Good Science

NSW Stage 4

Second edition

Authors

Amy Constant (series editor)

Rebecca Cashmere

Aleryk Fricker

Shirley Gilbert

Haris Harbas

Cristy Herron

Kiran Jagpal

John Ley

Rachael Riley

Contributing authors

Thomas Cosgrove

Emma Craven

Sarah Edwards

Nirupma Kumar

Vicki Maggs

Natalie Stinson

Ashley Suter

Investigations reviewer

Josh Aldridge

matilda education

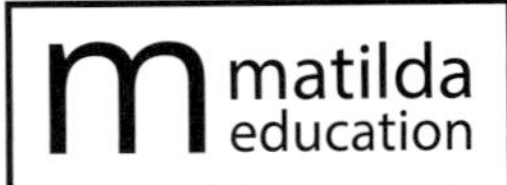

Good Science
NSW Stage 4
Second edition

Publisher: Melinda Schumann

Copy editors: Catherine Greenwood, Kate McGregor and Naomi Saligari

Project editors: Kate McGregor and Naomi Saligari

Proofreaders: Catherine Greenwood, Sally Woollett and Robyn Flemming

Cover and text design: Paul Ryan

Typesetter: Paul Ryan

Illustrator: QBS Learning

Warning: It is recommended that Aboriginal and Torres Strait Islander Peoples exercise caution when viewing this publication as it may contain images of deceased persons.

Acknowledgement of Country

We acknowledge the Aboriginal and Torres Strait Islander Peoples of this nation. We acknowledge the Traditional Custodians on whose unceded lands we have created this resource. We pay our respects to ancestors and Elders past and present.

Note about language

Please note that we use the terms 'Aboriginal and Torres Strait Islander' and 'First Nations' interchangeably.
We acknowledge that there is no one term that is universally accepted and we use these terms with respect.

First published in 2025 by **Matilda Education Australia**, an imprint of Meanwhile Education Pty Ltd

Melbourne, Australia

Telephone: 1300 277 235

Email: customersupport@matildaed.com.au

Web: www.matildaeducation.com.au

Publication data

Title: Good Science NSW Stage 4, Second edition

ISBN: 9780655093626

NATIONAL LIBRARY OF AUSTRALIA

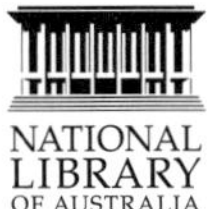

A catalogue record for this book is available from the National Library of Australia

Printed in Malaysia by Vivar Printing

December-2025

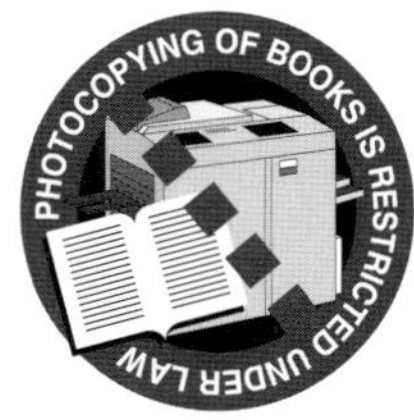

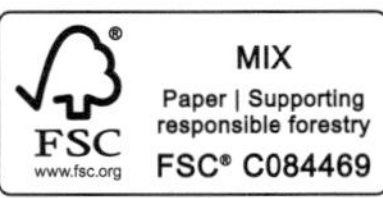

Contents

Learning Ladder for NSW Stage 4

Steps in progression	Observing the universe	Forces	Cells and classification	Solutions and mixtures	Living systems	Periodic table and atomic structure	Change	Data science 1
5	I can analyse how scientific advancement increases knowledge of the world.	I can investigate the impacts of forces in real-world contexts.	I can analyse how classification and understanding of organisms changes over time.	I can apply my understanding of the chemical properties of substances to real-world contexts.	I can predict the impact of changes to a living system.	I can investigate the use of elements and compounds based on their properties.	I can apply my knowledge of chemical changes to real-world contexts.	I can create and evaluate data-based models that explain scientific phenomena.
4	I can communicate the effect of changing variables in a range of first-hand investigations.	I can compare the role of different forces in real-world contexts.	I can compare the structural features of organisms using scientific conventions.	I can compare the properties of substances in different states.	I can compare features of different living systems.	I can explain how different elements and compounds are used.	I can communicate the processes involved in a range of changes.	I can explain how data is used to make predictions.
3	I can explain how observation can increase our understanding of scientific phenomena.	I can explain how forces are linked to real-world phenomena.	I can use scientific conventions to explain the organisation of organisms and cells.	I can explain how to use the properties of substances to separate or understand mixtures.	I can explain the structure and function of a living system.	I can compare the properties of elements and compounds, and relate them to the periodic table.	I can explain the role of energy in different types of changes.	I can recognise the relationship between data and scientific phenomena.
2	I can describe the importance of collaboration in science.	I can describe different types of forces.	I can describe the structure of cells and organisms.	I can describe the properties of different solutions.	I can describe the features of a living system.	I can describe the structures and properties of an atom, an element or a compound.	I can describe a scientific change.	I can describe the relevance of different types of data.
1	I can identify the branches of science.	I can identify multiple forces.	I can identify features of cells and organisms.	I can identify a range of solutions.	I can identify a living system.	I can identify that everything is made of different types of matter.	I can identify a scientific change.	I can differentiate between types of data sources.
Content	SC4-OTU-01	SC4-FOR-01	SC4-CLS-01	SC4-SOL-01	SC4-LIV-01	SC4-PRT-01	SC4-CHG-01	SC4-DA1-01

Steps in progression								
5	I can use observations and measurements to answer questions.	I can formulate testable questions and predictions, considering variables and controls.	I can construct an appropriate plan before conducting a first-hand investigation.	I can conduct first-hand investigations and accurately record the collected data.	I can process and interpret quantitative and qualitative data from a range of sources.	I can evaluate data and information for accuracy, reliability and validity.	I can evaluate problem-solving strategies used to solve an identified problem.	I can present scientific findings using appropriate conventions for specific audiences.
4	I can make inferences based on my observations.	I can make predictions to explain observations or phenomena.	I can develop appropriate scientific aims for a range of investigations.	I can collect accurate and reliable data from first-hand and second-hand investigations.	I can collect and present data in a range of appropriate formats.	I can draw conclusions based on patterns in data and information.	I can use given criteria to find solutions to scientific problems.	I can construct a range of appropriate scientific presentations based on first-hand and second-hand information and data.
3	I can record observations and measurements accurately.	I can construct questions to investigate scientific concepts or problems.	I can distinguish between controlled, dependent and independent variables.	I can implement safe practices when using scientific equipment.	I can construct appropriate graphs with headings and units for data.	I can explain relationships between datasets and information.	I can explain scientific problems and phenomena using cause-and-effect relationships.	I can use digital technologies to organise and present information and data.
2	I can use scientific tools to enhance observations.	I can select questions to investigate scientific concepts or problems.	I can describe ways to reduce risks for a range of investigations.	I can use scientific equipment to conduct investigations and gather first-hand data and information.	I can construct appropriate tables with headings and units for data.	I can describe trends from collected data and information.	I can suggest solutions to familiar scientific problems.	I can select appropriate ways to communicate information.
1	I can make observations using my senses.	I can make predictions based on prior knowledge and observations.	I can identify appropriate materials and technologies to conduct investigations.	I can identify correct processes for conducting investigations.	I can identify data from graphs, tables and digital sources.	I can identify trends within given data and information.	I can identify scientific problems.	I can recognise scientific information.
	Observing	**Questioning and predicting**	**Planning invest-igations**	**Conducting invest-igations**	**Processing data and information**	**Analysing data and information**	**Problem-solving**	**Commun-icating**
	SC4-WS-01	SC4-WS-02	SC4-WS-03	SC4-WS-04	SC4-WS-05	SC4-WS-06	SC4-WS-07	SC4-WS-08

Working scientifically processes

Stage 4 syllabus correlation grid

FOCUS AREAS	1.0 Observing the universe	2.0 Forces
Content		
SC4-OTU-01: Observing the universe Explains how observations are used by scientists to increase knowledge and understanding of the universe	✓	
SC4-FOR-01: Forces Describes the effects of forces in everyday contexts		✓
SC4-CLS-01: Cells and classification Describes the unique features of cells in living things and how structural features can be used to classify organisms		
SC4-SOL-01: Solutions and mixtures Explains how the properties of substances enable separation in a range of techniques		
SC4-LIV-01: Living systems Describes the role, structure and function of a range of living systems and their components		
SC4-PRT-01: Periodic table and atomic structure Explains how uses of elements and compounds are influenced by scientific understanding and discoveries relating to their properties		
SC4-CHG-01: Change Explains how energy causes geological and chemical change		
SC4-DA1-01 **Data science 1** Explains how data is used by scientists to model and predict scientific phenomena		

Working scientifically processes		
SC4-WS-01: Working scientifically: Observing Uses scientific tools and instruments for observations	✓	
SC4-WS-02: Working scientifically: Questioning and predicting Identifies questions and makes predictions to guide scientific investigations		✓
SC4-WS-03: Working scientifically: Planning investigations Plans safe and valid investigations		
SC4-WS-04: Working scientifically: Conducting investigations Follows a planned procedure to undertake safe and valid investigations	✓	
SC4-WS-05: Working scientifically: Processing data and information Uses a variety of ways to process and represent data		✓
SC4-WS-06: Working scientifically: Analysing data and information Uses data to identify trends, patterns and relationships and to draw conclusions		✓
SC4-WS-07: Working scientifically: Problem-solving Identifies problem-solving strategies and proposes solutions		✓
SC4-WS-08: Working scientifically: Communicating Communicates scientific concepts and ideas using a range of communication forms		

3.0 Cells and classification	4.0 Solutions and mixtures	5.0 Living systems	6.0 Periodic table and atomic structure	7.0 Change	8.0 Data science 1
✓					
	✓				
		✓			
			✓		
				✓	
					✓
✓				✓	
		✓			
	✓			✓	
✓	✓			✓	
		✓	✓		
			✓		✓
	✓				✓
✓		✓			

Focus area 1

1.0 Observing the universe

For thousands of years, scientists from different cultures have been observing the universe and asking big questions: How does it work? What does it mean? The way that science is understood, practised and taught is an outcome of culture. This means that different cultures have different ways of *doing* science. However, all scientists try to answer the questions of What? How? and Why?

Learning Ladder

The Learning Ladder contains the scientific content and processes you will learn in this chapter. Each area has five levels of progression. To move to higher levels, you need to practise activities at the earlier levels. This will help you develop the ability to complete tasks that are more complex.

Steps in progression	Content: Observing the universe	Working scientifically processes: Observing	Working scientifically processes: Conducting investigations	Steps in progression
5	I can analyse how scientific advancement increases knowledge of the world.	I can use observations and measurements to answer questions.	I can conduct first-hand investigations and accurately record the collected data.	5
4	I can communicate the effect of changing variables in a range of first-hand investigations.	I can make inferences based on my observations.	I can collect accurate and reliable data from first-hand and second-hand investigations.	4
3	I can explain how observation can increase our understanding of scientific phenomena.	I can record observations and measurements accurately.	I can implement safe practices when using scientific equipment.	3
2	I can describe the importance of collaboration in science.	I can use scientific tools to enhance observations.	I can use scientific equipment to conduct investigations and gather first-hand data and information.	2
1	I can identify the branches of science.	I can make observations using my senses.	I can identify correct processes for conducting investigations.	1

Figure 1.1: For thousands of years, Aboriginal and Torres Strait Islander Peoples have used observations of the universe – such as the Milky Way shown here – to thrive.

1.1 ► What is science?

Since humans evolved as a species, people have been making **observations** of natural events and using these observations to improve humans' knowledge and understanding of the universe.

Learning intention

At the end of this lesson, I will understand the different ways science is practised by Western and First Nations Peoples.

Key terms

branch: a type of science, such as biology or chemistry

collaboration: working cooperatively with other people to complete a task or solve a problem

Country: a specific area of a First Nations People that includes physical, linguistic and spiritual features

culture: a shared system of customs, habits, beliefs/spirituality, social organisation and ways of life that characterise different groups and communities

observation: noticing something you can see, touch, smell, hear or taste, and know to be true

phenomenon (plural **phenomena**): an observable fact or event

Investigation 1.1

Comparing observations, p.384

Content group: Nature of science

Science is the study of the universe

Science is the study of the universe and everything in it. The word 'science' comes from the Latin word *scientia*, which means to know or find out.

This links to the purpose of science, which is to build knowledge and understanding of the world (Earth) and the universe (everything: all of time and space and everything in it).

To do this, scientists work through a series of processes, which are:

- observing
- questioning and predicting
- planning investigations
- conducting investigations
- processing data and information
- analysing data and information
- problem-solving
- communicating findings.

Science can be divided into different areas or **branches**: astronomy, biology, chemistry, geology and physics. These are discussed in Section 1.3 on pages 8–9.

Science is an outcome of **culture**, which is a lens through which we see the world. There are different cultures and different ways of practising science.

Science is a necessary and important human activity. By practising science, we increase our knowledge. This allows us to solve problems, while enhancing our way of life and how we care for the environment.

First Nations and Western science have similarities

Science has been practised across the Australian continent and adjacent islands for as long as people have lived here. First Nations science and Western

Figure 1.2: For tens of thousands of years, First Nations Peoples have used fire to care for Country.

science are similar in many ways. Regardless of the science they practise, scientists observe natural **phenomena**, come up with explanations of these events, and conduct experiments to test these ideas.

The purpose of science for all scientists is also similar: to increase their understanding of the world.

First Nations Peoples also use this knowledge to care for **Country**. This role, this sacred responsibility, of managing the land, water, animals and plants is critical to Aboriginal and Torres Strait Islander Peoples' health and wellbeing.

The diverse range of activities involved in caring for Country requires First Nations Peoples to have a detailed understanding and knowledge of many branches of science, including those areas related to fire, animals, plants and the weather. This knowledge has been developed and passed down from generation to generation over thousands of years. This is a form of **collaboration**.

Learning Ladder

Observing the universe

1. Identify the branches of science that could explain how First Nations Peoples care for Country.
2. Explain how collaboration is related to First Nations Peoples' care for Country.
3. Propose reasons why making observations can improve our understanding of natural events.

Observing

see page 316

Go outside to complete the following tasks.

1. Use your senses to record observations about the weather. (Hint: What can you see, hear, feel and smell?) Record these observations in a table.
2. Use a thermometer to measure the temperature.
3. Record your temperature measurement, and the measurements of two other students, in a table. Include the unit of measurement.
4. Use your observations and the temperature data to explain what time of year or season it is.

In context

Give examples of how Western and First Nations Peoples have improved their knowledge of the world by using a scientific approach or idea.

Success criteria

- I can explain the purpose of science for Western and First Nations Peoples.

1·2 ▸ The world's oldest scientists

Learning intention

At the end of this lesson, I will understand the practice of First Nations science and the communication of this knowledge.

Key terms

Ancestors: the people we are descended from

Elder: a First Nations person who is a recognised custodian of knowledge, who has permission to pass on that knowledge, and who has specific responsibilities to young people in a community

spirits: the supernatural beings that many First Nations Peoples believe exist; can be associated with particular places, objects and rituals; are often connected with Dreaming stories

Content group: Nature of science

Aboriginal and Torres Strait Islander Peoples have lived on the Australian continent and the adjacent islands for at least 65 000 years. This means that First Nations cultures are the oldest continuous cultures in the world. Since the beginning, Aboriginal and Torres Strait Islander Peoples have practised science.

First Nations science is practised in the natural environment

We can understand the practice of First Nations science by looking at the idea of Country. Aboriginal and Torres Strait Islander Peoples' idea of Country includes many different things: the physical land, the waterways, the sky, the languages spoken on that land, family, **Ancestors**, **spirits**, identity and culture. Country is the close, interwoven relationship between all these things and more. None of these elements can be separated or isolated from the others; they all rely on each other.

The practice of First Nations science is similar across First Nations cultures. Aboriginal and Torres Strait Islander science practice does not isolate things (such as animals and plants) and study them away from their natural surroundings in a laboratory. Observations are made and experiments are conducted in the natural environment.

Figure 1.3: Aboriginal and Torres Strait Islander Peoples' knowledge can be communicated through art. This artwork was created by a Maluyligal and Wuthathi (Torres Strait) man, Brian Robinson. It shows the star cluster called Baidam Tithuyil (the Great Shark). When the shark touches the horizon, it is said to bring with it thunder and lightning and ushers in the monsoon season.

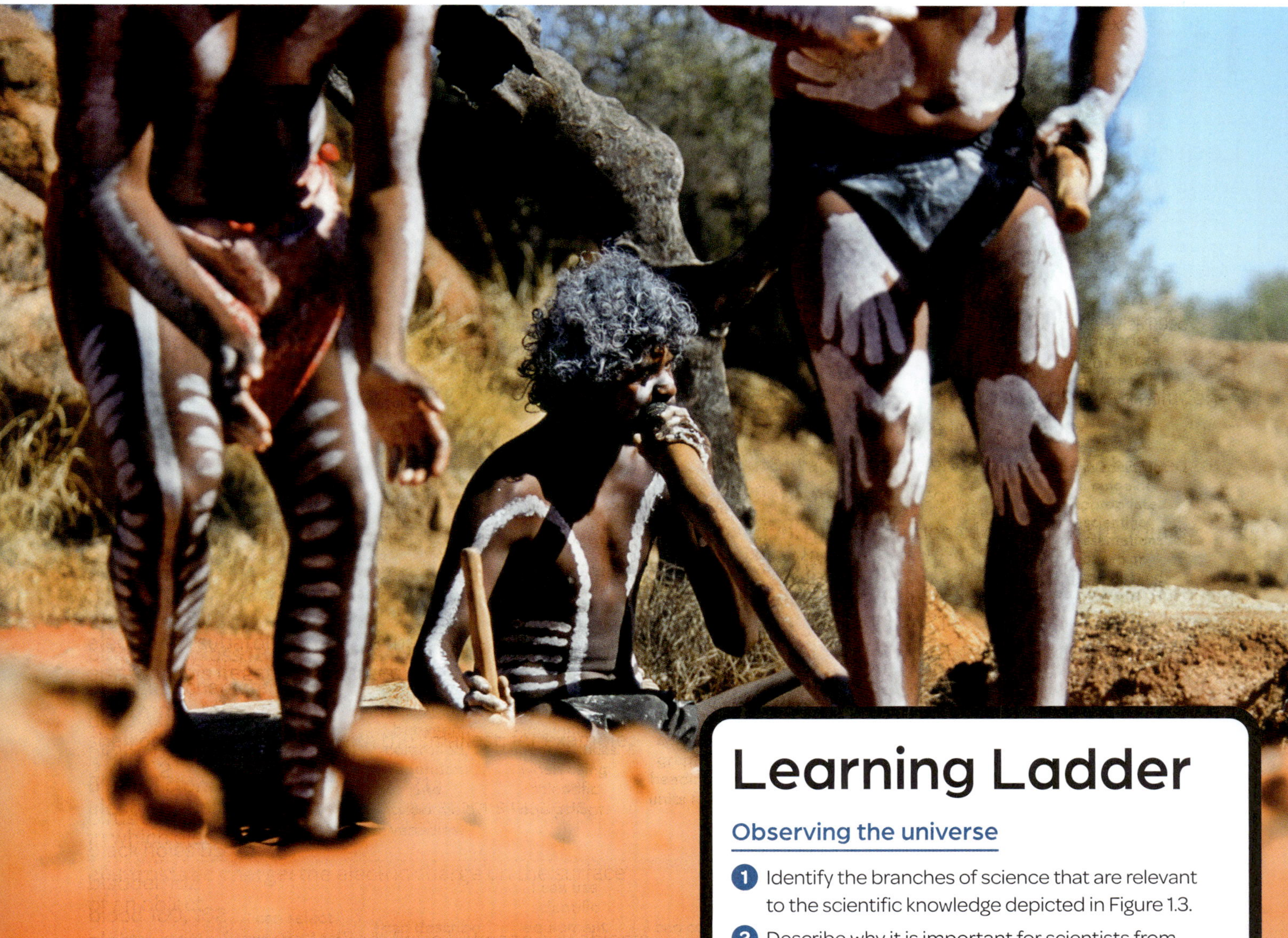

Figure 1.4: In Aboriginal and Torres Strait Islander cultures, scientific knowledge is often taught through dance.

First Nations scientific knowledge is communicated verbally and through art

Unlike the written reports favoured by Western scientists, the knowledge of First Nations science is passed on through oral traditions. The comprehensive scientific knowledge that First Nations Peoples have built – and continue to build – is passed from generation to generation through stories, song, dance and art (see Figure 1.3). Knowledge is often shared by the **Elders**.

Learning Ladder

Observing the universe

1. Identify the branches of science that are relevant to the scientific knowledge depicted in Figure 1.3.
2. Describe why it is important for scientists from different cultures to work together.
3. Explain how First Nations Peoples' observation of the stars increased their understanding of the weather.

Observing *see page 316*

1. Identify three senses that might be used in First Nations science to make observations of the environment.
2. Identify a piece of scientific equipment you could use to observe the Baidam Tithuyil star cluster as part of a first-hand investigation.
3. Describe how you could record your observations of the Baidam Tithuyil star cluster.

In context

Explain how First Nations artwork can be used to pass knowledge across generations.

Success criteria

- I can describe the practice of First Nations science and the communication of this knowledge.

1·3 ► The branches of science

Learning intention

At the end of this lesson, I will be able to:

- describe the different branches of science
- give examples of how the different areas of scientific knowledge can be used in the real world.

Key terms

interdisciplinary: combining more than one branch of knowledge

transdisciplinary: collaboratively working across multiple branches to learn or solve problems

Content group: Nature of science

Western science can be divided into different areas of knowledge. Each area is known as a branch of science.

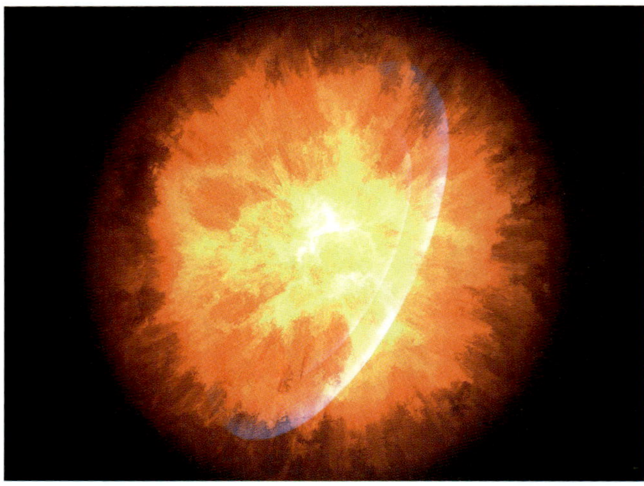

Figure 1.5: Astronomers observe and conduct investigations about phenomena in space, like this supernova. A supernova is an explosion of a star.

Western science has five branches

Western science is divided into five branches:

- **Astronomy** is the science of space. Astronomers observe the universe and investigate how things outside of Earth work. They also solve problems relating to space objects and travel. For example, how does the gravity of the Sun affect Earth and its position in the solar system?
- **Biology** is the science of living things. Biologists observe living things and investigate how they survive and thrive. They also solve problems relating to developing sustainable living practices. For example, what features does a polar bear have that help it to survive in a cold environment?
- **Chemistry** is the science of chemicals and matter. Chemists observe and investigate how chemicals and matter are structured and can be used. They also solve problems relating to energy and medicine. For example, what chemicals can treat deadly infectious diseases?
- **Geology** is the science of Earth. Geologists observe and investigate the structure of Earth. They also solve problems relating to human safety during natural disasters. For example, how can we predict the occurrence and severity of earthquakes?
- **Physics** is the science of matter and energy. Physicists observe and investigate how objects move and interact. They develop and test theories using modelling and other research to help solve problems in areas such as energy technologies and health care.

Within these main branches of science, there are many sub-branches. For example, ecology is a sub-branch of biology. Ecologists study how living things interact with their environments.

Figure 1.6: Physicists use machines like the Large Hadron Collider to conduct investigations. This machine allows scientists to observe how particles react under different conditions.

Solving some problems needs scientific knowledge from different branches

Many scientific problems are **interdisciplinary**, which means that more than one branch of scientific knowledge is needed to solve the problem. This means that science is **transdisciplinary**.

For example, climate change is a worldwide problem for humans and other living things. To tackle this problem, scientists from different branches of science are combining their knowledge and working together (see Table 1.1).

▲ **Figure 1.7:** Ecologists use tools like carbon dioxide meters to measure the impact of climate change in different environments.

Table 1.1: The roles of ecologists, meteorologists and physicists in tackling the problem of climate change

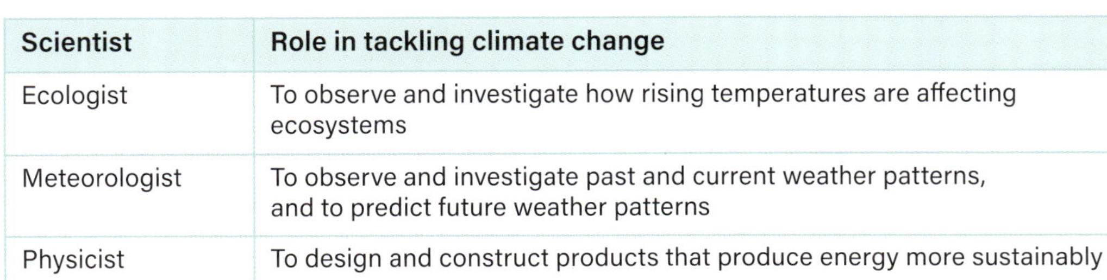

Scientist	Role in tackling climate change
Ecologist	To observe and investigate how rising temperatures are affecting ecosystems
Meteorologist	To observe and investigate past and current weather patterns, and to predict future weather patterns
Physicist	To design and construct products that produce energy more sustainably

Learning Ladder

Observing the universe

1. Identify the branch of science that would be used to study:
   a the features of an insect.
   b a star in a new galaxy.
   c how two chemicals interact.
2. Describe why it is important for science to be transdisciplinary.
3. Give examples of observations that can increase our understanding of climate change.
4. Describe changes that a geologist might look for that could indicate an earthquake is going to occur.
5. How might advancements like the Large Hadron Collider (see Figure 1.6) help scientists to better understand the world?

Observing

see page 316

1. **a** Identify three senses an ecologist might use to observe an environment.
   b Describe how an ecologist can use each of the senses you identified in Question 1a to learn about the environment.
2. Describe how scientists could use the Large Hadron Collider (see Figure 1.6) to improve their observations.
3. Describe how meteorologists record their observations about weather patterns.
4. Look at the picture of the supernova in Figure 1.5. What could you infer or suggest about the supernova based on your observations?

In context

Scientists are currently working to solve a range of problems in medicine. One problem is that some bacteria cannot be killed by antibiotics. Propose which three types of scientists (from which three branches of science) could work together to solve this problem. What could each type of scientist contribute to this investigation?

Success criteria

- I can describe the different branches of science.
- I can give examples of how the different branches of scientific knowledge can be used in the real world.

1·4 ► Collaboration in science

Learning intention

At the end of this lesson, I will be able to:

- explain what is involved in scientific collaboration
- explain why scientific research is usually collaborative.

Key term

telescope: a device that uses mirrors or lenses to focus on distant objects

Content group: Nature of science

The purpose of science is to build knowledge and understanding of the universe. Often, the best way for scientists to do this is to work together.

Collaboration is about working together

Collaboration is when scientists work together to tackle a task or develop understanding of a topic. Often, scientists who specialise in different branches of science work together. Scientists do this – build knowledge and understanding of the world and the universe – through observation, investigation and research. Scientific research involves collecting data, then analysing and interpreting it. Scientists throughout the world share or publish their results, which allows other scientists to access, use and build on this information. Working in this way is also a form of collaboration.

Collaboration is building our understanding of the solar system

The knowledge that shaped our modern understanding of the solar system started being developed in ancient Greece (12th to 1st century BCE). Since those ancient times, scientists have worked together and built on each other's work to improve our understanding of the solar system (see Table 1.2).

Teams of scientists are currently working with the National Aeronautics and Space Administration (NASA) on projects relating to space exploration. One of these is NASA's Artemis project, which aims to place a team of scientists on the Moon to collect information.

◄ **Figure 1.8:** This is the James Webb Space Telescope. Many scientists collaborated on the development and construction of this telescope.

Table 1.2: How astronomers have built on each other's work to develop our knowledge and understanding of the solar system

Scientist	Date	Findings	Findings built on
Plato	428–348 BCE	Everything in the solar system moves in circular orbits at a constant speed.	Plato's findings were based on his personal observations made with the naked eye – such as how the stars and planets move across the night sky.
Nicolaus Copernicus	1473–1543 CE	The Sun is at the centre of the solar system. Earth and the other planets orbit the Sun.	Copernicus's findings built on Plato's observations of the movement of the planets.
Galileo	1564–1642	Galileo used a **telescope** to observe the Sun, Moon, planets and stars. His observations of the solar system supported the idea that the Sun is at the centre of the solar system.	Galileo took Copernicus's model of the solar system and used a telescope to provide evidence that supported this model.
Isaac Newton	1642–1727	Newton provided more details about the moons and planets of the solar system. He did this with a telescope that used mirrors, which he developed.	Newton also built on the model of the solar system suggested by Copernicus and Galileo. He developed their ideas about the movement and locations of the planets.
Hubble Space Telescope team	1977 to the present	The Hubble Space Telescope allows us to observe the solar system in detail. The telescope was built by a team of scientists from different countries. This telescope was launched into space in 1990.	The team built on telescope and spacecraft technology developed over the preceding 400 years.
James Webb Space Telescope team	2004 to the present	The James Webb Space Telescope takes high-resolution pictures of the solar system. This telescope was launched into space by a team of scientists in 2021.	The team built on technology used to construct the Hubble Space Telescope.

Learning Ladder

Observing the universe

1. Identify which scientists (that is, from which branch of science) specialise in studying space.
2. Name two scientists who built on the work of others to improve our understanding of the solar system.
3. Explain how observation has improved our understanding of the solar system.
4. Discuss how changing the technology used to observe the solar system affected our ability to view objects in space.
5. Analyse how advancements in scientific equipment have improved our understanding of the solar system.

Conducting investigations

see page 326

1. Identify a step needed to investigate the solar system effectively.
2. Describe a piece of scientific equipment that is needed to conduct an investigation of the solar system.
3. Identify one safety practice that is relevant to gathering data about the solar system. Explain why this is needed.
4. Explain how data from a first-hand investigation of the solar system could be accurately collected.

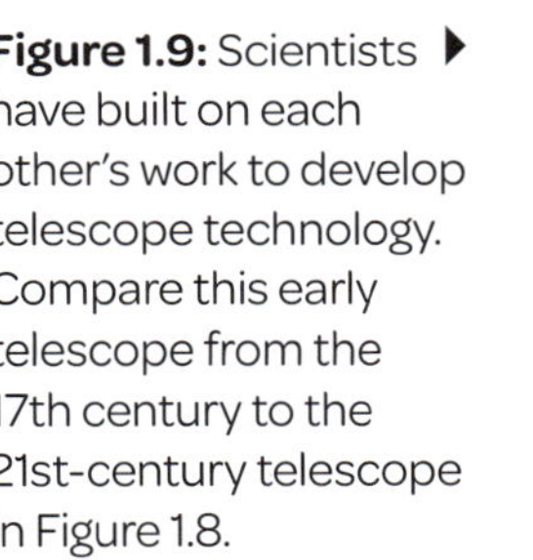

Figure 1.9: Scientists have built on each other's work to develop telescope technology. Compare this early telescope from the 17th century to the 21st-century telescope in Figure 1.8.

In context

NASA's Artemis project aims to place a team of scientists on the Moon to collect information. Explain how this would improve our understanding of the Moon.

Success criteria

- I can explain what is involved in scientific collaboration.
- I can explain why scientific research is usually collaborative and builds on the work of others.

1·5 ▸ Scientific theories and laws

Learning intention

At the end of this lesson, I will be able to:

- explain what a scientific theory is and what a scientific law is
- describe how scientific theories and laws are developed through observation and experimentation.

Key terms

experiment: a repeatable investigation carried out under controlled conditions to test a hypothesis

hypothesis: a suggested explanation or prediction of a scientific problem that can be tested with an investigation

scientific law: an explanation of a natural phenomenon that usually uses a mathematical equation

scientific theory: an explanation of a natural phenomenon that is supported by evidence and the results of repeated tests

Content group: Nature of science

To increase our knowledge and understanding of the world and the universe, scientists observe the natural phenomena around us, form hypotheses and repeatedly conduct experiments to test these ideas. It is through this process that scientific theories and laws are developed.

A scientific theory is an accepted explanation

A **scientific theory** is an accepted explanation of why a natural phenomenon occurs. Every scientific theory starts as a **hypothesis**, which is an untested idea, a suggested explanation for why an event in nature occurs. Before it can become a scientific theory, a hypothesis is tested and tested again. The results from many **experiments** must support the hypothesis before scientists consider the idea to be a scientific theory.

The big bang theory

One scientific theory is the big bang theory, which was first suggested by Belgian physicist Georges Lemaître (1894–1966) in the 1920s. According to this theory, the universe originated about 12 to 15 billion years ago as a result of a huge explosion. Since that moment, the universe has been expanding.

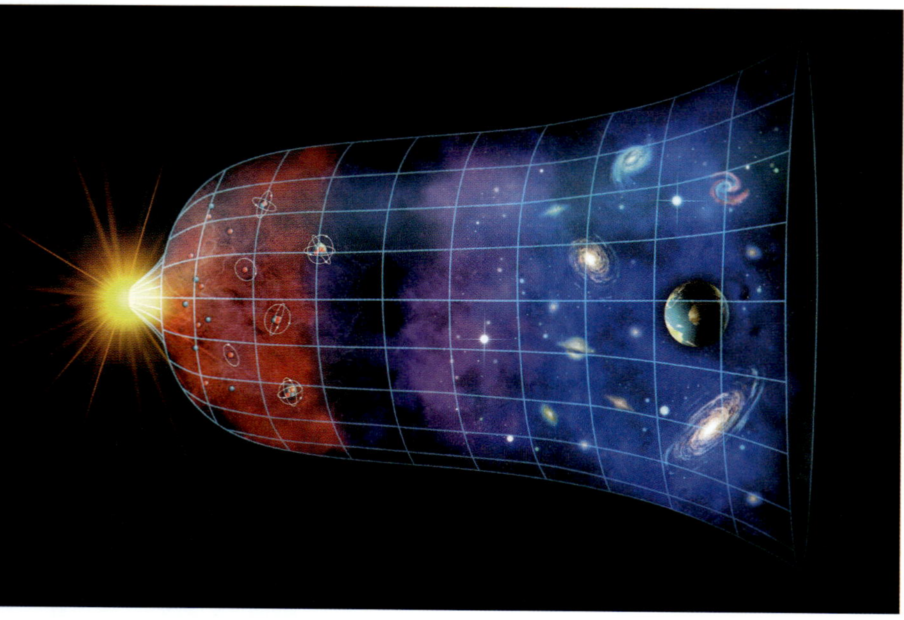

▲ **Figure 1.10:** This diagram represents the expansion of the universe from the big bang to the present. Many scientists' observations support the big bang theory as an explanation for how the universe came into existence.

Over the past 100 years, many scientists have conducted repeated experiments and observations to test the big bang theory. The data from these investigations supports Lemaître's theory. For example, numerous scientists have observed that objects in space are moving further away from one another. This means that the whole universe is expanding, which supports the theory.

Also, scientists have observed radioactive particles that are left over from the big bang. Scientists have found this radiation in the static from radios, and have seen the radiation in the universe by using telescopes.

A scientific law is based on maths

A **scientific law** is an explanation of why an event in nature occurs. Scientific laws have a mathematical basis. This means that repeated experiments have shown that, by using a particular mathematical equation, a natural phenomenon can be accurately predicted.

◀ **Figure 1.11:** This NASA astronaut is demonstrating that there is very little gravity in space.

Figure 1.12: Gravity is pulling this apple towards the ground. ▼

The law of gravity

One scientific law is the law of gravity. Gravity is the force of attraction between two objects. This law was developed by the English physicist and mathematician Isaac Newton (1642–1727).

Many scientists have conducted experiments to test this law. These experiments have consistently shown that Newton's law explains how gravity interacts between objects. This confirms a mathematical relationship for this law that says the force or weight of an object = the mass of the object × gravity. You will learn more about this law in Sections 2.6 and 2.7.

An experiment tests a hypothesis

A hypothesis cannot be proven (that is, an explanation of a natural phenomenon cannot be found to be true) without many scientists conducting repeated experiments to test the idea. Unless the results of multiple investigations support the hypothesis, the explanation will *not* be considered a scientific theory or a scientific law.

Learning Ladder

Observing the universe

1. Identify the branch of science that Isaac Newton specialised in.
2. Describe how scientists working together and building on each other's work has improved our understanding of how the universe was formed.
3. Explain how scientists' observations of the universe support the big bang theory.

Conducting investigations

see page 326

1. Outline what is required to ensure that the results of an experiment will be accepted.
2. Describe an example of how a specific piece of equipment was used to provide supporting evidence for a scientific theory based on experiments.
3. Outline one safety precaution that needs to be taken when investigating the law of gravity.
4. Propose data that would be collected from an investigation into the law of gravity.

In context

Scientific theories and laws explain natural phenomena. Discuss how a particular scientific advancement allowed scientists to provide evidence for a particular theory or law.

Success criteria

- I can explain what a scientific theory is and what a scientific law is.
- I can describe how scientific theories and laws are developed.

1·6 ▸ Safety in science

Learning intention

At the end of this lesson, I will be able to describe basic safety practices that allow investigations to be safely conducted in a science laboratory.

Key terms

Bunsen burner: a piece of scientific equipment that produces a single open gas flame

hazard: something that can harm living things, objects or the environment

heating flame: the blue (very hot) flame of a Bunsen burner (approx. 1500 °C); used for heating substances

risk: the chance that a hazard will cause harm

safety flame: the orange (cooler) flame of a Bunsen burner (approx. 300 °C); used between heating substances

Content group: Practice of science

Safety is critical in science because all investigations carry some degree of risk. Things can go wrong if you do not plan your procedure and act in a safe manner in the laboratory.

Good safety practices control risk

Whenever you step into a laboratory to undertake an investigation, it is important to identify **hazards** and use good safety practices to control the **risk** of them causing harm. (See 'Conducting investigations' and 'Using scientific equipment' in the Science how-to section on pages 326 and 375, respectively.)

It is important to use the safety equipment available in science laboratories, which includes:

- **safety glasses and face shields** to protect your eyes from fumes, particles and irritants
- **lab coats and aprons** to protect your clothes and body from chemical spills, flames and other hazards
- **gloves and hand protectors** to protect your hands when handling chemicals, biological materials and sharp objects
- **fire extinguishers**, which use dry chemicals (not water) to put out the flames if there is a fire

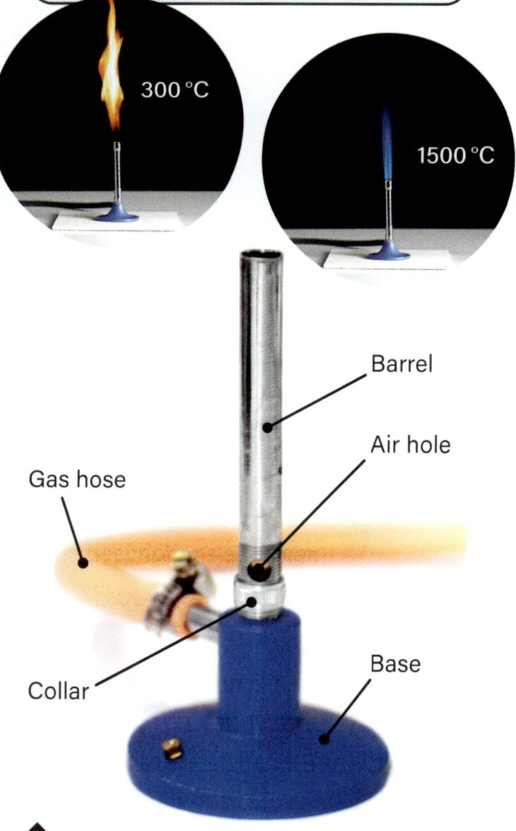

▲ **Figure 1.13:** Bunsen burners are used to safely heat materials in a laboratory.

▲ **Figure 1.14:** All laboratories have safety rules that you must follow.

- **eye wash stations**, to be used if you get something in your eye; never rub your eye
- **laboratory hoods**, to remove gases and fumes from the air; they are like exhaust fans in your bathroom or kitchen.

Figure 1.15: Wearing safety glasses, rubber gloves and a lab coat reduces risk in the laboratory.

Get to know Bunsen burner safety

A **Bunsen burner** is a common piece of scientific equipment that is used for heating things in a laboratory (see Figure 1.13). Learning how to set up and safely use a Bunsen burner is extremely important. Misusing this piece of equipment can result in severe burns or gas leaks.

A Bunsen burner has two flames:

- The **safety flame** is an orange flame that reaches temperatures of about 300 °C.
- The **heating flame** is a blue flame that reaches temperatures of about 1500 °C.

Setting up and lighting a Bunsen burner

1 Time for the safety check! If you have long hair, tie it up. Put on your lab coat and safety glasses. Make sure you know the location of the fire extinguisher and fire blanket. Check the gas hose for any cracks, holes or tears.

2 Place a heatproof mat on the bench. Place the Bunsen burner on top of the mat. Connect the gas hose tightly to the gas tap.

3 Turn the collar of the Bunsen burner so the air hole is closed.

4 Ask your teacher to check the set-up of your Bunsen burner and provide any feedback.

5 Lighting a Bunsen burner is best done by two people. Ask a partner to get ready to turn on the gas tap.

6 Light a match or taper before the gas tap is turned on. Position the match or taper over the top of the barrel.

7 Ask your partner to turn on the gas tap.

8 Your Bunsen burner should now be lit with a yellow flame. Move away from the burner and extinguish the match. Turning the collar to open the air hole creates a blue flame.

Learning Ladder

Observing the universe

1 Identify the branches of science in which safety precautions are important.

2 Describe how you can work together to maintain safety when lighting a Bunsen burner.

3 Explain how following the instructions in laboratory safety signs and labels can help you understand how to work in a laboratory.

4 Explain how changing the equipment you are using in an investigation may affect the safety measures you need to take.

Conducting investigations *see page 326*

1 Identify one safety rule that must be followed when working in a laboratory.

2 Identify a piece of equipment that can help keep you safe when conducting investigations in a laboratory.

3 Explain one safety precaution that needs to be taken when working with a Bunsen burner.

4 What data could you collect when testing the flame of a Bunsen burner?

In context

Scientists work together in laboratories all over the world. Discuss how scientists maintain safe practices when working in unfamiliar laboratories.

Success criteria

- I can describe basic safety practices in a laboratory.

1·7 ▸ Observations and inferences

Learning intention

At the end of this lesson, I will be able to explain how I can make observations using my senses and how observations can be used to make inferences.

Key term

inference: a conclusion that is based on evidence and observations

Content group: Practice of science

When conducting investigations, scientists make observations and draw inferences that explain natural phenomena.

Scientists observe using their senses

When conducting investigations, scientists can make observations using any of the five senses:

- **Sight**: Can a change be seen? For example, is a liquid bubbling or changing colour?
- **Touch**: Can a change be felt? For example, is a container warmer or colder?
- **Smell**: Is there a new scent? For example, can you smell something (like sulfur)?
- **Hearing**: Is there a new noise? For example, can you hear crackling or bubbling?
- **Taste**: Does a solution taste different? For example, does cake batter taste different before and after adding a specific ingredient? (Note: Do not taste chemical solutions in science unless you are explicitly instructed to do so by your teacher.)

Figure 1.16: By using your senses, you can make observations about what is occurring.

Inferences are educated guesses

Observations can be used to make inferences. **Inferences** are conclusions or educated guesses that are based on what has been observed that explain what has happened.

For example, a friend comes to school on crutches, so you *infer* that they have had an accident. At lunchtime, your friend bites into a sandwich and makes a face, so you *infer* they do not like their lunch.

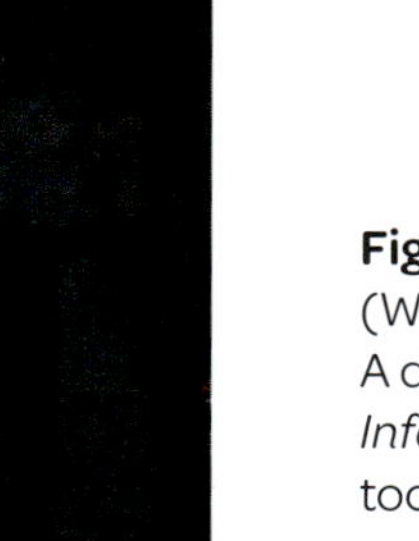

Figure 1.17: *Observation* (What do you see?): A drooping plant. *Inference:* The plant is getting too much or too little water.

Figure 1.18: *Observation* (What do you see?): A blackened landscape. *Inference:* There has been a bushfire in this area.

Table 1.3: A sample investigation, with observations and inferences

Investigation set-up	Observation	Inference
100 mL of cold water is in a beaker over the blue flame of a Bunsen burner.	After a certain amount of time, the water starts to bubble and steam forms.	The flame from the Bunsen burner is causing a change in the temperature of the water.
200 mL of cold water is in a beaker over the blue flame of a Bunsen burner.	After a longer amount of time, the water starts to bubble and steam forms.	The flame from the Bunsen burner is increasing the temperature of the water, but with more water it seems to takes longer.
300 mL of cold water is in a beaker over the blue flame of a Bunsen burner.	After an even longer amount of time, the water starts to bubble and steam forms.	The flame from the Bunsen burner is increasing the temperature of the water, but as the amount of water increases, so too does the time required for it to heat up and boil.

Figure 1.19: Making observations can allow us to make inferences about the impact of volume on heating time for water.

Learning Ladder

Observing the universe

1. Identify the branches of science that are relevant to heating water.
2. Explain how scientists work together to make observations during an investigation.
3. **a** Describe what you observe in Figure 1.19.
   b Infer what the flame is doing to the water.
4. Observe the the sample investigation in Table 1.3. Explain the impact of changing the volume of water before heating it.

Observing

see page 316

1. Identify the sense used to make each of the following observations.
   a A solution bubbles in a beaker.
   b A leaf crackles as it is crushed.
   c A glass cools when ice is added.
2. Identify which instrument might improve the accuracy of the observations made about the heat of the water in Figure 1.19.
3. Look at the sample investigation in Table 1.3. Explain how you could record your observations during this investigation.
4. Outline the inference drawn in relation to the water in the sample investigation in Table 1.3.

In context

Observe your classroom and ask a question about it. How could you investigate this question? How could you use observations to come to a conclusion?

Success criteria

- I can explain how I can make observations using my senses and how observations can be used to make inferences.

1·8 ▸ Using scientific equipment to make observations

Learning intention

At the end of this lesson, I will be able to describe the impact of using scientific equipment on the accuracy of my observations.

Key terms

accuracy: how close a measured value is to the true, exact value; how closely a recorded value matches the expected outcome of an investigation

data: facts and information collected for reference or analysis

meniscus: the curve seen at the top of a liquid in its container

qualitative data: data with qualities or characteristics that can be observed and described

quantitative data: information based on numbers that can be counted, measured or represented by numbers

reliable: provides consistent results when repeated

Investigation 1.8

Using scientific equipment to make observations, p.385

Content group: Practice of science

Making observations is an important part of investigations. From these observations, scientists draw inferences that explain natural phenomena. Observations that are made using only our senses lack accuracy, whereas observations made using scientific equipment are more accurate and reliable.

Measuring with equipment improves accuracy

During investigations, scientists make observations using any of the five senses: sight, touch, smell, sound and taste. They also use a variety of equipment to take accurate measurements and to collect **data**, such as thermometers, digital scales and stopwatches.

Data can be quantitative or qualitative:

- **Quantitative data** relates to quantities (that is, numbers) and can include the number of something, the volume, the length, time, or anything that can be measured or counted.
- **Qualitative data** relates to the qualities of something and can be written descriptions and observations. Qualitative data lacks **accuracy**.

In each of the investigations in Table 1.4, using measuring equipment enabled the scientist to translate the observations they made with their senses into accurate measurements.

This accuracy improved the scientist's observations. Accurate observations are more credible, valid and **reliable**, and are therefore more likely to be accepted by other scientists.

Figure 1.20: By using a thermometer, you can measure the exact temperature of liquids.

Figure 1.21: Scientists use equipment like microscopes to make observations.

Table 1.4: Making observations using the senses versus using measuring equipment

Investigation	Observation using the senses	Observation using measuring equipment
Heating water over a Bunsen burner	I can see that the water bubbles and produces steam when it gets hot.	I can use a thermometer to measure the temperature of the water. The water starts to bubble and produce steam when it reaches 100 °C.
Dissolving salt in cold water	I can see that a small spoonful of salt dissolves. I can see that a large spoonful of salt does not completely dissolve. I can see this because there are salt crystals at the bottom of the beaker.	I can use a digital scale to weigh accurate quantities of salt, so I can calculate the exact amount of salt that can be dissolved in the beaker of cold water.
Boiling different liquids	I can see that some liquids take longer to start bubbling than other liquids.	I can use a thermometer to measure the temperature of each liquid when it starts to boil. I can use a stopwatch to measure the exact amount of time it takes each liquid to start boiling.

Figure 1.22: Athletes use equipment like digital stopwatches and sensors to accurately measure their run times.

Figure 1.23: By using a measuring cylinder, you can measure solutions accurately. The base of the **meniscus** (curve) is the amount of solution in the cylinder.

Learning Ladder

Observing the universe

1. Outline which scientists (that is, from which branches of science) need to use accurate measurements in their investigations.
2. Describe how working with others may improve your ability to collect accurate observations during investigations.
3. Explain how making observations using measuring equipment can improve our understanding of phenomena (for example, the boiling points of different liquids).
4. Describe the effect of changing a variable in the dissolving salt investigation in Table 1.4.

Observing

see page 316

1. Identify which senses you use to make observations about the boiling points of liquids.
2. Identify two items of measuring equipment that can improve your observations of the boiling points of different liquids.
3. **a** Describe how you could record your observations of the boiling points of different liquids.
   b Explain how you could make sure your record is accurate.
4. Propose an inference you could make based on your observations of the boiling points of different liquids.

In context

Explain how a particular item of scientific equipment has assisted scientists to make improvements in the field of medicine. Provide examples to support your answer.

Success criteria

- I can explain why it is important to use scientific equipment to make observations during investigations.

1·9 ▸ Scientific variables

Learning intention

At the end of this lesson, I will be able to identify the independent, dependent and controlled variables of an investigation.

Key terms

controlled variable: a thing that needs to stay the same during an investigation

dependent variable: the thing that is measured in an investigation

independent variable: the thing that is deliberately changed in an investigation

Content group: Practice of science

To test their ideas, scientists conduct investigations. To be useful, a good scientific test needs to follow certain principles. Good scientific investigations have independent, dependent and controlled variables.

Variables are independent, dependent or controlled

Variables are the things that can be controlled, changed or measured during an investigation or experiment (see 'Questioning and predicting' in the Science how-to section on page 319).

The three main types of variables are:

- independent variables
- dependent variables
- controlled variables.

The **independent variable** is the one thing you change in an investigation.

The **dependent variable** is what you measure in an investigation. The dependent variable is altered by the independent variable. Examples of dependent variables are time in seconds and mass in grams.

The **controlled variables** are all the things you keep the same in an investigation. Examples of controlled variables are temperature, equipment, location and volume.

Figure 1.25: In this investigation into a person's allergies, the dependent variable (the person's reaction to different allergens) is being measured.

Change the independent variable and observe the dependent variable

When conducting your investigations, you will observe how changing the independent variable of the investigation can affect the dependent variable. In any investigation, only one factor (one independent variable) should be deliberately changed. All the other factors should be kept the same to make sure that the independent variable is responsible for any observed changes in the dependent variable.

For example, you put three plants in three different amounts of sunlight to see which plant grows the most. You make sure the plants are the same size, health and species, and only change the amount of sunlight each plant is getting; this is the independent variable. The dependent variable is the plants' growth (which could be their weight or their size, which you measure), and the controlled variables are all the other factors.

Figure 1.24: In this investigation, the scientist is observing how different nutrients in water affect the growth of garlic bulbs

Independent variable: The nutrients in the water (the water in each test tube contains different nutrients).

Dependent variable: The growth of the garlic bulbs (such as the roots and the sprouts)

Controlled variables: The type and size of the test tubes, the location of the bulbs, the temperature of the room, and the size of the bulbs at the start of the investigation

Learning Ladder

Observing the universe

1. Identify the branch of science relevant to Figure 1.24.
2. Describe how using variables when conducting investigations helps scientists to collaborate.
3. Explain how observing an investigation like the one in Figure 1.24 can help us to understand scientific phenomena.
4. Look at the investigation in Figure 1.24. Propose how changing the independent variable affected the dependent variable.

Observing

see page 316

Consider the plant investigation described on this page.

1. Using your senses, what observations could you make about the changes to the plants?
2. Identify a piece of equipment you could use to measure the growth of the plants.
3. Explain how you could record your observations of the plants.
4. After three weeks, you observe that the plant in full sunlight has grown by 15 cm and the plant in the dark has not grown at all. Propose an inference from this observation.

In context

Imagine you are conducting an investigation to measure the effect of water levels on plant growth. Identify the independent and dependent variables. Identify three controlled variables. Explain why the controlled variables are important.

Success criteria

- I can identify the independent, dependent and controlled variables of an investigation.
- I can explain the impact of each variable on an investigation.

1·10 ► Tabulating and graphing data

Learning intention

At the end of this lesson, I will be able to construct appropriate tables and graphs to represent data.

Key terms

primary data: data that you have collected from your own investigations

secondary data: data that has been collected by someone else

Investigation 1.10

Investigating how colour affects temperature, p.387

Content group: Practice of science

Data needs to be presented in a way that can be easily understood. There are many types of data, and scientists can construct different graphs and tables to present the information.

Scientists collect different types of data

There are many different types of data, including quantitative and qualitative data (see Section 1.8), and primary and secondary data.

- **Primary data** is first-hand data that you collect yourself through scientific investigation.
- **Secondary data** is second-hand data that has been gathered by someone else.

Figure 1.26: ► When you conduct an investigation, you collect primary data.

Data needs to be carefully collected and recorded

To make sure the data they collect is accurate, scientists record observations and measurements very carefully. Quantitative data might be recorded in a logbook or a table. Qualitative data might be recorded in a journal or workbook. Sometimes data is visual and can be recorded with a camera.

When measuring and recording quantitative data, use the appropriate units for physical quantities. (See 'Processing data and information' in the Science how-to section on page 329.)

Figure 1.27: This is an example of a good scientific data table. ▼

The descriptions are clear.

The table has a number and a title.

Table 1: Effect of sunlight on plant growth

Plant environment	Initial plant height (mm)	Growth after 1 week (mm)	Growth after 2 weeks (mm)	Growth after 3 weeks (mm)
Plant 1: no sunlight	181	0	−1	−3
Plant 2: indirect sunlight	175	1	2	4
Plant 3: direct sunlight	178	2	3	5

Units are provided at the top of each column.

The lines are ruled and the rows are easy to follow.

Data is organised and presented in tables and graphs

Collecting data is not the end of the process; the data needs to be analysed and evaluated. To effectively evaluate data, the information must be well organised and clearly presented.

Constructing a good scientific data table

One of the best ways to organise and present scientific data is in a table (see Figure 1.27). Always construct the table before you start your investigation.

Follow these steps to construct a table for scientific data:

1. Use a ruler to draw the lines so your table is clear and easy to read.
2. a. Give your table a descriptive and useful title.
   b. Give your table a number so you can easily refer to it in your investigation report.
3. Include the units of measurement in the column headings where needed (for example, mm).

Constructing a good scientific graph

A graph is an excellent way to visually present and explain data. Follow these steps to construct a graph for scientific data:

1. Pick a graph type that suits the data you are presenting (scatter plot, line graph or column graph). (See 'Processing data and information' in the Science how-to section on page 329.)
2. Use a ruler to draw the lines so the graph is clear and easy to read.

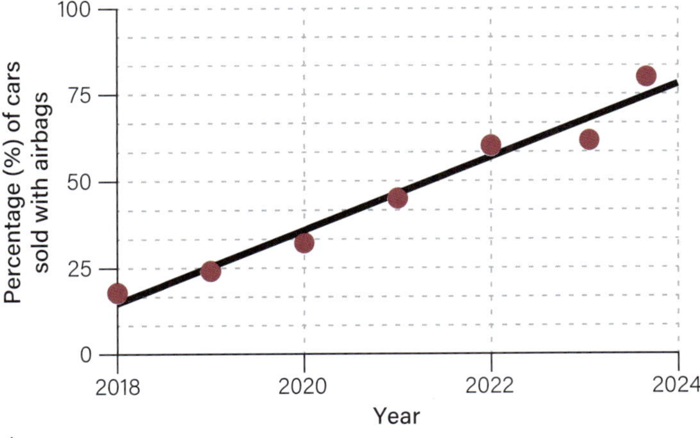

Figure 1.28: This scatter plot is a visual representation of the percentage of cars sold with airbags over a six-year period. This visual representation makes it easy to observe trends in the data.

3. Give your graph a descriptive and useful title.
4. a. Place the dependent (measured) variable on the *y*-axis (the vertical axis).
   b. Place the independent (changed) variable on the *x*-axis (the horizontal axis).
   c. Label the axes. Include the units of measurement, if appropriate.
   d. Have the numbers or information on each axis go up in an even scale and appropriate range.
5. Include a key or legend, if it is needed.
   This can be useful if your graph includes multiple lines or sections. You can use different colours to show different lines and use the key to say what each colour means.

Learning Ladder

Observing the universe

1. Identify which branch of science Figure 1.27 is relevant to.
2. Explain why using secondary data can be useful in scientific investigations.
3. Explain why making observations can be useful when collecting data.
4. Describe where you would place a changed variable in a scientific data table.

Conducting investigations *see page 326*

1. Outline why it is good practice to record data from a first-hand investigation in a table.
2. Identify the piece of equipment needed to draw scientific data tables. Why do we use this implement?
3. a. Outline the safety equipment you would need to use when gathering the data in Figure 1.27.
   b. Explain why you need to use this equipment to stay safe during this investigation.
4. Describe how you could make sure the data in Figure 1.28 is accurate.

In context

Tables and graphs are used to present a variety of scientific data. Propose a technology that could be used to improve the data collected in Figure 1.27. Construct a new scientific table with headings and units that could be used to represent the improved data.

Success criteria

- I can construct an appropriate scientific table.
- I can construct an appropriate scientific graph.

1·11 ► Observations of scientific phenomena

Learning intention

At the end of this lesson, I will be able to explain how making observations of natural events has allowed scientists to explain natural phenomena on Earth.

Key terms

axis: a real or imaginary line through the centre of an object

gravity: the force of attraction between two objects

orbit: the path of an object around a star or planet; for example, Earth around the Sun, or the Moon around Earth

pendulum: a mass that is attached to a fixed point by a string and can move backwards and forwards freely

revolve: move in a circular path around another object

rotate: spin on an axis

tide: the rise and fall of the waters in the ocean, caused by forces from the Moon and the Sun

Content group: Practice of science

By making observations of the Sun and the Moon, astronomers have worked out what causes everyday events on Earth, such as day and night, and the rise and fall of the water in the ocean.

Earth rotates on its axis

Earth is roughly a sphere, and it spins on its own **axis**. This was proven by the French scientist Léon Foucault (1819–1868). Foucault built on the work of other scientists, who had observed that objects seemed to move across the horizon. To prove this idea, Foucault built a **pendulum**. This device demonstrated that Earth **rotates** constantly as it **revolves** around the Sun in its **orbit**.

Day and night are due to Earth's rotation

A full day is 24 hours because Earth takes 24 hours to complete one full rotation on its axis. Viewed from the north pole, Earth rotates in an anticlockwise direction. This means that the Sun appears to rise in the east and set in the west. At any time, half of Earth is in daylight and half is in darkness. The side of Earth facing the Sun has day, and the side of Earth facing away from the Sun has night (see Figure 1.29).

▲ **Figure 1.29:** Scientists' observations have explained the phenomenon of day and night.

The Moon revolves around Earth

Moons are small bodies that revolve around planets. Some planets have more than one moon. Earth has one moon.

The Earth's Moon takes about 29.5 days to complete one revolution around Earth. Whenever the Moon passes over the side of Earth that is in daylight, it can be seen at the same time as the Sun. When the Moon is on the daylight side of Earth, the other side of Earth has night-time with no visible Moon.

The Earth's Moon can be seen with the naked eye. Scientists have observed the moons of other planets by using equipment such as telescopes.

Ocean tides are caused by gravity

A natural phenomenon that is commonly observable is the rise and fall of the water in the ocean. When it is high **tide**, the ocean reaches its highest level and the water comes the furthest up the beach. When it is low tide, the ocean is at its lowest level and the water is lower down the beach. Scientists' observations of the movement of the Sun, Moon and Earth have been used to explain the natural phenomenon of the tides.

Tides are caused by a natural force called **gravity**. The Earth, Moon and Sun all exert gravity. This means these bodies pull on each other. As the Moon pulls on Earth, it makes the oceans move.

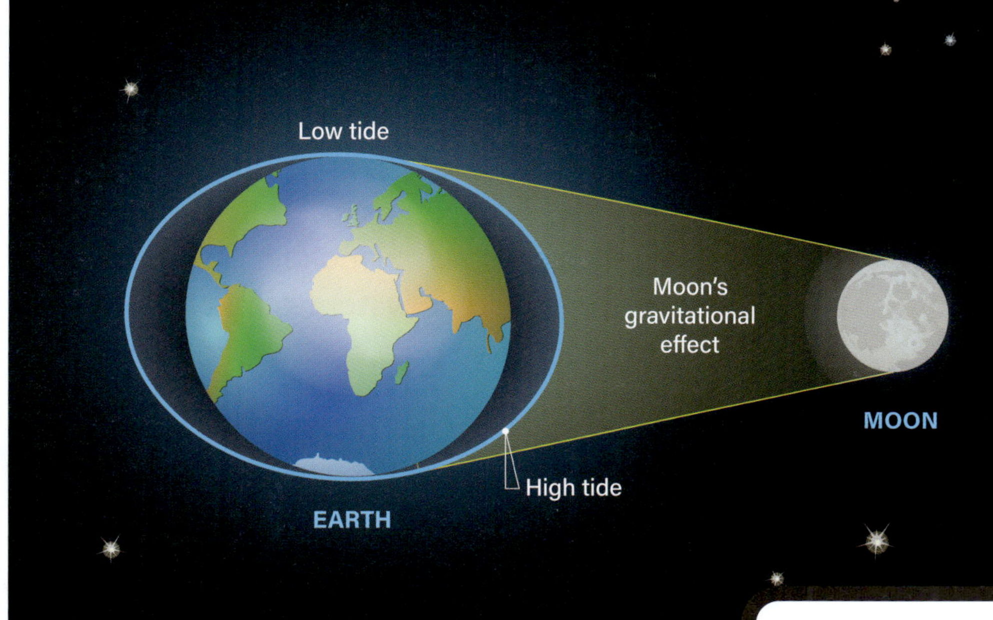

◀ **Figure 1.30:** The ocean's tides are caused by the movement and gravitational pull of the Moon.

On the side of Earth that is closest to the Moon, there is a high tide. At the same time, there is a high tide on the side of Earth directly opposite the Moon. In the areas of Earth between these two sides, there is a low tide (see Figure 1.30).

As Earth rotates on its axis and the Moon moves around Earth, the tides change. The tides go from high to low in a 24-hour period, depending on where the Moon is located in relation to the surface of Earth.

▲ **Figure 1.31:** Low tide (top) and high tide (bottom) at a beach in Mexico. The difference in the sea level is about four metres.

Learning Ladder

Observing the universe

1. Identify which scientists (that is, from which branch of science) study the objects of the solar system.
2. Outline how Léon Foucault built on the work of others.
3. Explain how scientists' observations of Earth's rotation relative to the Sun have increased our understanding of day and night.
4. Explain how the location of the Moon, relative to the Sun and Earth, affects the tides.
5. Analyse how the telescope has improved our ability to understand the natural phenomena occurring on Earth.

Observing

see page 316

1. Outline the senses people use to make observations about the movement of objects across the sky.
2. Identify a scientific tool that has improved our ability to observe the Sun and the Moon.
3. Explain how you could record the different heights of the ocean.
4. Scientists have observed that when the Sun and Moon are aligned, it affects the height of the tides.
   - **a** Outine what these tides are called.
   - **b** Explain why these tides occur.

In context

Surfing competitions occur at particular times of year and at specific times of day. Select a beach that is known for surfing in Australia. Propose reasons why specific times of year are selected for surfing competitions. Provide evidence to justify your reasoning.

Success criteria

- I can explain how making observations has allowed scientists to explain natural phenomena on Earth.

1·12 ► Models of the solar system

Learning intention

At the end of this lesson, I will be able to compare historical and current models of the solar system to show how models have been modified or rejected due to new scientific evidence.

Key terms

evidence: facts and observations that can be used to support or oppose a theory

geocentric model: a model of the solar system with Earth at the centre

heliocentric model: a model of the solar system with the Sun at the centre

model: a physical or mathematical representation of a scientific idea or concept

Content group: Space science

The ancient Greeks were known to make models of aspects of nature, like the solar system, to explain patterns they observed. Over time, these models were modified or rejected as a result of new scientific evidence.

The geocentric model put Earth at the centre of the solar system

One of the first **models** of the solar system was the **geocentric model** (see Figure 1.32). The word 'geocentric' comes from 'geo' (Earth) and 'centric' (centred).

According to the geocentric model:

- Earth was a sphere
- Earth was at the centre of the solar system
- Earth did not move
- the Sun and the planets revolved around Earth
- as the Sun and planets travelled around Earth, they also simultaneously moved around their own small circular paths.

Aristotle (384–322 BCE) argued in favour of the geocentric model. Aristotle was an ancient Greek philosopher who wrote about astronomy and whose observations were made with only the naked eye.

Another early scientist who agreed with the geocentric model was the Egyptian astronomer Ptolemy (100–170 CE). He agreed that Earth was at the centre of the solar system, but he suggested that the planets and the Sun revolved around a point outside Earth.

Figure 1.32: The geocentric model placed Earth at the centre of the universe with the Sun and the planets revolving around it.

Figure 1.33: Stonehenge is a circle of standing stones in England, built about 5000 years ago. This monument was built to align with the Sun's movement.

Aristotle's and Ptolemy's observations provided early scientists with **evidence** that supported the hypothesis that Earth was at the centre of the solar system. However, as time passed and mathematics and technologies developed, new evidence changed these initial hypotheses, and modified models of the solar system were developed.

The heliocentric model put the Sun at the centre of the solar system

In the 16th century, the **heliocentric model** replaced the geocentric model as the accepted model of the solar system. This model placed the Sun at the centre of the solar system (see Figure 1.34).

The heliocentric model was created by the Polish astronomer Nicolaus Copernicus (1473–1543), who used mathematics to explain the motion of objects in the heavens. His model proposed that:

- the Sun was at the centre of the solar system
- Earth and the other planets revolved in their orbits around the Sun.

Many people did not agree with the heliocentric model. For example, the early Catholic Church refused to accept this model of the solar system.

Galileo (1564–1642) was one of the first astronomers to use a telescope to observe the Sun, Moon, planets and stars. His observations of the solar system provided evidence that overwhelmingly supported the heliocentric model.

Eventually, the heliocentric model was accepted around the world. However, as telescopes improved, people noticed that stars did not orbit the Sun.

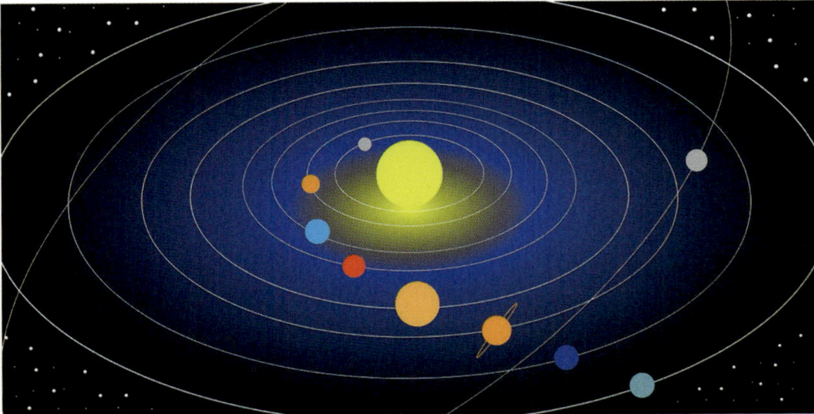

Figure 1.34: The heliocentric model placed the Sun at the centre of the solar system with the planets revolving around it.

By the early 19th century, scientists realised that – although the Sun is the centre of our solar system – the Sun is not the centre of the universe.

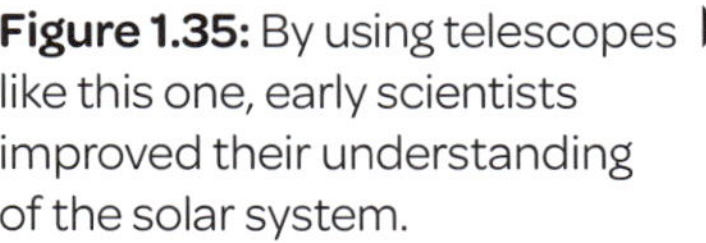

Figure 1.35: By using telescopes like this one, early scientists improved their understanding of the solar system.

Learning Ladder

Observing the universe

1. Identify which branches of science – other than astronomy – have contributed to our understanding of the universe.
2. Describe the contributions towards our current understanding of the solar system made by:
   a Ptolemy.
   b Aristotle.
   c Galileo.
3. Explain how Copernicus's heliocentric model improved our understanding of the universe and our place in it.
4. Explain how changing the perspective of our observations over time changed our understanding of the universe.

Observing *see page 316*

1. Identify the senses ancient astronomers used to make observations of the solar system.
2. Describe how the telescope enhanced our observations of the location of objects in our solar system.
3. Discuss how Copernicus recorded his observations of the solar system in a way that presented evidence for his heliocentric model.
4. Outline the conclusion that Galileo drew from his observations of the solar system using his telescope.

In context

Research a technology that is currently being used to observe the universe. Explain how it improves our understanding of a particular object.

Success criteria

- I can explain how historical models of the solar system changed over time as new evidence was discovered.

1.13 ▸ Seasons and eclipses

Figure 1.36: In Australia, spring features warming temperatures and many flowering plants, such as this red flowering gum.

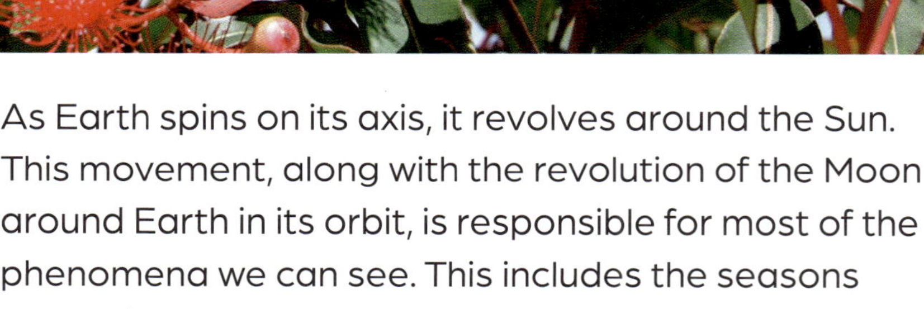

Learning intention

At the end of this lesson, I will be able to explain that predictable and observable phenomena on Earth – including seasons and eclipses – are caused by the relative positions of the Sun, Moon and Earth.

Key terms

eclipse: the blocking of the Sun's light from Earth

equinox: the two times of the year when night and day are about the same length

penumbra: the outer part of the Moon's shadow on Earth

solstice: the two times of the year when night and day are the most different in length

umbra: the inner part of the Moon's shadow on Earth

Investigation 1.13

Modelling eclipses, p.389

Content group: Space science

As Earth spins on its axis, it revolves around the Sun. This movement, along with the revolution of the Moon around Earth in its orbit, is responsible for most of the phenomena we can see. This includes the seasons and **eclipses**.

Most places on Earth have four seasons

Australia has four seasons: winter, spring, summer and autumn.

Table 1.5: Earth's seasons in the northern and southern hemispheres

Season	Temperature	Time of year	
		Southern hemisphere (e.g. Australia)	**Northern hemisphere (e.g. the United Kingdom)**
Winter	Cool to cold	June, July, August	December, January, February
Spring	Warming	September, October, November	March, April, May
Summer	Warm or hot	December, January, February	June, July, August
Autumn	Cooling	March, April, May	September, October, November

Earth's tilt causes the seasons

Earth's seasons are due to the tilt of its axis. Earth is tilted about 23.5 degrees on its axis as it travels around the Sun. The seasons are caused by the intensity of the sunlight: how much the light spreads on Earth's surface (see Figure 1.37). For example, in January, Australia directly faces the Sun and so the weather is hot.

Most areas of the world experience the same four separate seasons. These seasons do not occur at the same time of the year throughout the world. This is because the seasons in the northern hemisphere are the opposite of the seasons in the southern hemisphere (see Table 1.5). The closer a place is to the north and south poles, the greater the difference between winter and summer.

Equinoxes and solstices: equal and different lengths of day and night

The **equinox** is when day and night are the same length: 12 hours each. Equinoxes happen twice a year, once in spring and once in autumn. They happen at the same time around the world. When one hemisphere has spring equinox, the other hemisphere has autumn equinox, and vice versa. Equinoxes occur when the Sun appears directly over the equator. This is when Earth is tilted so that its axis is side-on to the Sun.

The **solstice** occurs when one of the poles of Earth is at its maximum tilt towards the Sun.

- On 21 December, the south pole is tilted towards the Sun, and it is the **summer solstice** in the southern hemisphere (longest day and shortest night) and the **winter solstice** in the northern hemisphere (shortest day and longest night).
- On 21 June, the north pole is tilted towards the Sun, and it is the summer solstice in the northern hemisphere and the winter solstice in the southern hemisphere.

▼ **Figure 1.37:** The tilt of Earth means that different countries have more or less direct sunlight, depending on where Earth is in its revolution around the Sun. Earth's tilt also results in equinoxes and solstices.

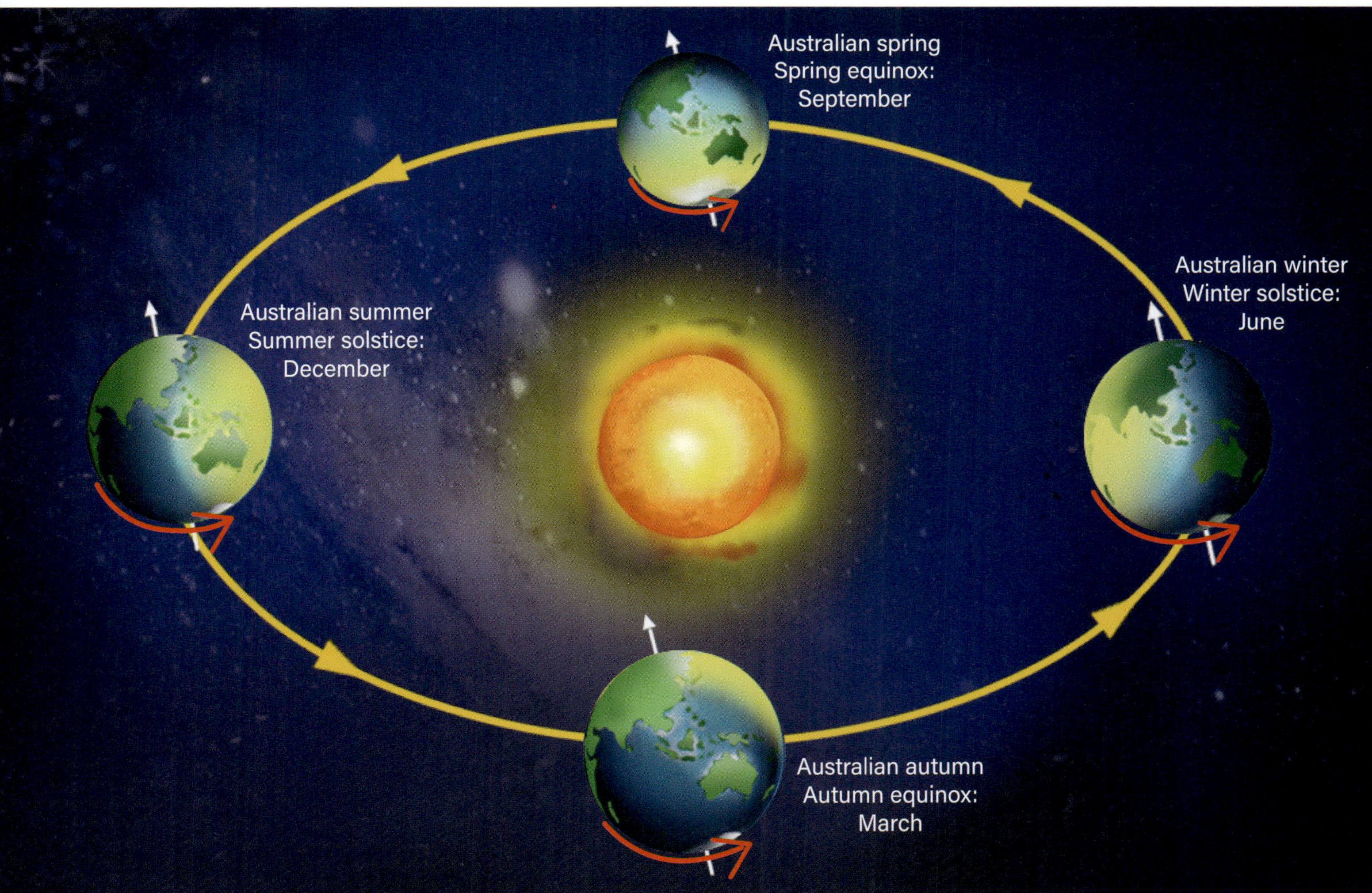

The Moon blocks the Sun's light during a solar eclipse

A solar eclipse occurs when the Moon passes between the Sun and Earth. The Moon blocks the Sun's light and casts a shadow on Earth. The centre of this shadow is the **umbra** and the outer ring of shadow is the **penumbra**.

There are three types of solar eclipse: total, annular and partial. They differ in how much light the Moon blocks, as seen by a viewer on Earth.

A **total solar eclipse** occurs when the Moon, Sun and Earth align (see Figures 1.38 and 1.40). People within the umbra see the Moon block *all* the Sun's light; those within the penumbra see the Moon block *part* of the Sun's light.

An **annular solar eclipse** occurs when the Moon is further away from Earth, making it appear smaller than usual. Because of this, the Moon only covers the centre of the Sun. During an annular eclipse, we can see the outer edges of the Sun. This is called an annulus or 'ring of fire'.

A **partial solar eclipse** occurs when the Sun, Moon and Earth are aligned but they are not positioned in a perfectly straight line. The Moon only blocks part of the Sun when this happens. There is no umbra in a partial solar eclipse; everyone who sees it is in the penumbra.

When there is a solar eclipse, there is a 'new Moon', which means the Moon can be very difficult to see in the night sky. This is because the Sun's light is hitting the side of the Moon that is opposite to the side facing Earth.

Earth blocks the Sun's light during a lunar eclipse

A lunar eclipse occurs when Earth passes between the Sun and the Moon. Earth blocks the Sun's light and casts a shadow on the Moon. Depending where you are on Earth during a lunar eclipse, some or all of the Sun's light is blocked.

Lunar eclipses always happen when the Moon is full, which is when the whole Moon can be seen in the night sky from Earth. There is a full Moon every 29.5 days, which is the time it takes for the Moon to complete one revolution around Earth. During most months, a full Moon occurs when the Moon does not line up with the Sun and Earth. During a lunar eclipse, the Sun, Earth and Moon are aligned.

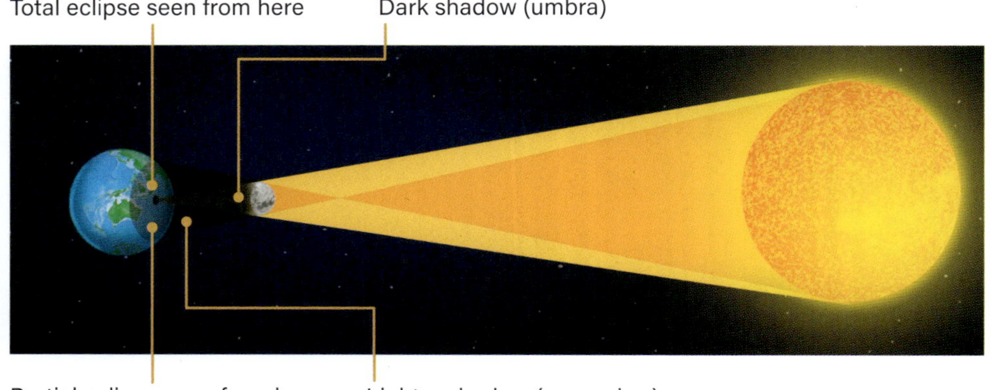

Figure 1.38: In a solar eclipse, the Moon casts a shadow over part of Earth.

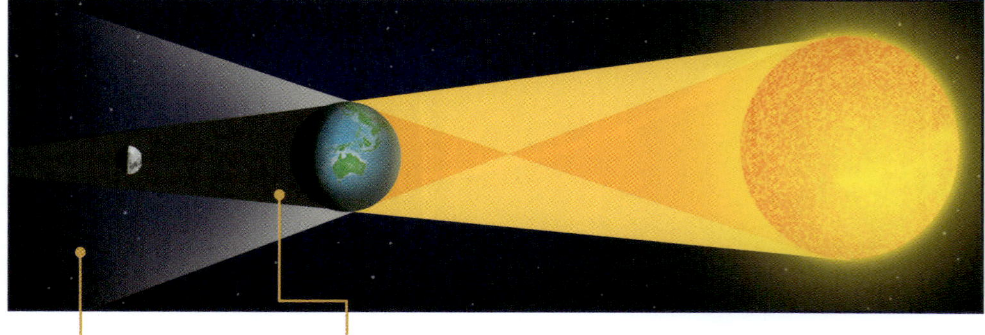

Figure 1.39: In a total lunar eclipse, the whole Moon is covered by Earth's shadow.

There are three types of lunar eclipse: total, partial and penumbral. Each type of eclipse is named according to how much light is blocked by Earth, as seen by a viewer on Earth.

A **total lunar eclipse** occurs when the Sun, Earth and Moon are perfectly in line. The whole Moon is covered by Earth's shadow (see Figure 1.39). Total lunar eclipses are known as 'blood moons' because the Moon changes to a striking red colour.

A **partial lunar eclipse** occurs when the Sun, Earth and Moon are not completely in line. Only part of the Moon is covered by Earth's shadow. During a partial lunar eclipse, you can see the curved shape of Earth's shadow on the Moon.

A **penumbral lunar eclipse** occurs when the Moon only passes through the penumbra, which is the outer edge of Earth's shadow. These eclipses are often not noticed because the Moon appears only slightly dimmer than a regular full Moon.

▲ **Figure 1.40:** The Moon is moving between the Earth and Sun, causing a total solar eclipse.

Learning Ladder

Observing the universe

1. Identify which scientists (that is, from which branch of science) study the Sun, Moon and Earth.
2. Describe how scientists from different countries could work together to study the Sun's effect on the seasons.
3. Explain how observing the orbits of Earth and the Moon can improve our understanding of eclipses.
4. Explain the effect that the position of the Moon has on the type of solar and lunar eclipses that occur.

Conducting investigations

see page 326

1. Identify which of the following processes is the correct way to collect data about the seasons.
   - **A** Record monthly average temperatures for one year. Identify which months are the warmest and the coolest.
   - **B** Record monthly average temperatures for three years. Identify which months are, on average, the warmest and the coolest.
   - **C** Record daily average temperatures for a year. Identify which days are hotter. Check for a pattern.
2. Describe why using a telescope is beneficial when tracking the movement of the Moon.
3. Identify one hazard of using a telescope. Propose how the risk of this hazard causing harm could be reduced.
4. **a** Outline how you would record data of the type described in Question 1.
   b Identify the unit of measurement you would use.

In context

The seasons are opposites in the northern and southern hemispheres. Propose a technology that could record evidence to support this fact.

Success criteria

- I can explain how the movement and position of Earth causes the seasons.
- I can explain how the positions of the Earth, Moon and Sun cause solar and lunar eclipses.

1·14 ▸ The Moon and the tides

Learning intention

At the end of this lesson, I will be able to explain how Aboriginal and Torres Strait Islander Peoples use observations of the phases of the Moon to predict tidal changes.

Key terms

Dreaming stories: important stories for First Nations Peoples; these stories often contain important knowledge of the world

ebb: the phase of the tide when the water flows towards the ocean

Elder: a First Nations person who is a recognised custodian of knowledge, who has permission to pass on that knowledge, and who has specific responsibilities to young people in a community

flow: the phase of the tide when the water flows towards the land

waning: when the Moon looks as though it is getting smaller

waxing: when the Moon looks as though it is getting bigger

Content group: Aboriginal and Torres Strait Islander Peoples' cultural knowledges of astronomy

Aboriginal and Torres Strait Islander Peoples have long recognised an alignment between what happens in the sky and what happens on Earth. This knowledge has been used for many purposes, including to accurately predict the timing and height of the tides.

The Moon moves through eight phases

The phases of the Moon are its different shapes we see in the night sky (see Figure 1.41). Over a month, we see a complete lunar cycle of eight different shapes, which then repeats. The phases of the Moon are described in Table 1.6.

Table 1.6: Phases of the Moon

Phase of the Moon	What we see
New Moon	The Moon is not visible in the sky.
Waxing crescent	A small crescent is illuminated on the left side of the Moon.
First quarter	The left side of the Moon is illuminated.
Waxing gibbous	Most of the Moon is illuminated. Only a small crescent on the right side of the Moon is dark.
Full Moon	The whole circular face of the Moon is illuminated, forming a full circle.
Waning gibbous	Most of the Moon is illuminated. Only a small crescent on the left side of the Moon is dark.
Third quarter	The right side of the Moon is illuminated.
Waning crescent	A small crescent is illuminated on the right side of the Moon.

Figure 1.41: The phases of the Moon

Western science uses telescopes to explain the Moon's phases

Western scientists' explanation of the phases of the Moon is based on observations made with telescopes. These observations have shown that the Moon does not give off light as it orbits the Earth. 'Moonlight' is actually the Sun's light reflected off the Moon's surface. As the Moon moves around Earth, the Sun lights up different parts of the Moon's surface. Although it looks as if the Moon is changing shape, it is actually our view of the Moon that is changing.

The Yolŋu story of the Moon's phases tells about Ngalindi

A First Nations account of the phases of the Moon is captured in the **Dreaming story** about Ngalindi (the Moon-man). This story is from the Yolŋu People.

This is how part of this story was told by a CSIRO astronomer, Ray Norris:

> 'In the Yolŋu story, [the Moon is] called Ngalindi and he was big and round and fat like the full Moon, and he was lazy.
>
> His wives and children got so angry because he did nothing to help, so they chopped off bits of him and he went from being a round, fat Moon and got thinner and thinner, which is why you get phases of the Moon.
>
> Eventually he died and stayed dead for three nights before he came back to life, as a new Moon.
>
> He cursed everyone and said that when he died, he would come back to life, but when others died, they would stay dead.'

The phases of the Moon are used to predict tidal changes

For thousands of years, Aboriginal and Torres Strait Islander Peoples have observed the positions and phases of the Moon and how these phenomena correlate with the rise and fall of the water in the ocean (which is called the **ebb** and **flow** of the tide).

This knowledge is captured in many Dreaming stories and passed on by **Elders**, including the Yolŋu story about the Moon-man, Ngalindi. In the story, Ngalindi gradually fills up with water from the ocean; when he is full (that is, when there is a full Moon), it is high tide. Then, as the water runs out of Ngalindi, the tides fall.

First Nations Peoples' knowledge of the cycle of the Moon's phases and how these align to the ocean tides enables accurate predictions of the timing and height of the tides. This can be seen in the tidal fish traps built by some First Nations Peoples. These traps consist of stone walls that create pools or pens. At high tide, the walls of the fish traps are submerged, allowing fish to swim into the area. As the tide ebbs and the water level drops, the stone walls sit above the water level and trap the fish in the pools.

Learning Ladder

Observing the universe

1. Identify which scientists study the movement of the Moon.
2. Explain why it is important to listen to and learn from the Elders.
3. Describe how the Yolŋu Peoples' observations of the Moon improved their understanding of natural events that occur on Earth.
4. Construct a diagram of the Moon, Earth and Sun, showing where the Moon is in its orbit around Earth when there is a full Moon. You will need to indicate where a person on Earth is viewing from.

Observing

see page 316

1. Thousands of years ago, the Yolŋu People observed something about the Moon when the tide was high. Outline what they observed.
2. Identify a scientific tool that could be used to observe the Moon.
3. Construct an illustration to accompany the Yolŋu story about Ngalindi. In your drawing, show high tide and low tide and the corresponding phases of the Moon, as described in the story.
4. Propose a conclusion that the Yolŋu People may have come to, based on their observations of the Moon and the tide.

In context

Propose ways in which First Nations Peoples' knowledge could be applied to modern Australian fishing practices.

Success criteria

- I can explain how First Nations Peoples use their observations of the phases of the Moon to understand, predict and use the ocean tides.

1·15 ▸ First Nations Peoples' use of the stars

Figure 1.42: Whirlwinds are common when the Scorpius constellation can be seen in the sky.

Learning intention

At the end of this lesson, I will be able to explain some of the many ways that First Nations Peoples use the stars.

Key terms

diffuse: when light becomes spread out as it passes through the atmosphere

sky Country: in First Nations knowledge, the aspects of Country that are not linked to the land; all the observable celestial phenomena and the spirits and lore linked to these events

songlines or **song series:** complex ways that First Nations Peoples record important information relating to many different topics; a body of songs sung sequentially and which are intended to convey knowledge

Content group: Aboriginal and Torres Strait Islander Peoples' cultural knowledges of astronomy

For thousands of years, Aboriginal and Torres Strait Islander Peoples have used their observations of the stars to predict weather, identify when the seasons will change, predict animal behaviour and identify the best time to source different foods.

First Nations astronomers know sky Country

The stars are an important part of First Nations Peoples' knowledge and understanding of the world. First Nations Peoples have used oral stories, songs, **songlines**, dances and art to transmit this knowledge from generation to generation.

First Nations astronomers

Aboriginal and Torres Strait Islander communities have their own astronomers. These are individuals who are selected to be trusted with knowledge of the stars. In the western Torres Strait, each of these astronomers is called 'Zugubau Mabaig', meaning 'star man'.

First Nations astronomers are responsible for carefully watching the sky and noticing the most subtle changes in the position and properties (brightness, colour) of the stars. These astronomers fuse their observations with their vast knowledge of the **sky Country**, and then advise the community on a range of issues, such as when to harvest certain foods. This practice is often referred to as 'reading the stars'.

Stars are used to predict the weather

Aboriginal and Torres Strait Islander Peoples use their observations of the colour, brightness, twinkling and positions of the stars to predict many different weather patterns.

Twinkling stars

In the eastern Torres Strait – which is off the north coast of Cape York in Queensland – Meriam Elders use the stars to predict rainfall and wind.

The Elders know that when they observe twinkling stars towards the end of the year (in December), this predicts the start of the northern Australian monsoon season (also known as the wet season) and the shift in the winds from cooler, drier south-easterlies to hotter, wetter north-westerlies.

This knowledge is preserved in the Meriam song 'Uier Naskaisreda' ('The Twinkling Stars'):

> Aipki- em pemut bapiti-e
> Nalugem pe-ueir Naskais-reda
> Ur Kakaper ise Bapri-eda
> Karim nowag-em e
> Ziai giru baz-pe aisli
> No-wabim MDW em di-kir
> Nalu-gem pe-wer Naskais-reda
>
> Why is it so calm tonight?
> Why are the stars twinkling like embers?
> Me, I think it's because of the big wind.
> The clouds are coming from the south
> And being swept to the north-west.
> Why are the stars twinkling like embers?
>
> – 'Uier Naskaisreda', song by George Passi

The stars appear to twinkle at this time of year because the hot, humid and windy conditions interfere with the light from the stars as it passes through the atmosphere, increasing how much the light shifts.

Blue, fuzzy stars

By carefully observing the stars and the weather over generations, Elders know that when the stars look blue and fuzzy, it will rain soon. The stars look blurry because higher levels of water in the atmosphere (which is called humidity) cause the light from the stars to **diffuse**.

The Scorpius constellation

The Kamilaroi and Euahlayi peoples know that when they can see the Scorpius constellation (see Figure 1.43), there are likely to be whirlwinds.

In the story, the dark patches in the Scorpius constellation relate to Wilbaarr, the spirit of the whirlwind. Scorpius is visible from May to August in the southern hemisphere. This is the windiest period of the year in the Kamilaroi and Euahlayi Countries, which are on the east coast of Australia, and whirlwinds are common at this time.

This same constellation can also be seen during the dry season, when scorpions are more common.

Figure 1.43: The Scorpius constellation looks like a scorpion in the sky, as shown here where an artist has drawn a scorpion over an image of the constellation.

Stars are used to predict animal behaviour and to identify when to source different foods

The rising and setting of particular stars at dusk and dawn are closely linked to Aboriginal and Torres Strait Islander Peoples' stories about the behaviour of animals (for example, when particular animals breed, give birth and migrate) and the availability of different foods.

The Emu in the Sky constellation

For the people who live in the Kamilaroi and Euahlayi Countries, the position of the Emu in the Sky constellation indicates different stages in the emu breeding season.

The shape of an emu can be seen in the dark patches in the Milky Way between the Coalsack Nebula (the emu head), the constellations of Scorpius and Sagittarius (the emu body) and the constellations of Ophiuchus and Aquila (the emu legs) (see Figure 1.44).

When the emu appears to rise in the sky at dusk (in April and May), this indicates that it is emu breeding season and it is the best time to collect emu eggs for food.

When the position of the emu in the sky changes (around August or September), collecting emu eggs must stop, for two reasons. First, there needs to be enough emu hatchlings to maintain an emu population. Second, the emu chicks have been developing in the fertilised eggs and as embryos are too developed to be suitable for eating.

▲ **Figure 1.44:** The shape of an emu can be seen in the dark bands of the Milky Way.

The Arcturus star

In Arnhem Land, the Yolŋu People harvest the corms (underground stems) of spike-rush plants (see Figure 1.45) when the Arcturus star can be seen in the sky just before sunrise.

Corms are usually at their biggest after several weeks of rainfall. Approximately 90 per cent of the annual rainfall in northern Australia occurs in the wet season (November to March). So, the optimal time to harvest spike-rush corms is late November, which is when the Arcturus star is around 10 degrees above the horizon at sunrise.

The Arcturus star is known as Marpeankurruk by the Boorong People, whose Country is near Lake Tyrell in north-west Victoria. The appearance of Marpeankurruk in the night sky in August coincides with the hatching larvae of the carpenter ant. These larvae can be found under logs and small rocks. They are high in protein and can be cooked or eaten raw. This ant's larvae can survive harsh droughts, so they are an excellent food source during dry periods.

◀ **Figure 1.45:** The spike-rush plant is an important food source for First Nations Peoples who live in northern Australia.

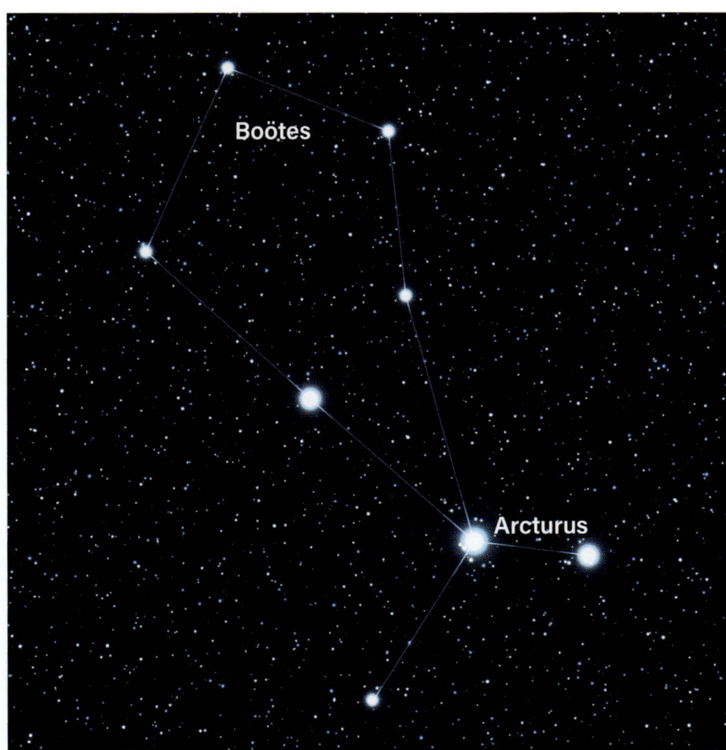

▲ **Figure 1.46:** The Arcturus star is the fourth-brightest star in the sky. It is part of the Boötes constellation.

Learning Ladder

Observing the universe

1. Identify two branches of science relevant to observing the stars and predicting the weather.
2. Explain why it is important for the Zugubau Mabaig to share his knowledge with others.
3. Select one example of how First Nations Peoples use their observations of the stars. Explain this example in detail.
4. Imagine you live in Far North Queensland and it is your job to carefully observe the stars. One night, you observe the stars twinkling. Propose what this means.
5. Discuss how observing the stars and understanding First Nations astronomy help with sustainable practices in Australian agriculture processes during droughts.

Observing

see page 316

1. Describe the observations that the Kamilaroi and Euahlayi Peoples use to determine when it is a good time to collect emu eggs.
2. Describe how First Nations Peoples pass on knowledge about how observations of the stars provide information about the weather.
3. Explain how you could accurately record your observations of the stars.
4. Outline what you can conclude about emu mating season when the Emu in the Sky constellation becomes visible.
5. Discuss how the observations of the Meriam Elders enabled them to pass down knowledge about predicting weather patterns.

In context

Explain how we can use First Nations Peoples' astronomical knowledge to develop Western scientific knowledge of Australia's weather patterns.

Success criteria

- I can explain some of the many ways that First Nations Peoples use the stars.

1·16 ▸ Predicting seasonal events based on observations of the stars

Learning intention

At the end of this lesson, I will be able to describe how Aboriginal and Torres Strait Islander Peoples predict seasonal events based on their observations of the stars.

Key term

brumation: the state of dormancy (inactivity) reptiles undergo in winter (similar to hibernation)

Content group: Aboriginal and Torres Strait Islander Peoples' cultural knowledges of astronomy

The stars have been important sources of information for First Nations Peoples for thousands of years. Aboriginal and Torres Strait Islander Peoples have long predicted seasonal phenomena based on their observations of the stars.

Stars are used to predict the weather and animal behaviour

Aboriginal and Torres Strait Islander Peoples use many stars and star clusters to predict seasonal phenomena, such as weather and animal behaviour. For example, the Pitjantjatjara People – whose Country is the Central Australian Desert, near Uluru – use the appearance and position in the night sky of the Seven Sisters (Pleiades) star cluster to predict the beginning of the winter frost.

The Pitjantjatjara People know that when the winter frost begins, many reptiles go into **brumation** and so these animals cannot be hunted at this time. The First Nations People of the Central Australian Desert also know that when the Seven Sisters star cluster rises in the sky at dawn (in June), dingos begin having their pups. This knowledge affects decisions about how to manage Country to make sure there are plenty of small animals for the dingos to hunt to support the nursing dingo mothers and their newborn pups.

Figure 1.47: This is a painting of the Seven Sisters star cluster by Gabriella Possum Nungurrayi.

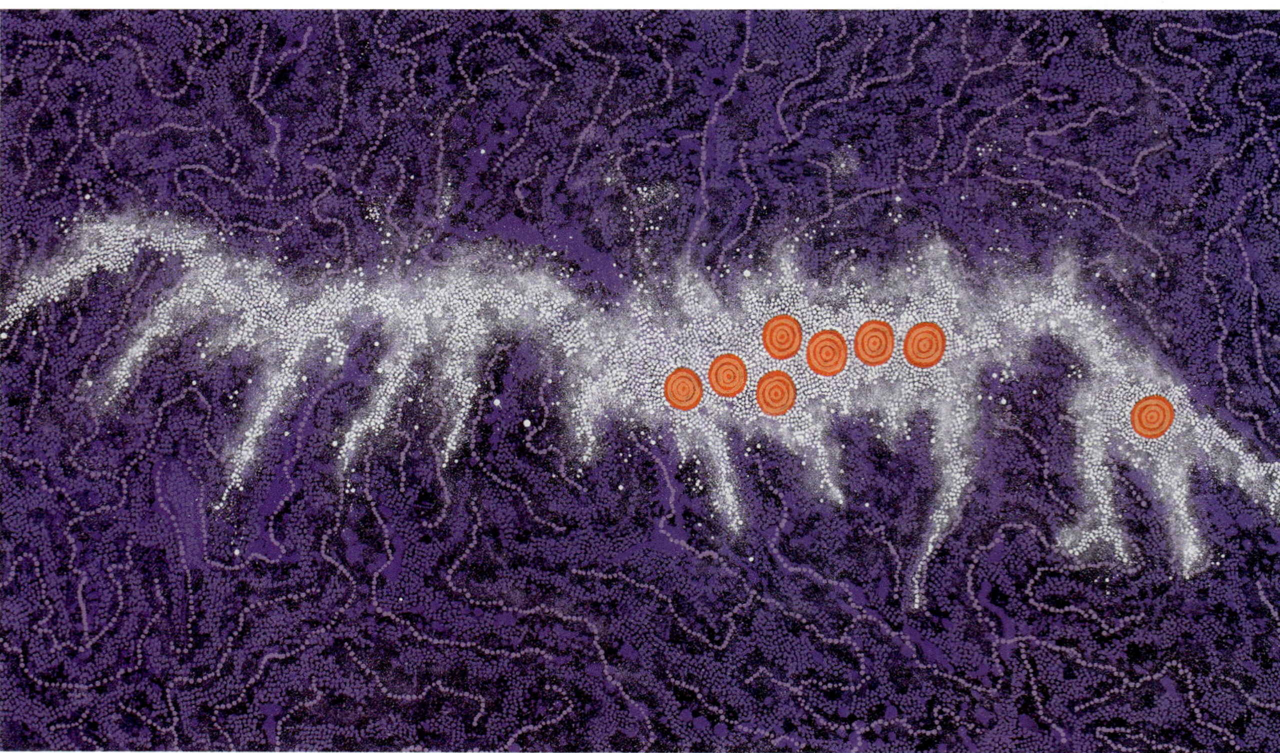

Gabriella Possum Nungurrayi, Anmatyerre People, Northern Territory, born Mt Allen, Northern Territory, 1967. *Seven Sisters, Milky Way Dreaming* 2002, Melbourne. Synthetic polymer paint on canvas 115.0 × 204.0 cm. Santos Fund for Aboriginal Art 2002 Art Gallery of South Australia, Adelaide 20022P8

Stars are used to predict plant cycles

Aboriginal and Torres Strait Islander Peoples also use celestial observations to predict the cycles of plants. For example, in the Torres Strait, the appearance of the Kek star – also called the yam star and Achernar by Western astronomers – indicates it is time to begin harvesting yams.

A yam is a root vegetable that grows underground, which means it is difficult to predict when yams are ready to be harvested without digging them up.

The best time to harvest yams is at the end of the wet season, when they have had plenty of water to help them grow and before the conditions of the dry season make them inedible.

The Kek star appears in late March, towards the end of the wet season. When Aboriginal and Torres Strait Islander Peoples see it, they know the dry season is coming.

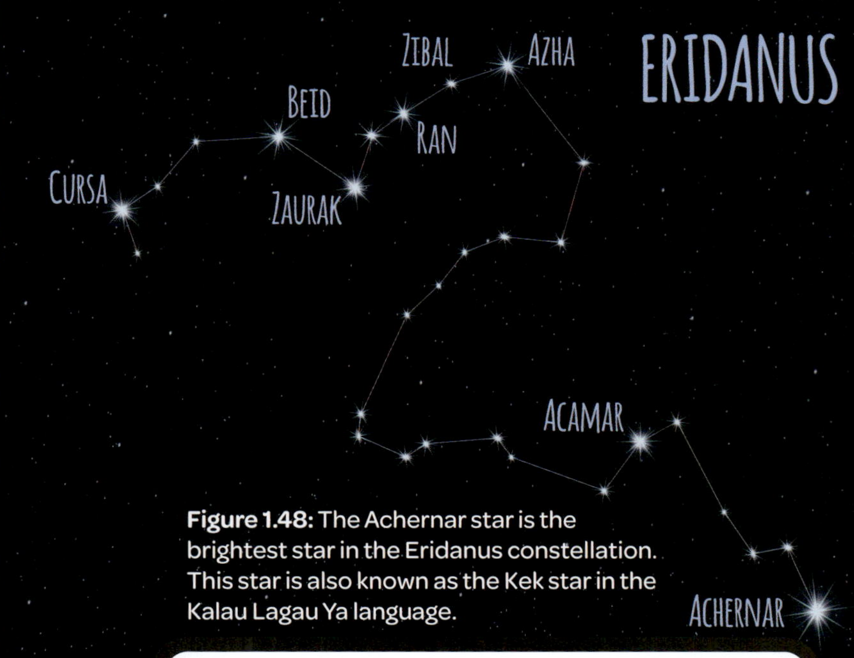

Figure 1.48: The Achernar star is the brightest star in the Eridanus constellation. This star is also known as the Kek star in the Kalau Lagau Ya language.

Figure 1.49: Dingos have pups once a year; June to August are critical feeding months for newborn pups. The start of this period corresponds to when the Seven Sisters star cluster first rises in the sky at dawn.

Learning Ladder

Observing the universe

1. Identify the branches of science that are relevant to knowing how to grow yams and when to harvest them.
2. Explain why it is important to share knowledge about the stars and how they connect to events that happen on Earth.
3. Explain how the Pitjantjatjara People use their observations of the Seven Sisters star cluster.

Observing

see page 316

1. Describe the senses that you think the Pitjantjatjara People use to determine whether it is the right time to harvest yams.
2. Identify a scientific tool that could enhance observations about a seasonal event.
3. Describe how you could accurately record your observations of a seasonal event.
4. Explain what can be inferred by the appearance of the Kek star in the sky in late March.
5. The Seven Sisters star cluster is approximately 442 light years from Earth. How far away are these stars in kilometres?

In context

Explain how we could use First Nations Peoples' understandings of astronomy to support ecological practices in contemporary Australia.

Success criteria

- I can describe how Aboriginal and Torres Strait Islander Peoples predict some seasonal phenomena based on their observations of the stars.

1·17 ▶ Observing the universe in context

Learning intention

At the end of this lesson, I will be able to explain how advancements in scientific technology have enabled scientists to increase our knowledge of the world and the universe.

Key terms

galaxy: a large system of stars

nebula: an enormous cloud of dust and gas in space

Content group: Observing the universe in context

Advancements in technology have made it possible for scientists to increase our knowledge of the world and the universe. These advancements include the improvement of scientific equipment, such as telescopes and rockets.

Figure 1.50: Astronauts took photos and collected lunar soil samples during the *Apollo 11* mission.

The telescope changed our view of the universe

Until the early 20th century, scientists working in the field of astronomy thought that the entire universe consisted of a single **galaxy**. They also thought that all the stars were relatively close to Earth.

This idea was challenged in the 1920s by the American astronomer Edwin Hubble (1889–1953). Hubble used a 2.5-metre telescope to study a star in what was then called the Andromeda Nebula. His observations led him to conclude that:

- Andromeda is not close to Earth
- Andromeda is not a **nebula** but a separate galaxy (now known as the Andromeda Galaxy)
- there are billions of galaxies in the universe
- each galaxy contains millions of stars.

Hubble's inferences completely changed our understanding of the universe.

Rocket technology put humans on the Moon

The *Apollo 11* mission

On 16 July 1969, a spacecraft carrying three astronauts – Neil Armstrong, Michael Collins and Edwin 'Buzz' Aldrin Jr – was launched into space using a rocket. Four days later, the spacecraft, with the men safely inside, landed on the Moon. That same day, 20 July 1969, Armstrong stepped out of the lunar landing module and onto the surface of the Moon. This was the first time any human had walked on the Moon. Since this historic mission, humans have landed on the Moon four times.

Landing on the Moon has increased our knowledge of the universe

During the missions to the Moon, astronauts have taken photos and collected rock samples from the Moon's surface (see Figure 1.50). Astronauts have collected over 382 kilograms of lunar samples.

Many scientists – such as geologists, who specialise in studying Earth – have examined these samples and developed theories about the origin of the Moon. The lunar samples show that some of the Moon's matter comes from Earth, and some comes from another source.

The current theory of the Moon's origin is that a very young Earth was hit by a stray body about half its size. This collision threw debris out around Earth, which eventually collected to form the Moon.

Australian astronomers have made many discoveries

Australian astronomers are collaborating with scientists from all over the world to improve our understanding of the universe.

Parkes Murriyang Telescope

The Parkes Murriyang Telescope (see Figure 1.51) is a radio telescope located in Parkes, New South Wales. For decades, this telescope has been used by astronomers from Australia and around the world to make observations and to assist in scientific endeavours.

For example, in July 1969, the Parkes Murriyang Telescope was one of many telescopes around the world that assisted scientists at NASA to maintain contact with the astronauts in the *Apollo 11* mission. Without the Parkes Murriyang Telescope, NASA would have lost communication with their astronauts when the Moon was on the same side of Earth as Australia.

Figure 1.51: The Parkes Murriyang Telescope in New South Wales has been used by astronomers from all over the world.

Learning Ladder

Observing the universe

1. Identify two types of scientists (that is, from two branches of science) involved in missions to the Moon and in studying lunar samples.
2. Describe how Australian astronomers have collaborated with scientists from other countries.
3. Explain how Edwin Hubble's observations improved our understanding of the universe.
4. Explain how changing the way we investigate the Moon (for example, by using robots) affects our ability to understand how the Moon is structured.

Observing *see page 316*

1. Describe some observations scientists can make about the Moon using just their senses.
2. Describe how Edwin Hubble used scientific equipment to enhance his observations of the universe.
3. Consider the astronaut in Figure 1.50. Explain how he is making sure his observations are accurately recorded.
4. Identify the inferences Edwin Hubble made based on his observations of the universe.

Success criteria

- I can explain how advancements in scientific technology have made it possible for scientists to improve our understanding of the world and the universe.

Observing the universe summary

Nature of science

The purpose of science is to build knowledge and understanding of the universe. This is achieved through observation, experimentation and analysis.

Science is divided into different branches of knowledge.

Scientists work together and build on each other's work.

Scientific theories and laws are based on repeated experiments and observations that describe or predict natural phenomena.

Practice of science

All investigations have a degree of risk. When practising science, it is important to use good safety practices to make sure you are not harmed.

Scientists make observations and draw inferences from those observations that explain natural phenomena.

Observations made using measuring equipment are more accurate and reliable than observations obtained using only our senses.

Good scientific investigations have independent, dependent and controlledvariables.

Space science

Predictable and observable phenomena on Earth – including seasons and eclipses – are caused by the relative positions of the Sun, Earth and Moon.

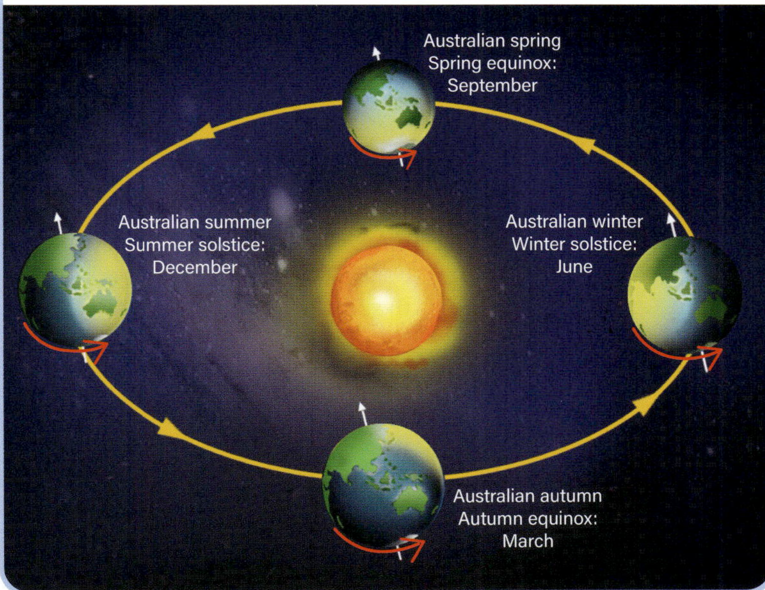

The Moon blocks the Sun's light during a solar eclipse.

Earth blocks the Sun's light during a lunar eclipse.

Aboriginal and Torres Strait Islander Peoples' cultural knowledges of astronomy

Some First Nations Peoples use their observations of the phases of the Moon to understand, predict and use the ocean tides.

Some First Nations Peoples use their observations of the stars to predict the weather and animal behaviour, and to identify when to source different foods.

Masterclass

Steps in progression

		1	2
Content	Observing the universe	Identify scientists from three branches of science who may collaborate to learn more about Mars.	Describe why it is so important for scientists to work together to learn about Mars.
Processes	Observing	While on the Mars research station, what is one observation scientists could make using their: **a** eyes? **b** ears? **c** hands?	Describe how two pieces of scientific equipment shown in Figure 1.54 could be used by the scientists to enhance their observations.
Processes	Conducting investigations	Identify three processes that would help scientists working on Mars to conduct investigations where they are looking at the soil structure and composition of the planet.	Describe three pieces of scientific equipment that would be necessary for scientists to take to Mars. Describe why each item is needed for scientific investigations.

Observing the universe: eyes on Mars

In the desert in Utah in the United States, a Mars research station has been set up. At this station, scientists can conduct first-hand investigations as if they were on Mars! For example, researchers only use the materials they bring with them, and when they head out to explore the surrounding desert and collect samples, they do so in special space suits.

In 2019 and 2020, a team of Australian scientists visited this research station as part of the Expedition Boomerang missions, to help them understand what living on Mars would be like. These scientists continue to look at opportunities to collaborate with others so that a research facility on Mars can become a reality.

Figure 1.52: The Mars desert research station is in a desert in Utah, United States.

Demonstrate your understanding

3	4	5	
Explain how learning more about living on Mars may help scientists to better understand our solar system.	Identify three investigations scientists could conduct on Mars. For each, identify the independent, dependent and two controlled variables.	Analyse how three scientific advancements have contributed to scientists' understanding of Mars.	
Which pieces of equipment do scientists need to accurately record measurements when they are collecting rock samples from the surface of Mars?	Explain what you can infer about the size and layout of the research station planned to be built on Mars, based on the bird's-eye view of the station in Utah in Figure 1.52.	Consider the investigations you identified in Step 4 of the knowledge ladder. For each investigation, explain what questions could be answered by the data the scientists collect.	Science how-to p. 316
Identify three safety practices that scientists would need to implement when working on Mars. Explain why each practice is important.	Analyse the processes scientists would need to use to ensure that the data they collect on Mars is accurate.	Explain in detail three processes you could implement to improve the accuracy of your investigations and data recordings.	Science how-to p. 326

Figure 1.53: Australian astronomers simulating conducting a first-hand investigation on Mars during an Expedition Boomerang mission

Figure 1.54: During the Expedition Boomerang mission, researchers practise working in laboratories, as they would on Mars.

Focus area 2

2.0 Forces

Figure 2.1: Forces keep planets and spacecraft in motion.

Forces are acting all around us. Forces keep Earth moving around the Sun, protect us from cosmic rays, and provide efficient transport. Forces are everywhere, all of the time.

Forces are also acting *on us* all of the time. They bring us back to the ground after we jump in the air and they make our hair stick up if we rub it with a balloon. Many different forces push and pull us in all directions. Our understanding of forces allows us to achieve incredible things, from constructing the pyramids to landing spacecraft on distant planets.

Learning Ladder

The Learning Ladder contains the scientific content and processes you will learn in this chapter. Each area has five levels of progression. To move to higher levels, you need to practise activities at the earlier levels. This will help you develop the ability to complete tasks that are more complex.

Steps in progression	Forces (Content)	Questioning and predicting (Working scientifically processes)	Processing data and information (Working scientifically processes)	Analysing data and information (Working scientifically processes)	Problem-solving (Working scientifically processes)	Steps in progression
5	I can investigate the impacts of forces in real-world contexts.	I can formulate testable questions and predictions, considering variables and controls.	I can process and interpret quantitative and qualitative data from a range of sources.	I can evaluate data and information for accuracy, reliability and validity.	I can evaluate problem-solving strategies used to solve an identified problem.	5
4	I can compare the role of different forces in real-world contexts.	I can make predictions to explain observations or phenomena.	I can collect and present data in a range of appropriate formats.	I can draw conclusions based on patterns in data and information.	I can use given criteria to find solutions to scientific problems.	4
3	I can explain how forces are linked to real-world phenomena.	I can construct questions to investigate scientific concepts or problems.	I can construct appropriate graphs with headings and units for data.	I can explain relationships between datasets and information.	I can explain scientific problems and phenomena using cause-and-effect relationships.	3
2	I can describe different types of forces.	I can select questions to investigate scientific concepts or problems.	I can construct appropriate tables with headings and units for data.	I can describe trends from collected data and information.	I can suggest solutions to familiar scientific problems.	2
1	I can identify multiple forces.	I can make predictions based on prior knowledge and observations.	I can identify data from graphs, tables and digital sources.	I can identify trends within given data and information.	I can identify scientific problems.	1

Figure 2.2: Forces allow us to ski down slopes and bring us back to the ground.

2·1 ▸ Direct and indirect forces

Figure 2.3: A soccer ball is set in motion when kicked by a player.

Learning intention

At the end of this lesson, I will be able to explain forces as either direct (contact) or indirect (non-contact).

Key terms

direct force: a force that acts on an object in contact with another object; also known as contact force

field: an area of space where objects are affected by an indirect force

force: a push or pull on an object when it interacts with another object

free-body diagram: a simplified diagram of the forces acting on an object

friction: a force between two surfaces that are sliding, or trying to slide, across each other

indirect force: a force that acts on an object without the need for physical contact; also known as non-contact force

interact: act upon or have an effect on

newton: the unit for measuring force, SI unit N; 1 kg on Earth is equivalent to 10 N

tension: a pulling force exerted by each end of an object

Investigation 2.1

Balloon rockets, p.391

Content group: Forces in action

A **force** is a push, a pull or a twist. Forces occur when objects **interact** with each other.

Forces happen when objects interact

A ball sitting on the ground will not move until a force is applied to it. If you apply a force by kicking the ball, several things happen.

- The ball changes shape for a moment.
- The ball moves in the direction of the force from the kick.
- The ball's speed changes. Initially, the ball is not moving, then it moves, hits the ground and stops, rolls or bounces.

Energy is required to apply a force. The person kicking needs energy to kick the ball. The energy is transferred to the ball at the same time as force is applied to it.

Direct forces are in contact

A **direct force** acts on an object that is in contact with another object. For example, kicking a soccer ball is a direct force because there is physical contact between the foot and the ball.

Direct forces are also known as contact forces and include:

- **friction**, when two objects slide past each other
- **tension**, a pulling force exerted by each end of an object such as a string or rope.

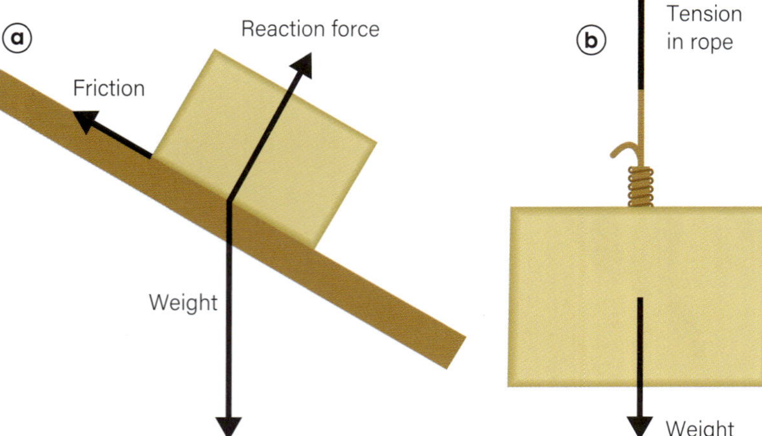

Figure 2.4: (a) Forces acting on an object moving down a ramp. (b) Forces acting on an object suspended by a rope.

We use free-body diagrams like Figure 2.4 to show the forces acting on an object. A **free-body diagram** is a simplified diagram of the forces acting on an object. The arrows represent the size and direction of the forces. Longer arrows represent bigger forces. Force is measured in units called **newtons** (N).

Indirect forces are not in contact

Some objects can exert forces on other objects even if they are not touching them. These forces are called **indirect forces** and act through a **field** – an area around the object.

The three main types of indirect forces are gravitational, magnetic and electrostatic. For example, Earth exerts a pulling force of gravity on the Moon because the Moon is in Earth's gravitational field. A magnet exerts a pulling magnetic force on iron filings in its magnetic field.

Figure 2.5: Planets in our solar system stay in their orbits around the Sun because they are in the Sun's gravitational field – the area where the Sun's force of gravity acts.

Learning Ladder

Forces

1. Identify each action force as a push, pull or twist.
   a Opening a door towards you
   b Moving your shopping trolley straight ahead
   c Using a key to unlock a door
   d Wringing out water from a wet towel
   e Playing tug-of-war
2. Identify an example of:
   a a direct force.
   b an indirect force.
3. Explain, in terms of forces, what happens when a paperclip is brought close to a magnet.

Processing data and information *see page 329*

1. Create a table like the one below to record your responses from Question 1 above. Include rows for parts **a–e** in your table.

Action force	Push	Pull	Twist
Opening a door towards you			

2. Investigate the forces required to open doors: push, pull and twist (turning a door handle). Which force is largest? Conduct three trials for each action and measure the force in newtons (N). Construct a table to record data for this investigation. (Hint: Include a column for the average force.)
3. Construct a column graph to represent the data from Question 2. (See the Science how-to section on page 370 for how to create a column graph.)
4. Construct a pie chart to represent the data from Question 1. Show the percentage of pushes, pulls and twists. (See the Science how-to section on page 371 for how to create a pie chart.)

In context

A soccer ball has a rubber bladder inside and a synthetic outer covering to prevent it from becoming waterlogged. This changes how the ball reacts to the force of a kick. Identify another piece of sporting equipment that is designed to optimise the forces applied to it during use, and explain how it achieves this.

Success criteria

- I can define forces.
- I can explain what direct and indirect forces are, using examples.

2·2 ▸ Balanced and unbalanced forces

Learning intention

At the end of this lesson, I will be able to use force diagrams to model balanced and unbalanced forces and predict the net effect acting in everyday situations.

Key terms

magnitude: the size of a force, an object or an amount of energy

net force: the resultant force when all forces acting on an object are added together and cancelled out

normal force: the force that acts from a surface pushing against an object

Investigation 2.2

Blowball, p.392

Content group: Forces in action

An object can be acted on by many forces at once. If these forces are the same size but acting in opposite directions, then they are balanced. Objects acted upon by balanced forces stay at rest or move at a constant speed in one direction. If the forces acting are not the same size, then the forces are unbalanced.

The forces on a stationary object are balanced

Even objects that are stationary (not moving) have forces acting on them. A ball at rest on the ground is not moving but is interacting with the ground. The ground applies a force on the ball, which is called the **normal force**. The normal force is the same size as the force of gravity, but it acts in the opposite direction, so the ball does not move. The forces acting on the ball are balanced, so the ball remains stationary. When forces are balanced, the overall force acting on the object is zero.

Figure 2.6 shows the forces acting on the ball as both a force diagram and a free-body diagram. The arrows show that the force of gravity acts downwards and the normal force (from the ground) acts upwards. The arrows are the same length, which means that the two forces are the same size (**magnitude**) and the ball is not moving up or down. In Figure 2.6b, the ball is represented by a point and the arrows representing the forces start at the centre of the ball.

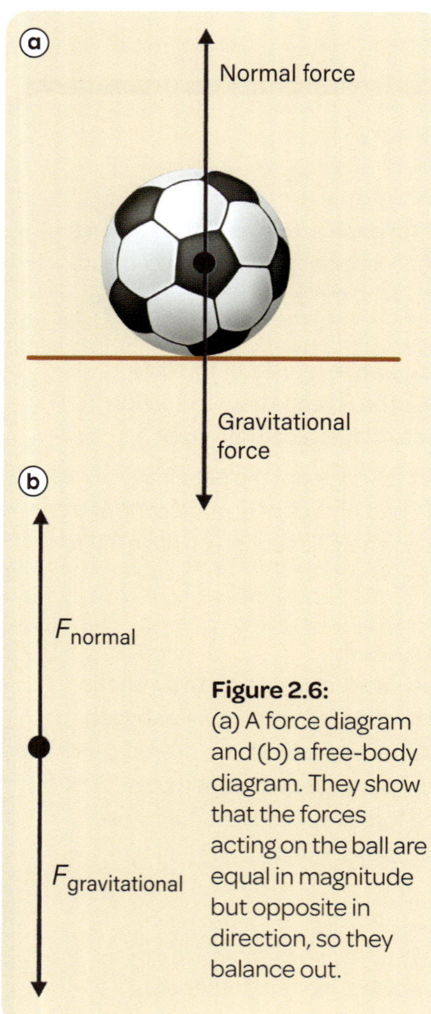

Figure 2.6: (a) A force diagram and (b) a free-body diagram. They show that the forces acting on the ball are equal in magnitude but opposite in direction, so they balance out.

Forces can become unbalanced

If someone kicks the ball, the forces become unbalanced. This means the ball will start moving. When the size of the net force is greater than zero, the result is a change in speed or direction (or both) of the object (Figure 2.7).

▼ **Figure 2.7:** Balanced and unbalanced forces on a ball

Balanced forces

Two forces are the same size but opposite in direction.

They cancel each other out.

The net force is equal to 0 N.

Unbalanced forces

Two forces are different sizes and opposite in direction.

They do not cancel each other out.

The force pushing the ball to the right is bigger than the force pushing to the left. The net force is equal to 5 N to the right.

Adding or removing a force can unbalance the forces on an object

The forces acting on objects, including humans, can become unbalanced in several ways. Figures 2.8 and 2.9 show two examples.

Figure 2.8: (a) A skateboarder stands still on his board – the forces acting on him are balanced and he is stationary; (b) then he pushes off, adding a force in the direction of his movement. The forces are no longer balanced and the skateboarder moves.

Figure 2.9: (a) A diver stands on a board – the normal force and gravity are balanced and he is stationary; (b) then he jumps off the board, removing the normal (upwards) force opposing the downwards force of gravity. The forces are no longer balanced and the diver moves down.

A moving object might also be affected by balanced forces

When forces on an object are balanced, the object keeps doing what it has been doing. The object can be stationary or moving.

When a car moves at a constant speed and direction, the forces acting on it are balanced. The thrust force from the car's engine is balanced by forces acting in the opposite direction (such as air resistance and friction). The normal force (pushing upwards from the ground) is balanced by the force of gravity (see Figure 2.10).

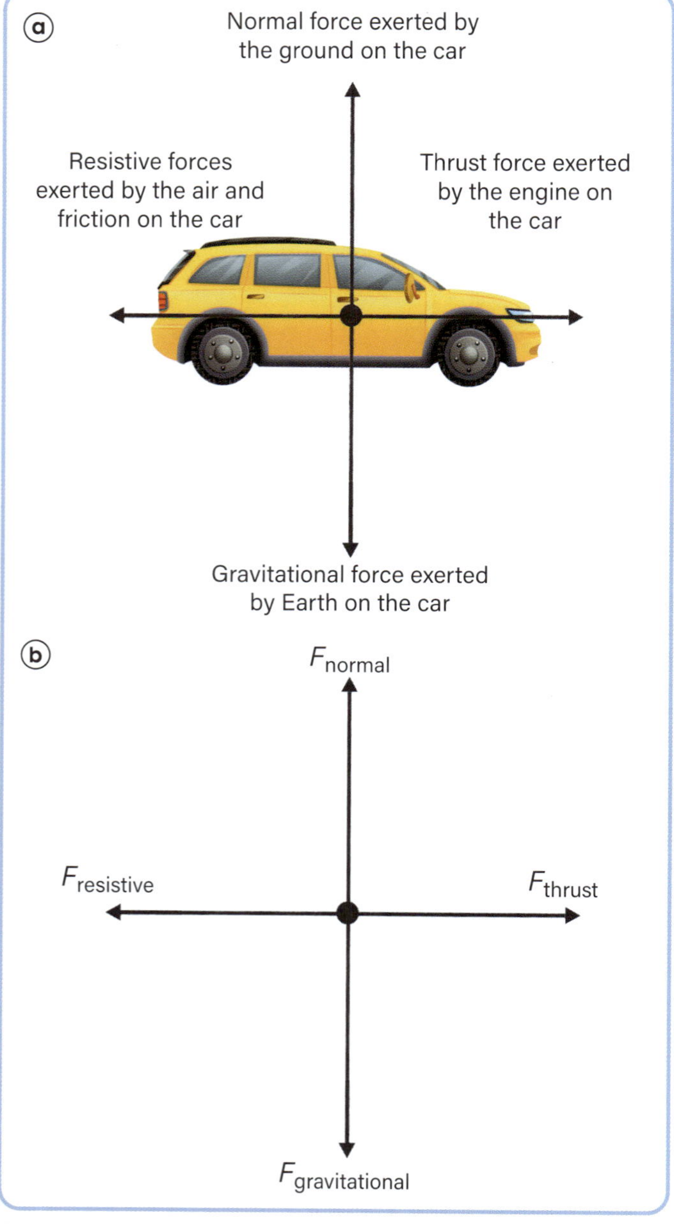

Figure 2.10: (a) A force diagram; and (b) a free-body diagram of the balanced forces acting on a car moving at constant speed

Forces can be added up to work out their effect

When many forces act on an object, you can calculate the direction and size of the overall force (**net force**) that is acting.

Imagine a car that has broken down and two people are pushing it to the side of the road (see Figure 2.11). They are applying a force to the car to get it moving, and they need to have energy to apply the force. The forces acting on the car become unbalanced, and it starts to roll forward.

You can work out the direction the car will move in, and the net force acting on it, as shown in Figure 2.11.

The two people apply forces of 275 N and 395 N, to oppose the resistive forces of 560 N. To calculate the net force, add the pushing forces and subtract the resistive forces because they act in the opposite direction:

$$(275 + 395) - 560 = 670 - 560$$
$$= 110 \text{ N}$$

So the car moves to the right with a net force of 110 N.

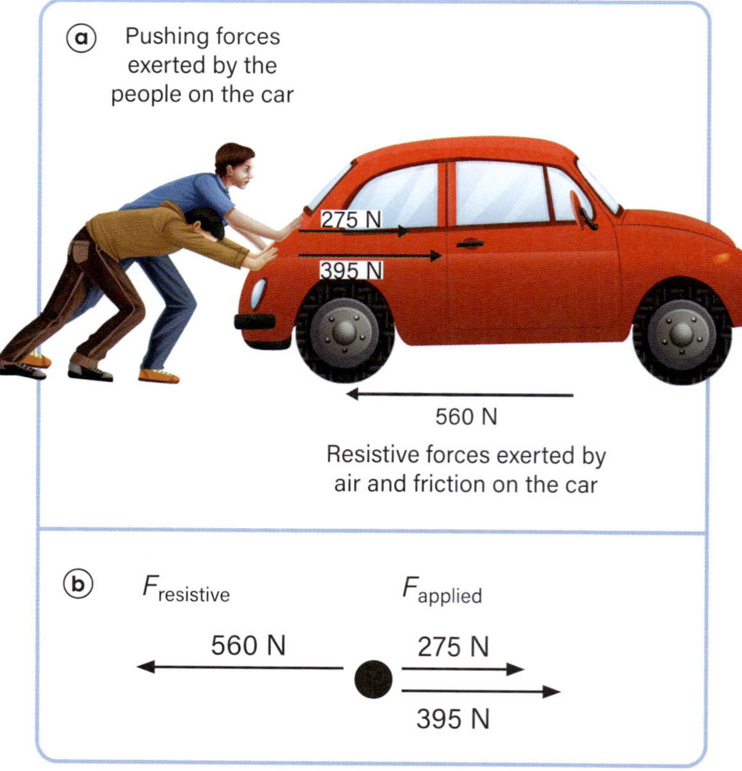

Figure 2.11: (a) The magnitude of forces acting when pushing a stalled car; and (b) a free-body diagram of forces acting on the car

Learning Ladder

Forces

1. Identify the force that:
   - **a** stops a person from falling through the ground.
   - **b** stops a person from floating off into space.
   - **c** slows a skateboarder down.
2. Identify two ways that unbalanced forces can change an object's motion.
3. Consider a stationary cyclist.
   - **a** Construct a free-body diagram of the forces acting on them.
   - **b** Describe two ways that the overall forces acting on them can become unbalanced.
   - **c** Draw a free-body diagram to show the forces in part b.
4. An object with a force acting on it in the opposite direction to its motion will slow down. Justify why.
5. Identify whether the following objects have balanced or unbalanced forces acting on them.
   - **a** A car that is speeding up
   - **b** A laptop resting on a desk
   - **c** A rolling ball that is slowing down
   - **d** An aeroplane travelling at a constant speed

Analysing data and information

see page 334

1. **a** Calculate the net force acting on each of the following objects and state which direction the object will move.

F = 75 N ← □ → *F* = 50 N

F = 200 N (up), *F* = 100 N (down)

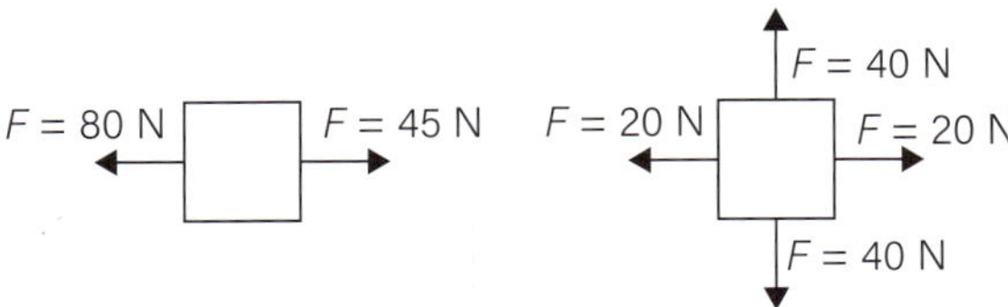

   b Identify the forces acting on this girl. Calculate the net force acting on her and state the direction she will travel.

2. Students conducted an experiment where they increased the mass of a cart and measured the force required to move the cart across a surface. Their results are shown in the table below.

Mass (kg)	Net force (N)
1	5
3	15
5	25
7	35
10	50

   - **a** Describe the trend in the data they have collected.
   - **b** Explain the relationship the students found in the data.

3. A box was pushed across the floor in a straight line for 10 metres. The net force exerted over this distance is shown in the graph below.

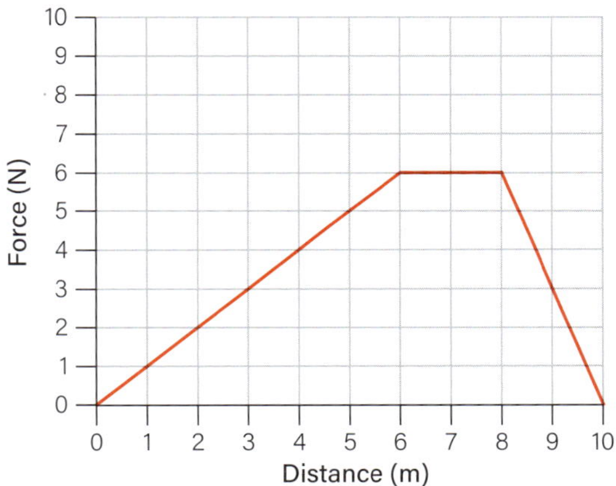

   - **a** Identify the forces acting on the box as it moved across the floor.
   - **b** What was the greatest net force applied, and between which distances was it applied?

In context

With the aid of force diagrams, explain the forces acting on a skydiver during each stage of their motion: jumping out of the aeroplane, falling at a constant speed, opening the parachute and then finally landing.

Success criteria

- I can explain what balanced and unbalanced forces are.
- I can draw force diagrams and free-body diagrams to show balanced and unbalanced forces.

2·3 ► Introduction to fields

Learning intention

At the end of this lesson, I will be able to use the term 'field' in describing forces acting at a distance.

Key terms

attractive force: a force that pulls objects towards each other

electric field: an area around a charged particle in which it exerts a force on other charged particles

electrostatic force: an indirect force between any objects with an electric charge

field line: a line used to show the direction of a force within a field

gravitational field: a region in which an object with mass can experience a gravitational force

gravitational force: an indirect force that attracts physical objects with mass towards each other

magnetic force: an indirect force that affects any object made of certain metals

radial field: a field in which the field lines radiate from a centre

repulsive force: a force that pushes objects away from each other

Investigation 2.3

Magnetic shielding, p.393

Content group: Forces in action

Indirect forces act through fields. Objects do not have to be touching each other.

Indirect forces act through fields

Indirect forces can work at a distance, without contact, as long as objects are in their fields.

- **Gravitational forces** affect objects containing matter.
- **Magnetic forces** affect objects made of certain metals, such as iron, nickel or cobalt.
- **Electrostatic forces** affect objects that have an electrical charge, even very weak charges.

We cannot always tell where a field ends.

The strength of a force varies within a field

Magnetic field lines

If you sprinkle iron filings near a bar magnet, most of them arrange themselves at the ends of the magnet (see Figure 2.12a). This shows that the strength of the magnetic force is not the same throughout the field.

Figure 2.12b shows the magnetic **field lines** around the bar magnet. The field is strongest where the field lines are closer together, at the ends of the magnet. You can see that the field lines match the pattern of iron filings in Figure 2.12a.

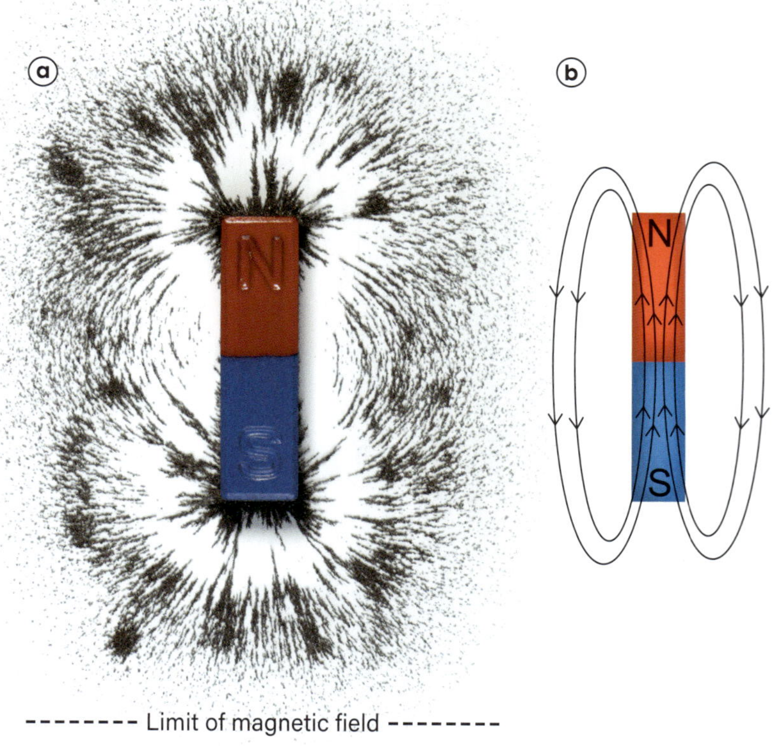

Figure 2.12: (a) Iron filings gather in the areas around a magnet where the magnetic field is strongest. Beyond the limits of the field, the filings are no longer attracted. (b) Field lines show how the different non-contact forces operate within the magnetic field.

Gravitational field lines

Any object that has mass has a **gravitational field** around it. This is an area in which other objects experience a force from the first object. Planets, people, even ants, all have gravitational fields, but the fields around smaller objects are too small to affect us. We only witness the effects of gravitational fields for larger objects such as moons, stars, planets and galaxies. Earth has a gravitational field – any object in its field experiences a gravitational pull towards Earth.

Figure 2.13 shows the shape of Earth's gravitational field. It is a **radial field** – all the field lines point towards the centre of Earth. The gravitational field gets weaker further from the centre of Earth. The arrows show that an object in the field would experience an **attractive force** because they are pointing towards Earth.

The Moon orbits Earth because it is in Earth's gravitational field and experiences an attractive force.

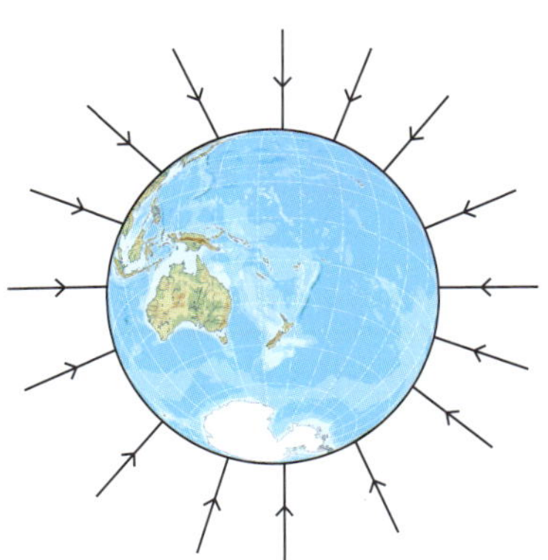

Figure 2.13: The direction of the forces in Earth's gravitational field

Electric field lines

An **electric field** exists around a charged object. The electric field radiates outwards from a positive charge and radiates inwards to a negative point charge, as shown in Figure 2.14.

Charged particles have a radial field. However, the electrostatic force within the electric field can be attractive (pulling) or **repulsive** (pushing), depending on the charged object.

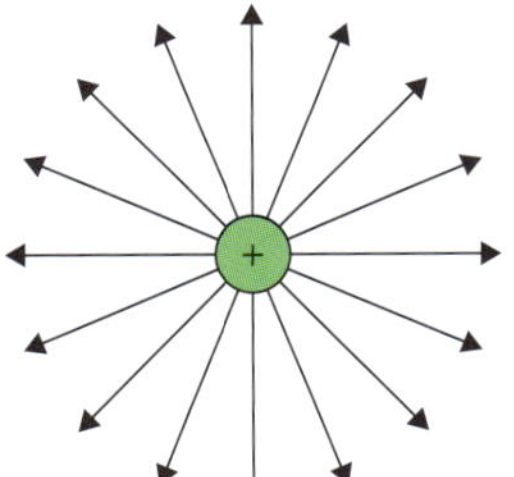

Electric field lines of a positive point charge point away from the charge.

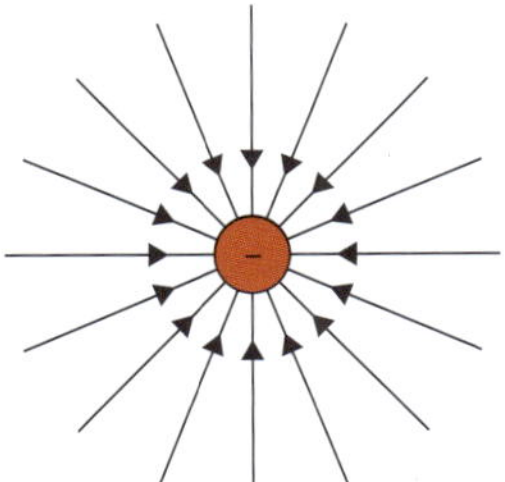

Electric field lines of a negative point charge point towards the charge.

Figure 2.14: Electric fields of individual charged particles (point charges)

Figure 2.15: A black hole is a massive amount of matter packed into a very small space. Black holes have such strong gravitational fields that nothing – not even light – can escape!

Learning Ladder

Forces

1. Outline what an indirect force is. Name three examples.
2. Describe how a named indirect force acts on a specific object.
3. Describe something that reduces the force on an object and is common between the three types of indirect forces.
4. Compare how Earth's gravitational field and a magnetic field would affect a nail as it moves away from them.
5. Outline the evidence that suggests fields generate forces of different sizes in certain areas.

Problem-solving

see page 339

1. Using Figure 2.12 to help you, draw some field lines around a horseshoe magnet.
2. Predict what would happen to the Moon's orbit around Earth if Earth's gravitational field was weaker. Suggest what problems, if any, this may cause.
3. Explain why objects eventually fall back towards Earth when they are thrown upwards.
4. Drivers are asked to turn off mobile phones while putting fuel in their car. Propose a reason why.

In context

Astronauts on a spacewalk are required to wear special clothing to protect their bodies. Research and describe what fields and forces may affect their bodies, and how the equipment helps to protect them.

Success criteria

- I can identify indirect forces as those that are caused by a field.
- I can describe the strength of the force within a field.

2·4 ► Electrostatic forces

Learning intention

At the end of this lesson, I will be able to describe the electrostatic forces exerted between charged objects.

Key terms

electron: a subatomic particle that orbits the nucleus of an atom; it is negatively charged

electrostatic charge: the electric charge on the surface of an object

neutron: a subatomic particle located in the nucleus of an atom; it is neutrally charged

nucleus: the centre of an atom, which contains protons and neutrons

proton: a subatomic particle located in the nucleus of an atom; it is positively charged

Investigation 2.4

Charging balloons, p.394

Content group: Forces in action

All matter is made up of atoms. At the centre of each atom is the **nucleus**. The nucleus is made up of uncharged particles called **neutrons** and positive particles called **protons**. Around the nucleus are one or more negative particles called **electrons**.

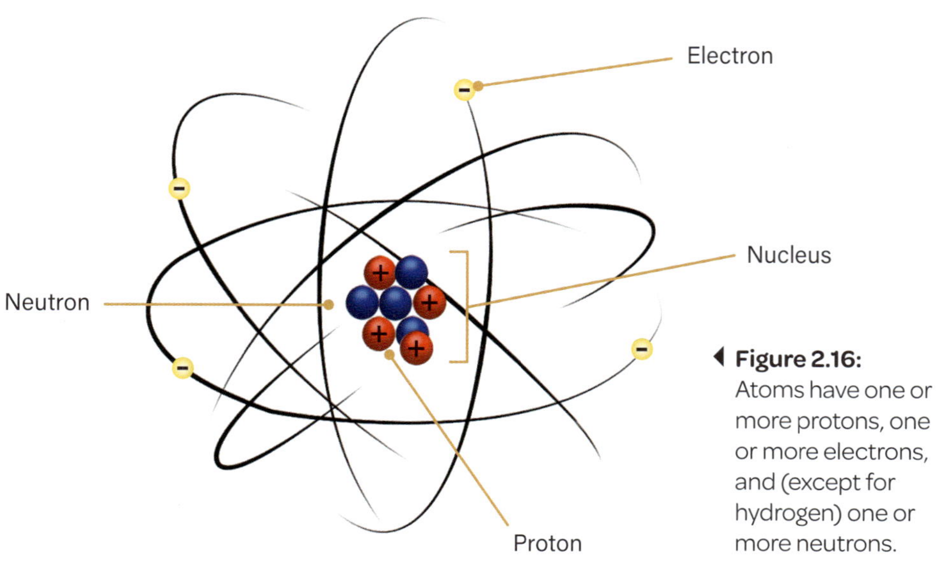

Figure 2.16: Atoms have one or more protons, one or more electrons, and (except for hydrogen) one or more neutrons.

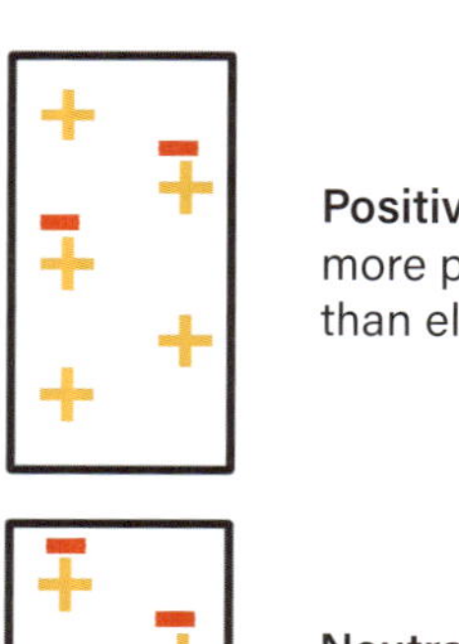

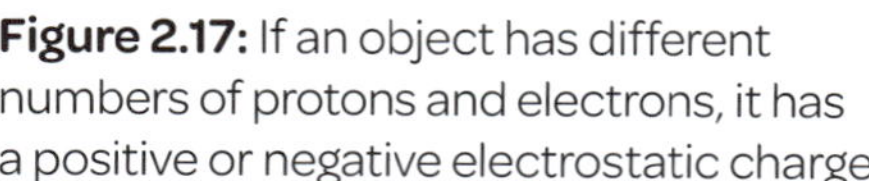

Figure 2.17: If an object has different numbers of protons and electrons, it has a positive or negative electrostatic charge.

Charged objects can have a positive or negative charge

Each proton and electron carries a tiny electric charge. The positive charge of a proton is equal but opposite to the negative charge of an electron. When the number of protons is the same as the number of electrons, the atom has no overall charge: it is neutral.

When an atom has more protons than electrons, it has a positive charge. When it has more electrons than protons, the atom has a negative charge. This gives an object an overall **electrostatic charge** and it can exert an electrostatic force on other objects, whether they are charged or uncharged (see Figure 2.17).

Like charges repel; opposite charges attract

When two objects with the same charge (positive or negative) are brought close together, they experience a repulsive force that pushes them apart. When two objects with opposite charges are brought close together, they experience an attractive force that pulls them even closer together. You can see this in the balloons in Figures 2.19–2.21. Balloons can be given an electrostatic charge by rubbing them with wool or rubber. Charged balloons will attract or repel other charged balloons that are close enough to be in their electrostatic field.

Figure 2.18: A Van de Graaff generator is a device that creates a build-up of electric charge. Here it has transferred electrons to the woman. Each strand of her hair has the same charge, so they repel each other. This is why her hair is standing on end.

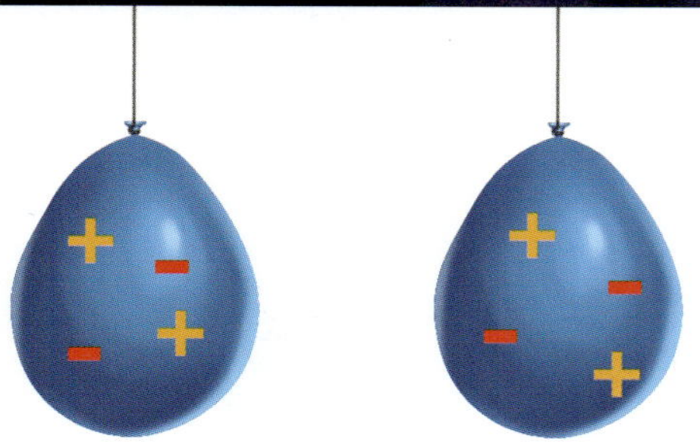

Learning Ladder

Forces

1. Identify the types of charges an object can have.
2. Describe how an object can become negatively charged.
3. Explain how electrostatic forces can cause two charged objects to have an attractive force between them.
4. Explain why your hair stands on its ends after you rub it with an inflated balloon.
5. Propose why electrons, and not other types of particles, move from the atoms of one material to another.

Questioning and predicting *see page 319*

1. Hundreds-and-thousands sprinkles often stick to the sides of their plastic container. Predict which two surfaces are rubbing together to cause this effect.
2. Rub a plastic comb vigorously with some wool, then hold it close to some tiny pieces of paper. You can make an observation about the paper's behaviour. Which of the following is an appropriate scientific question to investigate this phenomenon?
   - **A** How does the charge of a comb affect the movement of paper when they are brought closer together?
   - **B** Does a comb conduct electricity?
3. The more you rub a comb over paper, the greater the attraction between them. Construct a scientific question for this investigation.
4. Construct a hypothesis predicting what will happen when a comb is moved over paper.

In context

The triboelectric series is a list of objects ordered by how easily they gain or lose charge. Predict which of these materials would be most and least likely to gain a positive or negative charge: skin, wool, aluminium, paper, steel, rubber, plastic. Check your predictions by comparing your answers to an online triboelectric series.

Success criteria

- I can explain what it means for an object to be charged.
- I can describe the electrostatic forces exerted between charged objects.

Figure 2.19: Two balloons with neutral charge hanging close together won't attract or repel each other, so they hang straight down.

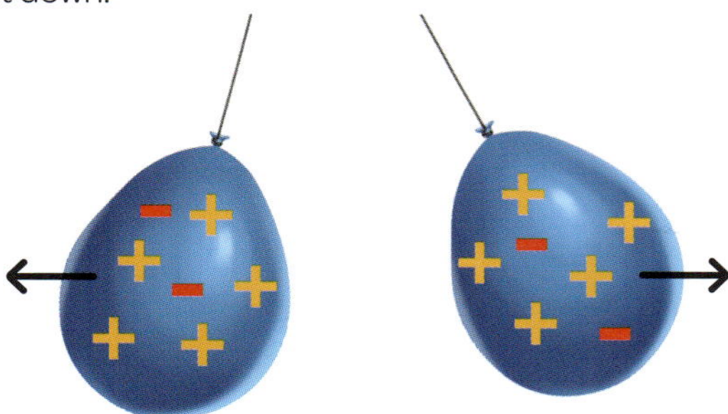

Figure 2.20: Two balloons close together and with the same charge repel each other, so they push away from each other. This is because they experience an electrostatic force. If the charged balloons are moved further apart, and are outside each other's electrostatic field, then they will not experience a force.

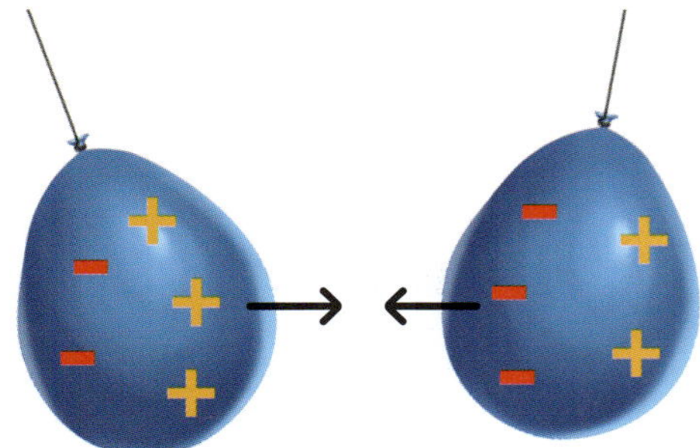

Figure 2.21: Two balloons close together and with opposite charges attract each other. The attractive forces pull them slightly together.

2·5 ▸ How charged objects behave

Learning intention

At the end of this lesson, I will be able to describe how charged objects behave.

Key term

electrostatic shock: the rapid movement of charged particles from one object to another

Content group: Forces in action

Figure 2.22: The opposite electrostatic charges in the atmosphere and the top of the building cause lightning to strike during a storm.

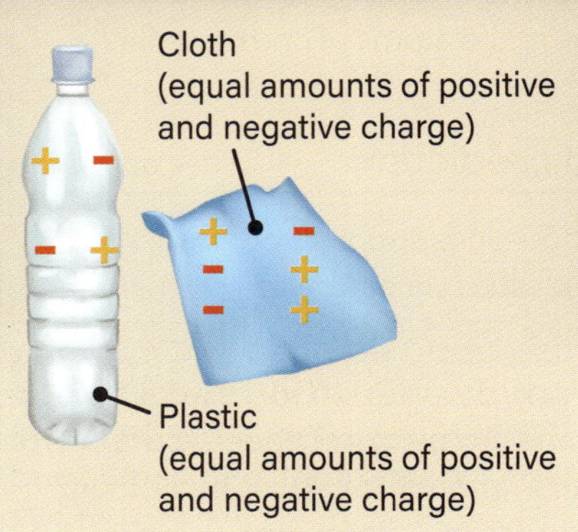

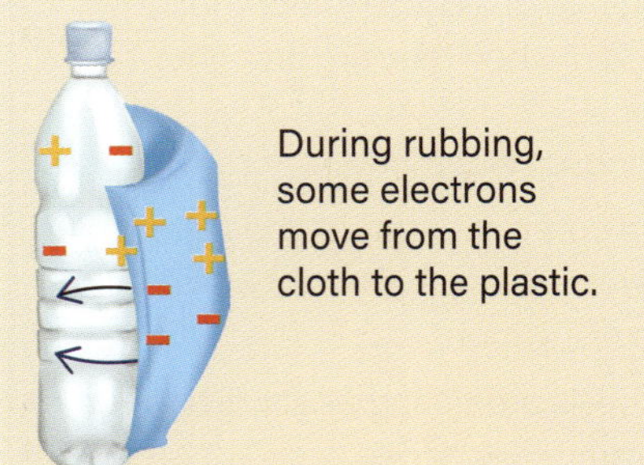

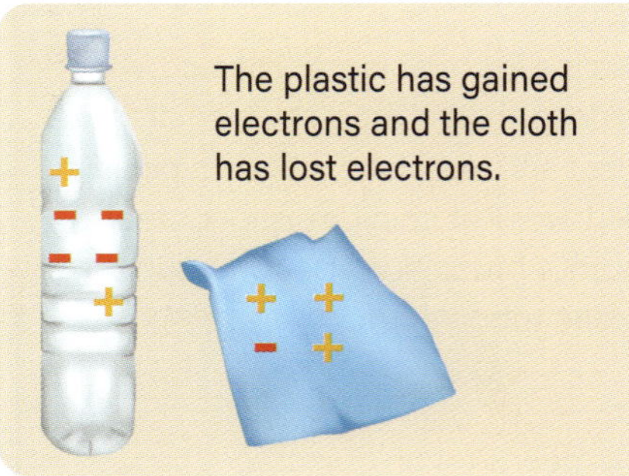

▲ **Figure 2.23:** When you rub plastic and cloth together, an electrostatic charge is made on both objects.

When two objects with the same charge are brought close together, the force between them pushes them apart, or repels them. When two objects with opposite charges are brought close together, the force between them pulls them together, or attracts them.

Objects gain and lose electrostatic charge

When electrons move onto or off an object, it becomes charged.

Some materials attract electrons more than other materials do. When these materials come into contact with a material that does not attract electrons as strongly, some electrons will move onto the electron-attracting material. For example, plastic attracts electrons more strongly than cloth does. So, if plastic and cloth are rubbed together, some of the electrons within the cloth transfer to the plastic. The plastic becomes negatively charged because it has more electrons than protons. The cloth becomes positively charged because it has more protons than electrons (see Figure 2.23).

Everyday items gain electrostatic charge

You can generate an electrostatic charge by walking around or touching items together. This is often called 'static electricity' and is caused by friction between certain objects.

An easy way to see this is to blow up a balloon and rub it on your hair. The material of the balloon attracts electrons more than your hair does, so electrons from your hair move into the material of the balloon, creating a slight negative charge. At the same time, your hair gains a slight positive charge because now it has more protons than electrons.

If the electrostatic charge is large enough, it can create a spark as the electrons move quickly back into the positively charged material. For example, if you take off a woollen jumper on a dry day, you may see small sparks.

Large-scale electrostatics can cause lightning

Lightning is an example of this movement of charge, or **electrostatic shock**, on a much larger scale. In a storm, two oppositely charged regions develop. The two charged areas can be within the storm clouds or in the clouds and the ground. These regions are separated by air, which acts as an insulator. When enough charge builds up, the air breaks down and there is a rapid movement of the charge. This discharge can be seen as lightning.

Figure 2.24: An electrostatic charge on your body may create a spark as it discharges onto a nearby metal surface.

Figure 2.25: Rubbing balloons on your hair will create a slight positive charge in your hair, causing the strands to repel each other.

Learning Ladder

Forces

1. Identify how objects can gain electrostatic charge.
2. Construct a diagram to show how you could positively charge an object. Show the charges on the object before and after charging it and how the charge moves.
3. Explain why you feel a 'zap' or see a small spark when you touch something that has been statically charged.
4. You and your friend take turns going down a slide and high-fiving each other. When you go down the slide, you both get a static shock when you do a high-five. But when your friend goes down, you don't get a shock when you high-five each other. Propose reasons for the difference.

Questioning and predicting

see page 319

1. An inflated balloon is rubbed on a person's hair. Predict what will happen to the hair.
2. Students want to conduct an investigation on how strong an electrostatic force needs to be to pick up small pieces of paper by rubbing a plastic pipe with a cloth. List the variables in this experiment, identifying which variables will need to be controlled, so that they could be used in a scientific question.
3. Construct a scientific question to investigate the concept in Question 2.
4. Construct a hypothesis for the investigation in Question 2.

In context

List all the times you have seen or felt static shocks. Identify the materials that were interacting to create the static charge. Compare your list with those of other students. Are there any materials in common?

Success criteria

- I can describe how charged objects behave.
- I can identify some ways that objects gain electrostatic charge.

2·6 ▸ Gravitational forces

Figure 2.26: This astronaut is floating in space, outside Earth's gravitational field.

Learning intention

At the end of this lesson, I will be able to use the concept of forces to describe the motion of objects in a gravitational field.

Key term

satellite: an object that orbits a planet; Earth has a natural satellite (the Moon) and many purpose-built satellites (referred to as artificial satellites)

Investigation 2.6

Measuring gravity, p.395

Content group: Forces in action

Gravitational forces act towards the centre of the object that produces them. That is why, wherever they are, all objects fall towards Earth's centre.

Gravity is always an attractive force

All objects with mass make a gravitational field. Within this field, any other object will have a force generated on it. The size of this force depends on the:

- mass of the object generating the force
- distance between the two objects.

The size of gravitational force varies, but the direction does not. Any force due to gravity will always be an attractive force. This means that it pulls objects towards the centre of the mass generating the gravitational field.

Wherever you are on Earth, if you drop something, it falls to the ground. This is because the attractive force is directed to the centre of Earth's mass. We are affected by Earth more than any other object because it is the largest and nearest object to us. The Sun is much larger, but it is so far away that it has little effect on an object that is dropped, compared to Earth's gravity.

You can draw field lines to represent the area where a gravitational force will be generated on an object:

- Stronger fields are shown by drawing field lines close together; weaker fields are shown with field lines further apart.
- Gravitational field lines point in the direction that the force is acting in.

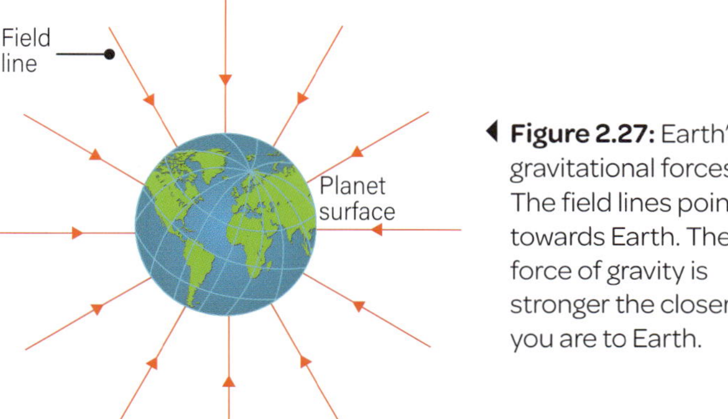

Figure 2.27: Earth's gravitational forces. The field lines point towards Earth. The force of gravity is stronger the closer you are to Earth.

In Figure 2.27, you can see that the gravitational field decreases in strength with increasing distance from the centre of Earth. The red field lines are closer together near Earth and further apart away from Earth.

Gravitational forces change with distance

On Earth's surface, gravity is constant because the distance from the centre of Earth is about the same everywhere. Further away from Earth, the gravitational force is smaller and the field lines are further apart.

Imagine a mattress with a bowling ball placed in the centre, making a hollow. If you place some marbles at different points around the mattress, eventually they will all roll into the hollow made by the bowling ball. Those at the very edges may roll slowly at first, but they will roll faster as they get closer to the bowling ball, where the slope becomes steeper. This is very similar to the way objects are affected by Earth's gravity: the force increases as the objects move nearer.

Figure 2.28: We can use gravity to have fun in many sports, such as skydiving.

Gravity holds objects in orbit

Gravity is the force that holds the planets in orbit around the Sun and **satellites** in orbit around Earth. The closer a planet is to the Sun, the stronger the Sun's gravitational pull on it and the faster the planet moves in its orbit. The further it is from the Sun, the weaker the Sun's gravitational pull on it and the slower it moves. Table 2.1 shows how long each planet takes to orbit the Sun.

You might wonder why gravity does not cause a planet to crash into the Sun. The Sun's gravitational force constantly pulls the planets towards it. However, the planets are travelling fast enough to balance the gravitational effect. This delicate balance causes the elliptical orbits of planets around the Sun, which is the centre of gravity for our solar system.

Table 2.1: The number of Earth days taken by each of the planets to orbit the Sun, in order of closest to furthest away

Planet	Number of Earth days
Mercury	88
Venus	225
Earth	365
Mars	687
Jupiter	4331
Saturn	10 747
Uranus	30 589
Neptune	59 800

Satellites can orbit planets, in the same way that planets can orbit the Sun. The only difference is the central object being orbited. Earth is the central object, and the centre of gravity, for satellites orbiting our planet. The distance at which the satellites orbit from the surface of Earth depends on how fast they travel.

Figure 2.29: The planets of our solar system, showing the order of distance of their orbits from the Sun

▲ **Figure 2.30:** Many satellites orbit Earth. Satellites have many uses, such as communication, weather forecasting and navigation. The Starlink satellites are a global network of communication satellites.

Learning Ladder

Forces

1. Identify the force that keeps you on the ground.
2. Describe what gravity is and what its strength depends on.
3. Explain the relationship between the speed at which a planet orbits the Sun and its distance from it.
4. Explain why satellites orbit Earth rather than crashing into the surface.
5. Predict what would happen to the motion of the planets if the Sun had a stronger gravitational field. Justify your answer.

Processing data and information *see page 329*

1. Refer to Table 2.1.
   - **a** How many Earth days does it take for Mars to orbit the Sun?
   - **b** Calculate how many seconds this is.
2. One day on Venus lasts approximately 243 Earth days. Using Table 2.1, construct a new table that shows how many 'Venus days' it takes planets to orbit the Sun. (Hint: Divide the current number of days by 243.)
3. Construct a graph to display the information in Table 2.1. (See the Science how-to section on page 331.)
4. Convert the data from your table in Question 2 into a pie chart that shows the difference in time between the planets' orbits around the Sun. (See the Science how-to section on page 371.)

In context

Does Earth exert a greater gravitational pull on the Moon than the Moon does on Earth? Explain your answer.

Success criteria

- I can use the concept of forces to describe the motion of objects in a gravitational field.

2·7 ▸ Weight and mass

Learning intention

At the end of this lesson, I will be able to:

- distinguish between the terms 'mass' and 'weight'
- calculate the weight of an object.

Key terms

acceleration: the rate at which speed changes over time, or the rate at which you fall due to gravity

mass: the amount of matter that a physical body contains

weight: the force of a gravitational field on the mass of a body

Content group: Forces in action

Although people talk about weight in terms of grams and kilograms, these are actually units of mass. What is the difference?

An object's weight depends on gravity

An object's **mass** is the amount of matter it contains. Mass is measured in grams and kilograms. A ball contains a specific amount of matter. Wherever that ball is, even if it is on the Moon, its mass does not change.

An object's **weight** is a measurement of how much gravitational force is pulling on the object. Weight is measured in newtons (N). Changing the gravitational force acting on an object will change its weight. The Moon exerts a small gravitational force, which is why astronauts on the Moon weigh less and can move around with huge jumps. It is also why astronauts on the International Space Station float, even though their bodies still have mass.

Trevor's mass on Earth is 60 kg. His weight on Earth is almost 600 N.

Trevor's mass on the Moon is also 60 kg. However, his weight on the Moon is almost 100 N.

▲ **Figure 2.31:** No matter where you are, your mass stays the same. However, your weight changes depending on gravity.

Figure 2.32: Astronaut Eileen Collins, weightless and floating in *STS-114* ISS Commander Space Shuttle Mission

Weight can be calculated with a formula

Your weight depends on where you are within the gravitational field. The further away you are from the centre of a mass, the weaker the gravitational field strength. This means you weigh less further away from Earth's surface.

How do you know what your weight is if your mass is 60 kilograms? We can calculate this by multiplying your mass (m) by the **acceleration** due to gravity (g) of the planet or moon you are on. This can be written in the following formula, where F_g is your weight (the force of gravity).

$$F_g = mg$$

where:

m = mass (kg)

g = acceleration due to gravity (m/s^2)

On Earth's surface, the acceleration due to gravity (g) is approximately 9.8 m/s^2. So, if you are 60 kilograms on Earth, then your weight is:

$$F_g = 60 \text{ kg} \times 9.8 \text{ m/s}^2 = 588 \text{ N}$$

Table 2.2: Some bodies of the solar system and their acceleration due to gravity, mass and the ratio of gravity on the body and gravity on Earth (g/g-Earth)

Body	Acceleration due to gravity, g (m/s^2)	Mass (kg)	$\frac{g}{g\text{-Earth}}$
Sun	274.13	1.99×10^{30}	27.95
Mercury	3.59	3.18×10^{23}	0.37
Venus	8.87	4.88×10^{24}	0.90
Earth	9.81	5.98×10^{24}	1.00
Moon	1.62	7.36×10^{22}	0.17
Mars	3.77	6.42×10^{23}	0.38
Jupiter	25.95	1.90×10^{27}	2.65
Saturn	11.08	5.68×10^{26}	1.13
Uranus	10.67	8.68×10^{25}	1.09
Neptune	14.07	1.03×10^{26}	1.43

Note: This table expresses some numbers in scientific notation. For an explanation of scientific notation, see the Science how-to section on page 366.

On the Moon, the acceleration due to gravity (g) is approximately 1.6 m/s^2. So, your weight would be:

$$F_g = 60 \text{ kg} \times 1.6 \text{ m/s}^2 = 96 \text{ N}$$

You can use the second and third columns of Table 2.2 to calculate your weight on different planets.

Learning Ladder

Forces

1. The force on an object due to gravity is also known as what force? What metric unit is it measured in?
2. Describe the difference between mass and weight.
3. Explain how an object's weight can change even though its mass remains constant.
4. Compare your weight on Earth and on the Moon.
5. On the Moon, an object weighs 90 N. Objects weigh six times more on Earth than they do on the Moon. Calculate the object's weight on Earth.

Analysing data and information *see page 334*

1. The final column of Table 2.2 shows the ratio of g for different bodies in our solar system and g on Earth. Identify the:
   - **a** body in our solar system where you would weigh the most.
   - **b** body in our solar system where you would weigh the least.
   - **c** relationship between the mass and radius of each body in our solar system.
2. Describe the relationship between weight and gravity, based on the weights of Trevor in Figure 2.31.
3. Use the information in Table 2.2 to compare your weight on the Moon and on Neptune.
4. Choose three other planets in our solar system, then use Table 2.2 to calculate your weight on each of these planets.

In context

Astronauts on a mission in space are often described as being 'weightless'. Research and explain what this term means and how astronauts need to prepare to live in these conditions.

Success criteria

- I can explain the difference between mass and weight.
- I can calculate the weight of different objects using $F_g = mg$.

2·8 ► Magnetic forces

Learning intention

At the end of this lesson, I will be able to describe how magnets attract or repel each other based on their polarity.
I will also be able to draw the magnetic fields of magnets.

Key terms

like poles: magnetic poles that are the same (both north or both south); they repel each other

magnetic pole: one of the two ends of a magnet

unlike poles: magnetic poles that are different (south–north or north–south); they attract each other

Investigation 2.8

Magnetic accelerator, p.396

Content group:
Magnets in everyday life

A magnet is made of a material that has two **magnetic poles**, north and south. The direction of the magnetic force in a magnet is from the north pole to the south pole. When poles are brought near each other, they either pull together or push apart. Magnetic forces only exist in objects that contain certain metals, such as iron.

Magnetic field lines show the direction of magnetic forces

The field lines around a magnet show the magnetic forces of the magnet's poles. These lines also show the direction of the magnetic force and how that affects objects within the magnetic field.

One way to examine the field lines around a magnet is to use a compass. When you put a compass in a magnetic field, the compass needle lines up with the direction of the field lines around the magnet (see Figure 2.34). The needle will always point from the magnet's north pole to its south pole. This shows that the direction of a magnetic force around a magnet is from north to south.

Figure 2.33: The direction of the magnetic field in a bar magnet is in a loop from the north pole to the south pole.

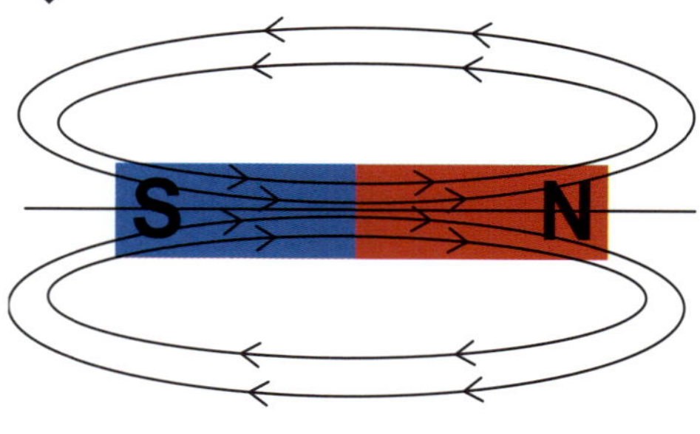

Figure 2.34: A compass can be used to show the direction of magnetic field lines.

In a magnetic field:

- the distance between field lines represents the strength of a field in any given area
- field lines can never cross one another
- field lines continue through the magnets.

Like poles repel each other

When the north poles of two magnets are brought close together, a repulsive force is made that pushes them away from each other. This is because the field from one north pole cannot enter or cross the field of the other north pole. Instead, the fields move away from each other and the field lines become more tightly packed (see Figure 2.35).

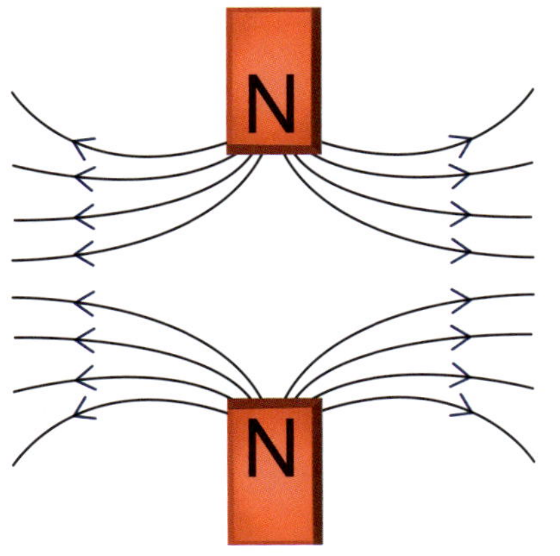

Figure 2.35: When two like poles are brought together, the repulsive force can be strong enough to cause one magnet to hover over the other.

Although the fields between two north poles have been squeezed into a smaller area, the distance between field lines still shows the strength of the field. The force pushing the two poles apart becomes stronger. A repulsive force also happens between the south poles of two magnets. This means we can say that, for all magnets, **like poles** repel.

Unlike poles attract each other

When the opposite poles of two magnets are brought close together, an attractive force is made. This is because the field lines from the north pole of one magnet can flow easily into the south pole of the other magnet (see Figure 2.36). In other words, **unlike poles** attract.

When two magnets contact one another, they act as one big magnet rather than two joined magnets.

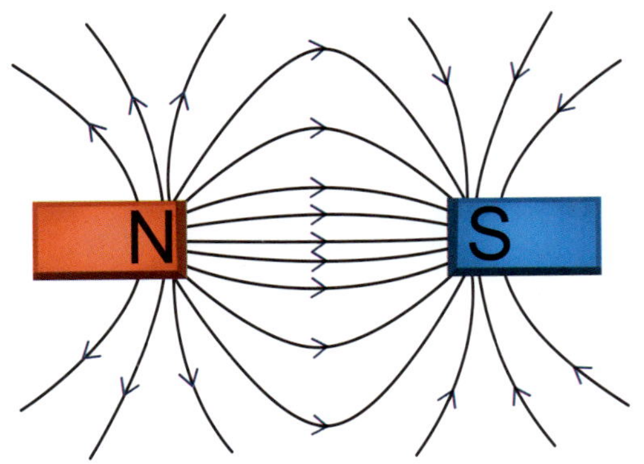

Figure 2.36: When two unlike poles are brought together, the forces pull the poles together.

Learning Ladder

Forces

1. Describe what happens when the like poles of two magnets are pushed closer together.
2. Describe the poles of a magnet, using examples.
3. Explain, using a field line diagram, how a magnet is attracted to another magnet.
4. Compare the density of the magnetic field lines in Figures 2.33 and 2.36. Suggest what they may mean in terms of their strength.
5. Propose what the impact would be if Earth did not have its own magnetic field.

Processing data and information *see page 329*

1. In the photo of iron filings around a bar magnet, the iron filings are not evenly distributed. Choose the most appropriate explanation for this observation.

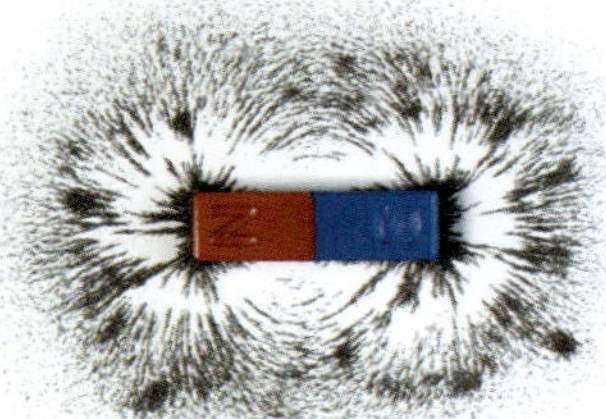

A The field lines are further apart at the poles, which is where the magnetic force is weakest.

B The field lines are closer together at the poles, which is where the magnetic force is strongest.

2. Some students were investigating if a magnet's size affects its strength. They collected four magnets of different sizes made of the same material. Then they used the magnets to pick up steel paperclips to form paperclip chains. The students found that:
   - a 2 cm magnet made a chain of 3 paperclips
   - a 4 cm magnet made a chain of 6 paperclips
   - a 6 cm magnet made a chain of 10 paperclips
   - an 8 cm magnet made a chain of 14 paperclips.

   Construct a table, with appropriate units and headings, to present the data.
3. Construct a graph to record the data in Question 2.

In context

Maglev trains use magnets to move very quickly. The train levitates (hovers) above the guide way. Propose how magnets could make this happen.

Success criteria

- I can describe how magnets attract or repel each other based on their polarity.
- I can draw the magnetic fields of magnets.

2·9 ▸ Uses of magnets and electromagnets

Learning intention

At the end of this lesson, I will be able to investigate how magnets and electromagnets are used in some everyday devices.

Key terms

electromagnet: a device that uses electricity to make a magnetic field

polarity: the direction of a force, such as a magnetic force

Investigation 2.9

Building an electric motor, p.397

Content group:
Magnets in everyday life

Magnets are very useful. **Electromagnets** are particularly useful because the strength and direction of their field can be easily changed. Electromagnets are in devices such as speakers, electric motors, medical equipment and computers.

Magnets can hold objects together

Ordinary magnets are sometimes called permanent magnets because they constantly have magnetic force. The most common use of magnets is to hold objects together. The constant attractive force acts on certain metals or on the opposite pole of another magnet. For example, some screwdrivers have a magnetic head that allows screws to 'stick' to the tool without needing to be held. Other everyday uses of magnets are as fridge magnets, and in building and construction toys, compasses, jewellery and machinery.

Electromagnets use electricity to make a magnetic field

A wire that has an electric current flowing through it produces a magnetic field. If the wire is coiled into a spiral, as shown in Figure 2.37, the field around the wire builds into one large field that is like the field around a bar magnet.

The field in the spiral can be strengthened by:

- placing an iron core in the middle
- increasing the amount of electricity
- making the electricity flow faster
- increasing the number of coils in the spiral.

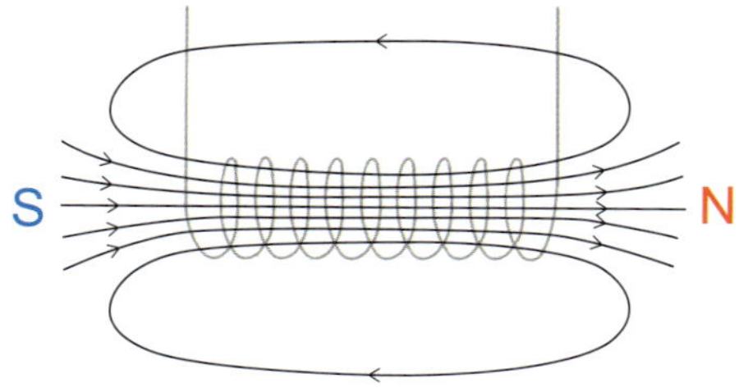

Figure 2.37: The magnetic field around an electromagnet has a north and south pole, like a bar magnet.

Figure 2.38: Electromagnets are used in scrapyards to lift scrap metal.

Electromagnets have many uses

Electromagnets are used in many devices and technologies. They are useful because we can change the strength of their magnetic field very precisely. We can also change the direction, or **polarity**, of their magnetic field.

When a speaker is connected to a powered device playing music, the electricity changes the coil in the speaker into an electromagnet. The direction of the electromagnet's field can rapidly flip back and forth. This attracts and then repels a permanent magnet inside the speaker cone. The permanent magnet moves back and forth, pushing the speaker cone back and forth, making sound waves or vibrations in the air.

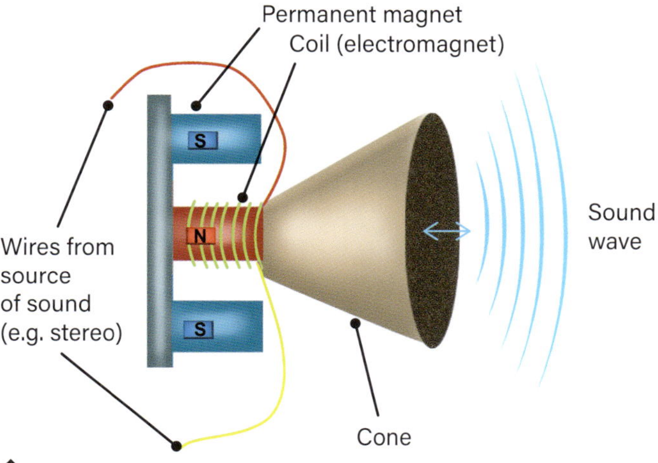

Figure 2.39: The coil inside a speaker is an electromagnet. It pulls closer or pushes further away from the permanent magnet, producing sound waves by moving the cone.

Very sensitive electromagnets are used in computer hard drives to read information stored on metallic strips. Electromagnets are also used in medical equipment. Magnetic resonance imaging (MRI) scanners generate magnetic fields that can be used to create images of the inside of the human body (see Figure 2.40).

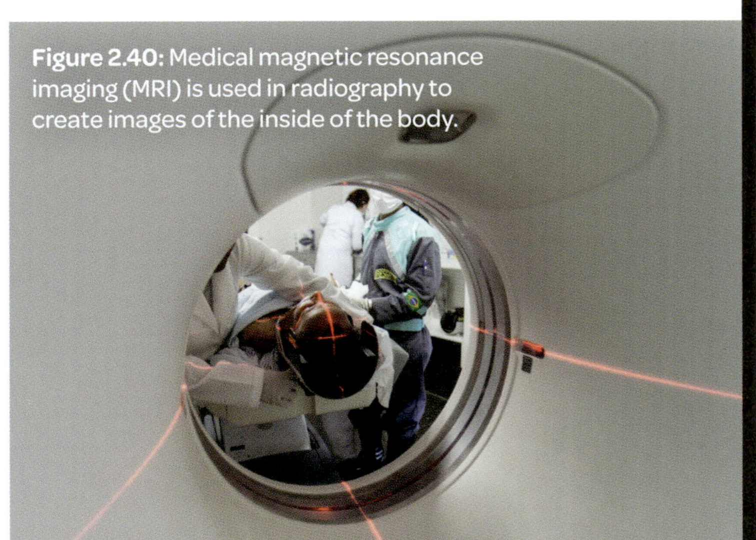

Figure 2.40: Medical magnetic resonance imaging (MRI) is used in radiography to create images of the inside of the body.

Learning Ladder

Forces

1. Identify the two ways a magnetic field can exist.
2. Describe how the strength of an electromagnet can be increased.
3. Explain the purpose of three magnets that you can find in your home or the classroom.
4. Compare the use of an electromagnet in a scrap yard with that of a magnet used in a construction toy.

Questioning and predicting

see page 319

A student is investigating the strength of an electro-magnet. The electromagnet is made of wire wrapped around a nail. The student measures the number of staples the electromagnet can pick up. They test different nails that are different widths. All other variables are kept the same.

1. Predict what will happen as the nail gets wider.
2. Identify a question for this investigation.
   A Does changing the width of a nail in an electromagnet affect how many staples it can hold up?
   B Does changing the number of coils of an electromagnet affect how many staples it can pick up?
3. Construct a scientific question to use if you change the type of metal in the electromagnet, rather than the width of the nail.
4. Propose a scientific hypothesis for the expected outcome of the investigation into nail width. Include the dependent and independent variables. (Hint: Can you reword the prediction you made in Question 1?)

In context

A common use of electromagnets is as transformers in electrical devices. Research what a transformer is and give some examples of how transformers are used.

Success criteria

- I can describe how magnets are used in everyday life.
- I can describe what an electromagnet is and how electromagnets are used in everyday life.

2·10 ▸ Uses of simple machines

Learning intention

At the end of this lesson, I will be able to:

- identify some simple machines, such as levers, pulleys, gears and inclined planes
- describe how simple machines reduce the amount of work needed to conduct everyday tasks.

Key terms

axle: the pin, post or rod that a wheel can spin around to transfer force

effort: the force applied to a simple machine to move another object

fulcrum: the point on which a lever turns when moving an object

inclined plane: a tilted flat surface; also called a ramp

lever: a bar acted upon at different points by two forces

load: an object that is moved or lifted by a simple machine

load force: the force from the object that is being moved

pulley: a wheel with a grooved rim for carrying a cable

Investigation 2.10A
Making a catapult, p.398

Investigation 2.10B
Modelling a seesaw, p.399

Investigation 2.10C
Pulley investigation, p.400

Investigation 2.10D
Using ramps , p.402

Content group: Simple machines in everyday life

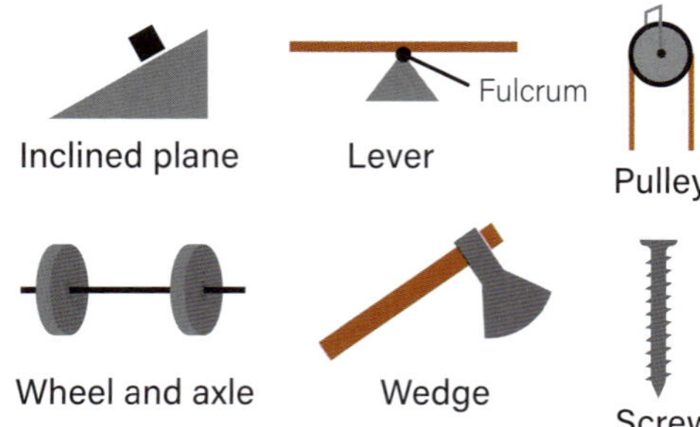

▲ **Figure 2.41:** Simple machines

Machines help make work easier by changing the strength or direction of a force.

Wheels can make simple machines to lift objects off the ground

A wheel with a post (**axle**) fixed to its centre can be used to move or roll things. Examples are steering wheels and doorknobs.

A **pulley** is a wheel with a rope threaded around it. A pulley changes the direction of the force applied to one end of the rope. The applied force is called the **effort**. When you pull the rope, you can lift a **load** attached to the other end. If you add more wheels to the pulley system, you can reduce how much effort force you need to lift the load. But you need to move the rope further to lift the load to the same height.

Levers reduce the force needed to move objects

Levers help us to lift objects with less effort. A **lever** consists of a rigid bar that pivots on a support called a **fulcrum**. When a force is applied to one end of a lever, the other end lifts. A seesaw is a lever with the fulcrum in the middle between the end where the effort force is applied and the **load force** on the other end.

There are three classes of levers, depending on the positions of the effort, fulcrum and load. First-class, second-class and third-class levers are summarised in Figure 2.43. A seesaw is a first-class lever.

Inclined planes help to lift and keep objects still

An **inclined plane** reduces the effort force needed to raise an object. Instead of lifting the object straight up, the object is moved up the sloping surface over a longer distance. Ramps and slides are inclined planes. A screw is a modified inclined plane. In the screw, the inclined plane is wrapped around a cylinder in a spiral pattern. It moves by twisting, usually with a screwdriver. A lot of force is used to overcome the friction between the screw and the item it is being screwed into. However, this means that the screw will hold the item in place!

A wedge is made of two inclined planes meeting at a point. This can be used to hold something in place or to cut something. You often see wedges holding doors open; they cause friction between the wedge and the base of the door.

Figure 2.42: Much less force is required to push a wheelchair up an inclined plane (the ramp) than to lift it.

Simple machines combine to make complex machines

A complex machine is made of two or more simple machines working together. A wheelbarrow is a combination of a wheel, an axle and a lever. The combined simple machines work together to help you lift and move a load.

First-class levers

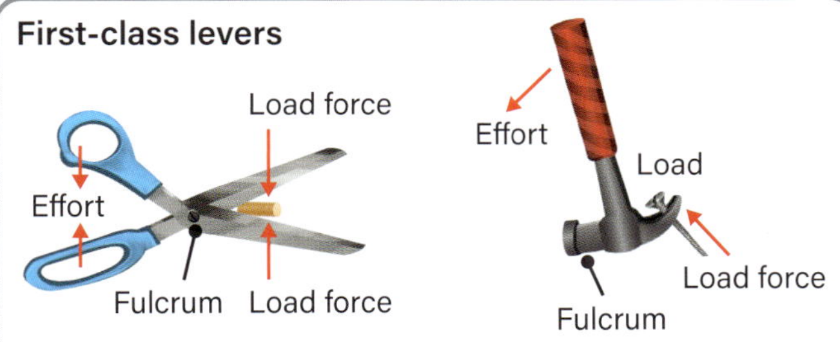

Second-class levers

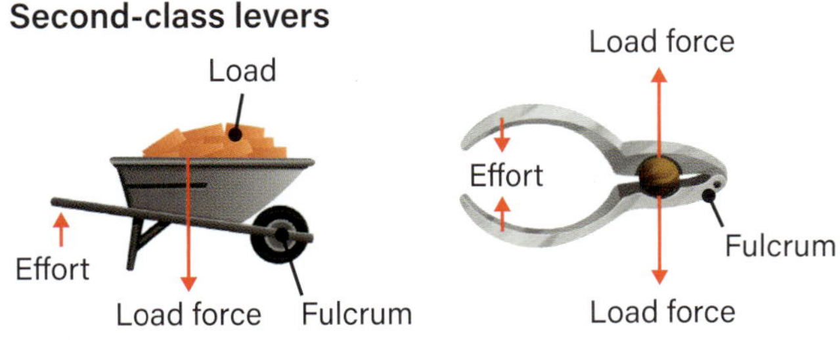

Third-class levers

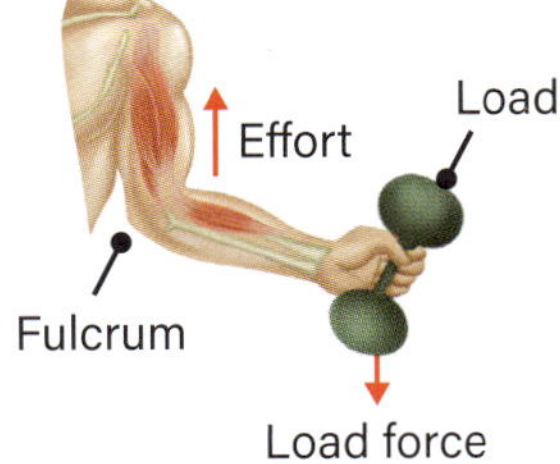

Figure 2.43: A summary of the different classes of levers. Except for first-class levers, the effort is in the opposite direction to the load force.

Learning Ladder

Forces

1. Draw a diagram of a pulley, labelling all the forces acting on the load.
2. Describe how a wheel and axle can transfer a force to a different point.
3. Explain how the location of an applied force on a lever can affect the motion of the other side.
4. Pulleys and ramps can both be used to lift a load to a height. Compare scenarios where an inclined plane or a pulley would be the most appropriate simple machine to use.
5. You want to lift some heavy objects and will need to use pulleys. Every pulley added will halve the force required to lift the objects. If you need two pulleys to lift one object, how many pulleys will you need to lift an object that is twice as heavy?

Problem-solving

see page 339

1. Propose why turning a tap on or off without its handle is nearly impossible.
2. A heavy goods vehicle parks in a loading bay. Propose a method for the goods to be loaded into the vehicle.
3. A farmer sells tomatoes at the local market. However, getting to the market involves a long walk down a mountain and across a river. Design a simple machine to move the tomatoes without squashing them. Explain why your design would be effective.

In context

A screw can be described as 'an inclined plane wrapped around a nail'. Explain why a screw can be considered stronger at holding than a nail, and why it takes longer to insert than a nail.

Success criteria

- I can identify the different types of simple machines.
- I can describe how simple machines reduce the work needed for different tasks.

2·11 ▸ Aboriginal and Torres Strait Islander Peoples' application of forces to tools

Learning intention

At the end of this lesson, I will be able to outline the main features of a range of tools and give examples of how they can be used by Aboriginal and Torres Strait Islander Peoples.

Key terms

grinding: the action of rubbing objects together to break up materials into smaller particles

knapping: the action of applying force to a specific type of rock to create a sharp shard

leverage: the action of a lever

rotation: turning around as if on an axis

Content group: Simple machines in everyday life

Aboriginal and Torres Strait Islander Peoples apply their knowledge of forces to create and use tools. Tools are used in food hunting and harvesting, in ceremonies and as items for trade.

Friction can be used to make useful items

Coastal and islander communities can access many shells to use as raw materials. Using friction, and applying forces such as pushing and **rotation**, people grind the shells carefully to make fishhooks, tools for cutting and scraping, and parts for ceremonial attire. Many coastal communities traded tools to peoples further inland who did not have this kind of raw material in their communities.

The **grinding** process to create a single fishhook requires enormous skill and patience, and takes significant time. The shell's outer surface is ground on a flat stone to make the shape, then it is fine-polished and sharpened.

Woomeras are simple machines

The woomera is a simple machine that uses **leverage** to help throw spears up to three times further. A woomera is a lever made of wood. One end is a handle attached to a paddle-shaped tool and the end usually has a bound piece of wood or stone that holds the spear securely in the launch mode. The woomera acts to increase the length of the thrower's arm, and therefore increases the launch speed of the spear.

Figure 2.44: ▸ A woomera is used to improve spear-throwing distance and accuracy.

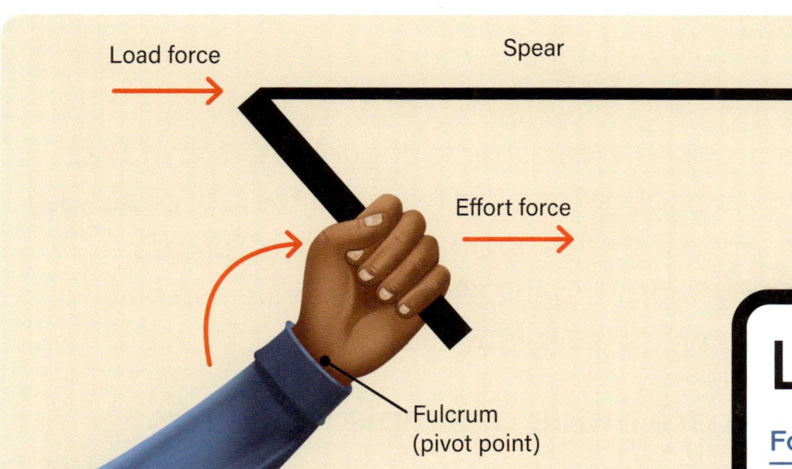

◀ **Figure 2.45:** A woomera is a first-class lever. The wrist of the spear-thrower acts as the fulcrum, the effort force is delivered through the hand and the load is the spear.

A well-constructed woomera allows the spear motion to stay constant and accurate. If an unstable force is applied to a spear by the spear-thrower, this can create unbalanced forces that cause the spear to wobble and slow down, reducing the distance it can travel.

Knapping uses force to create sharp tools

First Nations Peoples have used many stone tools for thousands of years. A common method of creating cutting tools requires the special skill of knapping. **Knapping** is where force is applied to a specific type of rock to create a sharp shard. The original rock is called a core, and the piece that is removed is a flake. Flakes can be very sharp and can cut into wood or even flesh.

Figure 2.46: Stone knapping

▲ **Figure 2.47:** A basalt Aboriginal axe head. These large rock flake pieces are often made into axe or spear heads, or sharpened by grinding into various other tools.

Learning ladder

Forces

1. Identify the kind of simple machine that a woomera spear thrower is.
2. Describe the force that grinding requires when used to shape shell hooks.
3. Explain how a woomera improves spear-throwing.
4. Compare the forces acting on a spear thrown by hand and by a woomera. How does this affect the distance the spear travels?

Questioning and predicting *see page 319*

1. Describe some possible uses for a woomera other than hunting.
2. Identify which of these two questions would be an appropriate scientific question to find out how a woomera works.
   A How does the use of a woomera change how far a spear can be thrown?
   B How many years of experience does a person need before they are able to use a woomera?
3. Describe how you could edit the question you selected in Question 2 so that it is in an appropriate scientific format.
4. Based on the information from this section, construct a hypothesis that you could test about how a woomera works.
5. Propose a scientific question that would allow you to test the efficiency of a woomera. Specify all relevant variables within your question.

In context

Summarise ways that First Nations Peoples' application of forces to tools inspired and supported contemporary technologies.

Success criteria

- I understand the principles of using force to create a range of tools.
- I can use questions and predictions to understand how First Nations Peoples have made and used tools.

2.12 ▸ Forces in context

Learning intention

At the end of this lesson, I will be able to describe how forces are used in different everyday contexts.

Key terms

force multiplier: a machine that amplifies the effect of an effort applied in a task

speed multiplier: a machine that increases the speed of a task

Content group: Forces in context

Forces are used in all aspects of our lives, from riding a bike to school, getting the train into the city or playing sports with our friends. Our knowledge of forces is continually improving the technologies we use and how we complete everyday tasks.

Bicycles are made of simple machines

▲ **Figure 2.48:** A bicycle is a complex machine made up of several simple machines.

A bicycle is a complex machine made up of several simple machines. The pedal arms are levers attached to a large gear. Applying force on the pedal causes the gear to turn.

Gears are wheels with teeth around their rims. When two gears are arranged so that their teeth interlock, the rotation of one gear causes the other gear to turn. The first gear is called the driving (or drive) gear. The second gear is called the driven gear. A set of gears connected together is called a gear train.

The gears help the chain to move. The chain acts as a pulley attached to gears or cogs at the axle, causing the rear wheel to turn. Moving the chain from a smaller rear cog to a larger one means less effort is required to achieve the same effect. The load force is the combined weight of the cyclist and the bicycle.

Wheels and gears can be used as a **force multiplier** or as a **speed multiplier**.

Magnetism is important for transport

Maglev (magnetic levitation) trains use magnetism to levitate above the tracks on which they travel. This reduces friction, so the trains can go faster and be more efficient.

Superconducting magnets suspend the train car above a guideway. These magnets repel one another when like poles face each other (see Figure 2.49).

Three types of loops are set into the guideway at specific intervals to do three important tasks. One type of loop creates a field that makes the train hover about 13 centimetres above the guideway. A second type of loop keeps the train stable horizontally. Both these loops use magnetic repulsion to keep the train car in the best position. The third type of loop is a propulsion system created by an electromagnet, where both magnetic attraction and repulsion are used to move the train car along the guideway.

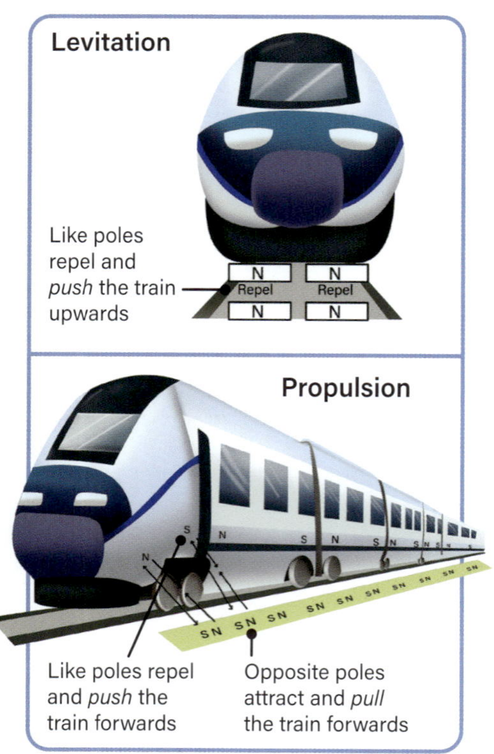

◀ **Figure 2.49:** The operation of a maglev train

Figure 2.50: An athlete performing a pole vault

Sports scientists use knowledge of forces to help athletes gain an edge

In sport, a slight advantage could mean the difference between winning and losing. For this reason, sports scientists apply scientific ideas and techniques to improve the performance of athletes.

For example, sports scientists apply their knowledge of forces to:

- help athletes reduce friction in the air or water, such as by wearing streamlined helmets when cycling or swimming with an efficient stroke that minimises splashing
- develop a method to throw a cricket ball or baseball with a desired curve or spin
- help athletes keep their centre of gravity low so they stay balanced when performing gymnastics, pole vaulting or high jumping.

Understanding the mechanical forces on the human body (biomechanics) helps to prevent injuries and to know the best way to recover from injuries.

Sometimes this knowledge can accidentally create an unfair advantage for athletes. For example, swimwear made from polyurethane has been banned in swimming events. The material does not let water soak through, so air gets trapped inside the suit, making the swimmer more buoyant. This allows the swimmer to swim closer to the surface, reducing water resistance and friction, and giving them an advantage over their competitors.

Learning Ladder

Forces

1. Identify the type of simple machine a wheel is.
2. Describe how a gear is different from a wheel.
3. Explain how wheels and gears can make a task easier.
4. Compare how magnets are used to repel and attract in maglev trains.
5. Discuss how forces could have negative effects on athletes.

Problem-solving

see page 339

1. **a** Identify ways in which forces may hinder an athlete's performance.
   b Outline ways in which forces may improve an athlete's performance.
2. Most trains have metal wheels, run on metal tracks and use diesel fuel. Identify a problem with these types of trains.
3. Propose a better type of train than the one described in Question 2, and justify your answer.
4. When bowling a cricket ball, a student found that the ball was often moving to the left of the stumps. Explain how they could overcome this problem by applying their understanding of the forces on the ball.

Success criteria

- I can describe how forces are used in different everyday contexts.

Forces summary

Forces in action

- A force can be a push, a pull or a twist.
- Forces can be direct (objects are in contact) or indirect (objects are not in contact).
- Forces occur when objects interact with each other.

- Balanced forces cause an object to continue what it is doing, such as staying still.
- Unbalanced forces cause an object to accelerate, like a skateboarder pushing off a walkway.

Balanced forces

Two forces are the same size but opposite in direction.

They cancel each other out.

5 N → ⚽ ← 5 N

The net force is equal to 0 N.

Unbalanced forces

Two forces are different sizes and opposite in direction.

They do not cancel each other out.

The force pushing the ball to the right is bigger than the force pushing to the left. The net force is equal to 5 N to the right.

- Force diagrams show the size of forces acting on an object and can be used to determine whether it is stationary or moving.
- Electrostatic forces are forces of attraction between positive and negative charges.
- Lightning strikes occur because of electrostatic forces.

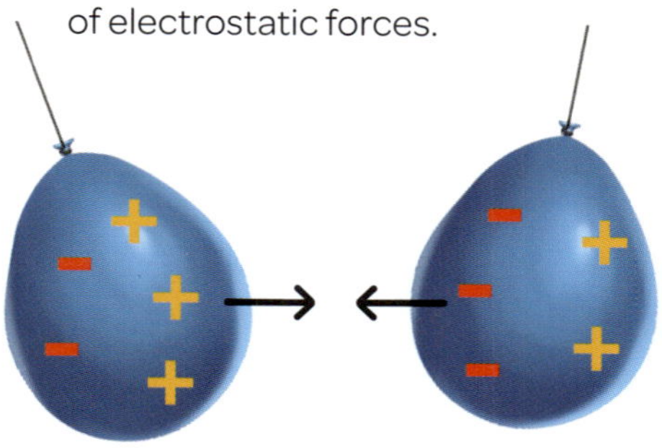

- Gravitational force is a force of attraction between objects with mass. The more mass the object has, the bigger its gravitational force.
- The Sun's gravity pulls on moving planets, which keeps them in their orbits.
- The mass of an object is the same no matter where it is in the solar system.

Forces in action CONTINUED

- An object's weight is its mass multiplied by the acceleration due to gravity: $F_g = mg$
- Weight is measured in newtons.

Magnets in everyday life

- A magnet has a positive and a negative pole. Like poles repel; opposite poles attract.
- Magnetic fields exist around magnets.

Simple machines in everyday life

- Simple machines reduce the amount of energy we use to carry out tasks.
- Wheels, levers and inclined planes are simple machines.
- A wheelbarrow uses a wheel and a lever to help move loads with less force.

Inclined plane

Lever

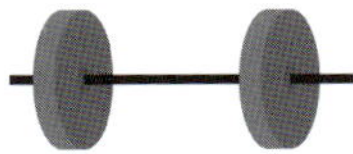

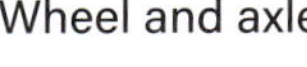

Wheel and axle

Wedge

Pulley

Screw

Aboriginal and Torres Strait Islander Peoples make and use a variety of tools or machines, including the woomera, which acts as a lever.

Masterclass

	Steps in progression	1	2
Content	**Forces**	Identify two forces that help to increase the speed of maglev trains.	Describe how each of the forces you identified in Step 1 works to increase the speed of a maglev train.
Processes	**Questioning and predicting**	Predict the impact of changing the shape of the front of a maglev train to make it look more like a truck.	Which question could be used to investigate electromagnet strength? **A** How does changing the metal in an electromagnet affect the strength of its magnetic field? **B** How does changing the size and shape of an electromagnet affect the strength of its magnetic field?
Processes	**Processing data and information**	Identify two proposed maglev routes from the data in Figure 2.52.	**a** What is the average speed a maglev train would travel between Newcastle and Sydney (see Figure 2.52)? **b** Use the average speed to construct a table showing the time it would take to travel between each set of destinations in Figure 2.52.
Processes	**Analysing data and information**	Identify a trend in the graph you constructed in Step 3 of the 'Processing data and information' row.	Consider the trend you identified in Step 1. Describe what this trend tells you about travelling between cities on maglev trains.
Processes	**Problem-solving**	Identify a problem with having a maglev train system in Australia.	Describe one problem that maglev trains could solve in relation to travelling in Australia.

Figure 2.51: Smooth, curved outer surfaces help to increase the speed of maglev trains.

Demonstrate your understanding

3	4	5	
Explain how the force of gravity is linked to the movement of maglev trains.	Compare the roles of friction, gravity, magnetic force and drag in relation to starting and stopping a maglev train.	Analyse the impact of reducing the strength of an electrical current supporting an electromagnet on a maglev track.	
Construct a question that could be used to investigate the relationship between the material a train is made from and the fastest speed the train can travel at.	Construct a scientific hypothesis that states the expected result of the investigation in Step 3.	**a** Identify the independent, dependent and controlled variables of the investigation in Step 3. **b** Are all the variables included in your question and hypothesis? Make any improvements based on the identified variables.	Science how-to 319
Convert the data in your table from Step 2 into a digital graph.	Convert the data in your table from Step 2 into another digital format that shows the difference in the time it would take to travel each of the routes on a maglev train.	Consider Figure 2.51. **a** What is one interpretation you can draw from this image? **b** Does this interpretation strengthen your confidence in the data you compiled in Step 2? Explain your answer.	Science how-to 329
Explain the relationship between the information in Figure 2.52 and the data you compiled in Step 2 of the 'Processing data and information' row.	Draw a conclusion about which route in Figure 2.52 would take the longest on a maglev train. Justify your reasoning with data.	Assess the data you compiled in Step 2 of the 'Processing data and information' row to determine if the information is accurate, reliable and valid.	Science how-to 334
Explain the link between magnetic force and train speed.	Suggest a solution to the problem of the large cost of implementing maglev trains in Australia. Consider which materials could be used and where they could be sourced from.	Assess the strategies you used to come up with a solution in Step 4. How effective were they?	Science how-to 339

Maglev trains set to glide into Australia

Maglev trains can travel more than 500 kilometres per hour. 'Maglev' is short for 'magnetic levitation', which explains how these trains work: they glide above magnetic tracks. Maglev trains have a streamlined design and smooth outer shell. Both of these factors minimise drag, which helps to increase the train's speed.

In Australia, scientists are exploring what would be involved in setting up maglev train routes between Brisbane, Newcastle, Sydney and Melbourne. Imagine a trip from Sydney to Newcastle taking only 25 minutes instead of two hours!

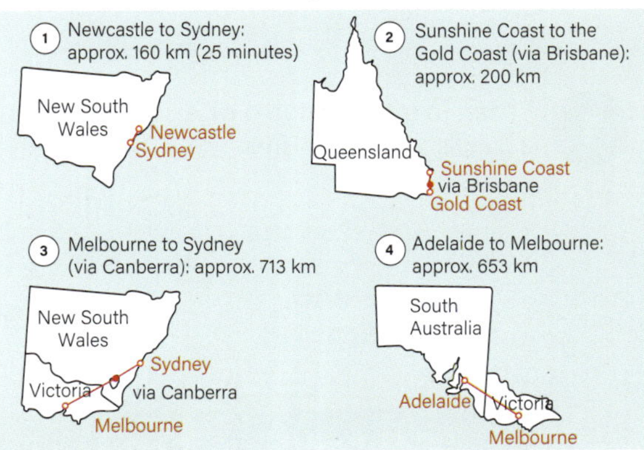

▲ **Figure 2.52:** Potential maglev train routes in Australia

Focus area 3

3.0 Cells and classification

Every living thing – including every animal and plant – is made up of cells. Humans are made up of 30–40 trillion cells! There is a wide variety of cells and a wide variety of living things. To order and organise the diversity of life on Earth, scientists use classification, where they group things that are similar in some way.

The Learning Ladder contains the scientific content and processes you will learn in this chapter. Each area has five levels of progression. To move to higher levels, you need to practise activities at the earlier levels. This will help you develop the ability to complete tasks that are more complex.

Steps in progression	Content	Working scientifically processes			Steps in progression
	Cells and classification	Observing	Conducting investigations	Communicating	
5	I can analyse how classification and understanding of organisms changes over time.	I can use observations and measurements to answer questions.	I can conduct first-hand investigations and accurately record the collected data.	I can present scientific findings using appropriate conventions for specific audiences.	5
4	I can compare the structural features of organisms using scientific conventions.	I can make inferences based on my observations.	I can collect accurate data from first-hand and second-hand investigations.	I can construct a range of appropriate scientific presentations based on first-hand and second-hand information and data.	4
3	I can use scientific conventions to explain the organisation of organisms and cells.	I can record observations and measurements accurately.	I can implement safe practices when using scientific equipment.	I can use digital technologies to organise and present information and data.	3
2	I can describe the structure of cells and organisms.	I can use scientific tools to enhance observations.	I can use scientific equipment to conduct investigations and gather first-hand data and information.	I can select appropriate ways to communicate information.	2
1	I can identify features of cells and organisms.	I can make observations using my senses.	I can identify correct processes for conducting investigations.	I can recognise scientific information.	1

Figure 3.1: This quokka and all the plants surrounding it are made up of cells.

3·1 ▸ Characteristics of living things

Learning intention

At the end of this lesson, I will be able to describe the characteristics all living things share.

Key terms

carbon dioxide: a colourless and odourless gas containing carbon and oxygen that is produced during respiration

living thing: an organism that displays specific characteristics of being alive

mass: the amount of matter that a physical body contains

organism: an individual animal, plant or other living thing

respiration: a chemical reaction that converts glucose to energy

stimuli: changes in the environment that cause responses in living things

Content group: Classification of living things

There are millions of different organisms on Earth. How do scientists classify things as living or non-living?

Living things share three key characteristics

All **living things**:

- are made of cells
- are made of molecules containing carbon
- have biological characteristics in common, such as being able to grow, move, reproduce and respond to **stimuli**.

Living things have many other characteristics and some organisms have the same features. Living things can also be very different from one another, such as dogs and octopuses!

Living things share biological characteristics

There are seven biological characteristics or features that **organisms** must have to be considered 'living' by scientists. These are:

- **Growth**: The organism gets bigger and develops over time. For example, frogs hatch from eggs as tadpoles and develop as they grow, and many plants grow from seeds.
- **Movement**: The organism can move part or all of itself. For example, humans can walk, and plants can move their leaves to face the sun.
- **Nutrition**: The organism takes in nutrients. For example, a bird eats insects, and plants absorb minerals from the soil.
- **Reproduction**: The organism can produce new offspring. This happens in many ways. Some organisms make identical copies of themselves, while others (like humans) require a mate.
- **Response to stimuli**: The organism can respond to changes in the environment. For example, kangaroos stay in shaded areas throughout the hottest parts of the day.
- **Respiration**: The organism can convert nutrients into energy. This happens in the organism's cells.
- **Waste removal**: The organism can get rid of waste products. In humans, this includes breathing out **carbon dioxide** gas and getting rid of urine and faeces.

If something does not have all seven of these characteristics, scientists will not classify it as living. For example, even though fire requires fuel to burn and spreads, it does not respond to stimuli in the environment, so fire is classified as non-living.

▲ **Figure 3.2:** Frogs develop from tadpoles.

We use scientific equipment to see the characteristics of living things

Some of the characteristics of living things can be hard to see. To see them, scientists use special equipment, such as:

- **Microscopes**: This technology allows scientists to see magnified images of different objects, including living things. This means that they can see the cells that make up living things.
- **Digital scales:** Scientists use digital scales to measure the **mass** of living things. Increasing mass means the organism is growing!
- **Carbon dioxide probes:** These special probes measure the level of carbon dioxide in the air. If there is an increase in carbon dioxide in the air of an enclosed environment containing an animal, it proves that the animal is respiring.

▲ **Figure 3.4:** Scientists can use microscopes to see the characteristics of living things.

Figure 3.3: Both bees and flowers have the seven biological characteristics of living things.

Learning Ladder

Cells and classification

1. Identify two characteristics shared by all living things.
2. Outline an example of each of the characteristics you identified in Question 1.
3. Imagine something that grows over time but does not get rid of waste. Would scientists classify this thing as living or non-living? Explain your choice.
4. Compare the bee and the flower in Figure 3.3. Identify two characteristics that they both have.

Observing

see page 316

1. Describe how you could observe a characteristic of a living thing using just your senses.
2. Identify a piece of scientific equipment that scientists use to improve their observations of living things.
3. You are using digital scales to measure the growth of a living thing over time. Explain how you would record your measurements.
4. What could you infer about an object in an enclosed space if a carbon dioxide probe registered no change in the carbon dioxide levels over time? Explain why you made this inference.

In context

Select a living thing from your school playground. Outline how it shows the seven biological characteristics of living things.

Success criteria

- I can describe the characteristics shared by all living things.

3·2 ▸ Classifying living things

Learning intention

At the end of this lesson, I will be able to explain what classification is and why classifying living things is useful.

Key terms

classification: the process of sorting things into groups or classes

species: a single, specific type of living thing

taxonomy: the science of classifying organisms

Investigation 3.2

Classifying items, p.403

Content group:
Classification of living things

Classification is the process of sorting things into groups. Similar objects are placed in the same group. Classification plays an important role in ordering and organising the diversity of life on Earth.

Classification means sorting things into groups

Things can be classified for many different reasons, depending on who is doing the arranging, sorting or grouping.

Different classification systems exist in many areas in your daily life:

- In your school library, books are classified according to the Dewey Decimal Classification system, which is useful for finding both subjects and authors.
- We can sort items that can be recycled into paper, glass and plastic.
- Animals can be classified into two groups: carnivores (which only eat meat) and herbivores (which only eat plants).

▲
Figure 3.5: Butterflies are classified by their colours and the shapes of their wings.

All scientists use classification

Biologists are scientists who study living things. The science of classifying organisms (sorting and arranging living things into groups) is called **taxonomy**. A biologist who specialises in classifying living things is called a taxonomist.

However, many scientists use classification:

- **Chemists** classify substances by their physical properties and their chemical reactions with other substances.
- **Geologists** classify rocks according to how they were formed and their mineral content, such as sedimentary, igneous and metamorphic.
- **Astronomers** classify objects in the universe by observing properties such as size, density and whether they give off light.
- **Meteorologists** classify different sections of the atmosphere surrounding Earth according to their temperature and how they affect the weather.

Classification is important

Classifying living things is important for many reasons:

- Classification helps biologists work out whether different living things (for example, different **species**) are connected and the relationship between different organisms.
- Classification helps biologists identify newly discovered organisms. New plants, insects, fish and bacteria are discovered every year.
- Using a classification system allows biologists from different countries to communicate their work. An Australian scientist may use different names for trees and plants than a Chinese, French or Indonesian scientist. However, if they use the same classification system, they can all easily identify an organism.
- Biologists can use classification to determine how species have evolved.

Without classification, none of these processes are possible.

Figure 3.6: Is a cat closely related to a lion? They have similarities but are also very different. Classification helps biologists identify how these two animals are related.

Figure 3.7: In our homes, we classify recyclable waste into different categories.

Learning Ladder

Cells and classification

1. Identify two features you can use to group living things.
2. Classify the following organisms by size: alpaca, beetle, echidna, horse, iguana, leopard, platypus, wasp, zebra.
3. Explain how taxonomists classify organisms into groups.
4. Explain how classification can help us identify which species are closely related.

Conducting investigations see page 326

A student collected the following objects: scissors, five coloured pencils, a lead pencil, a red pen, a blue pen, a glue stick, an eraser, a pencil sharpener and a ruler.

1. Identify which of the following steps should be included in the process of classifying these objects.
   - **A** Collect three of each object to compare them.
   - **B** Examine all the objects before beginning to classify them.
   - **C** Remove two of the objects because they are too hard to work with.
2. Describe an item of scientific equipment you could use to make observations of the objects listed above. How would you use this piece of equipment?
3. Explain how you can make sure you are safe while classifying these objects.
4. **a** Outline the data you would collect from the process of classifying these objects.
   b Explain how you would make sure the data is accurate.
5. Propose a table or other layout to collect and display this data.

In context

Identify five organisms that you can put into a group. Describe a feature that the five organisms share.

Success criteria

- I can describe what classification is and explain why it is useful.

3·3 ▸ The Linnaean classification system

Learning intention

At the end of this lesson, I will understand and be able to use the Linnaean system of classification.

Key terms

collaboration: working cooperatively with other people to complete a task or solve a problem

naturalist: someone who studies nature and its history

Content group: Classification of living things

It is important that scientists use the same system to classify living things, so their findings can be compared, and they can share information and **collaborate**.

Scientists use the Linnaean system to classify species

A system for grouping living things was first proposed by Swedish **naturalist** Carolus Linnaeus in 1753. It is known as the Linnaean classification system and has been used by scientists to classify species since the 18th century.

The Linnaean system has eight levels

Linnaeus's system classifies organisms by levels called taxa, and each taxa is divided into groups. These groups are based mainly on cell and structural characteristics, and on how the organisms behave and reproduce.

At the highest level, these groups are huge: one group contains all the plant life on Earth. As we go down the levels, the groups become more specific.

At the top of the system, the most general classifications are domains and kingdoms. The characteristics used to separate these groups are based on cell structure. At the bottom of the system are genus and species. This is where we can identify individual types of organisms.

Figure 3.8: The Linnaean classification system has six kingdoms. These kingdoms of life are broad classification levels for all living things.

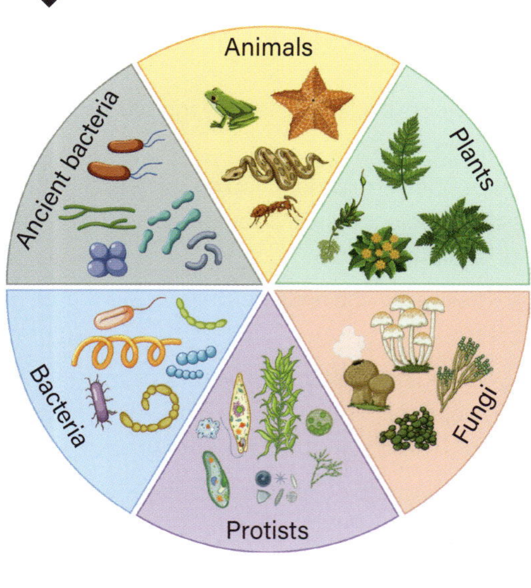

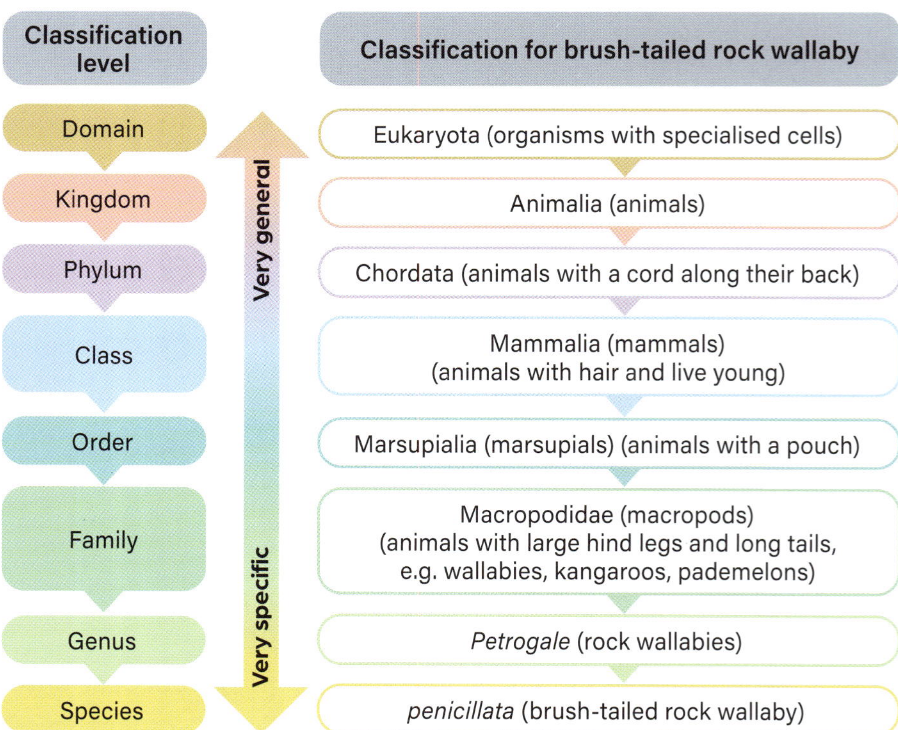

Classification level	Classification for brush-tailed rock wallaby
Domain	Eukaryota (organisms with specialised cells)
Kingdom	Animalia (animals)
Phylum	Chordata (animals with a cord along their back)
Class	Mammalia (mammals) (animals with hair and live young)
Order	Marsupialia (marsupials) (animals with a pouch)
Family	Macropodidae (macropods) (animals with large hind legs and long tails, e.g. wallabies, kangaroos, pademelons)
Genus	*Petrogale* (rock wallabies)
Species	*penicillata* (brush-tailed rock wallaby)

Figure 3.9: The classification of the brush-tailed rock wallaby, according to the Linnaean classification system.

Figure 3.10: The brush-tailed rock wallaby's scientific name is *Petrogale penicillata*.

See Figure 3.9 for how the brush-tailed rock wallaby is classified according to the Linnaean system.

Every species has a two-word name

All rock wallabies are part of the *Petrogale* genus, and the brush-tailed rock wallaby's species name is *penicillata*. This means that its full scientific name is *Petrogale penicillata*.

This naming is another important part of the Linnaean system: every species is given a two-word scientific name. These names are developed according to strict rules:

- The first word is the genus name and begins with a capital letter.
- The second word is the species name and begins with a lowercase letter.
- The name is shown in *italics* (or underlined when written by hand).

The words in these names are usually Latin or ancient Greek. Because these languages are no longer commonly used, they do not change over time as modern languages do. This is useful because the meanings stay the same. Even if scientists do not speak Latin, they can look up the words and find the same meaning each time.

Learning Ladder

Cells and classification

1. Identify three features of the brush-tailed rock wallaby that can help you to classify it.
2. Describe how the different levels of classification help us to better understand living things.
3. Classify the domestic dog from the least to the most specific:
   - Phylum: Chordata
   - Class: Mammalia
   - Kingdom: Animalia
   - Genus: *Canis*
   - Family: Canidae
   - Order: Carnivora
   - Domain: Eukaryota
   - Species: *Canis familiaris*
4. Compare the structural features of the plant and fungi illustrations in Figure 3.8.
5. Discuss how you could improve the Linnaean classification system. Provide reasons for the changes you suggest.

Communicating

see page 343

1. Identify the level of classification that is one step broader than class.
2. Write the scientific name for the great white shark.
   - Genus: *Carcharodon*
   - Species: *carcharias*
3. Create a flowchart to show how a short-beaked echidna is classified from domain to species.
4. Design a poster that explains how to classify a specific animal from the least to the most specific.

In context

Classifying living things can reduce the risk of them harming humans. Using common and scientific names, create an infographic that classifies five Australian snakes according to how venomous they are.

Success criteria

- I understand and can use the Linnaean system of classification.

3·4 ▸ Classification keys

Learning intention

At the end of this lesson, I will be able to use classification keys to categorise living things.

Key terms

classification key: a tool to help classify an organism by identifying important characteristics such as its features and behaviours

dichotomous: divided into two parts

Investigation 3.4

Classifying supermarket items, p.404

Content group: Classification of living things

Scientists use **classification keys** to help them categorise and identify living things. Classification keys may use pictures, words, numbers and instructions. There are different types of keys, including branching and dichotomous keys.

Some classification keys divide groups using branches

A common type of classification key is a branching key, which has two or more branches at each level to separate groups. As we move down the branches, we can identify characteristics until we determine the right organism. Consider a small group of animals: a bird, an earthworm, a lizard and a cat. You can use the branching key in Figure 3.11 to identify each of these organisms.

▾ **Figure 3.11:** An example of a branching key that classifies the bird, earthworm, lizard and cat

- Bird, earthworm, lizard, cat
  - Bones
    - Bird, lizard, cat
      - Feathers
        - Bird
      - No feathers
        - Lizard, cat
          - Fur
            - Cat
          - No fur
            - Lizard
  - No bones
    - Earthworm

Dichotomous keys use questions or statements

Another type of key is a **dichotomous key**. 'Dichotomous' means 'divided into two parts'. This classification key uses statements or questions with only two choices at each step. We can follow the directions until we identify the organism.

For example, a dichotomous key for identifying a bird, an earthworm, a lizard and a cat might look like Figure 3.12.

▾ **Figure 3.12:** A dichotomous key that classifies the earthworm, bird, lizard and cat

Step		Statement	Outcome
1	a	Skeleton of bone	Go to step 2
	b	Does not contain bones	Earthworm
2	a	Covered in feathers	Bird
	b	Not covered in feathers	Go to step 3
3	a	Covered with dry scales	Lizard
	b	Covered with fur	Cat

When using a dichotomous key, we do not always go to the next numbered step (see Figure 3.13).

▾ **Figure 3.13:** A dichotomous key that classifies the spider, scorpion, fly and dragonfly

Step		Statement	Outcome
1	a	4 pairs of legs	Go to step 2
	b	3 pairs of legs	Go to step 3
2	a	Fangs to inject venom	Spider
	b	Barbed stinger to inject venom	Scorpion
3	a	1 pair of wings	Fly
	b	2 pairs of wings	Dragonfly

Classification keys identify important characteristics

The most important part of designing a classification key is deciding which characteristics to include to identify organisms. The features chosen for a key depend on the organisms to be classified.

It makes sense for the first steps of a key to be easily observable features; for example, whether an organism has a backbone. When the classification becomes more specific – such as classifying different species of an animal – the key needs more detail; for example, whether an insect has spots on its wings.

The largest group of animals on Earth is the arthropods. 'Arthropoda' means 'jointed feet', and all arthropods have legs with joints as well as an exoskeleton, which is outside their bodies. Table 3.1 shows the five classes (subgroups) of arthropod, and the physical characteristics by which they are grouped.

Table 3.1: The five classes of arthropod

Class	Characteristics	Example(s)
Insects	• 3 body parts (head, thorax, abdomen) • 6 legs (3 pairs) • 0, 1 or 2 pairs of wings • 1 pair of antennae	Fly, mosquito, dragonfly
Crustaceans	• 3 body parts (head, thorax, abdomen) • 10 legs (5 pairs) • No wings • 2 pairs of antennae	Crab, lobster, prawn
Arachnids	• 2 body parts (cephalothorax, abdomen) • 8 legs (4 pairs) • No wings • No antennae	Spider, tick, scorpion
Centipedes	• Many body segments • 1 pair of legs on each segment • Flat body cross-section • No wings • 1 pair of antennae	Centipede
Millipedes	• Many body segments • 2 pairs of legs on each segment • Rounded body cross-section • No wings • 1 pair of antennae	Millipede

Learning Ladder

Cells and classification

1. Identify two features of an Australian funnel web spider.
2. Describe some structural features of humans that might be used to classify them.
3. You have discovered a new species of arthropod. It has 10 spiky legs, a head, a thorax and an abdomen, and it cannot fly. Which of the five classes of arthropod does your new species belong to? Explain your choice.
4. Scientists recently discovered a fossil of an arthropod that is 350 million years old. It had eight legs, three body parts, and contained silk glands but no spinneret (an organ that arthropods use to spin silky thread). Identify which modern-day arthropod is the most similar to this ancient arthropod.

Communicating

see page 343

1. Identify five terms that could be used to classify different types of insects.
2. Design a simple classification key to identify organisms in a local habitat of your choice.
3. Brainstorm 10 dog breeds. Construct a digital dichotomous key to identify the breeds.
4. Create a branching key to classify the shoes your classmates are wearing.

In context

Scientists classify new species as they discover them. Explain how a classification key may change if new species need to be added.

Success criteria

- I can use a simple classification key to categorise living things.

3.5 ▸ Classifying animals

Figure 3.14: Blue whales are the largest vertebrates in the world: they weigh up to 140 tonnes! A whale's skeleton could not support its weight on land.

Learning intention

At the end of this lesson, I will be able to explain the difference between a vertebrate and an invertebrate.

Key terms

invertebrate: an organism without a backbone or spinal cord

vertebrate: an organism with a backbone or spinal cord

Investigation 3.5

Investigating features of marine animals, p.405

Content group: Classification of living things

From the largest whale to the smallest insect, there are more than 800 000 different animal species in the kingdom Animalia. The first question to consider when classifying animals is whether they have backbones.

Vertebrates are animals with backbones

Some animals – such as humans – have a backbone or spine, which is made up of oddly shaped bones called vertebrae. Animals with backbones are called **vertebrates**. In the Linnaean classification system, vertebrates are members of the phylum Chordata. All the animals in this group have a flexible cord along their back (all vertebrates have a nerve cord inside their backbone), which sends and receives information between body parts.

Fossils show that vertebrates have existed for at least 518 million years. Over time, more and more vertebrate species have evolved. Today, there is a diverse range of animals with backbones.

Table 3.2: The five classes of the phylum Chordata

Class	Body covering	Temperature control	Reproduction	Structure for movement	Example
Fish	Slimy scales	Gains heat from surroundings	Lays eggs	Fins to move through water	Salmon
Amphibians	Naked skin	Gains heat from surroundings	Lays eggs in water	Young (tadpoles) have fins; adults have legs	Frog
Reptiles	Dry scales	Gains heat from surroundings	Lays leathery eggs	Legs for walking on land (except snakes)	Crocodile
Birds	Feathers and scales (on feet)	Produces own heat	Lays hard-shelled eggs	Legs for walking on land and wings for flying	Kookaburra
Mammals	Hair or fur	Produces own heat	Lays eggs	Legs with different shapes for walking, swimming or flying	Echidna
			Gives birth to underdeveloped young		Human
			Gives birth to fully developed young		Dog

Invertebrates are animals without backbones

The prefix 'in-' means 'not', so **invertebrate** means 'not a vertebrate'. These animals do not have backbones or any bones at all.

There are many different phyla (groups) of invertebrates. They can be grouped by structural features. Most invertebrates have an exoskeleton, which is a skeleton outside the body that is not made of bone. This supports the animal as it moves. Some invertebrates – such as worms – do not have an exoskeleton; these animals have inside structures that support them.

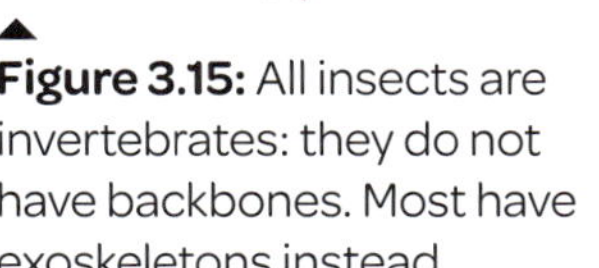

▲ **Figure 3.15:** All insects are invertebrates: they do not have backbones. Most have exoskeletons instead.

Table 3.3: Some of the many phyla (groups) of invertebrates

Phylum (group)	Characteristics	Example(s)
Cnidarians	Soft body with one opening	Anemone, coral, jellyfish
Annelids	Segmented worm, round cross-section, lives on land	Earthworm
Platyhelminths	Segmented worm, flat cross-section, freshwater or in another organism	Tapeworm, planarian worm
Nematodes	Unsegmented worm, round cross-section, often lives in another organism	Round worm, heartworm
Poriferans	Collection of cells, some with hair-like structures, one opening	Sponge
Molluscs	Shelled animal	Snail, clam, oyster
Echinoderms	Body parts arranged around a central point, has many 'feet'	Sea star, sea urchin, brittle star
Arthropods	Jointed legs, hard exoskeleton, segmented body	Insect, spider, prawn

There are more species of invertebrates than vertebrates

Invertebrates account for around 95 per cent of all the species in the animal kingdom. There is also a large range of invertebrate species. These diverse animals can be sorted into many different groups.

Learning Ladder

Cells and classification

1. Identify the main difference between an invertebrate and a vertebrate.
2. Classify the following animals into their phylum (group). Use Table 3.3 to help you.
   - **a** Earthworm
   - **b** Snail
   - **c** Mussel (a shelled animal)
   - **d** Lobster
3. Explain how you can tell that an organism is a vertebrate or an invertebrate just by looking at it.
4. Many vertebrates contain other similar bone structures, despite being completely different classes and species. Explain these similarities.

Observing

see page 316

1. Outline one observation you can make about the insects in Figure 3.15 that shows why they might be grouped together.
2. Consider the information in Investigation 3.5. Identify some scientific tools that are useful when making observations during a dissection. Explain why these tools are useful.
3. Discuss how you could accurately record your observations from a dissection.
4. Summarise your observations about the insects in Figure 3.15 that tell you how they survive in their environments.
5. Over time, changes might be made to the classification of invertebrates. Explain why this is.

In context

What categories can vertebrates be classified into? Classify three Australian vertebrates into more specific categories.

Success criteria

- I can explain the difference between a vertebrate and an invertebrate.
- I can give at least two examples of vertebrates and invertebrates.

3·6 ► Classifying plants

Figure 3.16: Ferns are bryophytes, which reproduce using spores.

Learning intention

At the end of this lesson, I will be able to classify a variety of plants based on some of their features.

Key terms

angiosperm: a plant that produces seeds in fruit or flowers

bryophyte: a plant that does not contain vascular tissue

gymnosperm: a plant that produces seeds in cones

spore: a tiny part of some plants that is used to reproduce

structure: the construction and arrangement of tissues, parts or organs

tissue: a group of cells with a similar structure and function

tracheophyte: a plant that contains vascular tissue

vascular tissue: tissue that transports fluid and nutrients throughout a plant

Content group: Classification of living things

There are almost 400 000 plant species in the kingdom Plantae. The incredible variety of plants includes ferns, mosses and flowering plants. Like other organisms, plants can be classified by their **structure**.

Plants can be classified by their tissue

Plants can be grouped according to their type of **tissue**:

- Vascular plants (**tracheophytes**) have **vascular tissue**.
- Non-vascular plants (**bryophytes**) do not have vascular tissue.

Vascular tissue transports sugars, water and minerals through tiny tubes in the plant. It also supports the plant structurally. Vascular tissue works in a similar way to the bones and blood vessels of humans and other mammals.

Tracheophytes (vascular plants)

Most plants are tracheophytes (vascular plants), which are supported from inside and can grow very tall. Water moves from the soil into the roots, the stem and then the leaves. Glucose is made in the leaves, and it travels to the rest of the plant to be used for energy or converted to starch and stored.

Most vascular plants reproduce using seeds. Seeds are small grains with hard outer coatings that contain the material plants need to reproduce.

Bryophytes (non-vascular plants)

Examples of bryophytes (non-vascular plants) are mosses and liverworts. Non-vascular plants do not have support inside them, so they cannot grow very tall. Instead, they appear as soft coverings on surfaces such as rocks and soil. Non-vascular plants do not have true roots or leaves. Bryophytes grow in areas where water is freely available, such as next to rivers or waterfalls, or in rainforests.

Figure 3.17: This Australian native plant, the kangaroo paw, is an angiosperm because it produces seeds in its flowers.

Non-vascular plants reproduce using **spores**. These are small, round spheres on the underside of leaves. Spores fall to the ground and grow into heart-shaped plants that produce male and female cells. Spores need water for fertilisation.

Seeds, fruits and flowers can be used to classify plants

Plants can also be grouped into those that produce seeds and those that do not. For example, ferns do not have seeds (they have spores), whereas flowering plants do have seeds.

Plants that produce seeds can be divided into those that produce seeds in:

- cones
- flowers and fruits.

Gymnosperms produce naked seeds that develop in cones. These plants have needle-like or thin leaves, and often grow into tall trees with woody trunks. Because they have large root systems, they can collect water from underground supplies and can grow in many different environments. Pines, firs, cycads and ginkgos are all gymnosperms.

Table 3.4: Some features used to classify plants

Class	Contains vascular tissue?	Produces seeds?	Method of producing seeds	Examples
Bryophytes	No	No	–	Mosses, liverworts
Ferns	Yes	No	–	Ferns
Gymnosperms	Yes	Yes	Cones	Pine tree
Angiosperms	Yes	Yes	Flowers and fruits	Orange tree

Figure 3.19: Norfolk Island pines are gymnosperms, which produce seeds in cones.

Angiosperms produce seeds in flowers or fruits. There are many different types of flowering plants, and this is the largest group of plants on Earth. Wattles, roses, grass, wheat and eucalypts are all angiosperms.

Figure 3.18: Mosses are bryophytes. These plants do not have inside support and grow outwards rather than up.

Learning Ladder

Cells and classification

1. Identify some features used to classify plants.
2. Describe the distinguishing features of ferns.
3. Is a cherry tree a gymnosperm or an angiosperm? Explain your answer.
4. Explain the difference between bryophytes and tracheophytes.

Communicating

see page 343

1. Identify an example of a:
   a tracheophyte.
   b bryophyte.
2. A plant is about 20 centimetres tall and has a pink flower blooming from a single stem. Which one of the following groups would you classify this plant into? Explain your choice.
   A Ferns
   B Gymnosperms
   C Angiosperms
3. **a** In your local area, identify a fern, a gymnosperm and an angiosperm.
   b Construct a digital table that shows how each plant is classified based on its features.
4. Select one of the plants from Question 3. Create a poster containing information about its classification. See page 349 of the Science how-to section for help with constructing your poster.

In context

Species are becoming extinct at an alarming rate. Write a persuasive article justifying why preserving plant species is as important as preserving animal species.

Success criteria

- I can classify a variety of plants based on some of their features.

3·7 ▸ Structural, physiological and behavioural adaptations

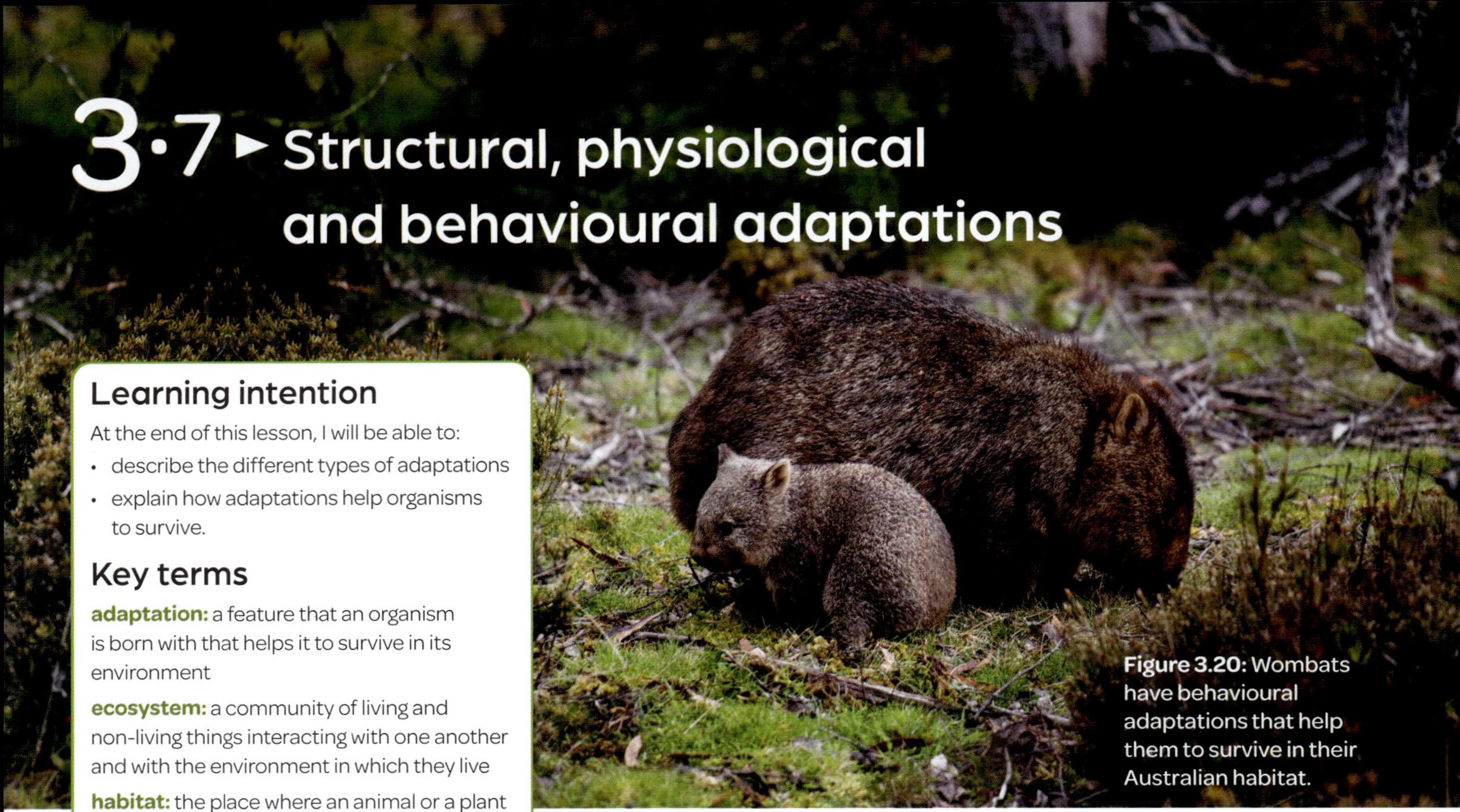

Figure 3.20: Wombats have behavioural adaptations that help them to survive in their Australian habitat.

Learning intention

At the end of this lesson, I will be able to:

- describe the different types of adaptations
- explain how adaptations help organisms to survive.

Key terms

adaptation: a feature that an organism is born with that helps it to survive in its environment

ecosystem: a community of living and non-living things interacting with one another and with the environment in which they live

habitat: the place where an animal or a plant naturally lives

Investigation 3.7

Investigating structural adaptations, p.406

Content group: Classification of living things

The characteristics that make organisms able to survive in their environments are called adaptations. Adaptations can be structural, physiological or behavioural.

Behavioural adaptations are something animals do

A behavioural **adaptation** is any behaviour that organisms exhibit that helps them to survive in their **habitat**. For example, a single organism collecting specific objects to build a habitat is a behavioural adaptation. A group of organisms choosing to live together is also a behavioural adaptation that helps them to survive.

In the Australian environment, many organisms display behavioural adaptations. For example, when a wombat is being chased, it runs towards its burrow and then stops suddenly. The predator runs into its back end, damaging its jaw on the wombat's reinforced tailbone.

Physiological adaptations are processes in the body

Physiological adaptations are processes and functions that help organisms to survive. They are triggered in response to the environment.

For example, pregnant female kangaroos can 'pause' the development of their joey if food is scarce. When food is plentiful again, the pregnancy continues. This reaction is controlled by hormones, and is not a conscious choice of the kangaroo, but it does improve the survival rates of joeys.

▲ **Figure 3.21:** The digestive systems of koalas allow them to eat the waxy, toxic leaves of eucalyptus trees without getting sick.

Structural adaptations are physical features of an organism

A structural adaptation is any physical feature of an organism that helps it to survive in its environment. Physical features include tissue, organs, and any visible features like tails, bodies and claws.

For example, the thick, brown fur that covers most of a platypus's body is waterproof and insulates against the cold, which allows these animals to hunt for food in cold rivers.

Figure 3.22: Platypuses have a structural adaptation – their thick, waterproof fur – that helps these animals to survive in cold water.

Australian organisms are adapted to their environments

Adaptations can be seen clearly in many Australian **ecosystems**, such as bushland areas (see Table 3.5).

Table 3.5: Examples of how Australian organisms are adapted to their environments

Organism	Adaptation	Type	Benefit
Eucalyptus tree	Leaves are coated in a waxy substance, which keeps moisture in the leaves	Structural	Helps the tree survive during droughts and in environments with little water
Koala	Digestive systems allow them to eat waxy, poisonous eucalyptus leaves without getting sick	Physiological	Provides a food source that other living things are not competing for
Musk lorikeet	Nests in eucalyptus trees	Behavioural	Provides a safe place to produce and raise their young

Learning Ladder

Cells and classification

1. Identify one structural adaptation that a platypus has that helps it to survive in its Australian habitat.
2. Describe how the structural adaptation you identified in Question 1 helps a platypus to survive in its Australian habitat.
3. Identify three features of a wombat that would help you to classify it.
4. Look at Figures 3.21 and 3.22. Compare the feet of the koala and the platypus.

Observing

see page 316

1. Outline some observations of eucalyptus leaves you could make using just your senses.
2. Identify a piece of scientific equipment that could improve your observations of the physiological adaptation of female kangaroos.
3. Draw a diagram of an Australian bush environment. Include examples of behavioural, physiological and structural adaptations of three organisms. Label the adaptations.
4. Propose an inference about a platypus's diet and habitat, based on the photo of this animal in Figure 3.22.

In context

Humans also have behavioural, physiological and structural adaptations. Propose two features or characteristics of humans that fall into each category. Justify each classification.

Success criteria

- I can describe the different types of adaptations.
- I can explain how adaptations help organisms to survive in an environment.

3·8 ▸ Aboriginal and Torres Strait Islander Peoples' systems of classification

Learning intention

At the end of this lesson, I will be able to explain some of the ways Aboriginal and Torres Strait Islander Peoples classify plants and animals.

Key terms

Ancestors: the people we are descended from

spirits: the supernatural beings that many First Nations Peoples believe exist; can be associated with particular places, objects and rituals; are often connected with Dreaming stories

Content group: Classification of living things

First Nations classification systems are complex webs with criss-crossing connections. These systems are based on Aboriginal and Torres Strait Islander Peoples' deep understanding and vast knowledge of living things.

First Nations classification systems are complex

The systems of classification used by Aboriginal and Torres Strait Islander Peoples are complex and sophisticated. Many factors are taken into consideration when grouping objects and organisms (see Table 3.6).

Table 3.6: Some of the factors used to classify objects and living things by Aboriginal and Torres Strait Islander Peoples

Living or non-living	Edible or non-edible
The organism's age	The organism's size
Venomous or non-venomous	The organism's sex
Status of the organism (e.g. within its herd)	The object or organism's observable features
Where the organism lives (e.g. coastal or inland)	Which stage of its breeding cycle the organism is in
The value of the object or the organism to a First Nations community	Use of the organism when alive and when dead (e.g. as food, medicine, tools, clothing)

Plants and animals are sorted by their use

One of the many ways that Aboriginal and Torres Strait Islander Peoples group living things is by how the organisms are used. For example, some trees (including swamp she-oaks, stringybarks and mulga trees) are classified as 'canoe trees' because First Nations Peoples made canoes by removing the bark from these trees (see Figure 3.23).

Figure 3.23: Canoe trees like this one all have large trunks and thick bark.

The Dhuwa and Yirritja classification system divides the universe into two halves

The Dhuwa and Yirritja classification system is fundamental to the culture of the Yolŋu People, who live in north-east Arnhem Land. In this system, *everything* in the universe is classified as either Dhuwa or Yirritja: the seasons, individuals, animals, plants, areas of land, waterways, clouds, **spirits**, **Ancestors**, winds.

▲ **Figure 3.24:** The Yolŋu People classify everything as either Dhuwa or Yirritja.

Learning Ladder

Cells and classification

1. Identify the names of the two groups in the Yolŋu classification system.
2. Describe how canoe trees are structurally similar.
3. Explain why swamp she-oaks, stringybarks and mulga trees are placed in the same group in a First Nations classification system.
4. Compare the Linnaean classification system to First Nations classification systems.

Communicating

see page 343

1. Examine Figure 3.24. Identify and describe two organisms that are:
   a Dhuwa. **b** Yirritja.
2. Describe how you could explain the differences between the Linnaean classification system and First Nations classification systems.
3. Create a digital presentation of the organisms you identified in Question 1.

In context

Research the classification system used by First Nations People in your area or nearby.

Success criteria

- I can explain some of the ways Aboriginal and Torres Strait Islander Peoples classify plants and animals.

3·9 ▸ What are cells?

Learning intention

At the end of this lesson, I will be able to explain what a cell is and outline the components of cell theory.

Key terms

cell: the smallest functional unit of an organism

hypothesis: a suggested explanation or prediction of a scientific problem that can be tested with an investigation

scientific theory: an explanation of a natural phenomenon that is supported by evidence and the results of repeated tests

Investigation 3.9

Examining cells under a microscope, p.407

Content group: Cells

Cells are the smallest structural and functional units of living things. This means that a cell is the smallest part of a living thing that can carry out the functions needed for life, on its own. All living things are made of cells.

All living things are made of one or more cells, which come from other cells

Cells were first identified in 1665 by English scientist Robert Hooke. He was using a simple microscope to study the bark of a cork tree. The tiny structures he could see through the microscope reminded him of the tiny rooms that monks sleep in, which are called cells.

Even after cells had been discovered, scientists were not sure whether all organisms were made of them. Then, in 1839, scientists developed a **hypothesis** about cells. All the evidence and discoveries made since then have confirmed this 19th-century idea, which is now recognised as a **scientific theory** called cell theory.

There are three parts to the theory of cells:

- All living things are made of one or more cells.
- Cells are the basic building blocks of life.
- Cells come from cells that already exist.

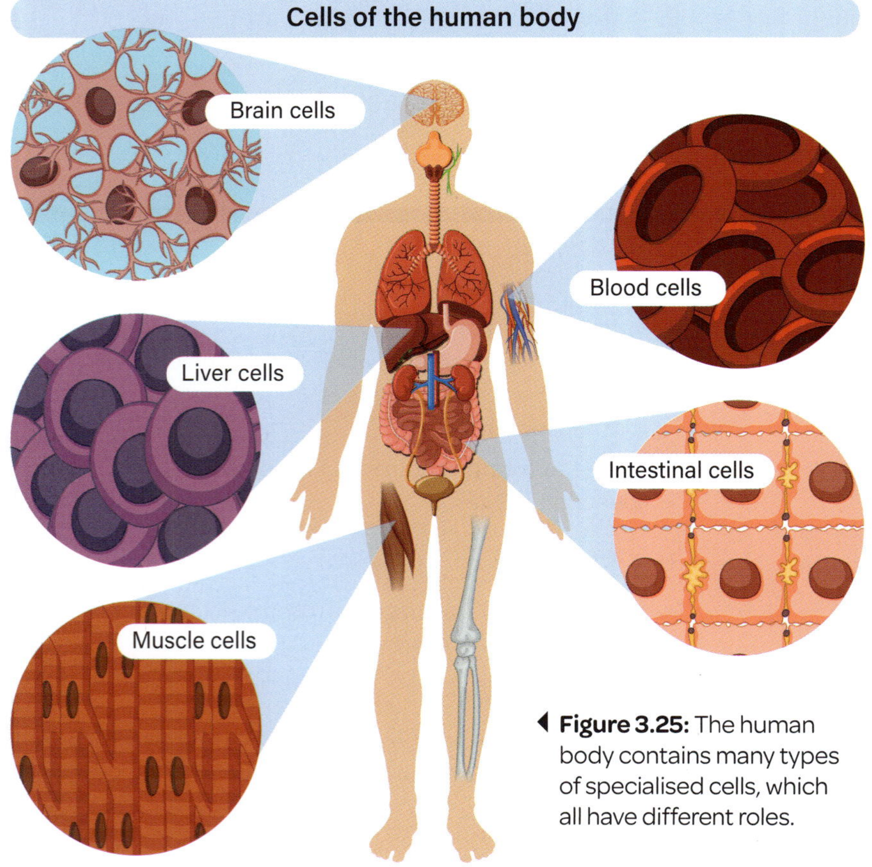

Figure 3.25: The human body contains many types of specialised cells, which all have different roles.

Cells are the building blocks of life

Cells are the basic building blocks of all living things. Sometimes a single cell is a complete – but very simple – organism (for example, a bacterium). More complex organisms (such as humans) are made up of many cells that work together to perform different functions (see Figure 3.25).

The human body is made up of trillions of cells. They provide structure, take in nutrients from food, change those nutrients into energy, and carry out other functions. Our cells contain our genetic material and can make copies of themselves. This allows us to grow new cells, such as for hair and fingernails.

Figure 3.26: An onion is made of many cells. The cells in a thin layer of onion membrane can be seen under a light microscope.

Microscopes allow us to see and identify cells

Cell size can vary a lot. The largest cells are ostrich eggs: a single unfertilised egg cell is about 15 centimetres wide and weighs more than one kilogram! Red blood cells are only 0.008 millimetres wide; a line of 125 red blood cells is only one millimetre long. Most cells are about this size, which is far too small for us to see with the naked eye.

Scientists use microsopes to see cells. The word comes from two ancient Greek terms: *micro*, which means 'small', and *scope*, which means 'to see'. A microscope is a tool for looking at small things.

The most common type of microscope in science laboratories is a light microscope. It allows scientists to study cells by shining a bright light through an extremely thin slice of tissue (usually 0.01 millimetres thick) from an organism (see Figure 3.26).

Learning Ladder

Cells and classification

1. Complete this sentence: 'Cells come from ...'
2. Approximately how many cells are there in the human body?
3. Explain why cells are described as 'the building blocks of life'.
4. In terms of cells, outline how humans are different from bacteria.
5. Analyse how cell theory changed scientists' understanding of living things.

Conducting investigations

see page 326

1. Outline how thin a slice of tissue has to be in order to be viewed under a light microscope.
2. Construct a step-by-step description of how you would use a microscope to view cells.
3. Identify one safety practice that you should implement when using a light microscope.
4. Explain how you can make sure that the information you collect when viewing tissue under a microscope is accurate.

In context

'All living things are made from one or more cells.' Propose how using technology allowed scientists like Robert Hooke to support this hypothesis with evidence. Present your information as an article on the history of cells.

Success criteria

- I can explain what a cell is and outline the components of cell theory.

3·10 ► Cell structures

Learning intention

At the end of this lesson, I will be able to:

- describe the common parts of plant and animal cells
- identify the similarities and differences in the structures (organelles) of plant, animal, fungus and bacteria cells.

Key terms

cell membrane: a thin layer around a cell that controls which substances enter and leave the cell

cell wall: a stiff layer around a plant cell that supports it

chloroplast: a small organelle in a plant cell that makes food for the plant

cytoplasm: a jelly-like fluid inside a cell

DNA: the genetic information inside a body's cells

nucleus: the control centre of a cell; contains DNA

organelle: a structure within a cell

Investigation 3.10

Examining cell structures, p.409

Content group: Cells

Cells are made up of different parts, and different cells have different roles.

Cells have different parts called organelles

Each cell has many different parts. Each part has a particular job that helps keep the cell healthy. However, not every cell has the same parts. For example, a plant cell has more parts than an animal cell. Scientists call the parts of a cell **organelles**.

Animal, plant and fungus cells have a nucleus, cytoplasm and cell membrane

The cells of all animals, plants and fungi (for example, mushrooms and moulds) have three organelles in common:

- A **nucleus**: This is the central part of a cell. It controls all the cell's activity. **DNA**, the genetic material, is located inside the nucleus.
- **Cytoplasm**: This is a jelly-like fluid in which cell organelles sit. Many of the chemical reactions in a cell happen in the cytoplasm.
- A **cell membrane**: This is a flexible, thin layer around a cell (like an envelope surrounding a letter). It controls substances coming in and out of a cell.

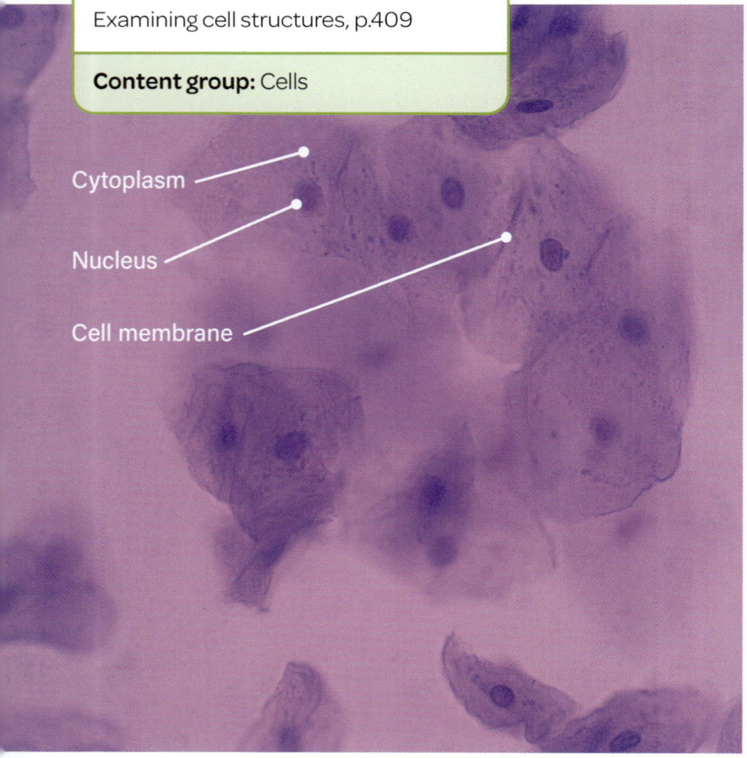

Figure 3.27: A group of typical animal cells, as seen under a microscope

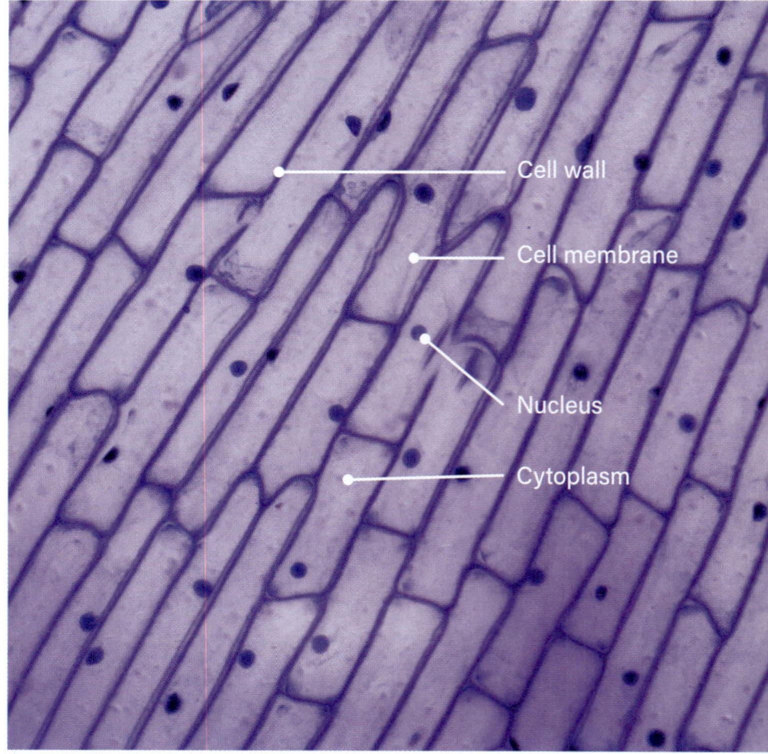

Figure 3.28: A group of typical plant cells, as seen under a microscope

Plant cells have a cell wall and chloroplasts

There are many obvious differences between plants and animals. No-one is likely to confuse a cat with a cactus, or a shark with seaweed!

However, some differences are not obvious, such as the organelles inside plant and animal cells. Like animal cells, plant cells have a nucleus, cytoplasm and cell membrane. However, they also have two structures that animal cells do not:

- A **cell wall**: This is a stiff layer around a plant cell, outside the cell membrane. It helps to protect and support the cell.
- One or more **chloroplasts**: Chloroplasts are organelles that act as energy producers. They are where plants make sugar using the Sun's energy.

Animal cells do not have these organelles because they do not need them. Animals have a skeleton to provide support and they can move around to get food.

Bacteria cells are simpler than animal, plant and fungus cells

Bacteria are different from plants, animals and fungi, which are all made of complex eukaryotic cells. Bacteria are made of simpler cells, called prokaryotic cells. These cells cannot join to form a bigger organism. Bacteria cells have DNA, but they do not have a nucleus to hold it in place. The DNA floats freely in the cytoplasm. However, there are some similarities between the different cells (see Figure 3.29).

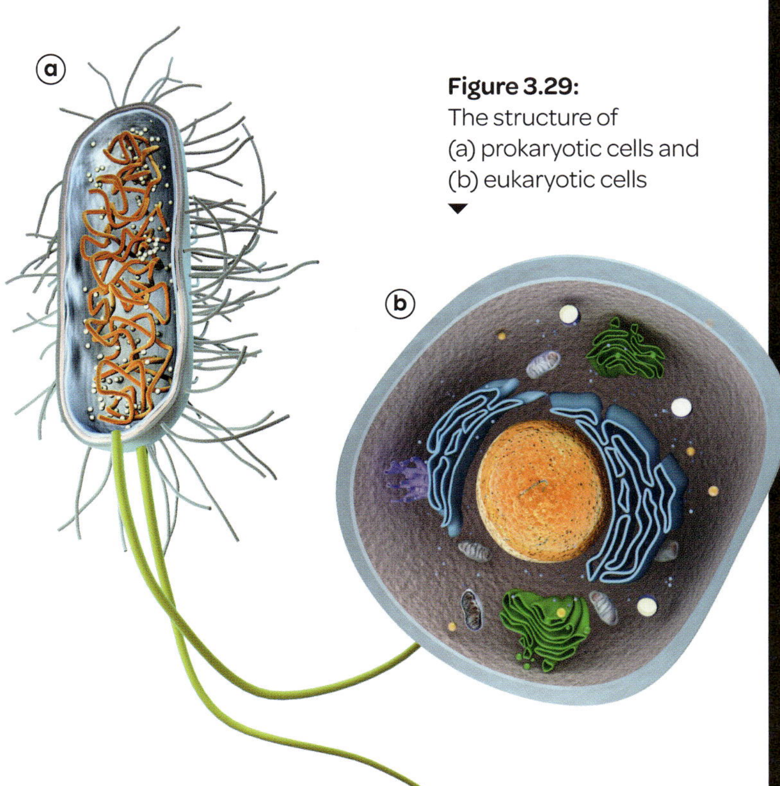

Figure 3.29: The structure of (a) prokaryotic cells and (b) eukaryotic cells

Staining shows the organelles of cells

Scientists use chemicals called stains to artificially colour some of the parts of cells. Staining makes them easier to see under a microscope. In Figure 3.28, a purple stain has been used to colour some of the cell parts.

Most cell structures are too small to be seen under a microscope. Parts of a cell that can be seen under a light microscope include the nucleus, cell membrane, cell wall and chloroplast. The cytoplasm cannot really be seen in Figure 3.28, but you can see floating substances: they look like a grainy background inside the cell membrane.

Learning Ladder

Cells and classification

1. Identify an organelle that is unique to plant cells.
2. Describe two organelles of plant cells.
3. Recall the names of three organelles in fungi cells.
4. Identify an organelle that is found in an animal cell that is not found in a bacteria cell. Outline the purpose of this organelle.

Observing see page 316

1. Can you use your senses to make observations about cell organelles? Explain your answer.
2. Identify which scientific equipment you could use to make observations of cells. Why would using this item improve your observations?
3. Outline how you could record your observations of cells.

In context

Draw diagrams of a bacteria cell, a plant cell and an animal cell. Label the diagrams by identifying the similarities and differences between the three cells.

Success criteria

- I can describe the common organelles of plant and animal cells.
- I can identify the similarities and differences in the cell organelles of animals, plants, fungi and bacteria.

3·11 ▸ Chemical reactions in cells

Learning intention

At the end of this lesson, I will be able to explain the processes of photosynthesis and respiration.

Key terms

aerobic respiration: the process of turning glucose into energy where the cells take in oxygen and release carbon dioxide

anaerobic respiration: how living things produce energy without oxygen

chlorophyll: the green pigment in plant cells; this pigment absorbs sunlight and enables photosynthesis

glucose: a type of sugar that is the energy source for cells

mitochondria: the organelles where respiration happens

photosynthesis: the chemical reaction, powered by sunlight, in plants that converts carbon dioxide and water into sugar and oxygen

respiration: a chemical reaction that converts glucose to energy

stomata: pores on the surface of a leaf; the site of gas exchange in plants

Investigation 3.11A

Investigating respiration, p.410

Investigation 3.11B

Energy from food, p.412

Content group: Cells

Animals and plants have many cells that carry out various functions to help the organism survive. To perform these functions, cells need a constant supply of energy. Animal cells get their energy from food, and plant cells make their food using sunlight.

All cells need energy to survive

Without a continuous supply of energy, cells cannot perform important functions and will die. The only type of energy that cells can use is chemical energy. Chemical energy comes from **glucose**, which is a type of sugar. Plant cells convert energy from the Sun into glucose. Animal cells get most of their energy from food.

Figure 3.30: Plant cells get energy from sunlight.

Photosynthesis is how plants make food

Plants rely on a chemical reaction called **photosynthesis** to make their food. **Chlorophyll** (a green pigment) in plant cells absorbs sunlight and uses it to change carbon dioxide and water into glucose and oxygen. The process of photosynthesis is summarised in Figure 3.31 and shown in Figure 3.32.

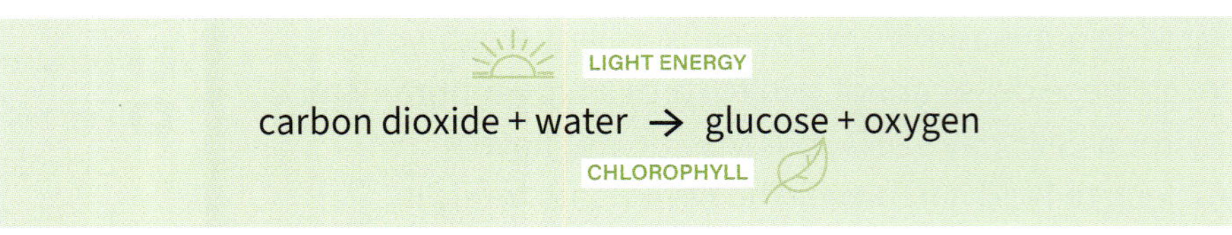

Figure 3.31: The process of photosynthesis

The glucose produced from this reaction is either used in respiration or stored for later use. The oxygen moves into the environment through special pores (openings) in the leaves called **stomata**, or is used during respiration.

Without photosynthesis, humans would not have the oxygen or glucose we need to survive. We breathe in oxygen released by plants, and consume glucose made in leaves. If plants did not exist, humans would not exist either!

Respiration is how cells make energy

Respiration is a chemical reaction that converts glucose to carbon dioxide and produces energy. Respiration occurs in all cells. There are two forms of respiration:

- **aerobic respiration**, which uses oxygen
- **anaerobic respiration**, which happens without oxygen.

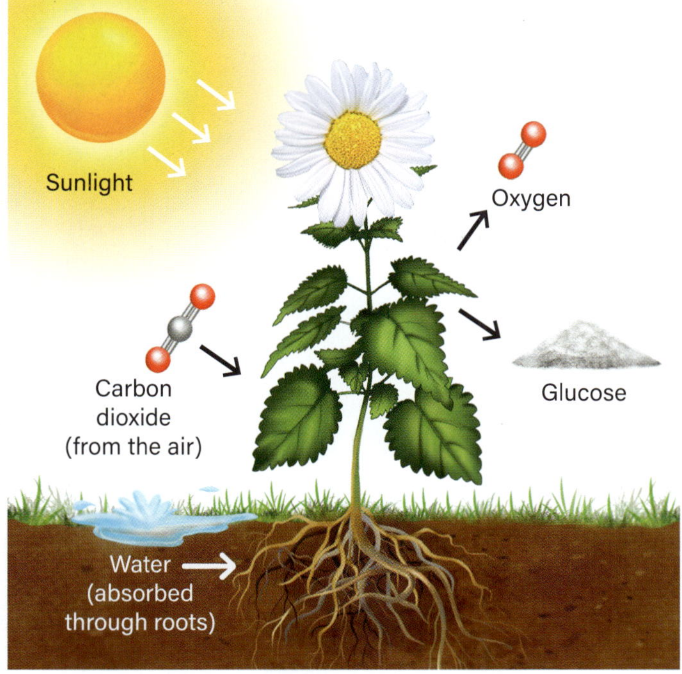

Figure 3.32: Photosynthesis happens in the leaves of plants.

Aerobic respiration provides animals and plants with most of their energy.

In animals, the process of aerobic respiration involves the organism taking in glucose (from food) and oxygen (from the environment via lungs or gills). When the glucose and oxygen combine, they react and form carbon dioxide, water and energy. Carbon dioxide and water are the waste products.

The chemical energy produced during respiration is transported to the cells that need it, while the carbon dioxide and water are removed from the cell. The process of aerobic respiration is summarised in Figure 3.33.

glucose + oxygen → carbon dioxide + water + energy

Figure 3.33: The process of aerobic respiration

Respiration happens in mitochondria

Respiration happens in organelles (cell structures) called **mitochondria**, in both plant and animal cells. Mitochondria are called the 'powerhouse' of cells because they take in nutrients, break them down, and change them into energy that cells can use. Animals and plants do not respire in the same way, and they gain energy from different nutrients, but mitochondria have the same purpose in all living things.

Mitochondria are very small. Some cells have several thousand mitochondria because they need lots of energy; for example, muscle cells in animals. Other cells have lower energy needs, so they have few or no mitochondria.

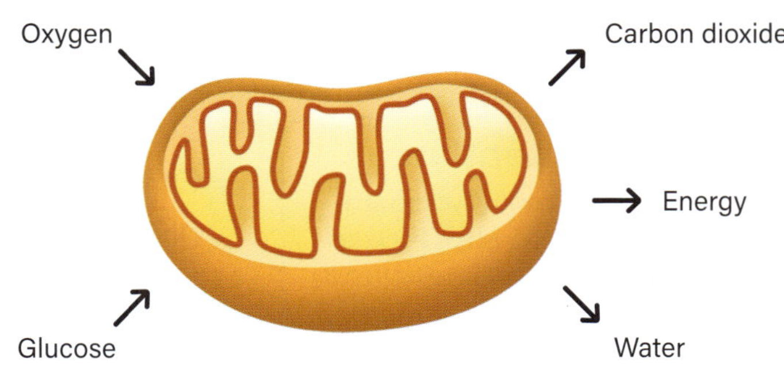

Figure 3.34: Mitochondria are the organelles responsible for respiration. Because their folded structure provides a lot of surface area, many reactions can occur at the same time.

Learning Ladder

Cells and classification

1. Identify one feature of a cell that allows it to conduct respiration.
2. Summarise what mitochondria are.
3. Explain why stomata are located on the underside of leaves.
4. Compare how animals and plants take in oxygen.

Conducting investigations

see page 326

Read the aim, materials and method of Investigation 3.11B on page 412, then answer the following questions.

1. Identify the correct way to secure the test tube.
2. Describe how you would use a thermometer to collect data.
3. Identify two hazards associated with this investigation. Outline how you could minimise the risk of harm occurring.
4. Propose ways that you could make sure that the data you collect from this investigation is accurate.

In context

Discuss the differences between respiration and breathing.

Success criteria

- I can explain the processes of photosynthesis and respiration.

3·12 ► Cells, tissues and organs

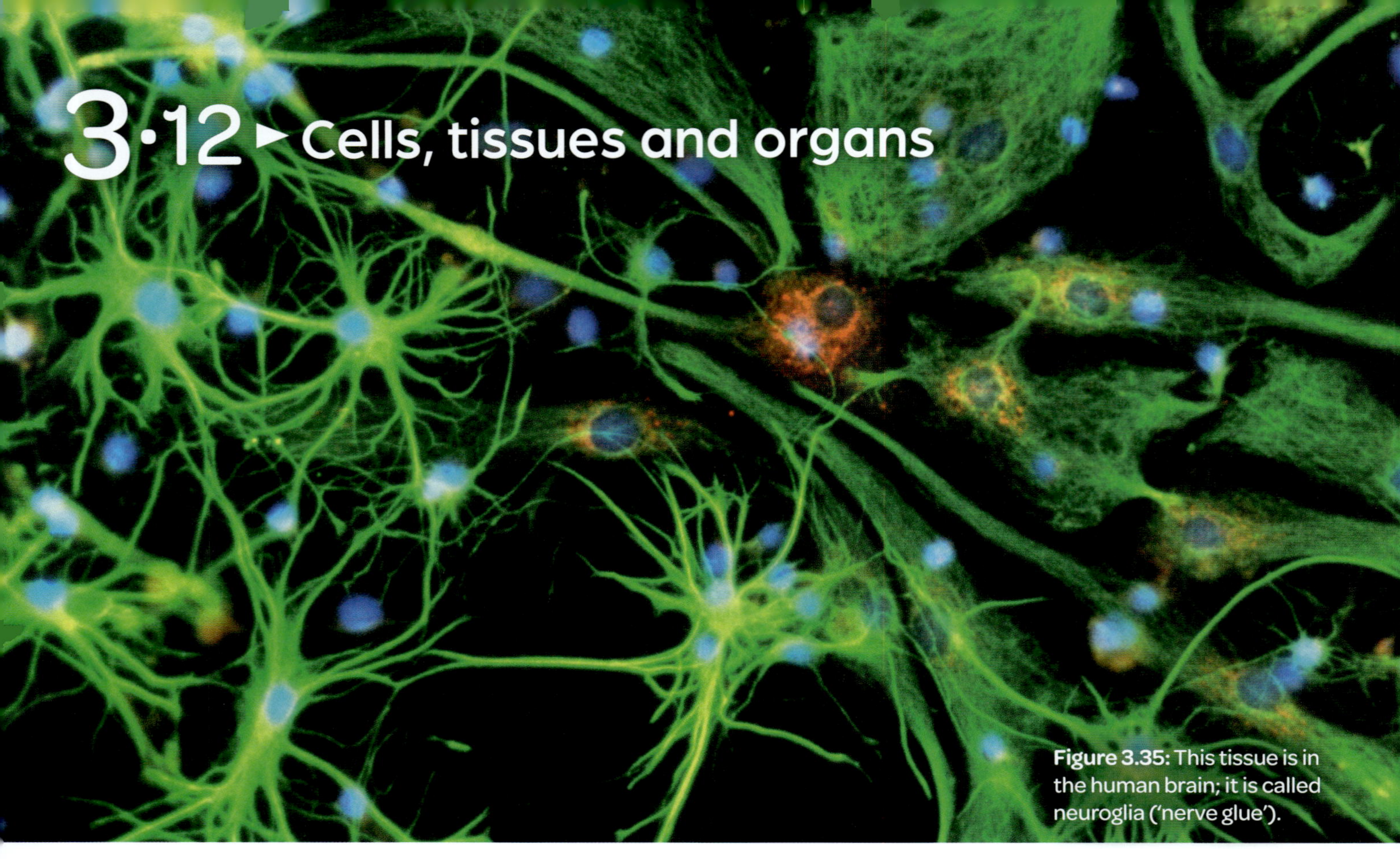
Figure 3.35: This tissue is in the human brain; it is called neuroglia ('nerve glue').

Learning intention

At the end of this lesson, I will be able to examine the relationships between cells, tissues and organs in living things.

Key terms

differentiate: to change to have a particular function

multicellular: consisting of many specialised cells that form tissues and organs, which together make one functional organism

organ: a structure made up of two of more tissues that has a specific function

unicellular: consisting of one cell that carries out all the functions itself

Content group: Cells

Some organisms are **unicellular** (such as bacteria). Other organisms are **multicellular** and have many different types of specialised cells.

Specialised cells perform specific tasks

A multicellular organism starts as a single cell. This one cell divides into many cells that are joined together. As the organism grows and more cells form, the cells **differentiate** so they perform different functions to help the organism work effectively.

Some of the specialised cells in plants and animals are listed in Table 3.7.

Table 3.7: Some of the specialised cells in plants and animals

Organism	Specialised cell	Function
Human	Epidermal cell	Forms a protective outer layer (skin)
	Nerve cell	Transmits messages around the body
	Red blood cell	Transports oxygen and nutrients around the body
Plant	Palisade cell	Contains chlorophyll to carry out photosynthesis
	Xylem cell	Transports water from the roots to the stem and leaves
	Phloem cell	Transports sugar around the plant

Specialised cells can form tissues and organs

Cells that specialise in the same function can group together to form tissues and **organs**. These structures are larger and more complex than individual cells. Tissues and organs are structured in specific ways to help organisms function efficiently.

The following flowchart summarises the levels of organisation in organisms.

organelle → specialised cell → tissue → organ → organ system → organism

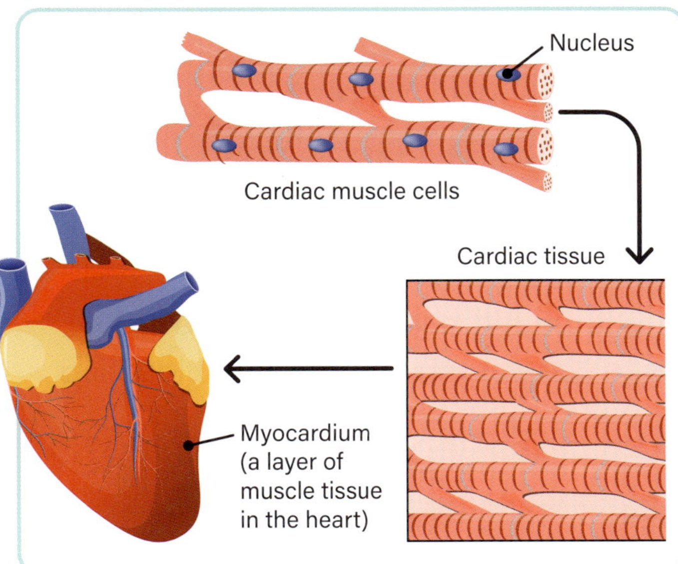

▲ **Figure 3.36:** The heart is an organ that has a muscular layer of tissue called the myocardium. The myocardium is made of cardiac tissue, which in turn is made of cardiac muscle cells.

A microscope is needed to view specialised cells

Cells are too small to be seen by the naked eye, so scientists use microscopes to view and make observations about cells.

Light microscopes can be used to view the overall size and shape of cells, and to see some of their larger organelles, such as the cell wall and the nucleus.

Electron microscopes have much higher magnifications than light microscopes. These microscopes can be used to look at the organelles in different cells, allowing scientists to see the differences between specialised cells. These differences can include which organelles are present, and how many there are.

Figure 3.37: Scientists need to use microscopes to see specialised cells.

Learning Ladder

Cells and classification

1. Identify a specialised cell found in:
   a humans. **b** plants.
2. For the cells you identified in Question 1a, describe their structure and function.
3. Explain how specialised cells form the basis of more complex structures in multicellular organisms.
4. Compare the complexity of a cell organelle to an organ in a human body.

Observing

see page 316

1. Outline the observations you can make about the specialised structures of plants by just using your senses.
2. Describe how a microscope would assist you to make detailed observations of the structure of the specialised cells in humans.
3. **a** Research a specific specialised cell.
   b Construct a digital flowchart that shows how this specialised cell fits in the structure of a functional organism. (Hint: What tissue and organ is it part of?)
4. Create an infographic about the heart as an organ. Include information about the specialised cells and tissues that combine to allow the heart to function.

In context

Research how improving our understanding of specialised cells could assist doctors to conduct organ transplants.

Success criteria

- I can describe the structures and functions of many specialised cells.
- I can explain how specialised cells form larger functional networks in living things.

3·13 ► Cells and classification in context

Figure 3.38: The scientific name of the quokka is *Setonix brachyurus*.

Learning intention

At the end of this lesson, I will understand the classification of the quokka.

Key term

nocturnal: active during the night

Content group: Cells and classification in context

The quokka is a **nocturnal** Australian animal. We can examine quokkas to understand how they are classified.

Why is a quokka classified as a living thing?

There are seven biological characteristics that organisms must have to be considered 'living' by scientists. The quokka has all seven of these features (see Table 3.8).

Table 3.8: Characteristics of quokkas that classify them as living things

Biological characteristic	Description	Quokka behaviours and features
Growth	The organism gets bigger and develops over time.	A baby quokka is the size of a bean and weighs around 35 grams; adult quokkas weigh between 2.5 and 5 kilograms and their bodies are around 40–54 centimetres long.
Movement	The organism can move part or all of itself.	Quokkas run and hop along the ground using their large hind legs and long tails.
Nutrition	The organism takes in nutrients.	Quokkas take in nutrients by eating plants, such as native grasses, leaves and berries.
Reproduction	The organism can produce offspring.	Quokkas give birth to live young.
Response to stimuli	The organism can respond to changes in the environment.	Quokkas shelter in dense vegetation when it is hot during the day, and are active at night when it is cooler.
Respiration	The organism can convert nutrients into energy.	Quokkas are made of cells; cells use respiration to transform into energy the nutrients in the plants that quokkas eat.
Waste removal	The organism can get rid of waste products.	Quokkas get rid of both urine and faeces.

Figure 3.39: The quokka is a herbivore, which means it eats plants to obtain nutrients such as glucose.

How do we classify the quokka?

The species of the quokka can be classified using the Linnaean system of classification (see Table 3.9).

Table 3.9: The classification of the quokka in the Linnaean system

Classification level	Quokka classification	Description
Domain	Eukaryota	Organisms with specialised cells
Kingdom	Animalia	Animals
Phylum	Chordata	Animals with a cord along their back
Class	Mammalia	Mammals: have hair and live young
Order	Diprotodontia	Marsupials: have a pouch
Family	Macropodidae	Macropods: have large hind legs and long tails
Genus	*Setonix*	Specific to the quokka only
Species	*brachyurus*	

What else can we classify?

The criteria for classifying living things and the Linnaean system of classification can be applied to anything you see or know of. Try applying these systems to organisms in your local area.

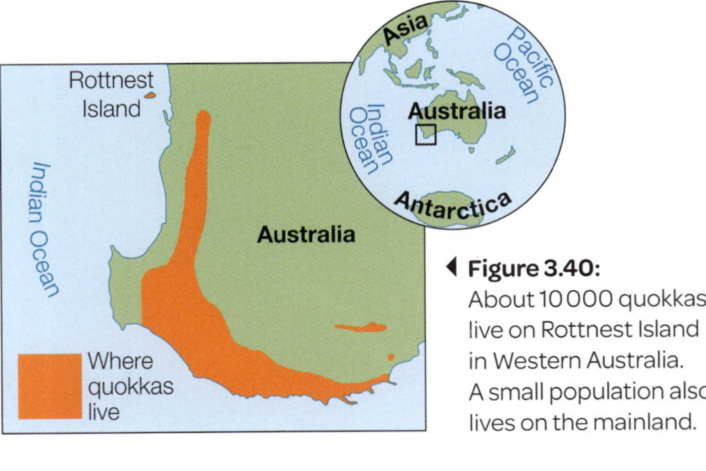

Figure 3.40: About 10 000 quokkas live on Rottnest Island in Western Australia. A small population also lives on the mainland.

Learning Ladder

Cells and classification

1. Summarise the seven biological characteristics of all living things.
2. Identify two physical features of macropods.
3. Quokkas are closely related to kangaroos and wallabies. Explain why this is the case, using the terms in the Linnaean system of classification.
4. Look at the photos of the quokka. Identify a similar organism and justify your answer.
5. The quokka was first described by Western Europeans as 'a kind of rat as big as a common cat'. Does this match the current classification of the quokka? Discuss what has changed since this statement was made.

Communicating

see page 343

1. Explain what respiration is.
2. Describe a way you could communicate information about the diet of the quokka.
3. Construct a digital flowchart that shows the classification of the quokka.
4. Create a poster that shows information about the quokka. Highlight the features that justify its classification as a living thing.
5. Select a plant or an animal from your local area. Use the Linnaean system to classify it, from domain to species.

Success criteria

- I understand the classification of the quokka.

Cells and classification summary

Classification of living things

All living things are made of cells and share biological characteristics: they can grow, move, reproduce, respond to stimuli, take in nutrients, convert nutrients to energy, and excrete waste.

First Nations Peoples use complex classification systems, which can include many factors such as the age, location and value of living things to specific communities.

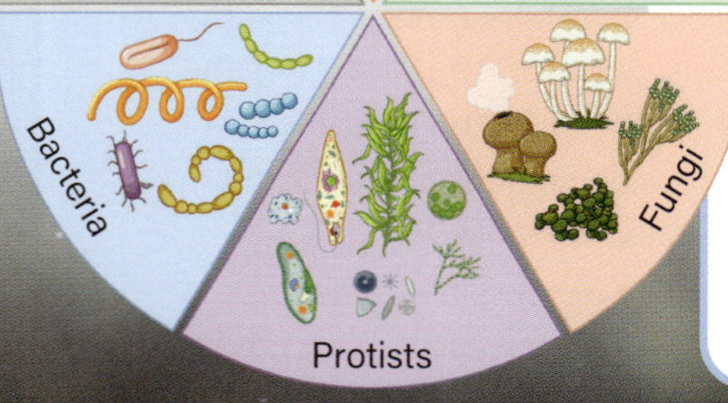

Classification level	Classification for brush-tailed rock wallaby
Domain	Eukaryota (organisms with specialised cells)
Kingdom	Animalia (animals)
Phylum	Chordata (animals with a cord along their back)
Class	Mammalia (mammals) (animals with hair and live young)
Order	Marsupialia (marsupials) (animals with a pouch)
Family	Macropodidae (macropods) (animals with large hind legs and long tails, e.g. wallabies, kangaroos, pademelons)
Genus	*Petrogale* (rock wallabies)
Species	*penicillata* (brush-tailed rock wallaby)

Very general → Very specific

In biology, taxonomists use the Linnaean system of classification to sort living things from domain (broad) to species (specific).

Classification keys – including dichotomous and branching keys – can be used to sort organisms by their specific features.

Animals have many adaptations, including structural features, that can help in classifying them into groups. These features also help them to survive, like the water-resistant fur of the platypus.

Cells

Cell theory tells us about how living things are made:

- All living things are made of cells.
- Cells are the basic building blocks of life.
- All cells come from cells that already exist.

Cells contain structures called organelles that allow them to function; organelles carry out specific roles.

Mitochondria enable cells to complete cellular respiration (a chemical reaction that converts glucose to energy that can be used by organisms).

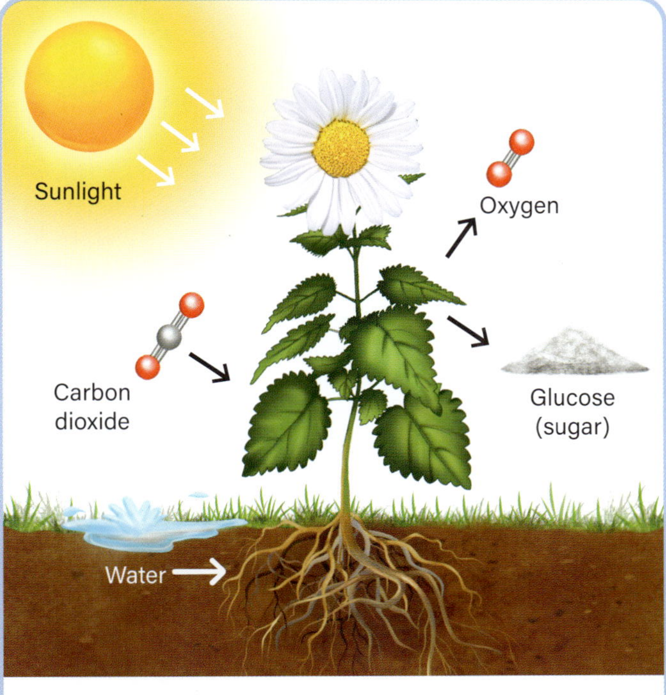

Photosynthesis occurs in chloroplasts. Photosynthesis is the chemical reaction powered by sunlight that converts carbon dioxide and water into glucose and oxygen.

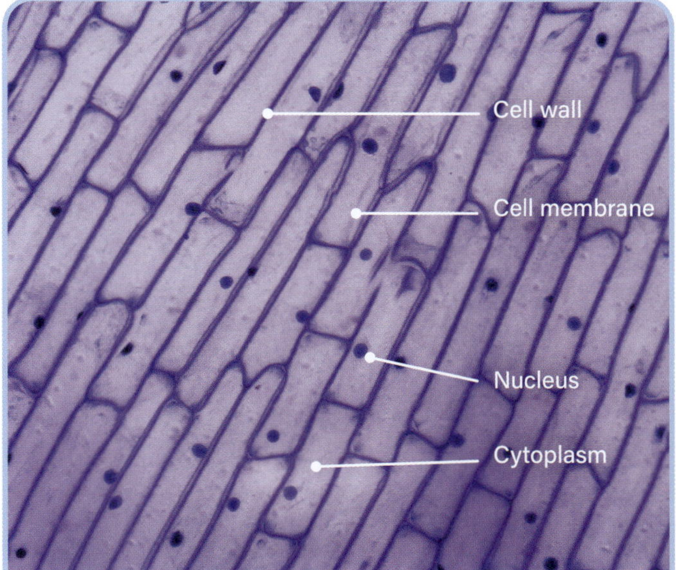

Animal, plant and fungus cells all contain a nucleus, cytoplasm, ribosomes and a cell membrane. Only plant cells also contain a cell wall and chloroplasts.

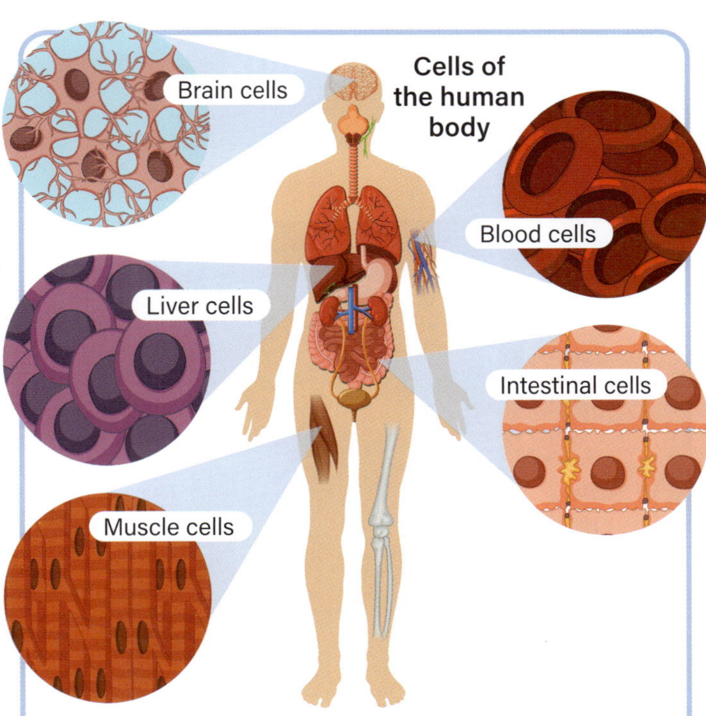

Cells can specialise to perform specific roles in multicellular organisms. Specialised cells form more complex structures in the multicellular organism.

organelle → specialised cell → tissue → organ → organ system → organism

Cells and classification in context

Our understanding of cells and classification can be applied to all living things so that we can better understand them.

Masterclass

Steps in progression	1	2
Content — Cells and classification	Identify three features of the mouse in Figure 3.41.	Describe the features of the delicate mouse in Figure 3.41.
Processes — Observing	What are two observations that scientists could make about these mice with their senses when conducting field studies?	Describe a scientific tool that has been used to enhance our observations of these mice species.
Processes — Conducting investigations	Identify a process that scientists could use to ensure they are correctly classifying the species of mice they see in the Australian desert.	Describe how a scientist could use a: • digital scale • digital camera to gather first-hand data about the delicate mouse.
Processes — Communicating	Identify two pieces of scientific information about the delicate mouse.	Describe a way you could communicate information about the delicate mouse to other scientists.

Figure 3.41: In 2024, scientists confirmed they had found two new species of the delicate mouse in Australia.

▲ **Figure 3.42:** Scientists conduct field work to help them find and classify new species of many organisms, like the delicate mouse.

Demonstrate your understanding

3	4	5	
Explain how classification conventions can be used to distinctly categorise the three species of the delicate mouse.	Compare the features of the delicate mouse to the features of another named organism that lives in the Australian desert. How could these features assist the mice to survive?	Analyse how using technologies has allowed scientists to change the classification of these species of mice.	
Explain how scientists could accurately record their observations of the mice species while working in the field.	What is an inference that Australian scientists have made based on their initial observations of the delicate mouse?	How did scientists use more advanced technologies to improve their observations of the delicate mouse and change their initial inferences?	Science how-to p. 316
Explain three safety practices required when conducting field work in the desert to observe the delicate mouse.	Suggest a way that scientists could ensure they collect reliable data about the delicate mouse, such as: • its habitat • the population sizes.	Analyse the processes scientists need to use to ensure that the data they record about the delicate mouse is accurate.	Science how-to p. 326
Explain how you would present data about the genetics of the different species of the delicate mouse using a named digital technology.	Use secondary source information to construct a scientific poster on the delicate mouse that shows its habitat, distribution and the distinct characteristics of each species.	Edit your poster into a written presentation that could be shown to a group of Year 7 students to teach them about the marsupials of the Australian desert.	Science how-to p. 343

▲ **Figure 3.43:** Many special laboratory technologies allow scientists to study and map the genetics of animals, which helps to classify them.

New Australian mice scamper around Australian deserts

In February 2024, Australian scientists confirmed they had found two new species of native mice in the Australian desert. These mice are types of the 'delicate mouse'. While it has previously been thought there was only one species of this Australian marsupial, scientists have now confirmed that there are at least three distinct species that inhabit different parts of the Australian desert: *Pseudomys delicatulus*, *Pseudomys pilbarensis* and *Pseudomys mimulus*.

Scientists have used specific tools and technology to make these new classifications, including mapping the genes of the mice to see how similar they are. What they found is amazing: although they look similar, these mice are quite different. They could not reproduce with mice from the other species!

Focus area 4

4.0 Solutions and mixtures

Figure 4.1: Magnets allow you to separate metallic materials from non-metallic materials.

All substances can be classified according to what they are made up of. A substance that contains only one type of particle is called a pure substance. A substance that contains different particles is a mixture. Mixtures can be separated.

A solution is a special kind of mixture made up of a solvent and a solute. Solutes dissolve in a solvent to make a solution. In addition, substances can be classified according to their states of matter: solid, liquid or gas. For example, water can be solid ice, liquid water or gaseous water vapour.

Learning Ladder

The Learning Ladder contains the scientific content and processes you will learn in this chapter. Each area has five levels of progress. To move to higher levels, you need to practise activities at the earlier levels. This will help you develop the ability to complete tasks that are more complex.

Steps in progression	Content: Solutions and mixtures	Working scientifically processes: Planning investigations	Working scientifically processes: Conducting investigations	Working scientifically processes: Problem-solving	Steps in progression
5	I can apply my understanding of the chemical properties of substances to real-world contexts.	I can construct an appropriate plan before conducting a first-hand investigation.	I can conduct first-hand investigations and accurately record the collected data.	I can evaluate problem-solving strategies used to solve an identified problem.	5
4	I can compare the properties of substances in different states.	I can develop appropriate scientific aims for a range of investigations.	I can collect accurate and reliable data from first-hand and second-hand investigations.	I can use given criteria to find solutions to scientific problems.	4
3	I can explain how to use the properties of substances to separate or understand mixtures.	I can distinguish between controlled, dependent and independent variables.	I can implement safe practices when using scientific equipment.	I can explain scientific problems and phenomena using cause-and-effect relationships.	3
2	I can describe the properties of different solutions.	I can describe ways to reduce risks for a range of investigations.	I can use scientific equipment to conduct investigations and gather first-hand data and information.	I can suggest solutions to familiar scientific problems.	2
1	I can identify a range of solutions.	I can identify appropriate materials and technologies to conduct investigations.	I can identify correct processes for conducting investigations.	I can identify scientific problems.	1

Figure 4.2: Mixtures are two or more substances combined that can be separated.

4·1 ► States of matter

Figure 4.3: Water as a solid and a liquid

Learning intention

At the end of this lesson, I will be able to describe the behaviour of matter in terms of particles that are continuously moving and interacting.

Key terms

compress: squash something into a smaller shape

condense: change state from gas to liquid

matter: any substance that has mass and takes up space

particle: a very small amount of matter

volume: the amount of space an object takes up

Investigation 4.1

Compressing liquids and gases, p.414

Content group: Properties of matter

Matter is anything that takes up space and has mass. Substances on Earth can exist in three different states of matter: as solids, liquids or gases. Water is the only substance on Earth that is found naturally in all three states. This is critical to the survival of all living things. Solid water is ice. If you heat ice, it melts to become liquid water. If you keep heating it, the water boils and becomes water vapour. This is because water is made up of **particles**. As you add heat, you increase how much the particles move, which will eventually change the water's state of matter.

The particles in a solid are packed closely together

Solids are materials such as wood, metal or plastic. The particles in solids are packed closely together, like bricks in a wall. However, the particles are not still; they are constantly vibrating.

A solid has a fixed shape and a fixed **volume**. Because the particles in a solid are highly attracted to each other, the solid keeps its shape and does not spread out or flow. A solid cannot be **compressed** easily, because there is no space between its particles for them to squeeze closer together. Water exists in its solid state as ice. This occurs at temperatures below 0 °C. In this state, water molecules are packed tightly together in an organised pattern.

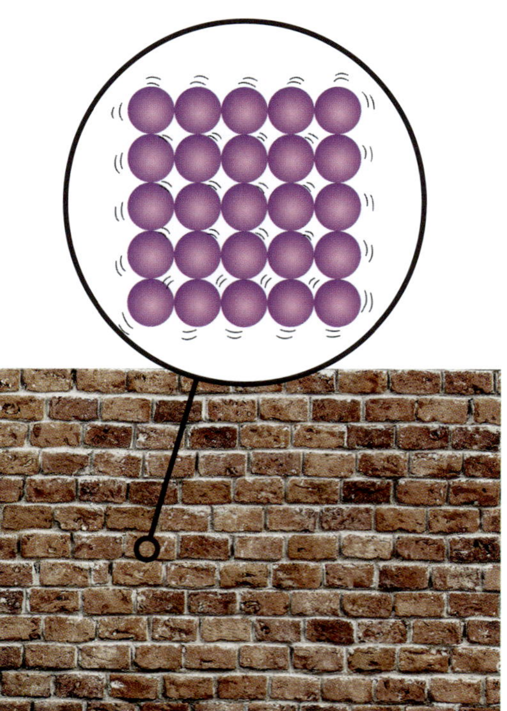

Figure 4.4: The particles in a solid are packed closely together, like bricks in a wall. They vibrate in their positions.

The particles in a liquid can move past each other

The particles in liquids are not as close together as they are in solids. The particles are still strongly attracted to each other, but they have room to move past each other.

A liquid has a fixed volume but not a fixed shape. A liquid takes on the shape of its container because the attraction between the particles is not strong enough to stop them from spreading out. A liquid cannot be compressed very much because there is not enough space between its particles for them to squeeze closer together. The liquid state of water is its most common form on Earth.

Figure 4.5: The particles in a liquid can move past each other. They take on the shape of their container.

The particles in a gas have large spaces between them

The particles in a gas are weakly attracted to each other, so they can move around a lot. The particles have large spaces between them, and they are constantly moving in all directions. Some common gases on Earth are oxygen and carbon dioxide. The air we breathe is a mixture of gases, including oxygen, nitrogen and carbon dioxide.

A gas has neither a fixed shape nor a fixed volume. Gases spread out to fill the container they are in. A gas can be compressed because there is space between the particles. In a smaller volume, the gas particles just have less room to move around.

Liquid water becomes a gas, known as water vapour, when it is heated. When temperatures rise above 100 °C, the water particles gain enough energy to break free from the liquid state. Water vapour is present in our atmosphere. It can **condense** to form liquid in clouds and may eventually fall back to Earth as rain or snow.

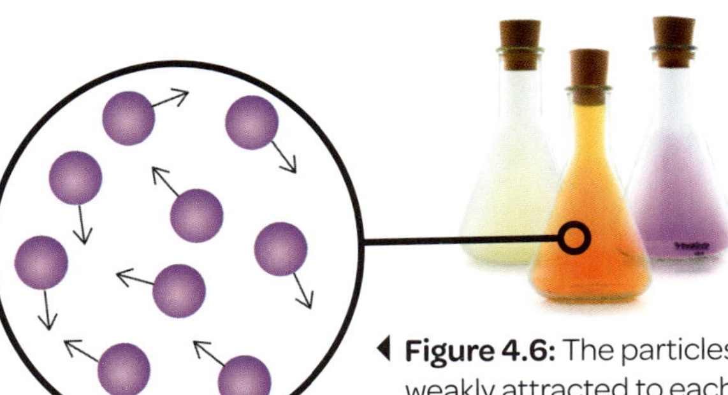

Figure 4.6: The particles in a gas are weakly attracted to each other. They can be squashed into a smaller space.

Learning Ladder

Solutions and mixtures

1. List three solids, three liquids and three gases you have come into contact with today.
2. Describe the properties of gas particles that make gases different from solids and liquids.
3. Explain why solids and liquids cannot be compressed but gases can. Refer to particles in your answer.
4. Compare the particle arrangement, energy level and movement of solids and liquids.

Planning investigations *see page 323*

Your teacher showed you an experiment called 'Crushing can' by heating some water in a can over a Bunsen burner. When the water started evaporating from the can, your teacher used metal tongs to flip the can upside down into a container of cold water. The can was crushed and made a loud sound.

1. Identify what was being investigated in the experiment.
2. Describe how the properties of the water particles may have caused the result you saw.
3. The gases inside the can were hot and had mostly expanded before it entered the water. Explain the link between the hot gas particles and the cold water that led to the final shape of the can.
4. Garbage takes up a lot of space in landfill. Discuss how this demonstration could be applied to provide a solution to this real-world problem.

In context

Plasma is sometimes called the fourth state of matter. Use the internet to research plasma and to understand how it differs from the other three states of matter.

Success criteria

- I can state the three main states of matter.
- I can describe the behaviour and arrangement of particles in each state.

4·2 ▸ Particle theory and water

Learning intention

At the end of this lesson, I will be able to compare the properties of solids, liquids and gases in terms of their particle arrangement and movement.

Key terms

atom: a particle that makes up all matter; made up of protons, neutrons and electrons

condense: change state from gas to liquid

deposition: changing state from gas to solid without going through the liquid state

evaporate: change state from liquid to gas

freeze: change state from liquid to solid

melt: change state from solid to liquid

molecule: a unit made up of two or more atoms chemically bonded to each other

particle theory: a model used to describe the properties of solids, liquids and gases

sublimation: changing state from solid to gas without going through the liquid state

Investigation 4.2

Changing states of water, p.415

Content group: Properties of matter

▲
Figure 4.7: The spaces between the joints in a bridge allow the metal to expand in hot weather without causing the bridge to buckle.

Particle theory says that all matter is made of tiny particles, which are constantly moving. The particles are individual **atoms**, or groups of atoms called **molecules**, and they are too small to see. Atoms cannot easily be divided any further.

Adding heat energy causes particles to move apart

When heat energy is added to matter, the energy is transferred to its particles. This causes the particles to vibrate and move more, which can result in the substance changing states. As the particles move more, they can overcome the forces holding them together.

We can see this when we add energy to ice (solid water). The ice **melts** and becomes liquid. If we continue to add energy, the liquid water will **evaporate** into a gas when the water boils at 100 °C.

Removing heat energy causes particles to move closer together

When matter loses heat energy, the particles lose energy and start to move less. The particles also become closer together because they no longer have enough energy to overcome the forces of attraction between them.

We can see this process with water. When we place a cold glass of water on a table, gaseous water vapour from the surrounding air **condenses** on the outside of the glass as liquid. If we put that same glass into the freezer, below 0 °C, then the water **freezes** into ice.

Table 4.1 shows how the particles of solids, liquids and gases behave.

Table 4.1: The arrangement of particles in the three states of matter

State of matter	Particle arrangement	Diagram	Forces
Solid	Closely packed; vibrate in place		Strong forces hold the particles together
Liquid	Loosely packed; can slide over one another		Moderate forces hold the particles near each other
Gas	Moving freely around the available volume; always moving and colliding		Weak forces between the particles allow them to move around

◀ **Figure 4.8:** Dry ice (solid) sublimes at room temperature, turning into a gas straight away.

Solids can become gases and gases can become solids

At room temperature, we exhale carbon dioxide gas into the air around us. At very low temperatures, carbon dioxide becomes a solid called dry ice. To stay solid, dry ice needs to be stored at −79 °C. When the temperature rises above −79 °C, the solid turns directly into carbon dioxide gas without going through the liquid state. It produces a white 'fog' (see Figure 4.8). This process of changing from a solid to a gas is called **sublimation**. The process of changing from a gas to a solid without going through the liquid state is called **deposition**.

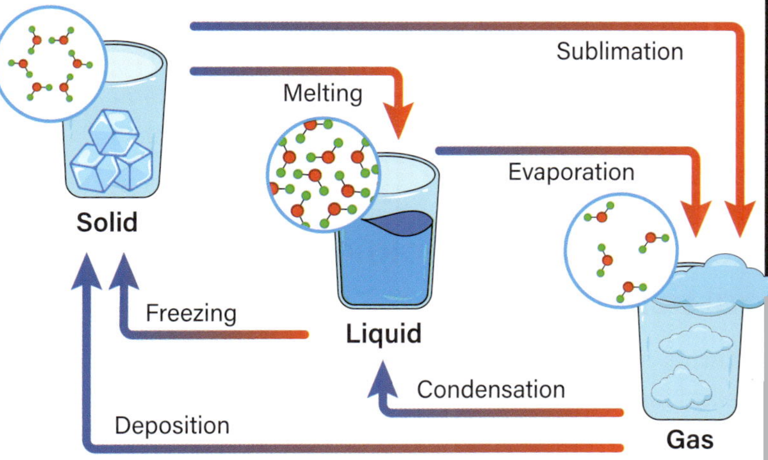

▲ **Figure 4.9:** Adding heat energy to water causes it to change from a solid to a liquid to a gas. Removing heat energy has the opposite effect: the gas changes to become liquid and then a solid.

Learning Ladder

Solutions and mixtures

1. Identify what happens to the particles in a substance when heat energy is:
   a added.
   b removed or when particles lose heat energy.
2. Describe how the particles of a solid substance would change if it was heated past its boiling point.
3. Explain, using scientific language, how water can change from a solid to a gas and then back to a solid. Use the words in Figure 4.9 to help you.
4. Compare the properties of solids, liquids and gases with reference to particle theory.

Problem-solving

see page 339

Imagine you are going to conduct an investigation to observe water changing between the three states of matter, from a solid to a liquid to a gas. Plan an investigation using the following instructions.

1. Identify a piece of scientific equipment you could use to add heat to the solid and liquid water.
2. Identify a risk associated with this piece of equipment. Describe two ways to manage the risk.
3. Describe three factors you would need to keep the same throughout this investigation, so you can compare your results to another group's.
4. Construct an appropriate aim for this investigation.

In context

Bridges have expansion joints to prevent buckling in hot weather, as shown in Figure 4.7. How else do engineers protect against thermal expansion when building other structures? Research and discuss two other strategies that can be used, with reference to the particles of the materials.

Success criteria

- I can describe how the particles are arranged in solids, liquids and gases.
- I can explain how energy affects how the particles of solids, liquids and gases move.

4·3 ▸ Physical properties of water

Figure 4.10: How is a heavy warship able to float on water?

Learning intention

At the end of this lesson, I will be able to explain other physical properties of water, such as density, buoyancy and surface tension.

Key terms

buoyancy: the upward force that opposes the downward force of gravity on an object in a liquid or gas

density: how heavy something is for its size; mass divided by volume

mass: the amount of matter that a physical body contains

surface tension: where the molecules at the surface of a liquid are more attracted to each other than to the air above the liquid

Investigation 4.3

Exploring density, p.417

Content group: Physical properties of water

What determines whether an object floats in water? The mass of the object is important but there must be more to it. Extremely heavy icebergs float, whereas much lighter coins sink. The answer has to do with a property of matter called density.

Water has unique and incredible physical properties. Buoyancy, density and surface tension allow small insects to walk on water's surface and large icebergs and ships to stay afloat.

Density is how heavy an object is for its size

You can calculate the density of an object by dividing its **mass** by its volume. In other words, **density** is a measure of how heavy an object is compared to its size.

Solids usually have higher densities than gases and liquids because their particles are packed more closely together. When comparing the density of two solids of the same size, the heavier one has a greater density. If you compare a piece of wood and a coin of the same size, the wood will be lighter and therefore less dense than the metal coin.

Figure 4.11: The particles on the left are more closely packed than the particles on the right. A substance with particles that are closely packed has a higher density.

Density can be explained by particle theory

In terms of particle theory, density is related to two things: the amount of space between particles and the mass of the particles (see Figure 4.11):

- The more tightly packed its particles are, the denser an object is.
- The heavier its particles are, the denser the object is.

Imagine a bike with an aluminium frame and an identical bike made of steel (iron metal). Although the bikes are the same size, the steel bike is heavier because iron atoms have more mass than aluminium atoms.

Figure 4.12: Liquids of different densities in a container will arrange so that the liquid with the highest density is at the bottom and the liquid with the lowest density is at the top.

Liquids and gases with lower density float on those with higher density

Liquids and gases with a lower density float on top of substances with a higher density (see Figure 4.12).

Have you ever heard that 'hot air rises'? This is because hot air is less dense than cool air. The spaces between the particles in the hot air are larger than the spaces between the particles in the cool air. This is what keeps a hot air balloon in the sky.

The warship in Figure 4.10 floats on water, so it must have a lower density than water. Even though the ship is made of metal, most of the inside of a warship is air. The air takes up a lot of volume but has very little mass, and so overall the ship has a lower density than water.

The different states of water have different densities

Water is a very unusual substance because its particles move further apart when it freezes. This means that ice is less dense than water and can float on water (see Figures 4.13 and 4.14). Water is in its most compact arrangement and at its most dense as a liquid at 4 °C. At lower temperatures, the water particles spread out and the water becomes less dense.

Figure 4.13: Ice is less dense than water, which is why icebergs float.

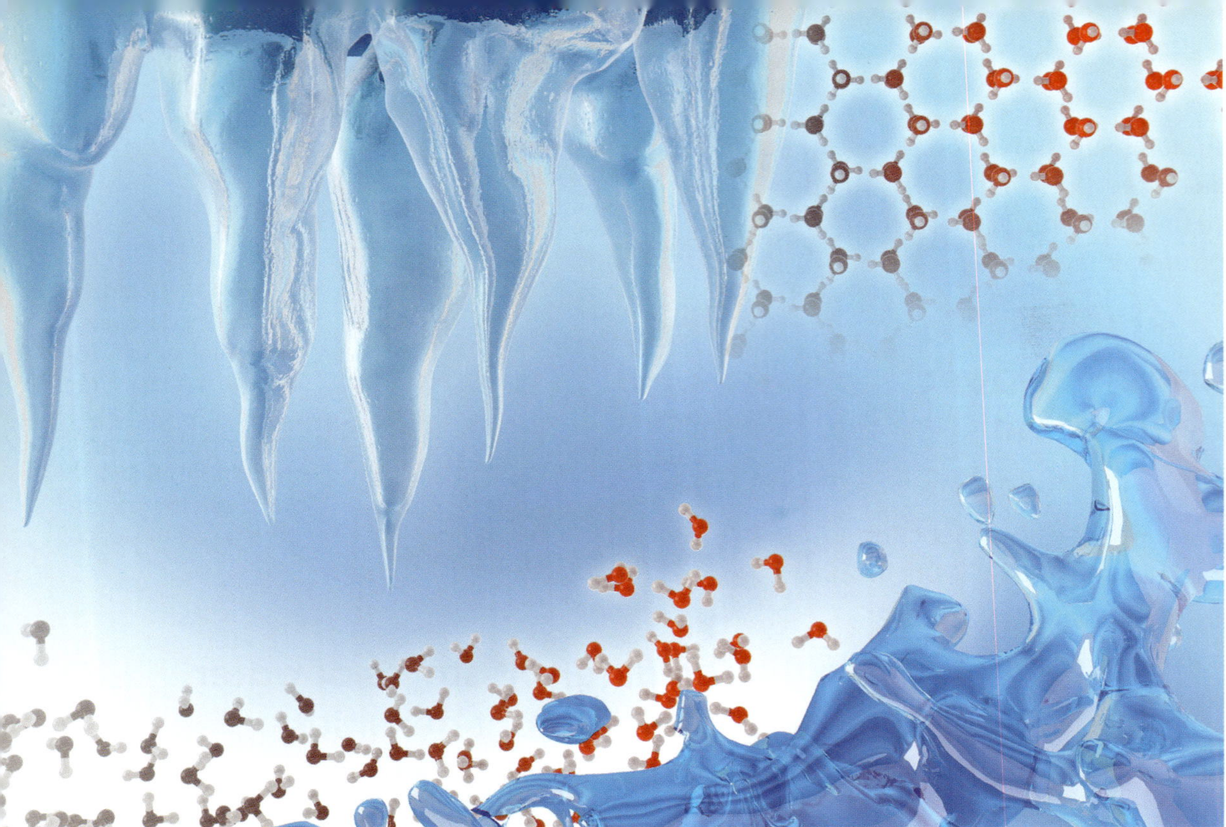

Figure 4.14: How ice can float on water. Water is an unusual substance because its solid form (ice) is less dense than its liquid form. The molecules in ice are arranged in an orderly lattice pattern and are further apart than the molecules in liquid water.

As water at the surface reaches 0 °C, it freezes. The water below the ice is at 4 °C and sinks because it has a higher density. This creates a pocket of air between the two layers. The air acts as an insulator, which is crucial for the survival of aquatic life during winter months.

Water vapour is the gaseous state of water, formed when liquid water evaporates into the atmosphere. Water vapour is much less dense than liquid water and ice. Water vapour is also less dense than air, which is why it rises in the atmosphere. This property is very important in the water cycle, as water vapour condenses to form clouds and eventually precipitates as rain or snow.

Buoyancy makes objects stay afloat

When an object is placed in water, it is pulled down by gravity. **Buoyancy** is the upward force that opposes the downward force of gravity on an object in a liquid or gas. This force counteracts the weight of the object, causing it to float or rise. If an object in water is less dense than water, then the upthrust is greater than the weight and the object rises. If the object is more dense than water, then the upthrust is less than the weight and the object sinks. If an object floats we say it is buoyant, and the upward thrust is equal to the downward force of gravity (see Figure 4.15).

Buoyancy is particularly important for aquatic life forms, from fish, octopuses and whales to seahorses. It allows them to stay afloat and move efficiently in their environments. Ocean water has a higher buoyancy force upwards than fresh water, because salt water is more dense than fresh water. This makes it easier to float in salt water. Buoyancy is what allows mega ships to stay afloat despite being large, heavy, and constructed from dense materials like metals.

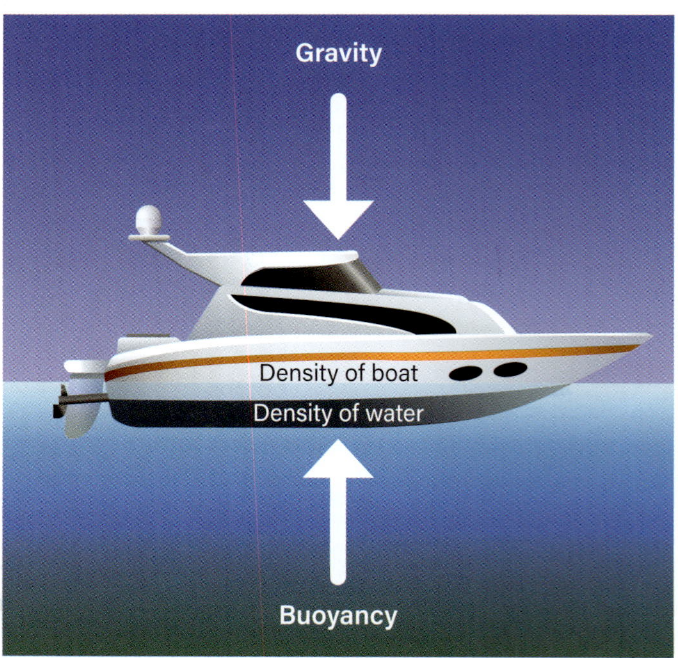

Figure 4.15: An object in water is pulled down by gravity and pushed up by buoyancy.

Surface tension holds water particles together

Surface tension is the reason a water strider, even though denser than water, can walk on the surface of water. **Surface tension** is a property of a liquid that allows it to resist external forces or resist the surface particles breaking apart. The reason water particles stick together is because of strong cohesive (bonding) forces. This force creates a 'skin-like' layer or film at the surface, causing surface tension.

Surface tension explains why water can flow against gravity through narrow spaces, such as a drinking straw. As the water molecules stick together and are also attracted to the sides of a narrow tube, the water can rise up the tube against gravity. This is known as capillary action. This occurs in plants, as water from roots moves against gravity towards leaves in narrow plant vessels, which is necessary for photosynthesis.

Detergents and soaps are used to reduce the surface tension of water by causing the water particles to break apart.

Figure 4.16: Surface tension is the reason a water strider can walk on water.

Learning Ladder

Solutions and mixtures

1. Identify which object has a greater density: a wooden log or a metal bar of the same size.
2. Describe the property of water that allows insects like a water strider to walk across it.
3. Explain why two objects can have different masses, even though they are the same size.
4. Compare the arrangement of particles in liquid and solid water and explain how this causes a change in density.

Planning investigations

see page 323

Imagine you are going to conduct an investigation to observe how adding salt to liquid water affects the density of water. Plan an investigation using the following instructions.

1. Identify a piece of scientific equipment you could use to add salt to the liquid water.
2. Describe a way that you could reduce one risk in this investigation.
3. Describe three factors you would need to keep the same throughout this investigation, so you can compare your results to another group's.
4. Construct an appropriate aim for this investigation.

In context

A type of star known as a neutron star contains some of the most dense material in the universe. Research the materials found inside a neutron star and suggest why they might be so dense.

Success criteria

- I can explain what density, buoyancy and surface tension are and relate these properties to water.
- I can draw a diagram showing high-density and low-density particle arrangements.

4·4 ▸ Calculating density

Learning intention

At the end of this lesson, I will be able to determine the volume and mass of regular-shaped and irregular-shaped objects and calculate their density using the formula $\rho = \frac{m}{V}$.

Key term

density formula: $\rho = \frac{m}{V}$ (unit: g/cm^3)

Investigation 4.4

Calculating density, p.419

Content group: Physical properties of water

Density is a measure of how much matter packs into a space. This affects whether a substance floats or sinks in water. Materials such as wood, feathers, rubber and Styrofoam float, whereas rocks, glass and metal sink. To calculate density, you need to know the volume and mass of objects.

Density is the amount of matter in a certain volume

Density is a physical property that measures how much matter packs into a specific volume. The volume is measured in millilitres (mL) or cubic centimetres (cm^3) and mass is measured in grams (g). One kilogram of cotton weighs the same as 1 kilogram of rocks. However, their densities are different because their volumes are different. The particles in cotton are more spaced out than the particles in the rocks, so the cotton has a higher volume and a lower density.

Therefore, density is the amount of mass per unit volume. In Figure 4.17, the object with lower density has fewer particles, so it has less mass, whereas the object with higher density has more particles, so it has more mass.

If an object is less dense than water, then it will float. If the object is denser than water, then it will sink.

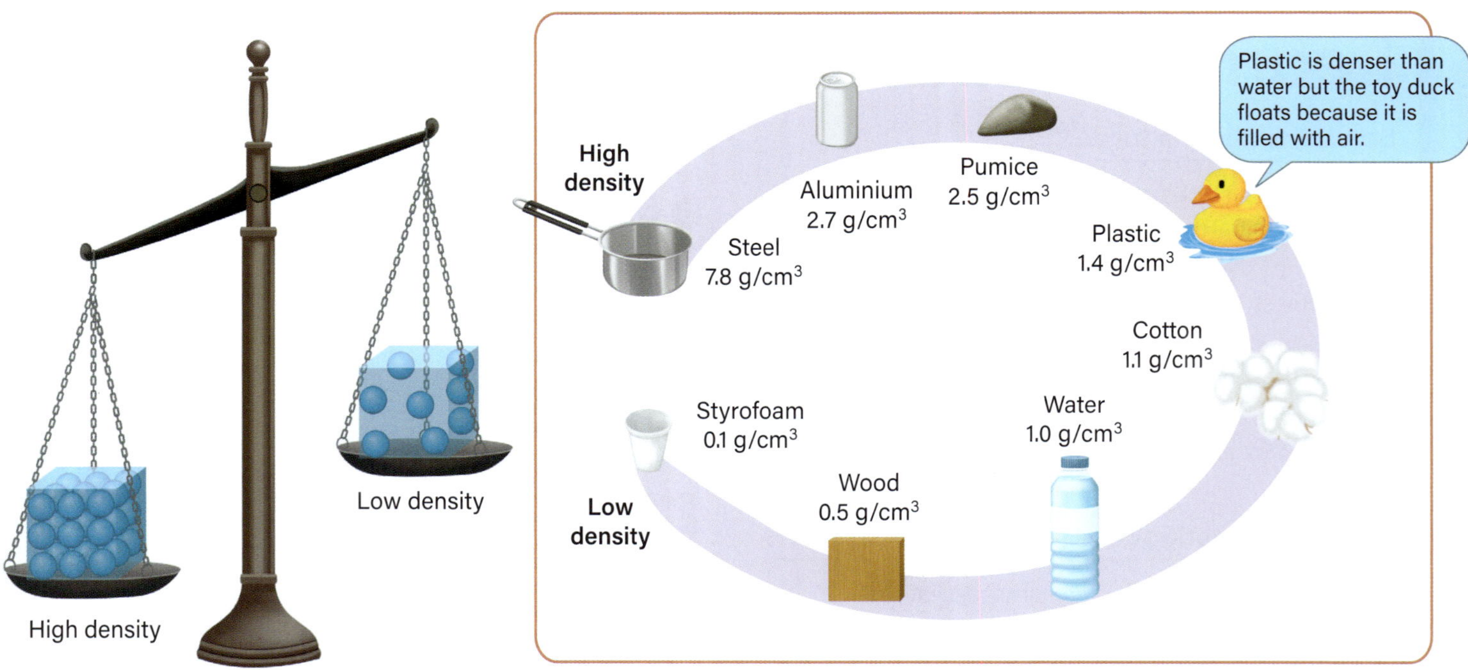

▲ **Figure 4.17:** The densities of two objects with the same volumes

▲ **Figure 4.18:** Density is a measure of how much mass is in a given volume of a material.

Density equals mass divided by volume

Density is measured in grams per cubic centimetre (g/cm^3). This means that the units for volume and mass determine the units for density. The mass in grams determines the amount of the substance, and the volume determines the space taken up.

In the laboratory, mass can be measured using electronic balances and volume can be measured using a measuring cylinder.

Density is independent of quantity and shape. This means that regular-shaped and irregular-shaped objects could have the same density even though their masses and volumes are different. For example, a steel cube that weighs 7.8 grams and has a volume of 1 cubic centimetre has a density of 7.8 g/cm^3. A steel nail could have the same density when its volume is 1.6 cubic centimetres and its mass is 12.5 grams.

Density can be calculated by using a formula

We can use the **density formula** to calculate density:

$$\rho = \frac{m}{V}$$

where ρ = density (g/cm^3)

m = mass (g)

V = volume (cm^3 or mL)

(If mass is given in kilograms and volume is given in litres, then convert to grams and millilitres by multiplying by 1000.)

Example:

Calculate the density of a cube with mass 5 g and volume 10 cm^3.

$$\rho = \frac{m}{V} = \frac{5}{10} = 0.5\ g/cm^3$$

For more information on calculating density, see page 373 of the Science how-to section.

Steel cube
Volume = 1.0 cm^3
Mass = 7.8 g
Density = 7.8 g/cm^3

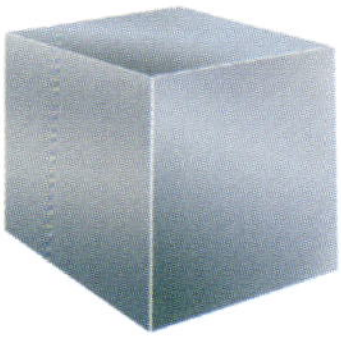

Steel nail
Volume = 1.6 cm^3
Mass = 12.5 g
Density = 7.8 g/cm^3

Figure 4.19: This steel cube and steel nail have the same density.

Learning Ladder

Solutions and mixtures

1. Identify which object has a greater density by using $\rho = \frac{m}{V}$.
   A A wooden log with mass 500 g and volume 2000 cm^3
   B A metal bar with mass 1.5 kg and volume of 500 cm^3
2. Suggest whether the log and metal bar from Question 1 would float or sink in water. (Hint: The density of water is 1 g/cm^3.)
3. Explain why it is easier to float in seawater (density 1.03 g/cm^3) than in fresh water (density 1 g/cm^3).
4. Compare the densities of a regular cube with volume 1 cm^3 and mass 4 g and an irregular rock with volume 12 cm^3 and mass 3 g. Explain your answer.

Planning investigations

see page 323

Imagine you are going to investigate the density of regular-shaped and irregular-shaped objects. Plan your investigation using the following instructions.

1. Identify a piece of scientific equipment that could be used to measure the volumes and masses of the regular-shaped and irregular-shaped objects.
2. Describe how you could reduce one risk in this investigation.
3. Identify the:
   a independent variable.
   b dependent variable.
   c controlled variable.
4. Construct an appropriate aim for this investigation.

In context

In 1912, the *Titanic* sank after hitting an iceberg. Research why icebergs can be dangerous for ships. Link your reasoning to density.

Success criteria

- I can explain the relationship between mass, volume and density.
- I can apply the formula $\rho = \frac{m}{V}$ to solve problems relating to the density of substances.

4.5 ► Solute, solvent, solution

Learning intention

At the end of this lesson, I will be able to describe mixtures in terms of solute, solvent and solution.

Key terms

concentration: the amount of a dissolved solute in a defined amount of solvent

dissolve: particles separating and spreading out, so that they seem to disappear, when added to another substance

insoluble: does not dissolve

mixture: two or more substances combined that can be physically separated

soluble: able to dissolve

solute: a substance that is dissolved by a solvent

solution: a mixture made up of a solvent and a solute

solvent: a substance that a solute dissolves in

Investigation 4.5

Solutes and solvents, p.421

Content group: Solutions

Figure 4.20: Cordial dissolves in water to make a solution.

Mixtures are made up of two or more substances that can be physically separated. A **solution** is a type of mixture that is made up of a solute and a solvent.

Solutes dissolve in other substances

A **solute** is any substance that can **dissolve** in another substance (the **solvent**) to form a solution. A solution is a homogeneous mixture in which one substance (the solute) is uniformly dissolved in another. Solutes can begin as any of the three main states of matter:

- **Solid**: For example, sugar, salt and instant coffee are all solutes that can dissolve in water.
- **Liquid**: For example, cordial can dissolve in water to make a sweet drink.
- **Gas**: For example, carbon dioxide can dissolve in soft drinks, which makes them fizzy.

Substances that dissolve are said to be **soluble**. Substances that do not dissolve are **insoluble**.

Solvents dissolve solutes

A solvent is any substance, usually a liquid, that can dissolve another substance (the solute). A soft drink is made up of cordial (a liquid), sugar (a solid) and carbon dioxide (a gas) dissolved in water (a liquid).

Some solutes will dissolve in one solvent, but not in others. For example, oil will not dissolve in water, but will dissolve in an alcohol-based solvent.

One way to determine if a mixture is a solution is to see if light passes easily through it. If you can see through the mixture (it is transparent), then it is a solution. For example, when copper sulfate solid dissolves in water, you get a blue, clear but coloured solution.

Some solutions are clear and colourless. For example, hydrochloric acid is a clear solution. Therefore, solutions can be coloured or colourless and are always transparent.

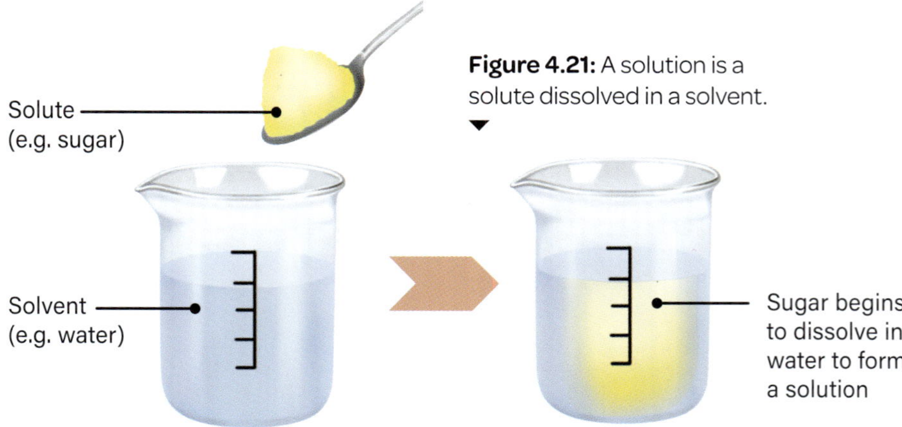

Figure 4.21: A solution is a solute dissolved in a solvent.

A solution is a solute in a solvent

A solution is a mixture of one or more solutes dissolved in a solvent. For example, a cup of black coffee is a solution made up of:

- solute(s): powdered coffee (and maybe sugar)
- a solvent: water.

We can calculate the concentration of a solution

We can calculate the **concentration** of a solution by dividing the mass of the solute by the volume of the solvent. Scientists commonly record concentration in grams per litre (g L^{-1}).

$$C = \frac{m}{V}$$

where: C = concentration (g/L)
m = mass (g)
V = volume (L)

If the solvent is a liquid, we can calculate the concentration of a solution as the percentage volume per volume (%v/v):

$$\%v/v = \frac{\text{volume of solute (mL)}}{\text{volume of solution (mL)}} \times 100$$

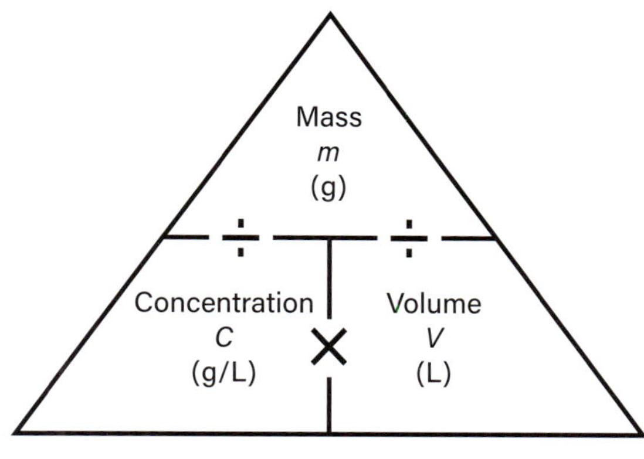

Figure 4.22: The formula triangle for calculating concentration, mass and volume. Cover up the value you are trying to work out. If the remaining two values are next to each other (e.g. C and V), multiply them. If one is on top of the other, divide the top one by the bottom one to get the answer.

Example:

What is the concentration (g/L) of a solution that contains 560 g of sugar dissolved in 2 L of water?

m = 560 g

V = 2 L

$$C = \frac{m}{V} = \frac{560}{2} = 280 \text{ g/L}$$

The concentration of the solution is 280 g/L.

Learning Ladder

Solutions and mixtures

1. In each of these solutions, identify the solute and the solvent.
   - **a** A cup of sweet, black tea
   - **b** A glass of orange juice made from powdered concentrate
   - **c** Seawater
   - **d** A glass of soda water
2. Describe how water can be used as a solvent to produce a variety of drinks.
3. Explain a property that can be used to determine if a liquid is a solution.

Planning investigations *see page 323*

Refer to Investigation 4.5 on page 421 to answer the following questions.

1. Identify a reason for using a measuring cylinder to measure and add the solvents to the test tubes.
2. Glassware can pose a risk during practical investigations. Describe two ways you can minimise the risk of breaking glassware during a practical investigation.
3. In this investigation, what is the:
   - **a** independent variable?
   - **b** dependent variable?
4. Was the aim of this investigation appropriate? Give a reason for your choice.

In context

Solutions are used all around the house for different purposes, including tasks like cooking and cleaning. Suggest two solutions that could be used for each of cooking and cleaning. Explain why you have classified them as solutions.

Success criteria

- I can describe the terms 'soluble', 'insoluble', 'solute', 'solvent' and 'solution'.
- I can identify substances that are soluble in water and substances that are insoluble in water.

4·6 ▸ Water as a solvent

Learning intention

At the end of this lesson, I will be able to describe the importance of water as a solvent in daily life, industries and the environment.

Key terms

aqueous: containing water

solubility: the amount of solute that will dissolve in a solvent

universal solvent: water, the solvent that dissolves most substances

Investigation 4.6A

Water as a solvent, p.422

Investigation 4.6B

Solubility and temperature, p.423

Content group: Solutions

Water is vital for life on Earth: without water we could not survive. Our bodies need water, and so do the plants that we eat. We also need water for daily activities such as cooking and washing.

We use water to extract minerals from underground and to make medicines. Water is known as the **universal solvent** because it dissolves more substances than any other liquid.

Water is a solvent in daily life

Water can dissolve many other substances: solids, liquids and gases. When these substances dissolve, they are called solutes. Any solution that has water as a solvent is called an **aqueous** solution.

Water acts as a solvent in daily life in many ways:

- **Human survival**: Water is an essential solvent for humans. Blood is mostly water, and it carries dissolved nutrients and oxygen around our bodies.
- **Preparing meals**: Water dissolves spices, flavours and other ingredients as food cooks.
- **Washing away dirt**: Water acts as a solvent for soaps and detergents when cleaning dishes, clothes and ourselves.

Figure 4.23: Water is a solvent for important minerals in our bodies. When we exercise intensely, we lose water and minerals dissolved in our sweat.

Figure 4.24: Water is often the solvent used to make medicines.

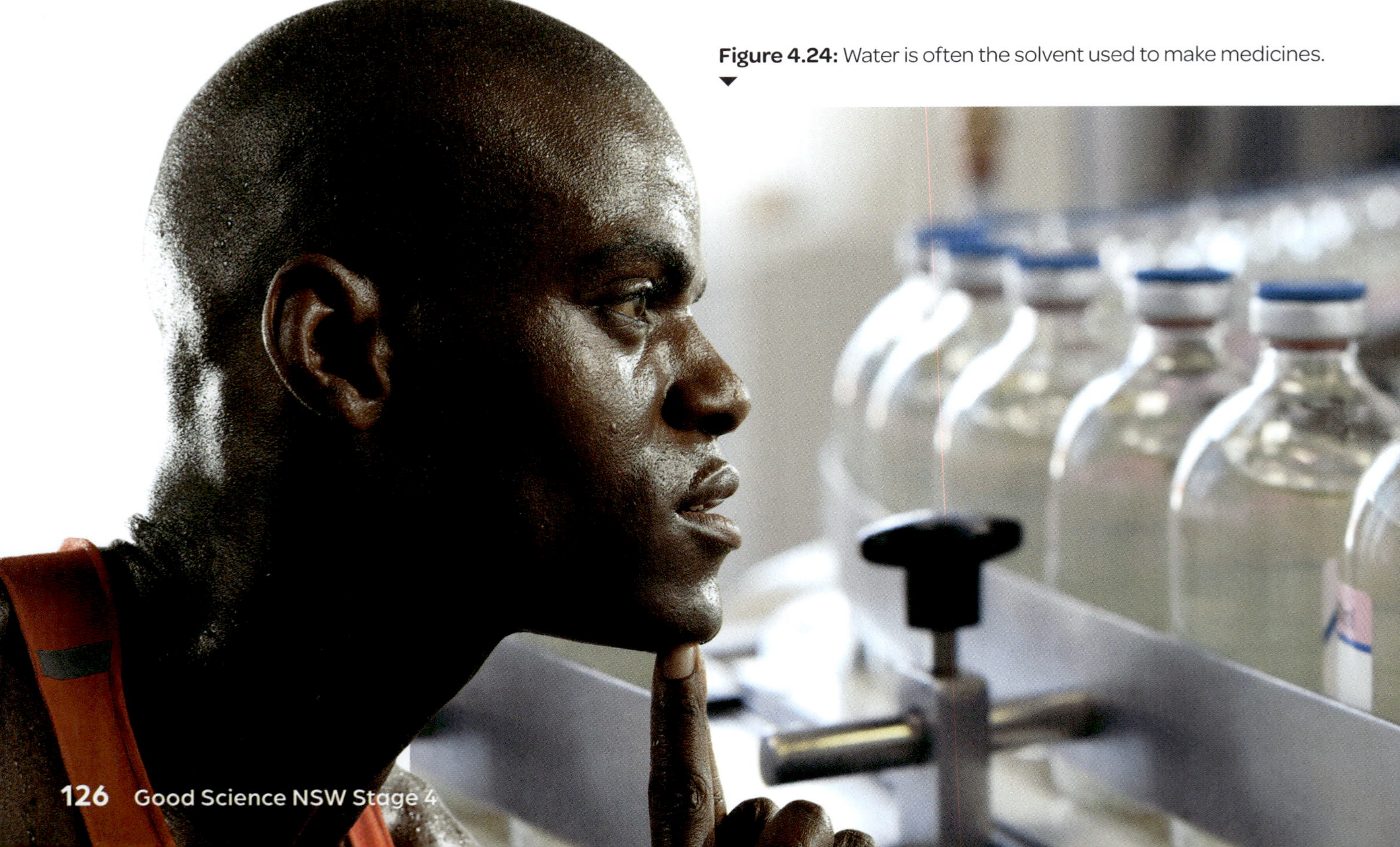

Water is used as a solvent in industry

Industries such as mining use water as a solvent to extract different minerals and elements. The element uranium is found in rocks deep underground. Uranium is used as a fuel in the reactors of nuclear power plants and to power nuclear submarines.

In Australia, some uranium mines use water as a solvent to remove the uranium from the rocks. Instead of digging up the rocks to remove the uranium, acids are dissolved into water that is then pumped into the rocks containing the uranium. The acidic solution dissolves the uranium, and the acidic water is then pumped back to the surface where it is processed to access the uranium.

Water is used as a solvent in the environment

Many essential elements, such as iron, zinc and calcium, are needed by plants and animals. They are found as minerals in rocks and soil. Water can dissolve many of these minerals, so plants can absorb them from groundwater. Animals eat the plants, receive these essential nutrients and are able to survive.

The water in rivers, lakes and oceans contains dissolved oxygen that is vital to fish and other animals. Similarly, water contains dissolved carbon dioxide, which marine plants need in order to photosynthesise.

The temperature of water affects how much solute dissolves

The properties of water change as the temperature changes. As heat is added, water molecules absorb more energy and can move more. This means that they become better at dissolving solutes: the warmer the water, the more solute it can dissolve. The **solubility** of the solute increases. Sometimes, solutes that do not dissolve in cold water will start to dissolve in warmer water.

Learning Ladder

Solutions and mixtures

1. List three uses of water as a solvent in daily life.
2. Describe how water can be used as a solvent in mining.
3. Explain why dissolved oxygen and carbon dioxide in water are important solutes.
4. Compare two ways that water is used in our daily lives.

Conducting investigations

see page 326

Refer to Investigation 4.6A on page 422 to answer the following questions. Read the aim, materials and method.

1. Explain why it is necessary to use the electronic balance accurately. How did you achieve this?
2. Describe how you would make sure that you are correctly measuring 50 mL of water in a 100 mL measuring cylinder.
3. Investigation 4.6A is relatively safe, but you should be able to identify three potential hazards. Suggest ways to minimise the risks of these hazards.
4. Propose ways to ensure that you accurately recorded your data while conducting this investigation.

In context

Water is often referred to as the 'universal solvent'. Discuss why it would be given this name, based on the information you have been given about solutes, solvents and solutions.

Success criteria

- I can describe a solvent.
- I can state how water is used as a solvent in daily life, industry and the environment.

4.7 ▸ Properties of solutions

Learning intention

At the end of this lesson, I will be able to describe how the concentration of solutions can be modelled.

Key terms

concentrated: when a solution has a large amount of solute in a certain volume

dilute: when a solution has a small amount of solute in a certain volume

saturated: a solution that cannot dissolve any more solute

saturation point: the point at which a solution has dissolved the maximum amount of solute

supersaturated: when a solution contains more than the maximum amount of solute it can normally dissolve

Investigation 4.7A

Calculating the concentrations of solutions, p.425

Investigation 4.7B

Preparing dilutions, p.427

Content group: Solutions

▲
Figure 4.26: Potassium permanganate forms a purple solution in water. The more concentrated the solution, the darker the colour.

Solutions are present all around us. The tea and coffee that we drink and the salt water that we swim in at the beach are all solutions! The particles of solutions interact in specific ways, meaning that they have determined properties. One of these is concentration.

Solutions can be concentrated or dilute

- A **concentrated** solution contains a lot of solute. The solute particles are close together within the solvent. For example, dissolving 20 tablespoons of cordial in a glass of water makes a concentrated (and unpleasantly sweet) solution.
- A **dilute** solution contains a very small amount of solute. The solute particles are widely spaced out within the solvent. For example, dissolving a quarter of a teaspoon of cordial in a glass of water makes a dilute (and not sweet enough) solution.

Concentration

Amount of solute present in a solution

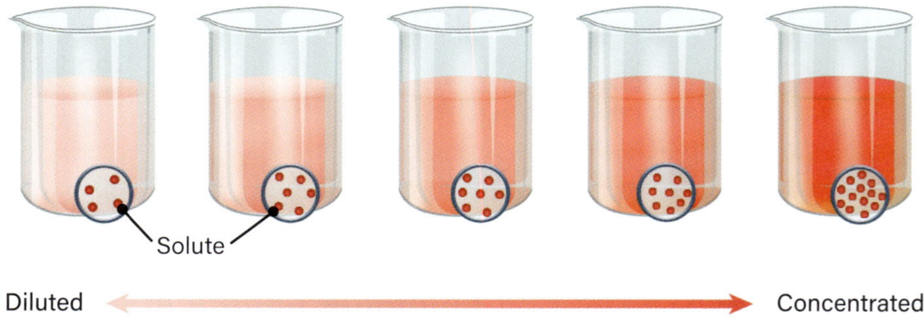

▲
Figure 4.25: The same volume of the concentrated solution has more solute particles than the dilute solution.

Saturated solutions contain the maximum amount of solute that can dissolve

A solvent can only hold a certain amount of solute. This means that at a certain point, called the **saturation point**, no more solute can dissolve into the solution, and the solution is described as **saturated**.

Heating a saturated solution can make a supersaturated solution

If we heat a saturated solution, we can dissolve more solute in it. Then, if we slowly cool the solution back to room temperature, the solute sometimes stays in solution, even at the lower temperature.

Figure 4.27: Honey is a supersaturated solution because it contains more sugar particles than can usually dissolve in water at room temperature.

These solutions are called **supersaturated** solutions because they contain *more* solute than the solution could otherwise hold if it were made at room temperature (see Figure 4.28).

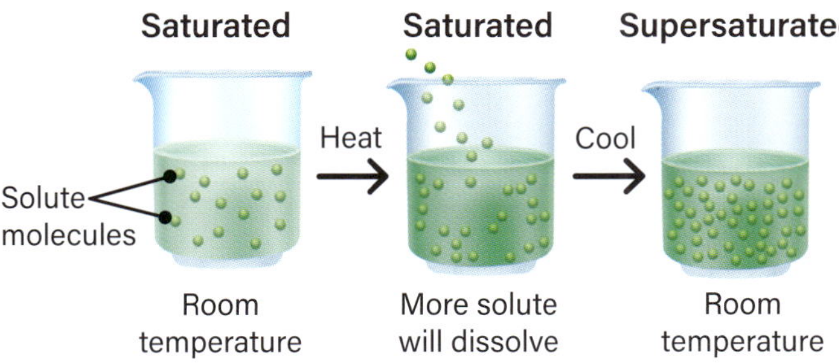

Figure 4.28: A supersaturated solution has the maximum amount of solute dissolved.

Honey is a supersaturated solution (see Figure 4.27). In making honey, the bees fan the honeycomb with their wings, which helps evaporate water from the honey. So, honey contains more sugars than could usually dissolve at room temperature. You might have noticed that after the honey has been stored for a while, the sugars crystallise out.

Learning Ladder

Solutions and mixtures

1. Identify an example of a supersaturated solution that you might find in your home.
2. Describe how the particles of a glass of dilute cordial would be arranged.
3. Explain how understanding the saturation point of a solution could help us to calculate its maximum concentration.
4. Compare the particle arrangement of dilute, saturated and supersaturated solutions. Construct a model of how the particles are arranged as part of your response.
5. Analyse the properties of dilute and concentrated solutions.

Conducting investigations

see page 326

Imagine you are going to conduct an investigation to make a supersaturated solution of sugar in water, after determining the maximum amount that needs to be added to a 20 mL sample. Plan an investigation using the following instructions.

1. Identify a piece of equipment that would allow you to correctly measure the amount of sugar being added to the solution.
2. Suggest a piece of equipment that would allow you to make the supersaturated solution from the 20 mL of water. (Hint: What do you need to add to allow more sugar particles to dissolve in the same amount of water?)
3. Explain a safety measure you would take to reduce the risk of using the piece of equipment you suggested in Question 2.
4. To obtain accurate data, you need to use precise quantities of materials. How could you ensure that you are obtaining precise measurements of sugar and water?

In context

In many local pools, chlorine is dissolved in water to form a solution. This helps to reduce the numbers of harmful germs in the water. Discuss why it would be important to maintain accurate levels of dissolved chlorine in pools.

Success criteria

- I can describe the concentration of dilute, concentrated, saturated and supersaturated solutions in relation to the amount of solute dissolved into the solvent.
- I can construct diagrams to demonstrate the amount of solute in different types of solutions.

4·8 ▸ Atoms, elements, compounds and mixtures

Figure 4.29: Gold is one of the few substances that exists in its elemental form in nature.

Learning intention

At the end of this lesson, I will be able to distinguish between elements, mixtures and compounds and classify them based on their properties by using particle theory and composition.

Key terms

chemical bond: a force that holds atoms together

colloid: a mixture made of tiny insoluble particles that remain dispersed evenly in a liquid

compound: a substance containing atoms of two or more elements chemically bonded together in a fixed ratio

element: a substance made of only one type of atom

heterogeneous: a mixture with an uneven (non-uniform) composition

homogeneous: a mixture with an even (uniform) composition

lattice: a three-dimensional shape made of a repeating pattern of atoms

suspension: a mixture made of large insoluble particles that settle to the bottom of the liquid

Investigation 4.8A

Classifying substances, p.428

Investigation 4.8B

The Tyndall effect, p.429

Content group: Separating mixtures

Substances can be elements, compounds or mixtures, which are all are made up of atoms. Elements are made up of one type of atom. Elements bond together to form compounds made up of more than one type of atom. A mixture can contain many types of substances or particles, but the parts of the mixture are not bonded together and can be separated.

Matter can be divided into pure substances and impure substances

Matter can be divided into pure substances and impure substances (see Figure 4.30).

A pure substance is composed of only one type of particle. Pure substances include elements (made up of one type of atom) and compounds. Compounds are made of two or more types of elements chemically joined together into particles that cannot be physically separated.

An impure substance is made up of two or more substances that can be separated. Impure substances include mixtures, which can be physically separated into their different components. Air (a mixture of different gases), milk (made of water, fat, protein and other substances) and salt water (salt mixed with water) are all examples of mixtures.

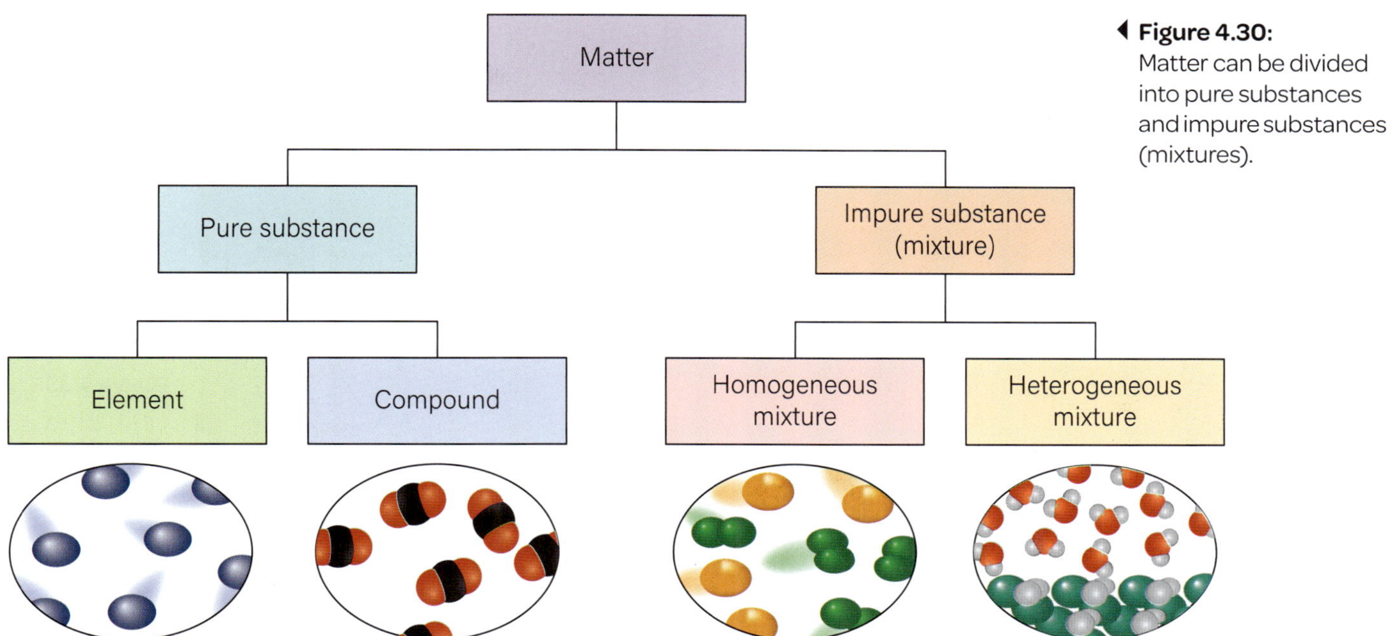

Figure 4.30: Matter can be divided into pure substances and impure substances (mixtures).

An element contains only one type of atom

An **element** is a pure substance that is made up of only one type of atom. Gold is an element because it consists only of gold atoms. Likewise, aluminium is made only of aluminium atoms, and hydrogen gas is made of only hydrogen atoms. Elements cannot be broken down further by ordinary chemical means.

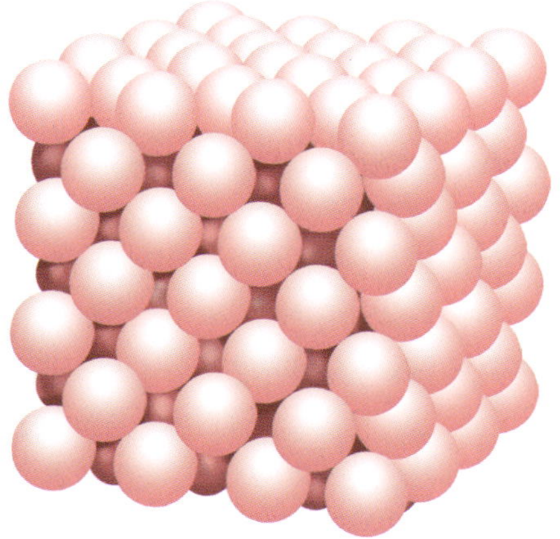

Figure 4.31: Pure gold (Au) is made of only gold atoms.

A compound is made of different elements bonded together

A **compound** is a substance made up of more than one type of atom **chemically bonded** together in a fixed ratio. Compounds can be broken down into simpler substances through chemical reactions.

Water is a compound: it is made up of water molecules, which each consist of one oxygen atom chemically bonded to two hydrogen atoms (see Figure 4.32).

Other compounds form a **lattice**, where the atoms are bonded together in a repeating three-dimensional pattern. Sodium chloride (table salt) is an example of a compound that exists as a lattice in which the sodium and chlorine atoms are in a 1:1 ratio (see Figure 4.32).

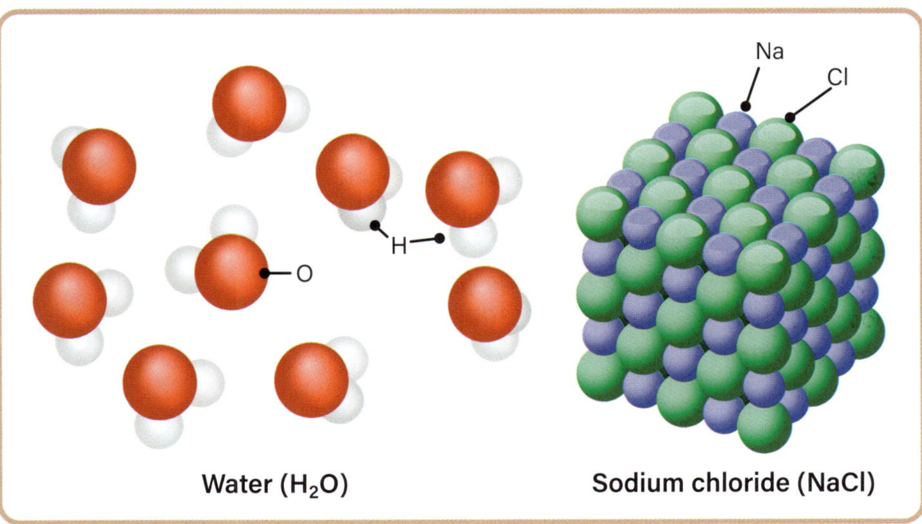

Figure 4.32: Compounds are usually molecules like water (H_2O) or lattices like salt (sodium chloride (NaCl)).

A mixture is made up of different substances that are not bonded together

Impure substances are mixtures because they consist of more than one type of particle. A mixture contains two or more elements and/or compounds that can be separated. Although a compound contains atoms of different elements, these cannot be easily separated because they are bonded together.

Seawater is a mixture of salt and water. If you boil seawater, the water escapes as a gas, leaving the salt behind.

Mixtures can be classified into heterogeneous and homogeneous mixtures

A **homogeneous** mixture has the same composition throughout. It is usually made up of a solvent, with one or more dissolved solutes. As the solutes dissolve, they become evenly distributed throughout the mixture. An example of a homogeneous mixture is salt water.

In a **heterogeneous** mixture, the substances are unevenly distributed. This is usually because one or more substances of the mixture are insoluble (do not dissolve). This means that different parts of the mixture have different compositions. For example, if a beaker contains a mixture of sand, water and pebbles, the sand and pebbles will settle to the bottom rather than dissolving evenly through the water. The mixture separates into layers (see Figure 4.33). So, different parts of the mixture have different compositions of sand, pebbles and water.

Suspensions and colloids are types of mixtures

Two other types of mixtures are suspensions and colloids. Their particles behave differently from the particles in solutions.

Figure 4.33: This mixture of sand, water and pebbles has separated into layers. The particles are not bonded together.

- A **suspension** is a mixture made up of large insoluble particles. At first, the particles are spread out evenly but eventually they settle to the bottom of the container. The mixture in a snow globe is a suspension: it is made up of plastic 'snow' particles mixed in water (see Figure 4.34).
- A **colloid** is a mixture made up of tiny insoluble particles. The particles are spread out evenly and never settle to the bottom of the container. Milk is a colloid: it is tiny droplets of fat mixed in water.

Figure 4.34: The mixture in a snow globe is a suspension of plastic particles in water.

There are three different types of colloids, depending on whether they are made up of solids, liquids or gases. They are summarised in Table 4.2.

Table 4.2: The three types of colloids

Colloid type	Description	Example
Sol	Solid in liquid	Milk
Emulsion	Liquid in liquid	Oil in water
Foam	Liquid in gas	Whipped cream
Aerosol	Gas in liquid or solid	Smoke, steam

You can tell whether a mixture is a solution, a colloid or a suspension by shining a light at it. Colloids and suspensions look cloudy because they scatter the light beam, whereas a solution does not. This is known as the Tyndall effect.

Figure 4.35: The Tyndall effect involves light scattering by particles in a colloid or suspension. Here, a laser beam is being shone on a glass of water containing red dye and another glass containing milk (a colloid).

Learning Ladder

Solutions and mixtures

1. Categorise these substances as elements, compounds or mixtures.
   - **a** Gold
   - **b** Carbon dioxide
   - **c** Milky tea
   - **d** Soil
   - **e** Oxygen
   - **f** Water
2. Describe the difference between homogeneous mixtures and heterogeneous mixtures.
3. Explain how you could tell if a substance is a compound or a mixture.
4. Compare how the particles of an element, a compound and a mixture are arranged. Give an example of each type of material.
5. A substance is added to a beaker of water. Initially the substance makes the water blue and cloudy. The substance then settles on the bottom of the beaker. What type of mixture is this? Provide a reason.

Problem-solving

see page 339

1. Propose reasons why it is difficult to classify some mixtures, such as blood, based on their composition.
2. Outline one way you could determine if a solution is an element or a heterogeneous mixture, just by looking at it.
3. If compounds such as salt and sugar are mixed in water, what new solution do they form? Explain how this relates to particle theory.
4. Use particle theory to draw a diagram for Question 3, in order to classify the type of mixture formed.

In context

The air in Earth's atmosphere is a mixture of elements and compounds. Research to find out the percentage of each of the elements and compounds in air. Present this data in a suitable graph. For more information on constructing graphs, see 'Graphing' in the Science how-to section on page 368.

Success criteria

- I can distinguish between atoms, mixtures and compounds and explain their properties using particle theory.
- I can classify matter into pure substances and impure substances based on their particle composition.

4·9 ▸ Separating mixtures: centrifugation and paper chromatography

Learning intention

At the end of this lesson, I will be able to explain how the physical properties of substances are used to separate the components of mixtures by centrifugation and paper chromatography.

Key terms

centrifugation: a technique for separating the components of a mixture on the basis of their different densities

paper chromatography: a technique that separates the components of a mixture based on their solubilities

Investigation 4.9

Separating colours in dyes by paper chromatography, p.430

Content group: Separating mixtures

Mixtures can be separated into their individual components by physical means. The choice of separation method depends on the physical properties of the components and the desired purity of the separated substances.

Mixtures can be physically separated

In science, a mixture is two or more substances mixed together that can be physically separated. Mixtures are all around us: concrete, air, mayonnaise, muddy water and seawater are all mixtures.

Some things are not mixtures. Oxygen, pure water and gold are not mixtures; they are pure substances. Mixtures can contain:

- soluble substances, which will dissolve, such as salt in water
- insoluble substances, which will not dissolve, such as sand.

Some substances partly dissolve, meaning that some, but not all, of the substance dissolves into the solvent.

Scientists use information about solubility to help decide which technique to use to separate a mixture.

Centrifugation separates the components of a mixture by density

Centrifugation is a technique that uses centrifugal force to separate components of a mixture on the basis of their densities. The mixture is spun at high speeds in a centrifuge, causing the denser components to move to the bottom of the container. For example, we can use centrifugation to separate blood into its different components.

Figure 4.36: Centrifugation of blood separates it into red blood cells, plasma, and white blood cells and platelets

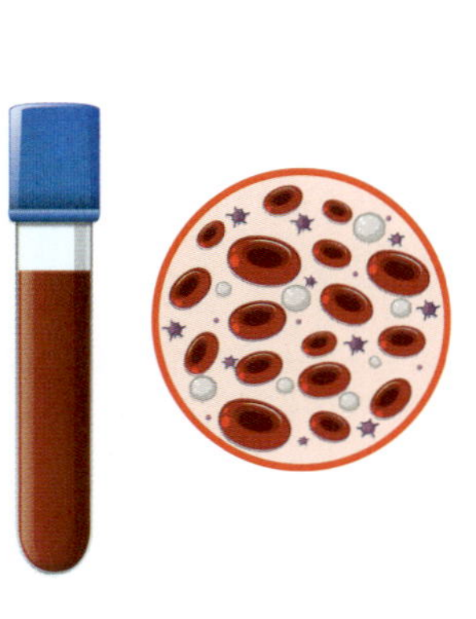

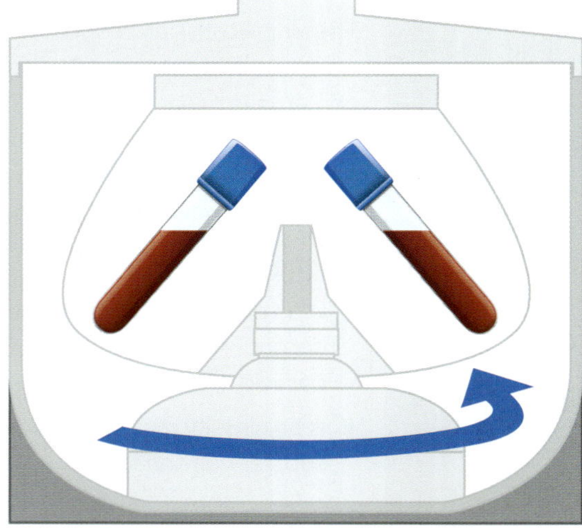

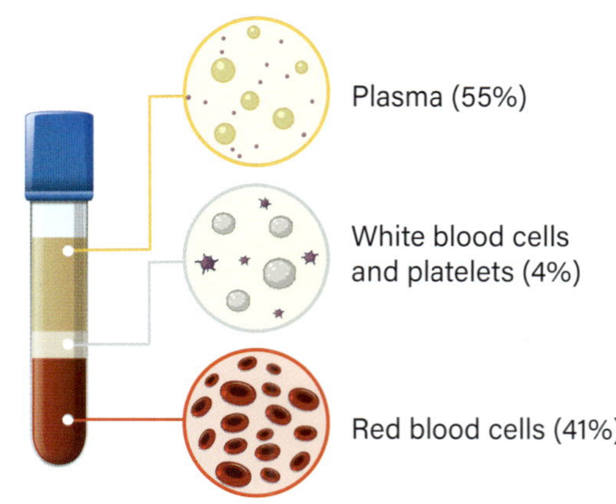

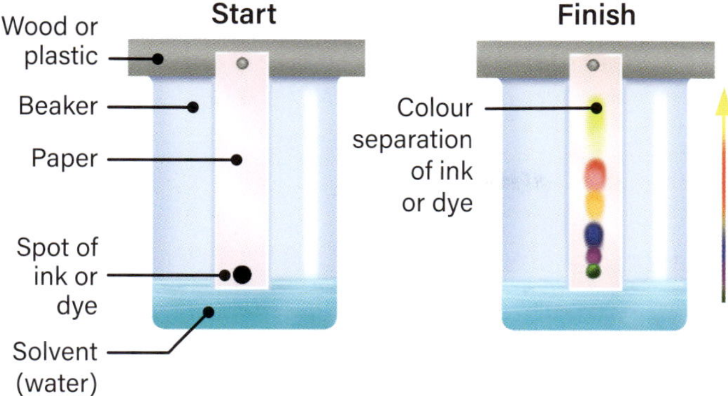

Figure 4.37: Chromatography on paper separates the black ink into its coloured components. The yellow spot moves up the furthest. The green component moves the least distance because it is the least soluble.

Paper chromatography separates dyes based on solubility

Paper chromatography is used to separate the components of mixtures of soluble substances on the basis of their solubility. The components are often coloured, such as inks or dyes (see Figure 4.37). A mixture is dissolved in a liquid to make a solution. A drop of the solution is placed at one end of a piece of chromatography paper, which is then placed in a solvent (for example, water). The solvent moves up the paper, carrying the dissolved substances with it. The most soluble substances move the furthest. Other substances are less soluble so do not move as far. In this way, the components of the mixture are separated.

Chromatography is used in industry to analyse substances found in oils and gases and to identify pollutants in the environment. In the pharmaceutical industry, chromatography can be used to test the effectiveness of medicinal substances that originated from animals and plants.

Figure 4.38: Forensic scientists can use chromatography to match pen and ink samples, or to compare dyes in fibre samples.

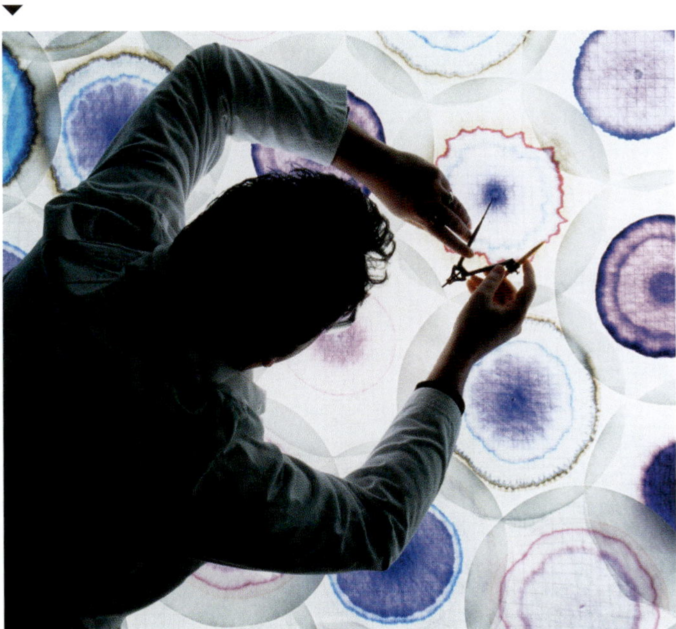

Learning Ladder

Solutions and mixtures

1. Identify which mixture could be separated by centrifugation.
   A Alcohol and water
   B Blood
   C Sugar and water
   D Muddy water
2. Describe the physical properties of substances that allow them to be separated by:
   a centrifugation.
   b chromatography.
3. Explain how using paper chromatography can help us to understand the different colours present in pen ink.
4. Compare the properties of blood cells that allow them to be separated by centrifugation.

Problem-solving

see page 329

1. Summarise reasons why it is difficult to classify some mixtures based on their composition, such as salt water or a sugar solution.
2. Propose a way you could use a separation technique to determine who had written a particular note in pen.
3. Explain the link between the solubility of substances and their ability to move along paper in chromatography.

In context

Blood is a homogeneous mixture made up of many parts (fractions) such as red blood cells, platelets, plasma and white blood cells. White blood cells can be separated by centrifugation. Research and explain why this process can be useful to society.

Success criteria

- I can describe the physical properties of substances.
- I can distinguish between the separation techniques of centrifugation and paper chromatography.
- I can explain how centrifugation and paper chromatography can be used to separate a mixture.

4·10 ► Separating mixtures: filtration and decantation

Learning intention

At the end of this lesson, I will be able to explain how the physical properties of substances are used to separate the components of mixtures by filtration and decantation.

Key terms

decant: to carefully pour off the liquid from a mixture, leaving the sediment behind

filtrate: the liquid that passes through a filter in filtration

filtration: the process of separating a mixture of solid particles from a liquid using a filter

immiscible: cannot be mixed

residue: the solid that does not pass through a filter in filtration

sediment: the solid that settles to the bottom of a liquid in a mixture

Investigation 4.10

Purifying muddy water, p.431

Content group: Separating mixtures

Imagine you are stranded in the bush and the only source of water is a muddy puddle. You scoop some water out with a container, and you notice how dirty it is. After a while, the mud sinks to the bottom of the mixture. You place a cloth over another container and, without disturbing the mud, carefully pour the liquid part of the mixture into the second container.

You have used decantation and filtration to separate the water and mud. You survive!

Filtration separates liquids from solids

Filtration is used to separate insoluble substances from liquids. The filter is a sieve: small holes allow the liquid to pass through but trap larger particles. So, filtration separates substances on the basis of particle size.

To make a cup of tea, we can use a strainer to filter the tea leaves. We can also use a tea bag where the tea leaves are inside the filter. Water can pass through, but the tea leaves cannot.

Figure 4.39 shows how to use filter paper to separate a mixture of chalk and water. The solid chalk particles get trapped in the filter paper but the liquid water passes through it. The chalk left in the filter paper is called the **residue** and the liquid that passes through the filter paper is called the **filtrate**.

Figure 4.39: Chalk particles can be separated from water by filtering the mixture through filter paper.

Decanting separates sediments from liquids

Decantation can also be used to separate mixtures of insoluble solids from liquids. After an insoluble solid in a mixture settles to the bottom of a container, the liquid can be poured out carefully, leaving the solid behind (see Figure 4.40).

Think back to the scenario at the start of this unit. The solid mud from the puddle water settled to the bottom of the container as a **sediment**. When the water was poured out from the top, the mud was left behind.

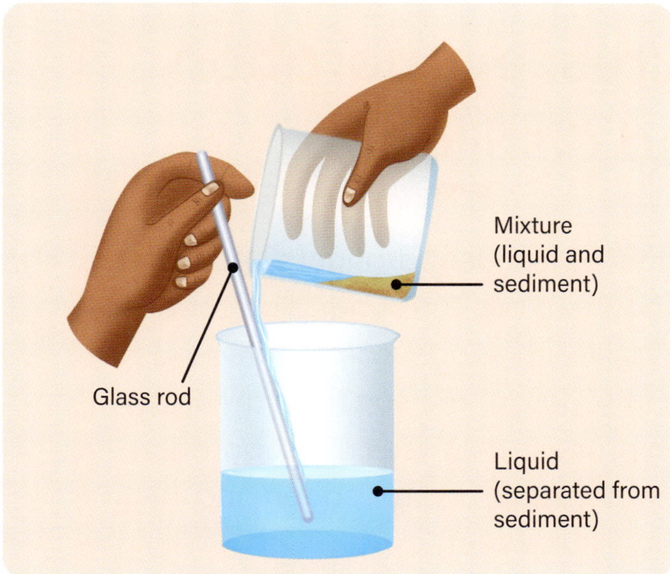

◀ **Figure 4.40:** After the solid sediment has settled, the liquid component of a mixture can be carefully decanted off the top. The glass rod assists when decanting by reducing spills and spattering.

Decanting is used in the process of making cheese. As the cheese forms, the remaining milk 'floats' to the top and is skimmed off.

If two liquids are **immiscible** (cannot be mixed), they can also be separated by decanting. Two immiscible liquids of different densities (for example, oil and water) will form two layers. The lighter liquid can be 'skimmed' off. The property that causes this separation is density. Another way to separate the immiscible liquids is by using a separating funnel (see Figure 4.41).

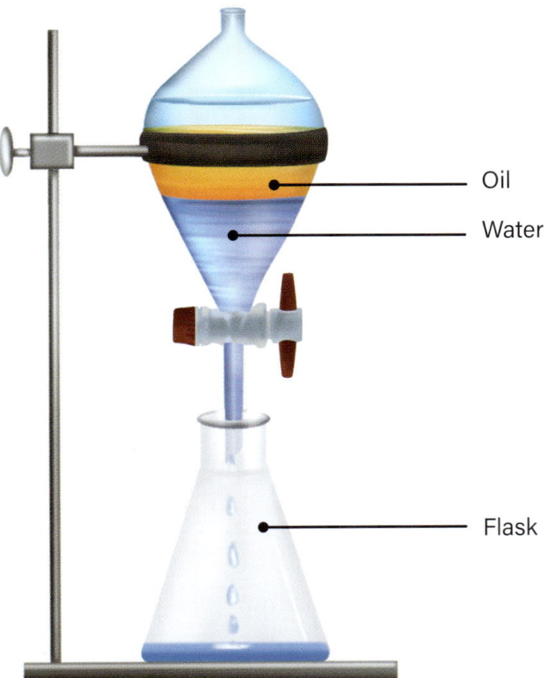

▲ **Figure 4.41:** Two immiscible liquids, such as oil and water, can be put into a separating funnel and left to settle into layers. When the tap is opened, the liquid at the bottom runs into the flask.

Learning Ladder

Solutions and mixtures

1. You use filtration to separate a mixture of sand and water. Identify the:
   a residue. **b** filtrate.
2. Describe a property of a chalk and water mixture that makes it suitable for separation by filtration.
3. Use words and a diagram to explain filtration.
4. Compare the physical properties that allow filtration and decantation to work.
5. Suggest ways that filtration and decantation could be used in the real world.

Conducting investigations

see page 326

Refer to Investigation 4.10 on page 431 to answer the following questions. Read the aim, materials and method.

1. Identify which of the following statements is correct when undertaking the process of decanting.
   A Stir the mixture before pouring it into another beaker.
   B Pour the mixture between the beakers as quickly as you can.
   C Allow the mixture to sit before decanting, so that the sediment can settle to the bottom of the beaker.
2. Describe how you would use one piece of scientific equipment while conducting this investigation. How would it help you to separate the mixture?
3. Investigation 4.10 is relatively safe, but you should be able to identify three potential hazards. Propose ways to minimise the risk of the three hazards.
4. Identify whether the data collected is qualitative or quantitative. Describe how you could ensure that you accurately recorded your data while conducting this investigation.

In context

LifeStraw™ is a brand of straw filter marketed for people going hiking or camping so they can safely drink water from contaminated sources. Explain why filtration would be an important part of the LifeStraw design.

Success criteria

- I can describe filtration and decantation.
- I can identify the appropriate separation method to use for different mixtures.

4·11 ▸ Separating mixtures: evaporation, crystallisation and distillation

Learning intention

At the end of this lesson, I will be able to explain how the physical properties of substances are used to separate the components of mixtures by evaporation, crystallisation and distillation.

Key terms

condensation: a change of state from gas to liquid

condenser: a glass tube cooled by water that cools a gas to become a liquid

crystallisation: the separation of a solution by evaporating the solvent, leaving behind solute crystals

desalination: the process of removing dissolved salts from seawater

distillation: the separation of liquids with different boiling points by evaporation and condensation

evaporation: a change of state from liquid to gas

Investigation 4.11A
Evaporating a solution, p.433

Investigation 4.11B
Growing crystals, p.434

Investigation 4.11C
Demonstrating distillation, p.435

Content group: Separating mixtures

Australia experiences frequent droughts and water shortages. Although we are surrounded by oceans, we (like many other living things) can't survive on salt water. But we can separate the salt from the water. Desalination is a process that uses the separation techniques of evaporation, crystallisation and distillation to purify salt water.

Solids can be crystallised out of solutions

Evaporation is the change of state from liquid to gas. It can be used to separate a mixture such as salt water by evaporating the liquid to leave the salt behind. This happens slowly at room temperature. Evaporation can be sped up by heating the solution in an open container such as an evaporating basin.

When most of the liquid has evaporated, the salt will start to form crystals; this is called **crystallisation**. Smaller crystals form if the liquid evaporates quickly, and larger crystals form if it evaporates slowly.

Crystallisation is used on a large scale to separate sea salt from seawater. The water is evaporated off to leave behind crystals of sea salt (see Figure 4.42). The salt crystals can then be used in cooking, in products such as body scrubs, or as a preservative so that food can be stored for longer periods of time without spoiling.

Distillation can purify water

Salt water can also be separated by **distillation**. First, the salt water is heated in a glass flask and the water evaporates off. The water vapour is captured and cooled in a tube called a **condenser**, which has cool water flowing around it. This causes the gas to change back to a liquid.

Figure 4.42: Salt evaporation ponds are shallow artificial ponds designed to extract salt from seawater. Natural salt ponds are geological formations that form when water evaporates and leaves behind a salt flat.

This change is called **condensation**. The condensed liquid (purified water) is collected in a different container, while the salt crystals remain in the flask.

You can see the equipment used in distillations in Figure 4.43.

Distillation can also be used to separate two liquids, such as water and alcohol. The physical property that allows these substances to be separated is boiling point. Every substance has a different boiling point and the substance with the lower boiling point boils off first. Alcohol has a lower boiling point than water, so it evaporates off and condenses in the flask, leaving the water behind, as shown in Figure 4.43.

Some everyday items contain solutions produced by distillation. Cars run on refined oil, which must be distilled as part of its purification process. Many perfume scents are collected through the process of distillation.

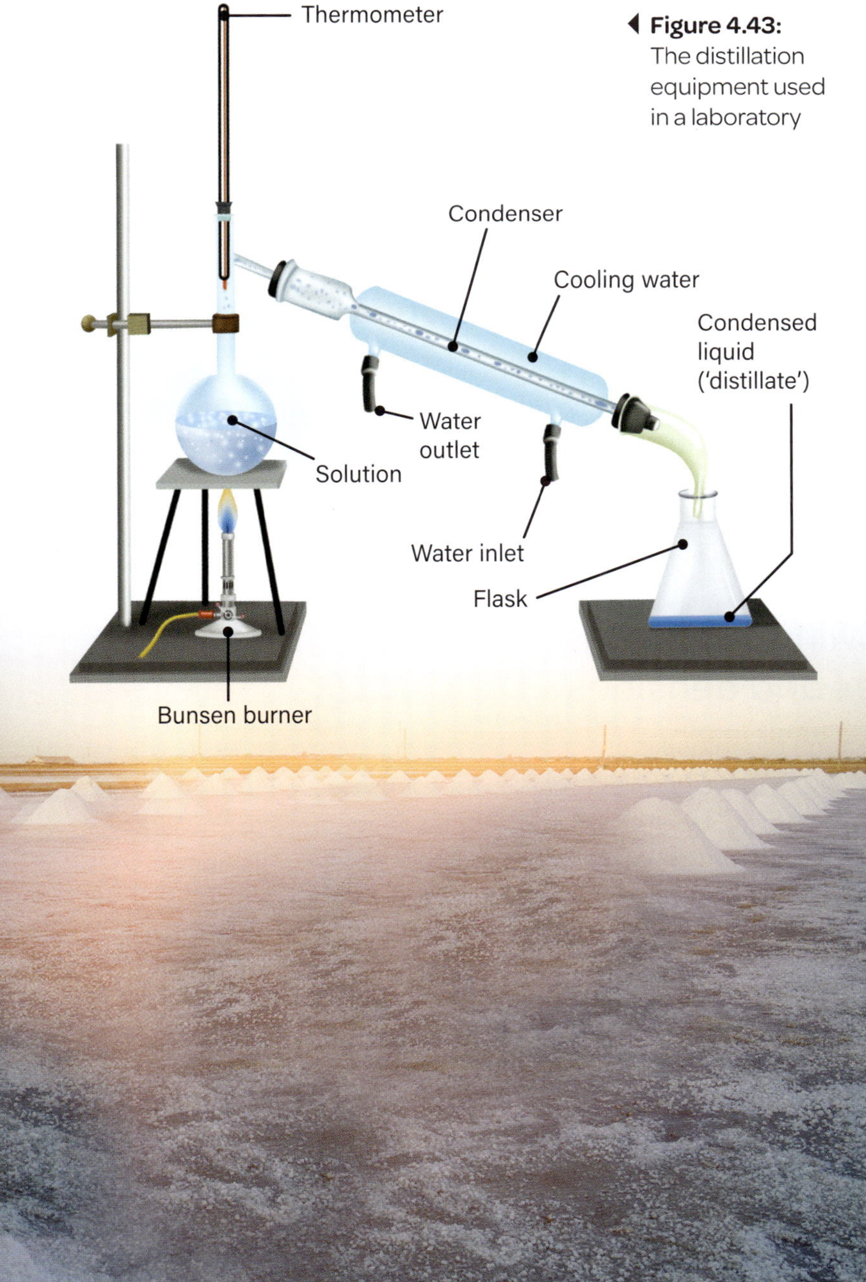

Figure 4.43: The distillation equipment used in a laboratory

Learning Ladder

Solutions and mixtures

1. Identify the physical property of a substance that is used to separate mixtures in distillation.
2. Describe the relationship between evaporation and crystallisation. Give an example of a mixture that can be separated by evaporation and crystallisation.
3. Explain how you would try to produce large crystals when separating a mixture by evaporation and crystallisation.
4. Compare the properties of liquids and solids that allow liquids to be separated by distillation.

Conducting investigations

see page 326

Imagine you are going to conduct an investigation to separate a mixture of salt water and alcohol. Plan an investigation using the following instructions.

1. Identify the correct separation techniques you would use in this investigation.
   A Distillation, evaporation and crystallisation
   B Distillation only
   C Evaporation only
   D Decantation and filtration
2. Describe how you would need a specific piece of scientific equipment to separate the alcohol from the salt water.
3. Identify a risk associated with this piece of equipment. Describe two ways that the risk can be managed.
4. What data would you collect from this investigation? Explain how you could ensure that it is accurate.

In context

At desalination plants, distillation is used in the separation process. Explain what a desalination plant is. Explain why distillation is an important step in separating out the desired components from a mixture.

Success criteria

- I can describe the difference between evaporation, crystallisation and distillation.
- I can give examples of situations in which each technique could be used to separate substances.

4·12 ▸ First Nations Peoples' separation techniques

Learning intention

At the end of this lesson, I will be able to understand the different ways that mixtures have been separated by First Nations Peoples for thousands of years.

Key terms

coolamon: a shallow dish traditionally used by First Nations Peoples to carry water and food and for winnowing seeds

hand picking: a simple method of separating a mixture of objects

leach: to remove soluble substances by the action of water passing through the material

sieving: using a sieve to separate larger particles from smaller particles

toxin: a poisonous substance produced by a living thing

winnowing: separating a mixture of seeds and husks by blowing air through the mixture

yandying: separating less dense particles from denser ones

Content group: Separating mixtures

In Australia, First Nations Peoples have used ecology, agriculture, aquaculture and animal husbandry to care for Country and to grow foods that are easy to cultivate and prepare for eating.

First Nations Peoples have used their knowledge of science and specific techniques to separate edible parts from inedible – and sometimes toxic – parts of plants or food.

Hand picking is a simple method of separating objects

First Nations Peoples have used eucalyptus leaves for thousands of years to ease coughs and colds. Some species are considered to have antifungal, antibacterial and mosquito-repellent properties. When brewing a gum leaf tea for medicine, the desired leaves are usually a lighter green and smaller than the older leaves. The leaves can be easily separated by **hand picking**.

Sieving separates particles of different sizes

Sieving is a common process of using a sieve or another object to separate larger particles from smaller particles. First Nations Peoples sometimes use sieving to catch small freshwater fish and yabbies in creeks and rivers. They hold vegetation and foliage while walking in a line across a creek to sieve the water. The water passes through the vegetation, and the fish and yabbies are caught in the vegetation.

Winnowing separates objects of different masses and densities

This separation technique is especially effective when there are two or more objects that have different masses and densities. **Winnowing** has been commonly used by First Nations Peoples to separate the heavier edible seeds of native grasses from their lighter and inedible husks. The tool used is a **coolamon**. The seeds and husks are thrown into the air. The wind blows away the husks, and the heavier seeds fall back into the container (see Figure 4.44).

Winnowing has been a key part of the process for First Nations Peoples to make flour and bake bread.

▲ **Figure 4.44:** This woman is using a coolamon to winnow seeds from other material. A coolamon is a traditional wooden object used for carrying various things.

Figure 4.45: This coolamon is being used to yandy seeds and chaff (seed coverings) through movement. Their different densities mean that they will form two groups with the movement and be easier to separate.

Yandying separates objects of different masses and densities

Yandying also relies on objects having different masses and densities. It involves moving a coolamon in particular ways to separate the objects. The coolamon is held with one corner raised up. It is then gently shaken to force the smaller and denser particles to collect at the bottom. The larger and less dense particles stay higher up.

Some First Nations People use yandying (such as the people of the Pilbara region of Western Australia). A coolamon can be modified to separate particles of tin from soil.

Leaching uses running water to remove toxins

Leaching is a less common separation technique for making food sources safe to eat by removing soluble **toxins**. For example, across the northern parts of Australia, some First Nations Peoples eat the seeds from the cycad plant. The seeds must be treated to remove toxic substances before being eaten.

A common way of treating the seeds is to **leach** out the toxins in running water. The seeds are placed into slow-moving water. Over time, the toxins leach from the seeds and dissolve in the water. The toxins are washed away in the moving water and the seeds can be cooked and eaten.

Figure 4.46: Cycad seeds on the tree ready for harvesting. The seeds contain toxic substances that can be removed by leaching.

Learning Ladder

Solutions and mixtures

1. In Figure 4.44 (winnowing), identify the objects in the mixture being separated.
2. Describe a common use of hand picking for separation.
3. Explain the properties a mixture must have for winnowing to be a successful separation technique.
4. Describe the process used to prepare many seeds to be separated. What are the edible parts of the seeds typically being separated from?
5. Consider products you use every day that need to be separated. Identify two that could be separated by each of the following techniques.
   - **a** Hand picking
   - **b** Sieving
   - **c** Winnowing
   - **d** Yandying
   - **e** Leaching

Problem-solving

see page 339

1. Identify one scientific process used to support separation. How does this help to make the separation process more effective?
2. Identify two important features of leaching that make it an effective method of removing toxins from cycad seeds.
3. Explain why winnowing has been an important process to enable the baking of bread by First Nations Peoples in Australia.
4. Describe some important properties of objects that enable them to be separated by winnowing and yandying.
5. You have been given a container with a mixture of soil and rocks of different sizes and densities. What would be the best way or ways to separate the mixture?

In context

How can knowledge of First Nations Peoples' ways of separating objects help other Australians consume native foods and medicines?

Success criteria

- I can understand the different techniques of First Nations Peoples for separating mixtures.
- I can consider the differences between the separating processes and why different processes are used with different mixtures.

4·13 ▸ Industrial separation techniques

Learning intention

At the end of this lesson, I will be able to describe the industrial applications of water purification and sewage treatment.

Key terms

disinfection: a method of destroying bacteria, often using special light or chlorine

sewage: semi-liquid human waste

sewerage: a system of pipes that carry sewage

Content group: Separating mixtures

Many separation techniques are used on a large scale by industry. This means large quantities of materials can be separated at one time. These separation techniques are often applied to water, in order to purify contaminated water sources, so that the water is able to be reused or released safely back to the environment.

Purification makes water safe to drink

Before water is safe for us to drink, it must be purified. This involves killing bacteria and removing harmful substances. In Australia, there are six main steps in water purification (see Table 4.3). When the water is safe to drink, it is distributed through pipes to homes, schools, hospitals, businesses and many other locations.

Table 4.3: The steps involved in water purification

Step	Description
1 Screening	Water passes through mesh screens to remove objects such as twigs and leaves.
2 Flocculation	A chemical called alum is added to the cloudy water to make the small floating particles clump together. These clumps are called floc.
3 Sedimentation	The floc is heavy and settles to the bottom of the tank to form a sediment. This sediment is collected as sludge.
4 Filtration	The water flows through tightly packed beds of different-sized pebbles, sand and crushed coal to trap and remove any remaining floc and other impurities.
5 Chemical treatment	Chemicals are added to the water, such as: • fluoride to help prevent tooth decay • chlorine to kill harmful organisms such as bacteria • other chemicals to reduce the acidity of the water.
6 Aeration	Oxygen is added to improve the smell and colour.

Figure 4.47: A water treatment plant has many different sections for the different stages, such as chemical treatment and aeration.

Sewage is processed to reduce harm to the environment

Sewage is semi-liquid human waste. When we flush a toilet, the sewage goes into a **sewerage** system and is treated.

Some countries do not have proper sewerage systems (see Figure 4.48). Untreated sewage is harmful to human health and the natural environment, so in more economically developed countries such as Australia, it is processed before release. In Australia, there are six main steps in sewage treatment (see Table 4.4).

▲ **Figure 4.48:** In some less economically developed countries, human waste goes into open sewers and is not treated.

Table 4.4: The steps involved in treating sewage

Step	Description
1 Sewerage	A network of pipes moves sewage from homes and businesses to sewage treatment plants.
2 Screening	Screens at the plant act as a sieve and catch large objects, which are physically removed.
3 Aeration	Air is pumped into tanks that hold the sewage. This feeds bacteria, which break down the sewage.
4 Settling	Other chemicals are added that cause the bacteria and solids to settle to the bottom of the tank as thick sludge. This sludge is removed and is used in soil and fertiliser products.
5 Filtration	The sewage passes through a filter made of pebbles. This traps more solids, which are removed.
6 Disinfection	Ultraviolet light or chlorine is used to kill harmful bacteria in the sewage.

After the final **disinfection** stage, the water is considered to be 'safe'. This water is then released back into the environment, where it re-enters the natural water cycle.

Learning Ladder

Solutions and mixtures

1. Identify the solutes, solution and solvent in the treatment of water.
2. Describe the properties of floc.
3. Explain how the first step of water purification is a separation technique. How does it help the purification process?
4. Compare the processes of screening and settling in sewage treatment. How do they help separate components of sewage that have different properties?
5. Analyse how the treatment and purification of water are important processes and critical to a healthy environment.

Problem-solving

see page 339

1. Identify problems that could occur if:
   a. flocculation is not completed effectively during water purification.
   b. the aeration process is ineffective during sewage treatment.
2. Propose solutions to the problems you identified in Question 1.
3. Explain why the process of filtration helps the purification of water, based on the particles in the solution.

In context

As global temperatures rise, there is a constant change in the weather patterns, which brings drought and reduced rainfall to parts of some countries like Australia. This affects the amount of water available for plants, animals and humans. Explain how processes such as sewage treatment could help to manage drought.

Success criteria

- I can describe water purification and its importance as an industrial technique.
- I can describe sewage treatment and its importance as an industrial technique.

4·14 ▸ Solutions and mixtures in context

Learning intention

At the end of this lesson, I will be able to apply my understanding of separation techniques to purify contaminated water.

Key terms

crude oil: oil that has not been separated into usable petroleum products

oil slick: a thin layer of oil on the surface of water

Investigation 4.14

Purifying oily water, p.437

Content group: Solutions and mixtures in context

Oil is one of the most important materials used today. When refined, it fuels the engines of vehicles, powers factories to produce electricity and is an ingredient in the manufacture of plastics. However, its extraction, transport and use have an impact on the environment. The accidental release of crude oil into the environment is called an oil spill. Oil spills in the ocean are very harmful to ecosystems. Understanding the properties of oil and how it mixes with water and other chemicals has allowed scientists to work out methods to clean up oil spills. This ensures that we do not contaminate water sources in the environment for extended periods of time.

Oil spills can be caused by different incidents

Oil spills in the marine environment can have different causes.

Crude oil is extracted by drilling it from reservoirs underground. Often this is done from oil rigs in the oceans. If an accident happens during the drilling or pumping process, crude oil can be released into the ocean.

Ships use oil as fuel. If a ship breaks down, runs aground, collides with another ship or has its tanks damaged, the oil can leak into the water.

Oil is transported by massive ships called oil tankers. If these ships are damaged, the oil they are carrying can spill into the environment.

Figure 4.49: Floating booms are used to contain oil spills on the surface of water. The oil can then be skimmed from the surface.

Oil spills can damage the environment

When oil is spilled into the ocean, it forms a thin layer on the surface known as an **oil slick**. Over time, the oil slick spreads out to cover a large area. If it reaches the shoreline, the oil will stick to and mix in with sediments, like sand.

The oil on the surface of the ocean stops oxygen and carbon dioxide gas in the atmosphere from dissolving into the water. If the levels decrease enough, this can kill marine plants and animals.

Oil spills can be contained and filtered

Oil slicks float on top of water because oil is less dense and so does not mix with the water. There are several ways that oil slicks can be cleaned up by taking advantage of these properties.

Floating booms are used to contain oil slicks. Devices called skimmers are then used to scoop or suck oil from the surface of the water within the enclosed areas. This oil can then be processed in a factory to remove any seawater that was also collected.

Figure 4.50: Birds caught in oil spills can lose their ability to fly, swim and float. They can also be poisoned by the oil when they try to clean themselves.

Sorbent booms are a special type of boom that work in a similar way to a disposable nappy, but they only absorb oil. The booms can then be removed and disposed of.

Oil on the shoreline can be difficult to clean up. One way to clean oil from the shoreline is to wash the oil back into the water so that it can be easily skimmed off the top. Another way is to remove the contaminated sediment. This is either disposed of in landfill or processed in factories to separate the oil from the sediment.

Learning Ladder

Solutions and mixtures

1. Identify whether crude oil is a pure substance or a mixture. Provide a reason.
2. Describe the properties of water and oil that make them unable to mix.
3. Explain how oil spills can be contained and cleaned.
4. Compare the properties of oil and water that allow them to be separated.

Planning investigations

see page 323

Refer to the contaminated water sample in Investigation 4.14 on page 437 to answer the following questions.

1. Identify one material you could use to assist you when attempting to purify the water sample.
2. Describe a risk that could be present during the purification process, and how you would reduce the likelihood of it occurring.
3. Suggest the independent variable in this investigation. What is its effect on the dependent variable?
4. Construct an appropriate aim for this investigation that you could write on your investigation scaffold.
5. What steps would you need to follow to completely purify the water sample? Write down these steps as your initial investigation plan.

Success criteria

- I can describe how a body of water can become polluted.
- I can describe how these pollutants such as oil spills can be removed.

▸ Solutions and mixtures summary

Properties of matter

- Matter is the particles that make up everything. There are three states of matter: solid, liquid and gas. Matter changes state as heat energy is added or removed.
- Density is how heavy something is for its size; this can be calculated by dividing an object's mass by its volume, using the formula $\rho = \frac{m}{V}$. Different objects have different densities.

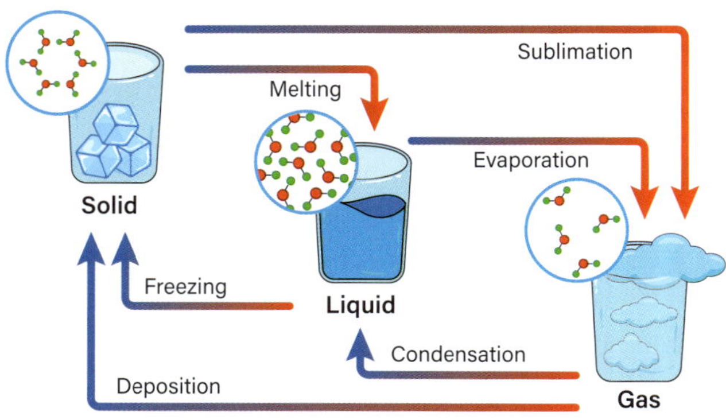

Physical properties of water

Water has many physical properties, including buoyancy and surface tension.

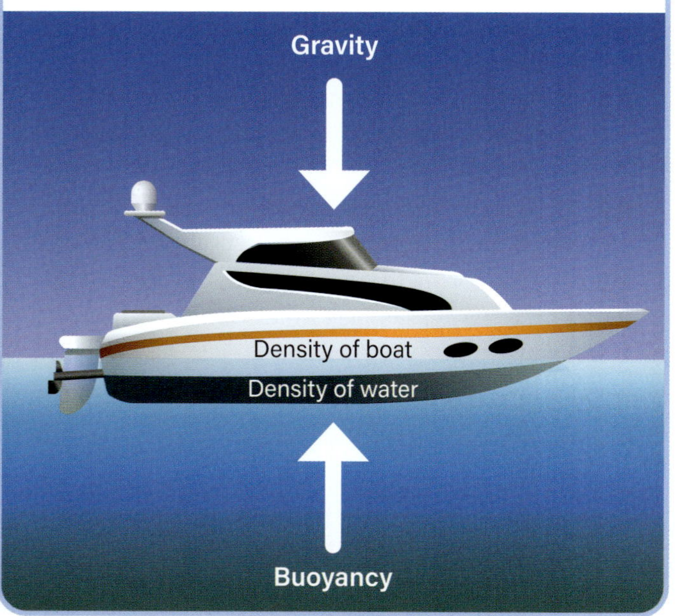

Solutions

- A solution is made when one material (solute) is dissolved into a liquid such as water (solvent).
- Materials that cannot be dissolved are called insoluble.
- The higher the temperature of a solution, the more solute it can hold.
- Solutions can be dilute (contain low amounts of solute in a large volume) or concentrated (contain high amounts of solute in a smaller volume).
- Saturated solutions have dissolved the maximum possible amount of solute at a given temperature. Supersaturated solutions are heated and then slowly cooled, so the solution contains more solute than it could normally hold at room temperature.
- The concentration of solutions can be calculated using $c = \frac{m}{V}$.

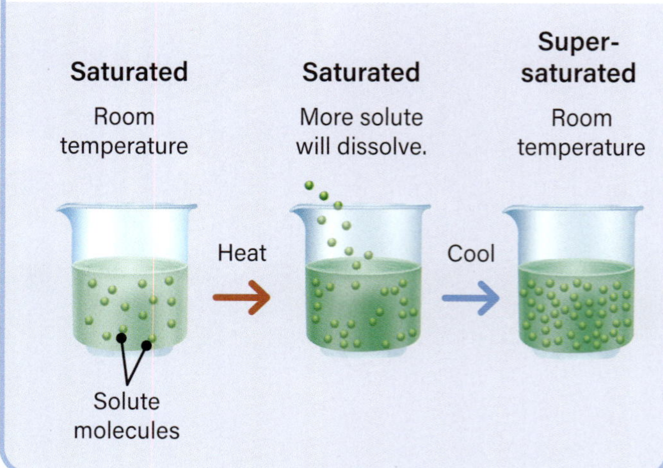

Separating mixtures

- **Elements** are pure substances made of one type of atom.
- **Compounds** are pure substances containing atoms of two or more elements. The atoms are chemically bonded together and cannot be physically separated. Atoms are present in a compound in a fixed ratio.
- **Mixtures** are impure substances, made of a combination of elements and compounds, and can be physically separated.

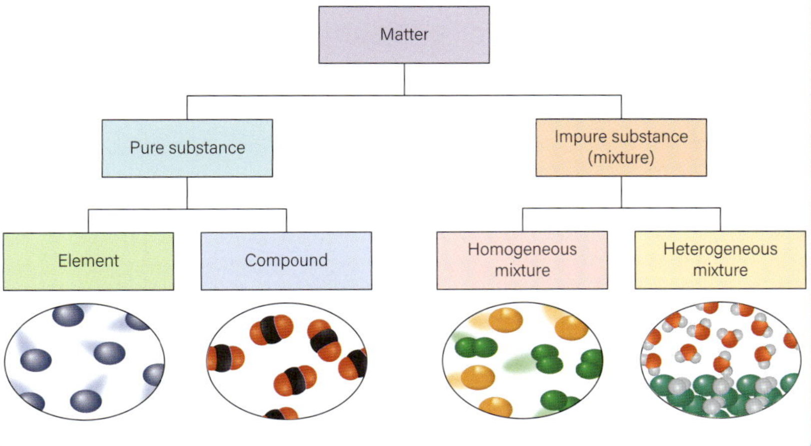

- Aboriginal and Torres Strait Islander Peoples use techniques such as winnowing, yandying and sieving to separate materials based on their sizes.
- Leaching can detoxify foods such as the cycad seed.

Industrial processes – such as purifying water for drinking – rely on separation techniques.

- **Decanting and filtering** can separate insoluble mixtures by using the density and size of the particles.
- **Evaporation and crystallisation** can separate dissolved solids from water or other liquids.
- **Distillation** can separate two liquids with different boiling points.
- **Centrifugation** can separate particles based on their mass.

Masterclass

Steps in progression

		1	2
Content	Solutions and mixtures	Identify one chemical that is added to water when it is being treated to help make it a safe solution.	**a** Describe a property of recycled water that is used in the purification process. **b** Name a technique used to clarify water.
Processes	Planning investigations	Identify a material being used in Figure 4.52 to test water samples.	**a** Describe three risks associated with purifying recycled water. **b** Conduct a risk assessment to help manage these risks.
	Conducting investigations	Identify a process being conducted in Figure 4.52.	Describe how you would use a named piece of scientific equipment to measure the salt levels in a water sample taken from a water-purification plant.
	Problem-solving	Identify one problem with using recycled water.	Suggest two solutions to the problem you identified in Step 1.

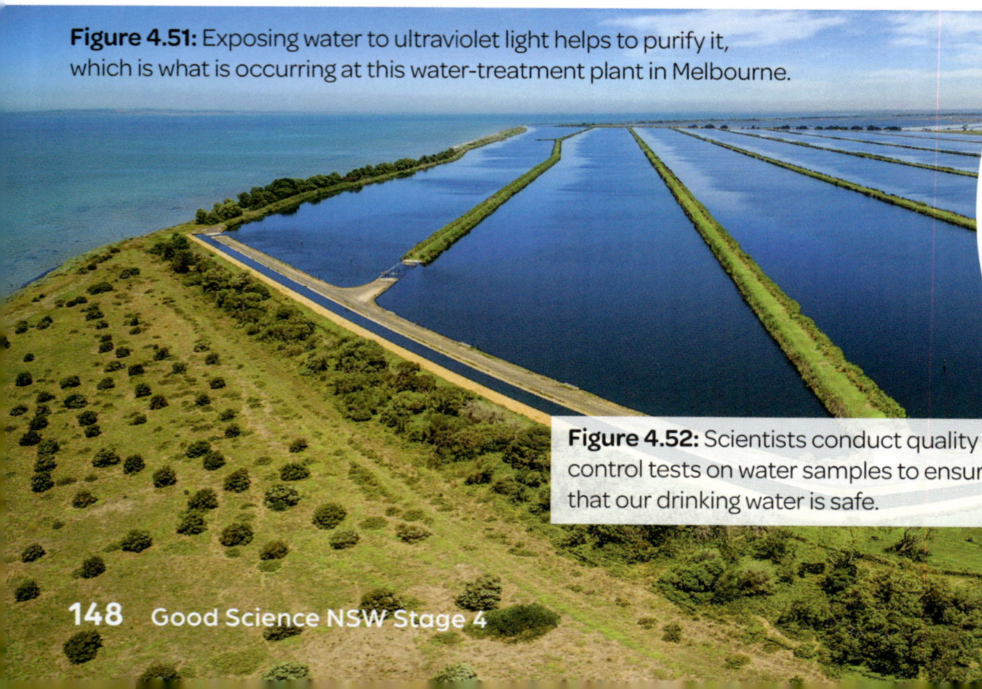

Figure 4.51: Exposing water to ultraviolet light helps to purify it, which is what is occurring at this water-treatment plant in Melbourne.

Figure 4.52: Scientists conduct quality control tests on water samples to ensure that our drinking water is safe.

Demonstrate your understanding

3	4	5	
Explain how the process of ultrafiltration works when purifying a water sample.	Compare the properties of recycled water used in home showers and toilets to the properties of water collected from stormwater runoff.	Analyse the use of recycled water as an option to help protect New South Wales' water supplies during droughts.	
A scientist wants to test the clarity of a water sample that had been exposed to different chemicals. Identify: • the independent variable. • the dependent variable. • three control variables.	Construct a scientific aim that could be used to address the question: How does exposure to ultraviolet light (UV) affect the microorganisms in water?	Consider the test in Step 3 and the question in Step 4: **a** Write a list of the equipment you would need to conduct each of these investigations. **b** Write the method for one of the investigations.	Science how-to p. 323
To test water samples, they have to be collected. Explain three safety procedures you would implement when collecting water samples.	Explain how scientists who conduct tests on purified water ensure that the data they collect is accurate and reliable.	Recall your investigation method from Step 5 in the 'Planning investigations' row. **a** Construct a results table, with units, that you could use to record the data. **b** How would you ensure that this data is accurate?	Science how-to p. 326
Explain the link between issues with water purification processes and community health.	The pH (presence of acid) of drinking water must be between 6.5 and 8.5 (i.e. close to neutral). How could the problem of a water sample having too much acid be solved?	One problem scientists have faced with water purification is controlling bacteria levels. To solve this problem, water is exposed to high levels of UV light and treated with chlorine. Evaluate the success of these solutions, giving evidence from New South Wales' water resources.	Science how-to p. 339

Purifying water in New South Wales takes recycling to the next level

The Australian climate means that New South Wales experiences drought, which can jeopardise the supply of water. Sydney Water wants to build infrastructure to help prevent this from happening, by enabling us to use purified, recycled water in our homes.

This would mean that water used in our homes – even water from our showers and toilets – would be diverted to water-recovery facilities, then moved to special purification plants, before it is directed back into the dams that hold our water supplies. Recycled water would be treated using normal water-treatment processes, but would also undergo ultrafiltration.

This means that all water would be passed through extremely fine filters and membranes, before special chemical processes are performed that cause new compounds to be formed in the water that are not harmful to humans.

All water would still have the usual chemical components like chlorine and fluoride added, so that it could be classed as being safe for consumption. The final water solution would be tested by scientists before being released back into water systems.

Focus area 5

5.0 Living systems

In multicellular organisms, cells make up organs that form complex systems. These living systems work together so that animals and plants can perform important functions, such as gaining energy and transporting it to new places, and removing waste products that might otherwise cause harm. Ecosystems behave in a similar way, but instead of internal organs forming complex systems, the organisms themselves and their surrounding environment are what make up the system. In this chapter, we will look at how living systems function normally, as well as how they can be affected by a variety of factors in and out of the control of the organism.

Learning Ladder

The Learning Ladder contains the scientific content and processes you will learn in this chapter. Each area has five levels of progression. To move to higher levels, you need to practise activities at the earlier levels. This will help you develop the ability to complete tasks that are more complex.

Steps in progression	Content: Living systems	Working scientifically processes: Questioning and predicting	Working scientifically processes: Processing data and information	Working scientifically processes: Communicating	Steps in progression
5	I can predict the impact of changes to a living system.	I can formulate testable questions and predictions, considering variables and controls.	I can process and interpret quantitative and qualitative data from a range of sources.	I can present scientific findings using appropriate conventions for specific audiences.	5
4	I can compare features of different living systems.	I can make predictions to explain observations or phenomena.	I can collect and present data in a range of appropriate formats.	I can construct a range of appropriate scientific presentations based on first-hand and second-hand information and data.	4
3	I can explain the structure and function of a living system.	I can construct questions to investigate scientific concepts or problems.	I can construct appropriate graphs with headings and units for data.	I can use digital technologies to organise and present information and data.	3
2	I can describe the features of a living system.	I can select questions to investigate scientific concepts or problems.	I can construct appropriate tables with headings and units for data.	I can select appropriate ways to communicate information.	2
1	I can identify a living system.	I can make predictions based on prior knowledge and observations.	I can identify data from graphs, tables and digital sources.	I can recognise scientific information.	1

Figure 5.1: These flowers from *Eucalyptus calophylla* are a living system: flowers contain the reproductive parts of this plant.

5·1 ▸ Body systems basics

Learning intention

At the end of this lesson, I will be able to explain the interrelationships between cells, tissues and organs.

Key terms

body system: two or more organs that are connected and working together

cell: the smallest functional unit of an organism

differentiate: to change to have a particular function

organ: a structure made up of two or more tissues that has a specific function

organism: an individual animal, plant or other living thing

tissue: a group of cells with a similar structure and function

Content group: Body systems

In **organisms** made of more than one **cell**, cells with similar functions group together to form tissues, organs and body systems.

Cells can perform different jobs

In unicellular organisms, the single cell must carry out all the major life functions. In multicellular organisms, the cells share the workload, and different cells have different functions.

Before an organism starts to develop, its cells are not specialised. These cells are called stem cells. As an organism grows, the cells **differentiate** and specialise to carry out different functions.

Tissue is a group of specialised cells

Cells that specialise in the same function group together to form **tissue**. There are four main types of tissue in animals:

- **Epithelial tissue** forms the skin, as well as the body's inner linings.
- **Connective tissue** transports substances (such as nutrients) to where they are needed.
- **Muscle tissue** contracts and relaxes to carry out different functions.
- **Nerve tissue** transmits information between the brain and other organs.

Cells work together as tissues, organs and systems

Different tissues working together are **organs**. Two or more organs that are connected and working together form a **body system**. The organs and tissues in each system are specialised to perform specific roles.

Figure 5.2: Muscles are made up of cells that are specialised to contract and relax.

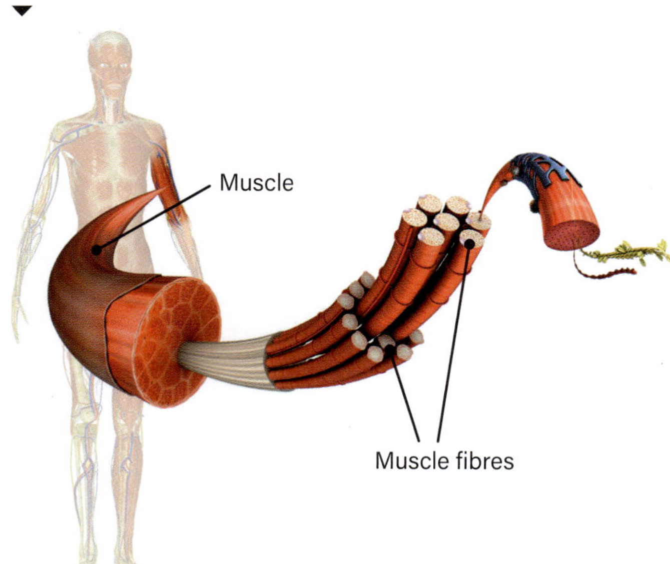

Organs perform an organism's main functions

An organ is a group of two or more tissues. The brain, heart, kidneys, liver and lungs are the main organs of the human body. Each organ has a specific function; Figure 5.3 shows how one is formed.

Each organ has a different structure. For example, in the heart, epithelial tissue lines the blood vessels. Connective tissue forms the valves of the heart and is in the walls of the blood vessels. Nerve tissue inside the heart controls the beating of the heart muscle, which pumps the blood out of the heart. All these tissues work together to keep the heart working efficiently.

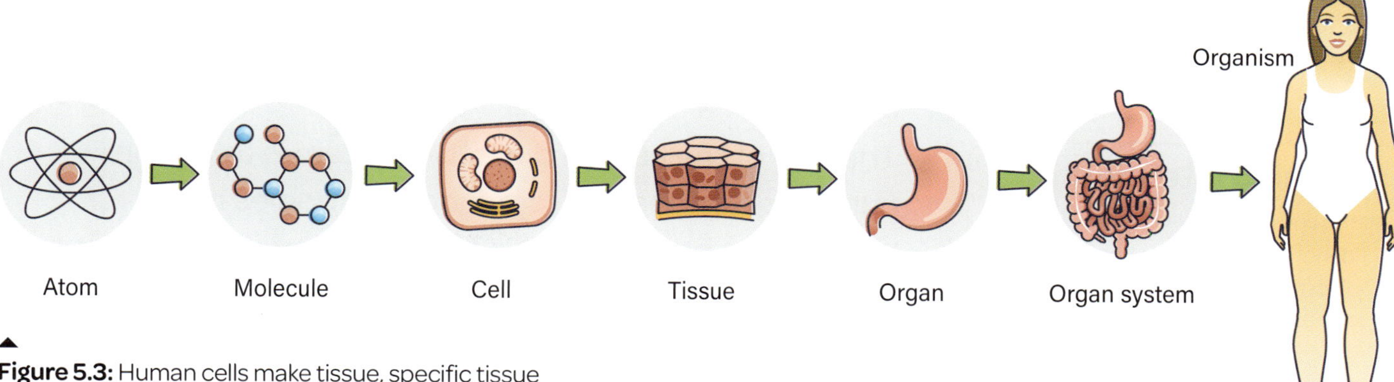

Figure 5.3: Human cells make tissue, specific tissue makes organs, and sets of organs make body systems.

Organs work together in systems

An organ is not useful on its own. In an organism, each organ is part of a body system, along with other necessary tissue such as blood. An organism's different body systems work together so it can function. The human body has several different body systems, each with its own function, as shown in Table 5.1.

Table 5.1: Some of the systems in the human body

Body system	Function	Major organs and tissues
Cardiovascular and circulatory	Transports oxygen and nutrients to cells	Heart, blood
Nervous	Transmits nerve impulses between parts of the body	Nerves, brain, spinal cord
Digestive	Breaks down food so nutrients can be absorbed	Oesophagus, stomach, liver, large and small intestines
Respiratory	Allows exchange of gases	Lungs, trachea, larynx, nasal passages
Excretory and urinary	Removes waste formed from digestion	Kidneys, ureters, urethra, bladder
Reproductive	Produces gametes (sex cells) and sex hormones	Female: ovaries, vagina, uterus, fallopian tubes Male: penis, testes, seminal glands
Integumentary system	Provides a barrier, helps regulate temperature, detects stimuli	Skin

Learning Ladder

Living systems

1. Order these terms from the smallest to the largest structure: tissue, cell, system, organ.
2. Classify each of the following as a tissue or an organ:
   - **a** lungs
   - **b** heart
   - **c** blood
   - **d** muscle
   - **e** brain
   - **f** skin.
3. Explain the difference between an organ and a tissue. Use examples in your explanation.

Communicating

see page 343

1. Identify four terms that could be used to describe tissue in multicellular organisms.
2. Describe some ways you could show or represent the tissues in the heart to communicate how it works.
3. Figure 5.2 is a diagram of cells that make up muscles. Outline the advantages and disadvantages of using an image like this one to present information about body systems.
4. Construct a diagram like Figure 5.2 to show the organisation of cells in the nervous system. Include nerve cells, neural (brain) tissue, the brain and the nervous system (brain and spinal cord).

In context

Conduct research online to answer these questions.

a Construct a list of the organs that are made up of each of the four types of tissues (epithelial, connective, muscle, nerve tissue).

b Discuss what each list of tissues or organs has in common in relation to structure and function.

Success criteria

- I can explain the interrelationships between cells, tissues and organs.

5·2 ▸ The digestive system

Learning intention

At the end of this lesson, I will be able to:

- identify the role of the human digestive system and name the major organs in this system
- describe how the structures and specialised features and cells of the digestive system enable it to carry out its functions.

Key terms

bile: a salty solution stored in the gallbladder, which aids digestion in the upper small intestine

chemical digestion: breaking down food in a chemical reaction that forms new molecules

chyme: a soupy mixture of partially digested food in the stomach and small intestine

digestion: the physical and chemical processes that break down food in the body

enzyme: a substance that enables or speeds up a chemical reaction

mechanical digestion: physically breaking down food into smaller pieces

peristalsis: the involuntary muscle action that pushes food through the digestive tract

villi: the tiny finger-like projections that line the walls of the small intestine

Content group: Body systems

The role of the human digestive system is to obtain nutrients and energy from food. Our bodies need nutrients to be delivered in a certain way so our cells can absorb them. The digestive system breaks down food both physically and chemically to allow nutrients to be absorbed into the body.

The digestive system breaks down food

Humans need to consume food to obtain nutrients. Nutrients are essential for energy production, growth, tissue repair and all other cellular processes. However, food cannot move directly into the cells. It first needs to be broken down into smaller molecules during **digestion**.

Digestion happens in two ways:

- **mechanical digestion**: food is physically broken down into smaller pieces (for example, by chewing) but the food material has not changed into new substances
- **chemical digestion**: digestive **enzymes** trigger chemical reactions, which break down large food molecules into new, smaller molecules that can be absorbed by the body's cells.

In the mouth, food is physically broken into small pieces by the teeth and mixed with saliva, which contains enzymes that chemically break down carbohydrates into simple sugars. When you swallow, this mix of food and saliva moves down into the oesophagus and then to the stomach.

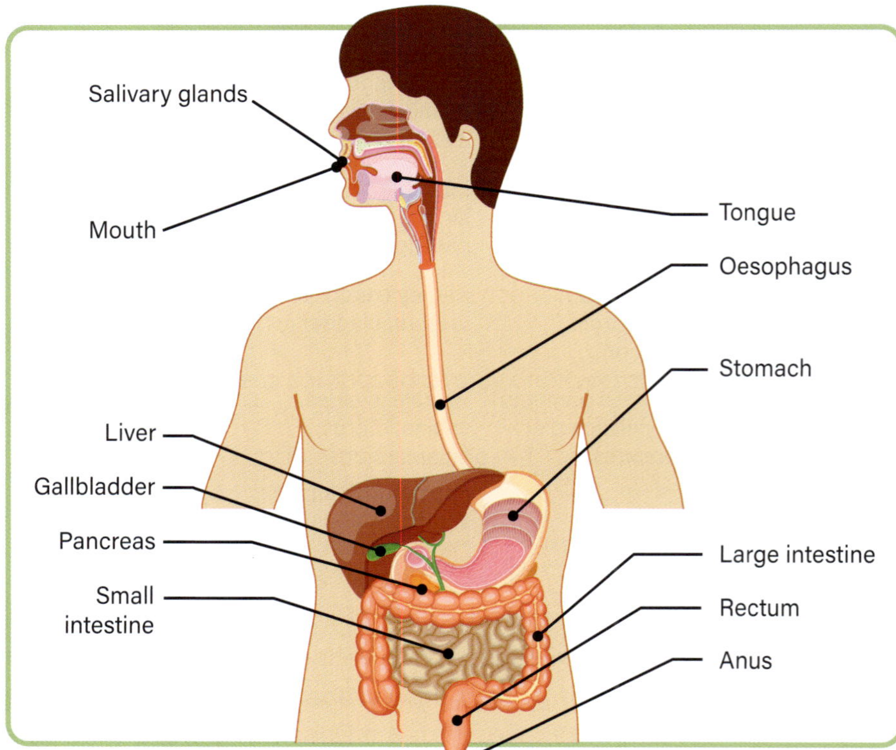

Figure 5.4: The process of digestion takes about 24 hours and involves many processes and organs.

Figure 5.5: The digestive system breaks down food to get the essential nutrients.

Within the stomach, the food is broken down further and combined with gastric juices that contain enzymes and hydrochloric acid. This soupy mixture (called **chyme**) is slowly released into the top section of the small intestine. Here, the chyme mixes with **bile** – a salty solution stored in the gallbladder – and pancreatic juices, which both contain more enzymes to chemically break down and neutralise the chyme.

As this mixture moves through the small intestine, the nutrients begin to be absorbed and sent around the body via other body systems. A lot of what we eat is left undigested as it passes through the small intestine; some nutrients are absorbed, but much of the material remains. This moves into the large intestine, also known as the bowel, where bacteria act on it, releasing important vitamins and minerals that can be absorbed and producing gas as a by-product.

Although water can be absorbed in the stomach and small intestine, the remaining absorption of water happens in the large intestine. The remaining solid waste collects in the rectum and is finally expelled as faeces.

The action that keeps things moving down your digestive tract is called **peristalsis**. Peristalsis is an involuntary muscular movement that creates wave-like contractions through your digestive tract. Once you have swallowed your food, peristalsis keeps it moving down your oesophagus and into your stomach. When your stomach 'gurgles' after eating, this is peristalsis accompanied by abdominal noises as food is squeezed from your stomach into the small intestine.

The epiglottis is a specialised component of the digestive system

The epiglottis is a flap of tissue in the throat (see Figure 5.6). The epiglottis covers the trachea (also called the airway) when we swallow so that food can pass into the oesophagus. Although it is not directly involved with the breakdown or absorption of food, without the epiglottis, food would be able to enter and block the airway.

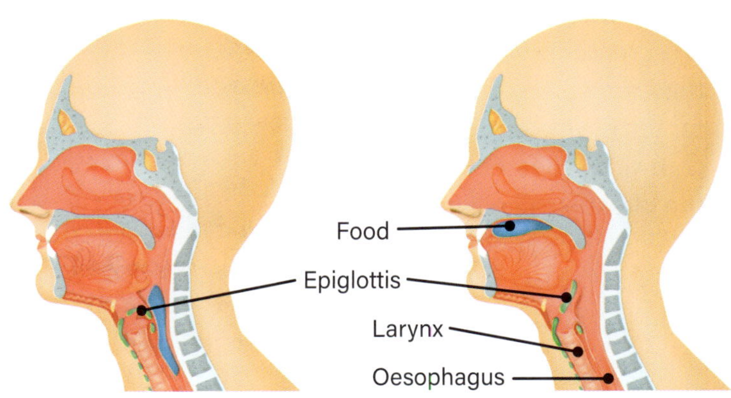

Figure 5.6: The epiglottis shifts to cover the trachea when we swallow so that food can pass into the oesophagus rather than block our airway.

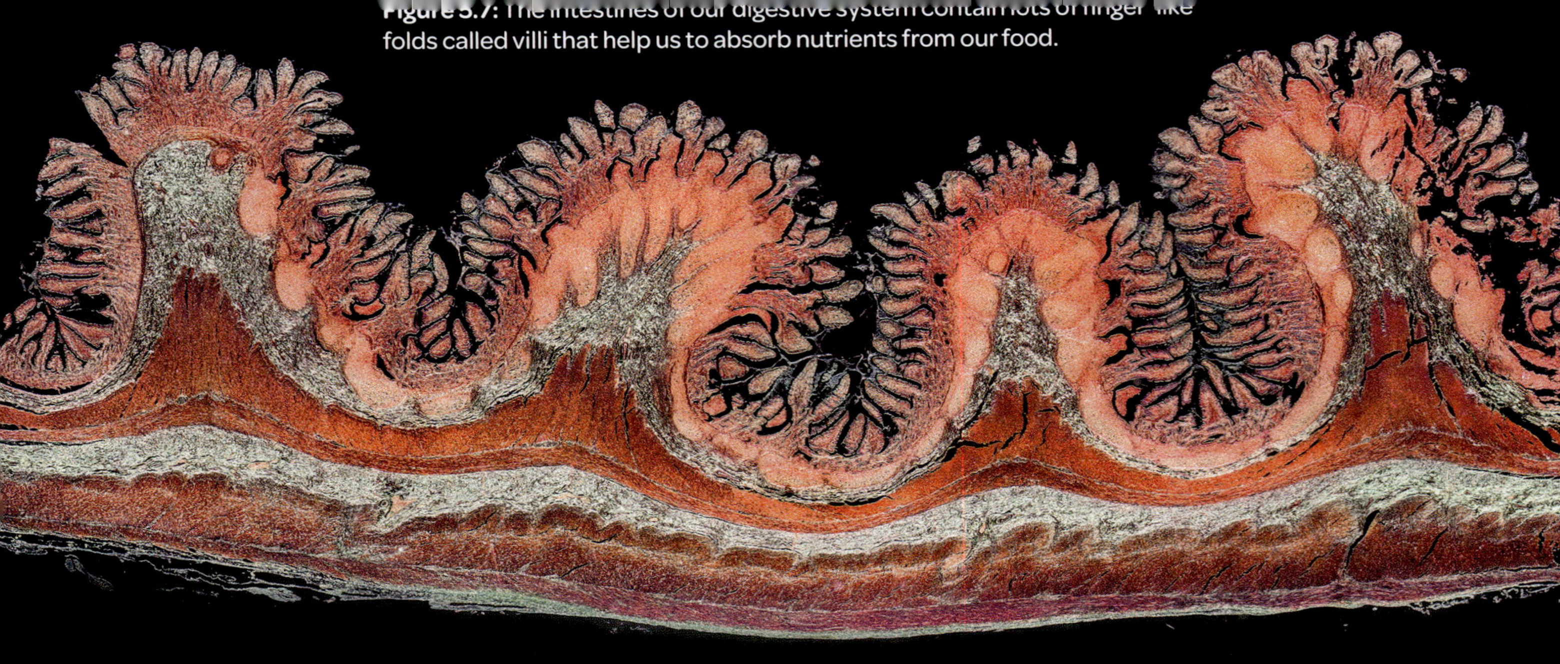

Figure 5.7: The intestines of our digestive system contain lots of finger-like folds called villi that help us to absorb nutrients from our food.

Specialised cells in the digestive system help it carry out its functions

The small intestine is lined with tiny projections called **villi**. Villi have specialised cells that help absorb nutrients during digestion. Villi contain capillaries, which are the smallest types of blood vessels. These capillaries allow nutrient molecules to pass through and to be dispersed in the body where needed. The shape of the villi maximises the amount of nutrients that can be absorbed, because the finger-like projections increase the surface area of the small intestine wall (see Figure 5.7). The greater the surface area, the greater the opportunity for the chyme mixture to come into contact with the intestinal capillaries.

Reflux is a disorder of the digestive system

Reflux is a disorder of the digestive system that can result in a person feeling pain or a burning sensation in their upper abdomen or chest. The condition is caused by stomach acid leaking from the stomach and moving up into the oesophagus. Antacids are bases that neutralise or cancel out the acid, and help stop the symptoms of reflux (see Figure 5.9).

Figure 5.8: This pH scale shows the acidity and basicity of different foods and substances.

▲ **Figure 5.9:** Antacids such as milk of magnesia can ease the symptoms of reflux by neutralising the acidic fluids in the oesophagus. Stomach acid has a pH of 1 and turns universal indicator red. Adding milk of magnesia causes the colour to change to green.

Learning Ladder

Living systems

1. **a** Summarise the purpose of the digestive systems in one sentence.
   b Identify the main organs that make up the digestive system.
2. Construct a step-by-step flowchart to describe the process of digestion, from the moment you take a bite of food to when you expel faeces.
3. Explain the difference between mechanical and chemical digestion.
4. 'Bolus' is food that has been chewed and swallowed into the oesophagus. Compare bolus and chyme by identifying their similarities and differences.
5. Discuss how a person would be affected if the walls of their small intestine were flat due to the absence of villi. Suggest how the person could modify their lifestyle to adjust to the lack of villi.

Questioning and predicting

see page 319

1. Lactase is the enzyme required to break down lactose in milk. Predict what would happen if a person were unable to produce lactase in their body.
2. Identify which of the following is the *most* suitable question to investigate a scientific concept.
   A How long does it take a person to digest a sandwich?
   B How do amylase enzymes in saliva start to break down complex carbohydrates in the mouth?
   C How does the amount of water you drink with a meal affect how long it takes to digest?
3. Construct a scientific question to investigate how the surface area of digestive structures (such as villi) affect the rate of chemical reactions, using marble chips and hydrochloric acid to simulate the reaction.
4. Propose a hypothesis that predicts the outcome of the investigation question constructed in Question 3. Use an 'If ... then' statement to structure your hypothesis.
5. Propose a scientific question to investigate the digestive systems of a carnivore (eats meat, such as a wolf), a herbivore (eats plants, such as a cow) and an omnivore (eats meat and plants, such as a human). (Hint: Consider how their diets might relate to the different structures in the digestive systems of these animals.)

In context

Select one of the following organs involved in the digestive system: gallbladder, pancreas, appendix, spleen. Research how the removal of the organ would impact the overall functioning of the digestive system. Share your findings with a classmate. Be sure to include interesting facts or real medical cases to engage your peer.

Success criteria

- I can identify the role of the human digestive system and name the major organs that make up the digestive system.
- I can describe how the specialised features and cells of the digestive system enable it to carry out its functions.

5·3 ▸ The respiratory system

Learning intention

At the end of this lesson, I will be able to:

- identify the role of the human respiratory system and name the major organs in this system
- describe how the structures and specialised features and cells of the respiratory system enable it to carry out its functions.

Key terms

aerobic respiration: the process of turning glucose into energy where the cells take in oxygen and release carbon dioxide

cellular respiration: the process of turning glucose into energy for cells to use

diaphragm: the band of muscle under the lungs that enables the physical action of inhalation and exhalation

diffuse: to move from an area of high concentration to an area of low concentration

gas exchange: the exchange of oxygen and carbon dioxide between an organism and the environment

hiccups: when the diaphragm contracts involuntarily and repeatedly to produce a vocal 'hic' sound

Investigation 5.3

Breathing rate and exercise, p.438

Content group: Body systems

The role of the human respiratory system is to bring oxygen into the body and remove waste gases.

The respiratory system is responsible for gas exchange

The major parts of the human respiratory system are the nose, mouth, trachea, bronchi and lungs. The lungs contain smaller, more specialised parts such as the bronchioles and alveoli (see Figure 5.10).

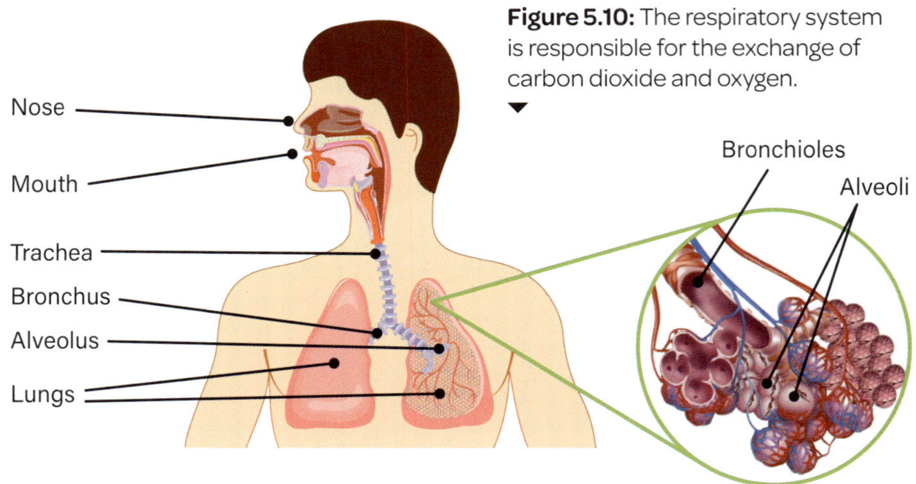

Figure 5.10: The respiratory system is responsible for the exchange of carbon dioxide and oxygen.

The human respiratory system gathers and processes oxygen. You breathe in (inhale) oxygen with air, and release carbon dioxide and water vapour when you breathe out (exhale). This **gas exchange** is only possible because of the special structures of the respiratory system.

During inhalation, air enters the nostrils or mouth and moves into the trachea (windpipe). It then travels into two branching bronchi – one for each lung – before passing into smaller and smaller passages called bronchioles. At the end of the bronchioles are clusters of tiny sacs called alveoli. Each alveolus is moist, thin and surrounded by many microscopic blood vessels called capillaries. Oxygen from the air **diffuses** through the thin cell membranes of the alveoli and into the capillaries. From there, it is transported by the blood to cells throughout the body for **aerobic respiration**.

At the same time, carbon dioxide and water that collect in the blood as products of **cellular respiration** are taken back to the lungs for removal. Carbon dioxide and water vapour diffuse back into the air within the alveoli. When you exhale, the air in the alveoli is swept back through the lungs, taking the reverse route to exit the respiratory system.

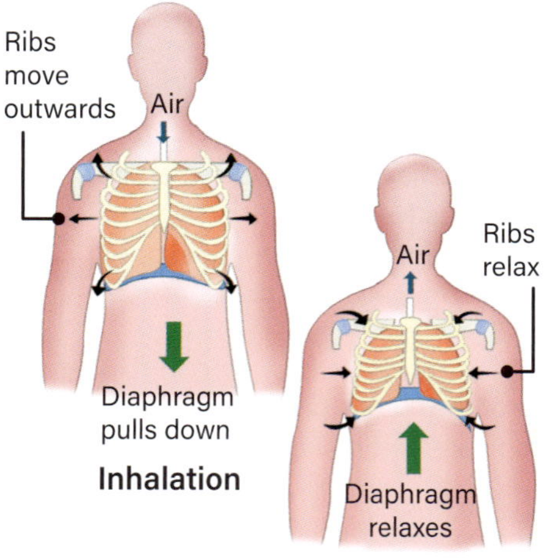

Figure 5.11: When the diaphragm contracts and relaxes, the chest expands and constricts.

The diaphragm is a specialised part of the respiratory system

The **diaphragm**, a band of specialised muscular and connective tissue, separates the chest cavity from the abdomen cavity. It plays a crucial role in the respiratory system despite not being directly involved with gas exchange. As shown in Figure 5.11, when you inhale, your diaphragm contracts to move downwards. This causes your ribs to move outwards, which increases the space in your chest cavity. This creates suction in your airway, drawing air into your lungs from outside your body through your nose or mouth, trachea and bronchi. When your diaphragm relaxes, the opposite occurs. It returns to its upward position, pushing air out of the lungs, meaning you exhale.

When the diaphragm suddenly contracts without warning, it closes off the vocal cords, producing a 'hic' sound. When this occurs repeatedly over a relatively short amount of time, we call it the **hiccups**. While usually harmless, persistent hiccups could signal an underlying medical issue.

Asthma is a disorder of the respiratory system

Asthma occurs when a person's airways become inflamed and narrow, making it difficult to breathe.

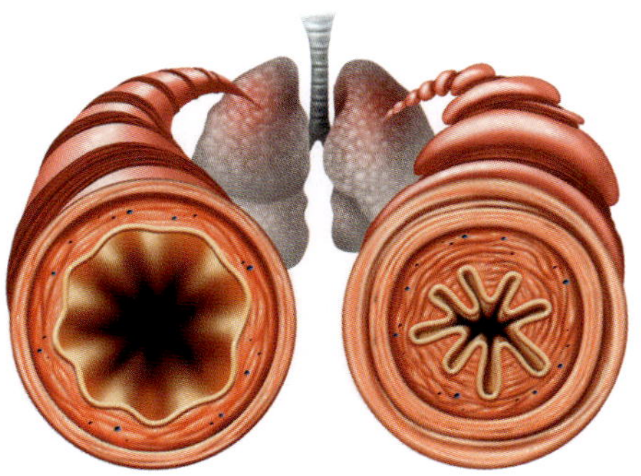

Figure 5.12: A healthy bronchial tube (left) and an unhealthy bronchial tube (right). This shows how the airway of a person with asthma can be restricted.

Symptoms of asthma include wheezing, coughing, tightness in the chest and shortness of breath. It is like trying to breathe through a thin straw. Many things can trigger an asthma flare-up, such as allergens, stress, illness or exercise. Medical treatment such as a quick-relief inhaler dilates the bronchi and eases the airway restriction.

Learning Ladder

Living systems

1. **a** Summarise the purpose of the respiratory system.
   b Identify the main organs of the respiratory system.
2. **a** Describe the role of the alveoli in the respiratory system.
   b Suggest the advantage of having thousands of tiny alveoli in the lungs.
3. Explain why carbon dioxide in blood needs to be exchanged with oxygen from the air.
4. **a** Evaluate the following statement: Breathing and respiration are pretty much the same thing.
   b Do you agree with the statement? Explain your answer.
5. Consider a person who is paralysed from the chest down but can still move their head, neck and face. Discuss how this might affect the functioning of their respiratory system. Use as many key terms from this section as possible.

Processing data and information *see page 329*

1. Assume a person exchanges half a litre of air per breath at a rate of 12 breaths per minute. Identify the volume of gas that is exchanged in a full day. Show your workings.
2. You are asked to investigate the difference in resting breathing rates between children and adults over a 15-minute period. Construct a table with appropriate headings that could be used to collect the necessary data.
3. Fill in your table from Question 2 with predicted data. Construct a graph to represent the 'results'. Follow the scientific conventions (see the Science how-to section on page 368).
4. Represent the data from Question 3 in a different format (for example, a chart or an infographic) to clearly communicate the results.
5. Select two lung diseases from cystic fibrosis, asthma, bronchitis, tuberculosis and pneumonia. Research the diseases to gather a variety of related statistics. Create an infographic poster to display the data.

In context

Find out about the structures of the respiratory systems in other animals such as fish and amphibians. Construct a table to compare them with those of a mammal. To extend yourself, create an infographic to display your findings.

Success criteria

- I can identify the role of the human respiratory system and name its major organs.
- I can describe how the specialised features and cells of the respiratory system enable it to carry out its functions.

5.4 ▸ The circulatory system

Learning intention

At the end of this lesson, I will be able to:

- identify the role of the human circulatory system and name the major organs in this system
- describe how the structures and specialised cells of the circulatory system enable it to carry out its functions.

Key terms

aorta: the main artery that leads from the heart to the rest of the body

artery: a type of blood vessel that carries blood away from the heart

atria: filling chambers of the heart

blood vessel: a tube such as a vein or artery that carries blood in the body

capillary: the smallest type of blood vessels

cardiovascular: related to the heart (cardio) and blood vessels (vascular)

vein: a type of blood vessel that returns blood to the heart

vena cava (plural **venae cavae**): the main vein that leads to the heart from the rest of the body

ventricles: pumping chambers of the heart

Investigation 5.4

Dissecting a heart, p.440

Content group: Body systems

The circulatory system includes an intricate network of vessels that allow blood to be continuously pumped throughout all parts of the body.

The circulatory system transports substances around the body

Your heart, blood vessels and blood make up your circulatory system. This system delivers oxygen, nutrients and other substances to every tissue in your body. It also helps your body to remove waste products, such as carbon dioxide.

The three main types of **blood vessels** are arteries, veins and capillaries (see Figure 5.13). **Arteries** take blood from the heart to the body. They have thick, muscular walls that expand and contract as the heart beats. **Veins** take blood from the body back to the heart. Their walls are not as thick as those of arteries, and they contain valves to keep the blood flowing in the right direction. **Capillaries** are the smallest blood vessels. They take blood to the body's cells. The walls of capillaries are one cell thick, which allows nutrients and oxygen to move easily into the cells from the blood, and waste products to move out of the cells into the blood.

The heart and blood vessels pump oxygenated blood from the lungs to the body's cells and deoxygenated blood from the body back to the lungs. The heart is made up of four chambers: two **atria**, which are 'filling' chambers; and two **ventricles**, which are 'pumping' chambers. Blood enters the atria and is pumped out through the ventricles (see Figure 5.14).

The process is as follows:

1. At the lungs, the blood picks up oxygen and flows through the pulmonary veins into the left atrium of the heart.
2. The oxygenated blood moves down into the left ventricle and is pumped out of the heart through the **aorta**, which is the main artery that leads to the rest of the body.
3. Once the blood has delivered oxygen to cells all over the body and collected waste material, it travels back to the heart.
4. The deoxygenated blood enters the right atrium of the heart from the superior (top) and inferior (bottom) **venae cavae**, which are the main veins that lead to the heart from the rest of the body.
5. The blood then flows down into the right ventricle, before being pumped back to the lungs for gas exchange via the pulmonary artery.

This is a cyclical process, so there is no start or finish.

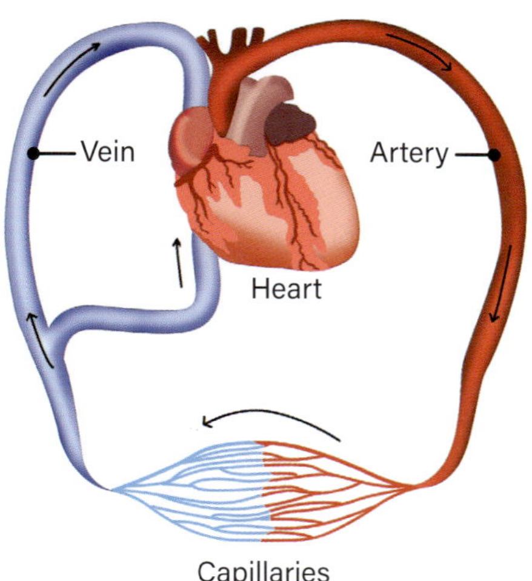

Figure 5.13: Arteries carry blood away from the heart and veins carry blood towards it.

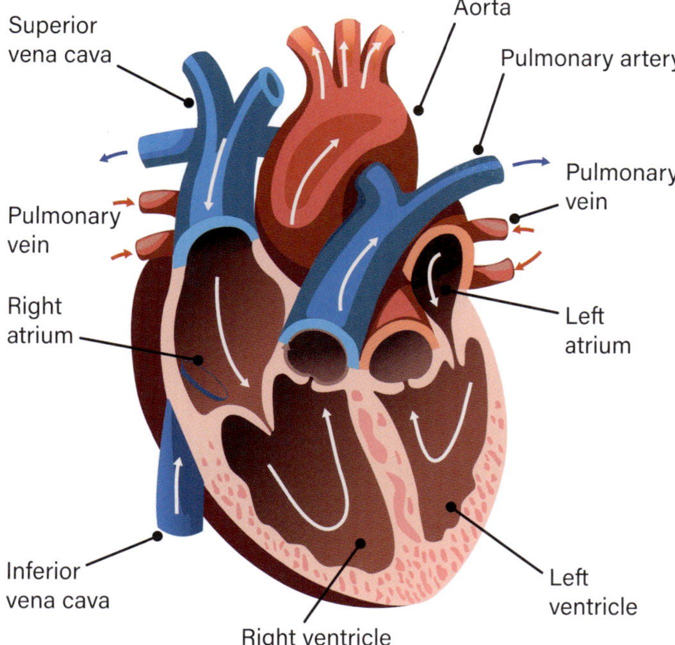

Figure 5.14: The heart pumps blood in specific directions through a complex series of vessels, chambers and valves. The blue vessels contain deoxygenated blood and the red vessels contain oxygen-rich blood. Viewpoint: inside the body looking out.

Heart valves are specialised parts of the circulatory system

The heart is the hardest working muscle in the body, with an adult's heart beating more than 115 000 times a day. **Cardiovascular** health is reliant on the heart's valves and tendons working efficiently. Heart valves allow blood to flow in one direction during the heartbeat cycle. They are highly organised connective tissue structures, consisting of several types of specialised cells.

The mitral valve, also called the bicuspid valve, regulates flow from the left atrium into the left ventricle. The aortic valve opens through to the aorta. When deoxygenated blood returns to the right atrium, the tricuspid valve lets it through to the right ventricle. Finally, the pulmonary valve promotes blood flow back to the lungs to be reoxygenated.

The mitral and tricuspid valves are assisted by a series of tough yet flexible tendons called chordae tendineae, commonly known as 'heart strings'. If these valves or tendons are damaged or malfunctioning, this can cause serious conditions such as a heart murmur, valve stenosis or mitral regurgitation, each of which can lead to heart failure.

Learning Ladder

Living systems

1. State the differences between a vein and an artery.
2. Construct a diagram of the circulatory system.
   - **a** Annotate the diagram to describe the parts of the circulatory system.
   - **b** Label the directional flow of blood throughout all structures in your diagram, including whether the blood is oxygenated or not.
3. Explain the importance of heart valves and heart strings to the overall function of the circulatory system.
4. Compare and contrast the functions of the mitral valves and the aortic valves.
5. **a** Identify the advantage of having separate right and left chambers of the heart.
   - **b** Suggest what problems would arise if they were not separated.

Questioning and predicting

see page 319

1. Predict the effect of a blockage of a major artery as a result of high cholesterol. In your answer, refer to the purpose of the circulatory system.
2. Identify which of the following is the *most* suitable question to investigate a scientific concept.
   - **A** How many litres of blood can the heart pump in a day?
   - **B** How does fitness level affect the time it takes for heart rate to return to normal after 10 minutes of high-intensity exercise?
   - **C** How many times does a person's heart beat a day?
3. Construct a scientific question to investigate the resting heart rates of humans, cats, sheep and horses.

In context

Conduct a search for the common misconception that deoxygenated blood in humans is blue. Write an evidence-based argument to convince someone that human blood is never blue. To extend yourself, investigate animals that do have blue blood, and find out why.

Success criteria

- I can identify the main structures of the heart and differentiate between the main types of blood vessels.
- I can describe, using annotated drawings, how the circulatory system transports and delivers oxygen to all parts of the body.

5·5 ▸ The excretory system

Learning intention

At the end of this lesson, I will be able to:

- identify the role of the human excretory system and name the major organs in this system
- describe how the structures and specialised cells of the excretory system enable it to carry out its functions.

Key terms

excretion: the elimination of cellular waste from the body through urine

glomerulus: a bundle of capillaries; the filtering unit within each nephron

nephron: a microscopic filtration structure in the kidneys that removes waste and excess water

renal: an adjective that means 'related to the kidney'

tubule: a small tube made of epithelial cells

urea: a product of the chemical breakdown of food, specifically proteins

Content group: Body systems

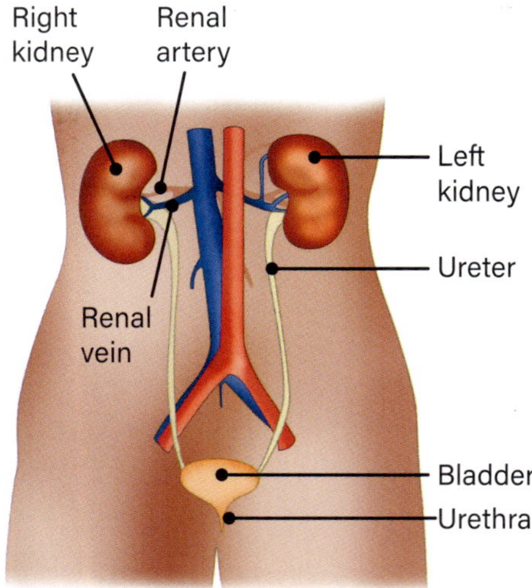

▲ **Figure 5.15:** The excretory system is made up of the kidneys, ureters, bladder and urethra. Viewpoint: inside the body looking out.

The processes occurring in our bodies produce wastes. If they are not removed, wastes can build up and cause harm. The body removes some forms of waste by exhaling or through sweat and faeces. Excretion refers mainly to the elimination of cell waste from the body through urine.

The excretory system removes waste from the body

Many specialised organs are involved in **excretion**, or the removal of waste, including the skin, lungs and liver. Specific organs also produce and eliminate urine. Four main organs make up the excretory system:

- **Kidneys** filter waste from the blood in the form of urine.
- **Ureters** carry urine from the kidneys to the bladder.
- The **bladder** is a muscular reservoir that stores urine.
- The **urethra** carries urine from the bladder out of the body .

The kidneys are two bean-shaped organs at the back and upper portion of the abdomen. The kidneys filter harmful wastes from the blood and ensure that the body retains the right amount of water. One major waste product is **urea**, a type of nitrogenous (containing nitrogen) waste produced in the liver. Urea is a by-product of reactions that break down proteins, which are large food molecules that contain nitrogen. Urine contains urea and any other substances that have been removed from the blood. Urine leaves the kidneys through the ureters and is stored in the bladder until it is expelled through the urethra. The blood that has been filtered then returns to the circulatory system through the renal vein.

Nephrons are structures within the kidneys

When blood moves into the kidneys through the **renal** arteries, it meets several pyramid-shaped structures made up of microscopic nephrons. **Nephrons** are filtration **tubules** that remove waste products and excess water. Each kidney contains about a million nephrons.

A nephron is made up of several types of highly specialised cells. Each type carries out a specific part of the filtration process. The main filtering unit within the nephron is a bundle of capillaries called the **glomerulus**, enclosed in a sac called the Bowman's capsule. From there, a network of tubular structures absorb and reabsorb components of the clean blood, ensuring that substances the body needs, such as glucose, remain in the blood. Collecting ducts from each nephron combine the waste to form urine (see Figure 5.16).

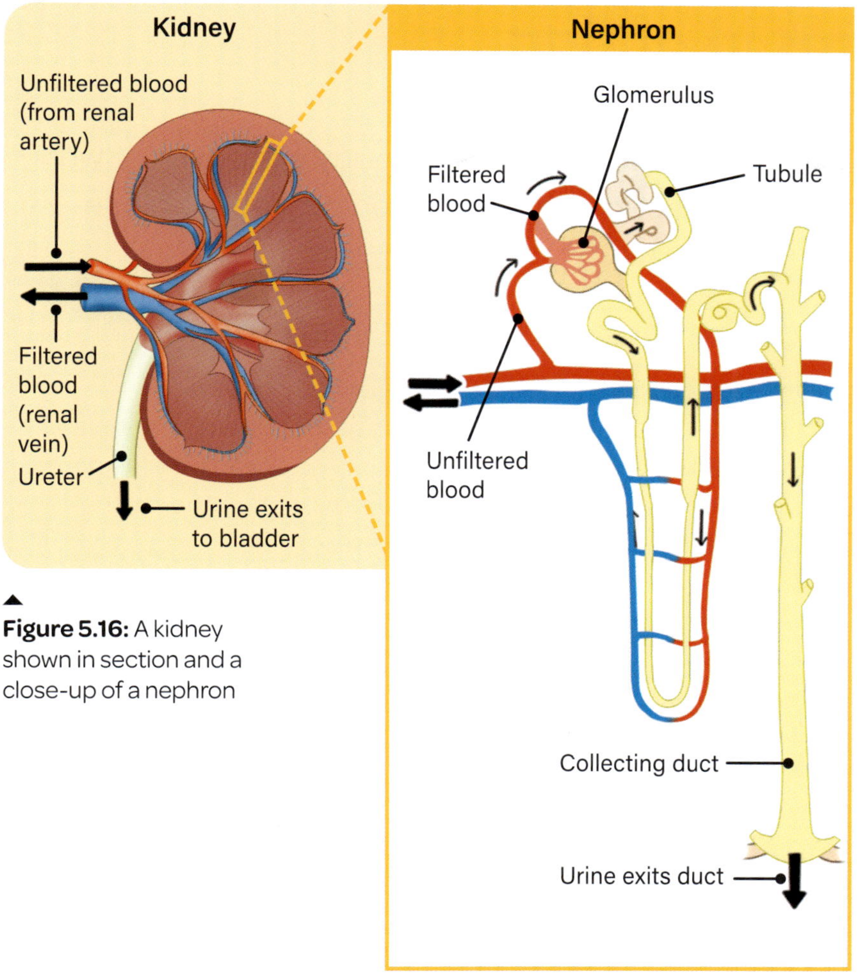

Figure 5.16: A kidney shown in section and a close-up of a nephron

Kidney disease is a disorder of the excretory system

Renal disease is any disorder of the kidneys. If the kidneys do not work properly, toxins build up, causing harm.

Kidney stones are like tiny rocks that form in the kidney, made up of minerals and salts from the urine. Usually, they are passed through urination, but large stones can get stuck and cause problems. They can be very painful and block urine flow, which can cause infection and lead to renal failure. It is possible to break the stones into smaller pieces so they can be passed into the ureter. Otherwise, they can be removed surgically. To avoid kidney stones, it is important to drink plenty of water and eat a healthy diet.

If the kidneys fail, the blood can be cleaned by dialysis, a process that mimics filtration by the nephrons. Eventually, though, it may be necessary to have a kidney transplant.

Learning Ladder

Living systems

1. Identify the role of the kidneys, ureters, bladder and urethra in removing waste from the body.
2. Describe the role of the nephron in excretion.
3. Explain how the excretory system ensures the human body can function effectively.
4. Defecation is the process of expelling faeces from the rectum. Evaluate the following statement: Defecation is a type of excretion because it removes food waste from the digestive system. Do you agree with the statement? Explain your answer.
5. People can live a normal life with only one kidney.
   - **a** Outline how this is possible.
   - **b** Propose some things you might recommend to someone with only one kidney.

Processing data and information *see page 329*

Healthy kidneys can filter about 120 mL of blood per minute.

1. Propose what it might mean if a person could only filter 50 mL of blood per minute.
2. Construct a table to show how much blood a healthy person's kidneys could filter in one hour in 10-minute intervals, as compared to the person described in Question 1.
3. Construct a graph to represent the data in the table in Question 2.
4. Conduct research to gather a variety of statistics related to kidney stones, kidney disease, renal failure, dialysis and other conditions of the kidney. Create an infographic poster to display the data.

In context

Investigate the cause, symptoms and treatments for urinary tract infections, and how explain they can be prevented.

Success criteria

- I can identify the major organs of the excretory system.
- I can use annotated drawings to describe how the excretory system rids the body of waste.
- I can explain how disorders of the kidneys can affect the overall functioning of the excretory system.

5·6 ▸ Body systems in action

Learning intention

At the end of this lesson, I will be able to explain that the systems in multicellular organisms work together to provide cell requirements, including gases, nutrients and water, and to remove cell wastes.

Key term

hormone: a chemical substance produced by the body that controls the activity of certain cells or organs

Content group: Body systems

For an organism to survive, each body system must work with the others, with materials often passed from one system to the next.

Animals have several body systems

Most animals, such as humans, have 11 major body systems (see Table 5.2). Some of these systems can be identified in multiple ways. For example, the body's skeleton and muscles can be considered separately as the skeletal and muscular systems, or together as the musculoskeletal system.

Table 5.2: The 11 major body systems

Body system	Function
Circulatory	Transports important materials such as oxygen and nutrients in blood through a vast network of blood vessels. Arteries carry blood away from the heart and veins carry blood towards the heart.
Nervous	Detects, processes and sends electrical signals
Respiratory	The oxygen in the lungs is moved into the blood and delivered to the cells for respiration. The carbon dioxide produced by cells is taken to the lungs to be removed.
Digestive	Breaks down food into nutrients, which are absorbed into the blood and transported to the cells
Skeletal	Provides support and structure to the body and organs
Muscular	Allows movement through the use of muscles
Excretory	Waste products from cellular processes move from cells into the blood, to be removed by the organs of the excretory system.
Endocrine	Produces hormones, which control growth and development
Reproductive	Produces sex cells, and supports pregnancy and birth
Immune	Makes white blood cells, which fight diseases and infections
Integumentary	Protects the body from damage

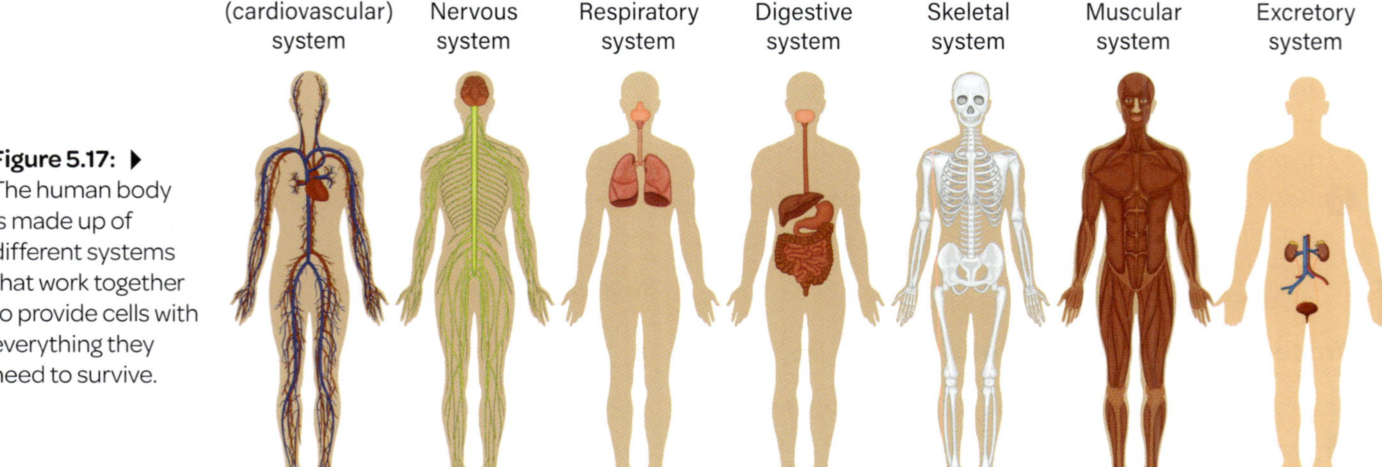

Figure 5.17: ▸ The human body is made up of different systems that work together to provide cells with everything they need to survive.

Body systems work together

Body systems must work together to provide cells with everything they need to function and survive, such as gases, nutrients and water. For example, in the human body the circulatory system is connected to every other system in the body. It transports nutrients, dissolved gases and waste products between cells. Without the circulatory system, other systems would not be able to function.

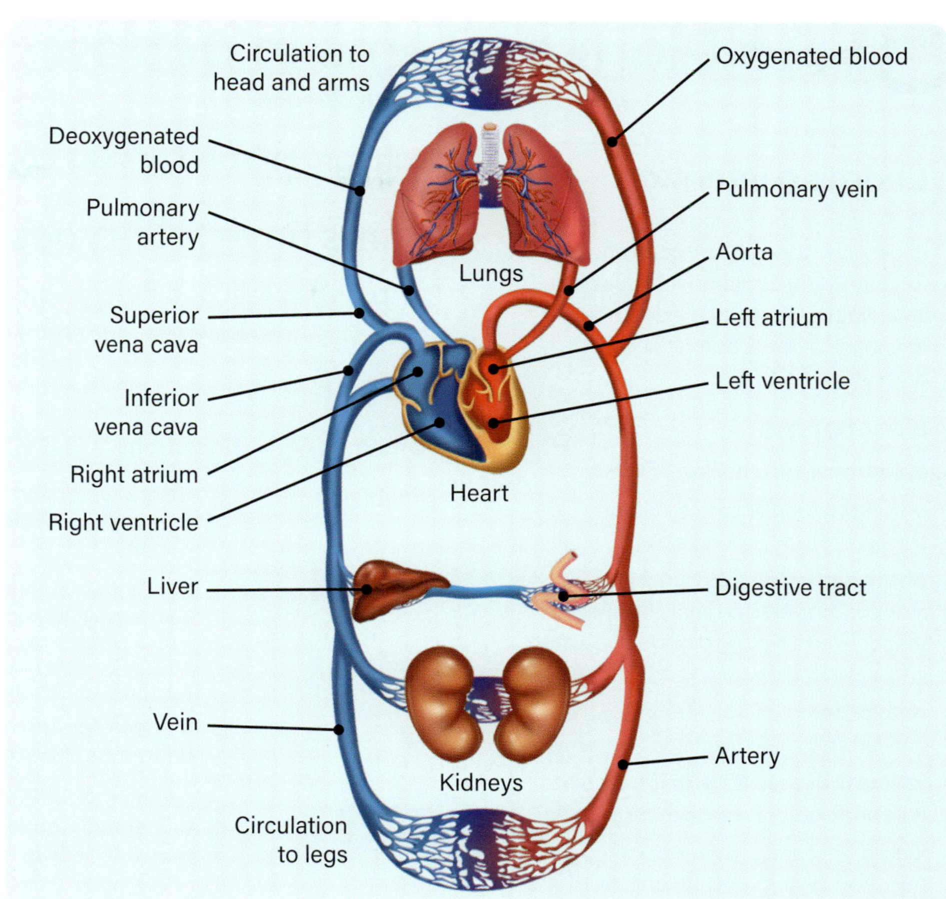

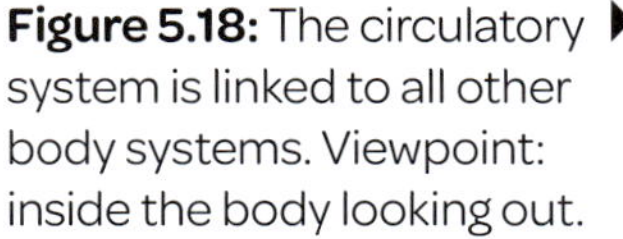

Figure 5.18: The circulatory system is linked to all other body systems. Viewpoint: inside the body looking out.

Learning Ladder

Living systems

1. Identify the system that helps to remove wastes in the body.
2. Identify and describe one feature of at least two body systems that allows them to interact.
3. **a** Explain how the musculoskeletal system provides structure for at least two other body systems.
   b Discuss how the two body systems you identified in part a would be affected if there were no skeletal system.
4. Both the circulatory and respiratory systems are responsible for ensuring that oxygen gets to our cells. Compare how they are responsible.
5. Cells require oxygen and water to survive. Discuss which body systems may assist cells to obtain these materials.

Communicating

see page 343

1. Identify why 'body system' means the same as 'organ system' in the human context.
2. Describe why each body system is represented on a separate human figure, rather than having all body systems shown on a single human diagram.
3. Choose at least four body systems and create a digital table that distinguishes their cells, organs, functions and primary purpose.
4. Construct a cyclical flowchart that shows the interconnectedness between the body systems from Question 3.
5. Annotate your flowchart from Question 4 with scientific drawings and information to highlight the interconnectedness of body systems that enable survival of organisms.

In context

Compare a human body system with the body system of a different animal. Display your findings using visual creativity; for example, a flowchart with annotated images.

Success criteria

- I can identify components of body systems that interact between systems.
- I can explain how body systems work together to enable the efficient functioning of an animal.

5·7 ▸ Plant systems in action

Learning intention

At the end of this lesson, I will be able to describe the role, structure and function of the components of a plant, including the xylem and phloem, in maintaining plants as multicellular organisms.

Key terms

epidermis: the outer layer of cells

phloem: a tubular structure that transports food around a plant

photosynthesis: the chemical reaction, powered by sunlight, in plants that converts carbon dioxide and water into sugar and oxygen

respiration: a chemical reaction that converts glucose to energy

vascular tissue: tissue that transports fluid and nutrients throughout a plant

xylem: a tubular structure that transports water and nutrients from the roots to various parts of a plant

Investigation 5.7

Water transport in plants, p.442

Content group: Plant systems

▲
Figure 5.19: Plants have two organ systems: the root system and the shoot system

Plant organs and tissues carry out vital functions.

Plants have four types of organs

Plants only have four different types of organs, as shown in Table 5.3. These organs are simple, but some carry out multiple functions.

Table 5.3: Plant organs and their functions

Organ	Functions
Roots	• Absorb water and minerals from the soil; plants need water for photosynthesis and to dissolve and move substances around • Anchor plant in the ground so that it does not fall over in strong winds or in ocean currents
Stem	• Forms the main body of the plant (the trunk of a tree is the same as the stem of a rose) • Provides support: helps a plant to stand up and bear the weight of leaves, flowers and fruit • Transports nutrients: connects the root and shoot systems • Aids growth: buds grow from the stem and form new branches, leaves or flowers
Leaves	• Where photosynthesis takes place • Sunlight is absorbed through the surface of the leaves, and products such as oxygen are released, or re-used during respiration.
Reproductive organs	• To produce offspring • Examples include flowers, fruits and seeds

Plants have a small number of tissue and organ systems

Plants have two organ systems: the root system and the shoot system (see Figure 5.19).

The **root system** consists of all the organs underground, such as the roots and any underground reproductive organs of the stem. This system absorbs nutrients and water from the soil. Despite it being called a root vegetable, when you eat a potato, you are eating part of the stem; it grows underground as part of the root system.

The **shoot system** mostly consists of the organs that grow above ground. These parts of the plant absorb sunlight and are where photosynthesis happens. The stem, fruit, flowers and leaves generally form the shoot system.

Plants also have four tissue systems. These are made up of specialised cells that perform important functions but do not combine to form organs.

- The **epidermis** is like the skin of the plant. These tissue cells form the outer surface of the leaves and the plant body.
- **Vascular tissue** transports fluids and nutrients through the plant, much like blood vessels do in your body.
- **Ground tissue** is the cells that make nutrients during photosynthesis and store them for later.
- **Meristematic cells** develop into various organs of a plant and are responsible for growth.

Plants have two primary vascular tissues

The **xylem** and **phloem** are complex tissues that form tubular structures to transport water, nutrients and food throughout a plant. The xylem only allows a flow of water and dissolved nutrients in one direction, from the roots to the shoot system. The phloem allows a two-way flow of sugars, proteins and other organic substances. The xylem and phloem are located next to each other as a cluster of vessels that make up the vascular bundle (see Figure 5.20).

Vascular bundles can be arranged in different ways, depending on the plant and its functions.

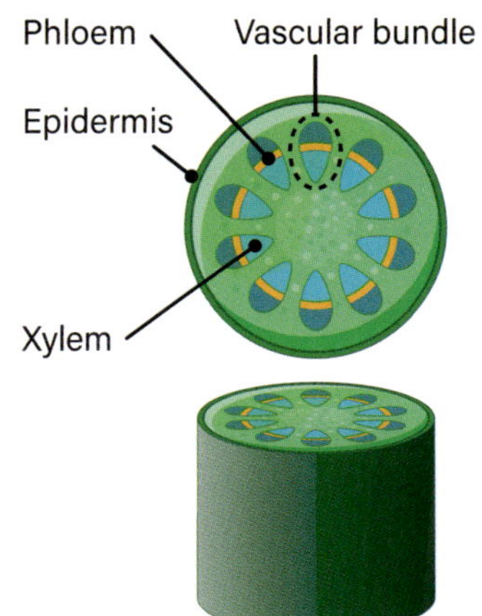

Figure 5.20: Xylem and phloem make up vascular bundles.

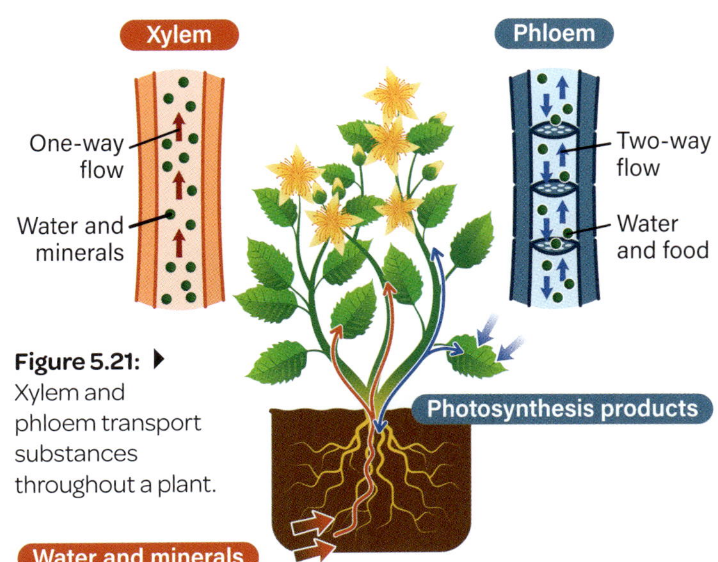

Figure 5.21: Xylem and phloem transport substances throughout a plant.

Learning Ladder

Living systems

1. Identify:
   - **a** how many types of organs plants have.
   - **b** how many organ systems plants have.
   - **c** the tissues that make up the transport system in plants.
2. Identify two important plant organs and describe their functions.
3. Explain the structure and function of each of the two organ systems in plants.
4. Construct a Venn diagram to compare and contrast the role and functions of the xylem and phloem.
5. Discuss how a plant would be affected if either of its vascular tissues were damaged.

Questioning and predicting

see page 319

1. Predict what will happen to a plant when you cut its stem to put it in a vase.
2. Identify which of the following is the *most* suitable question to investigate a scientific concept.
   - **A** Will flowers die if you do not give them water?
   - **B** How does the length of the stem affect how long it survives in a vase with water?
   - **C** How long will a flower survive in a vase without water?
3. Construct a scientific question to investigate the variety of vascular bundle patterns in different types of plants.
4. Refer to Investigation 5.7 on page 442 and read the aim, materials and method. Predict what will be observed.
5. Formulate a scientific question for Investigation 5.7 that considers relevant variables.

In context

Conduct an online search to compare the vascular arrangement or patterns of the xylem and phloem of at least three types of plants. Display your findings in annotated diagrams. Try to link the patterns to the type or purpose of each selected plant.

Success criteria

- I can give examples of at least one type of plant cell, tissue and organ.
- I can explain the relationship between plant cells, tissues and organs.

5·8 ▸ Flowering plants

Figure 5.22: The main purpose of flowers is for reproduction.

Learning intention

At the end of this lesson, I will be able to explain the role of the flower and leaf in maintaining flowering plants as functioning organisms.

Key terms

chlorophyll: the green pigment in plant cells that absorbs sunlight and enables photosynthesis

pistil: the female reproductive part of a flower (the stigma, style and ovary)

pollen: the fine powder produced by the male part of a flower; contains male sex cells

pollination: the movement of pollen from the male part of a flower to the female part (from the anther to the stigma)

stamen: the male reproductive part of a flower (the anther and filament)

stomata: pores on the surface of a leaf; the site of gas exchange in plants

Investigation 5.8

Dissecting a flower, p.443

Content group: Plant systems

Flowers may look and smell beautiful, but their main purpose is reproduction. Many flowers contain both male and female reproductive parts.

Flowers contain reproductive organs

The male reproductive part of a flower is called the **stamen** (see Figure 5.23). Each stamen is made up of the:

- **anther**: the organ that produces **pollen**, a fine powdery substance that contains the male sex cells of the plant
- **filament**: the stalk that supports the anther.

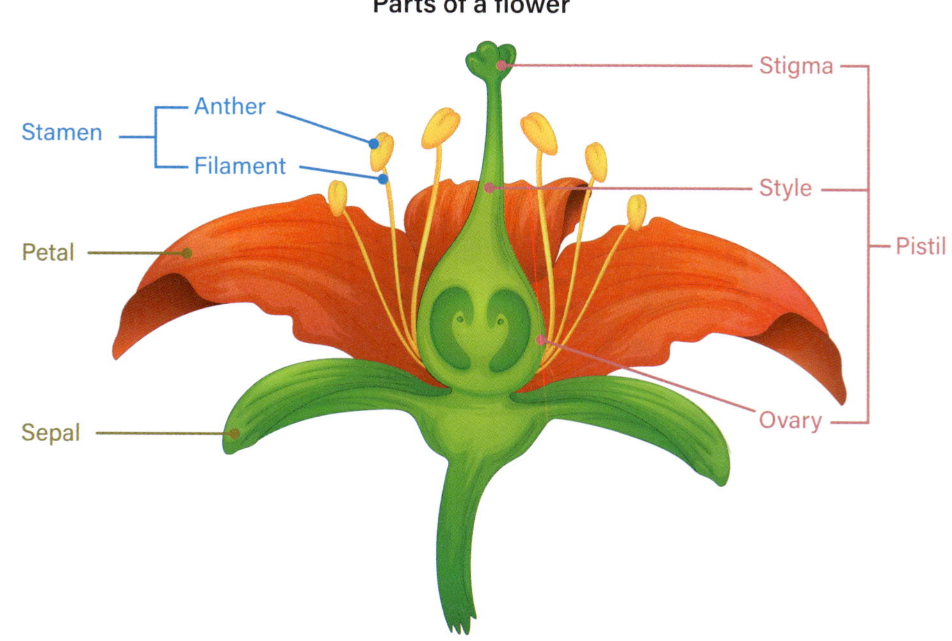

Figure 5.23: Flowers contain reproductive organs. The pistil is the female reproductive part and the stamen is the male reproductive part.

The female reproductive part of a flower is called the **pistil** (see Figure 5.23). Each pistil is made up of the:

- **stigma**: the organ where pollen germinates
- **ovary**: the organ that stores the female sex cells (ova or eggs)
- **style**: the stalk that connects the stigma and ovary.

For plants to reproduce, pollen needs to be moved from the anthers to the female part of the flower to fertilise the ovum (egg). This is called **pollination**.

Pollination can be assisted by wind, rain, birds, and insects such as bees. Birds and insects are attracted to bright and colourful flowers that smell nice and contain sugary nectar to eat.

Once fertilised, the ova (eggs) become seeds and the ovary swells and enlarges to become a fruit. The seeds in fruit can grow into new plants when conditions are suitable.

Photosynthesis happens in leaves

Leaves could be called the solar panels of plants because their main role is to perform photosynthesis. Leaves have many features that make them perfect for carrying out this process. They are often flat, which increases their surface area, allowing them to absorb more sunlight. They are thin, so carbon dioxide can travel easily into the cells from the environment. Plants contain green pigments called **chlorophyll**, which absorb light energy from the Sun to produce food. Veins in plants allow water and other substances needed for photosynthesis to travel to the leaf cells.

If you look at a leaf under a microscope, you will see round pores called **stomata** (see Figure 5.24). Stomata open and close to allow the exchange of gases, such as oxygen and carbon dioxide, with their environment. Water can also pass through these pores, and water loss is sometimes an unwanted consequence of gas exchange.

Figure 5.24: Stomata are tiny pores on the surface of leaves that allow plants to exchange gases with their environment.

Figure 5.25: Plants need pollinators such as birds in order to reproduce.

Learning Ladder

Living systems

1. Identify the parts that make up the:
   - **a** stamen.
   - **b** pistil.
2. Describe how a fruit is formed.
3. Suggest what would happen to a plant if it did not have a stem.
4. Photosynthesis needs carbon dioxide, water and sunlight. How does chlorophyll compare to stomata in helping a plant to obtain these resources?
5. Roots are usually found underground, so they do not receive the sunlight necessary to carry out photosynthesis.
   - **a** Discuss why roots are important to the process of photosynthesis.
   - **b** Predict what might happen if erosion caused roots to be exposed above ground.

Communicating

see page 343

1. Classify the following terms as either male or female features of a plant: anther, filament, ovary, pistil, pollen, stamen, stigma, style.
2. Construct a chart identifying the parts of a plant categorised in Question 1.
3. **a** Construct a digital diagram of a flower that is similar to Figure 5.23.
   b Annotate the diagram by labelling all of its parts, including a brief description for each term.
4. Construct an annotated flowchart to show the sequence of steps that take place when plants reproduce.

In context

How do insects and birds help plants to reproduce?

Success criteria

- I can describe the main roles of the flower and leaf in the body system of a flowering plant.
- I can sketch a diagram of a flowering plant to represent the main features and functions.

5·9 ▸ Ecosystem basics

Learning intention

At the end of this lesson, I will understand that ecosystems include communities of interdependent organisms.

Key terms

abiotic: non-living

biotic: living

community: a naturally occurring group of animals, plants and other organisms

consumer: an organism that gains energy by consuming other living organisms

decomposer: an organism that breaks down and recycles decaying matter

ecosystem: a community of living things interacting with one another and the non-living things in their environment

habitat: the place where an animal or a plant naturally lives

producer: an organism that makes its own food using energy from the Sun

Investigation 5.9

Identifying community members in a terrestrial ecosystem, p.444

Content group: Ecosystems

An **ecosystem** is an area that contains living and non-living things and their environment. Ecosystems can be very small, such as a pond, or very large, such as an island.

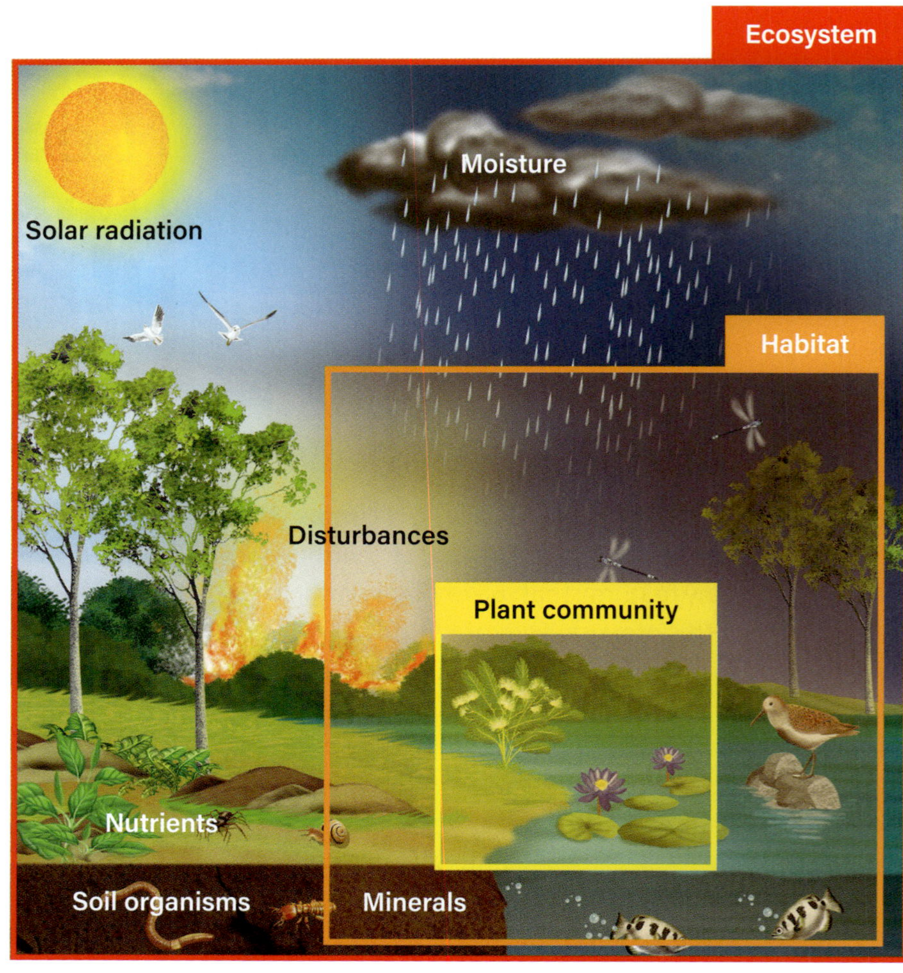

Figure 5.26: An ecosystem is an area that contains living and non-living things and their environment.

Communities are made up of different species

A **community** is made up of populations of all the different species that live in the ecosystem. All organisms within a community depend on each other for energy, nutrients and survival. **Producers** convert the energy from the Sun into chemical energy that can then be used by other organisms. **Consumers** obtain their energy by eating other organisms.

Decomposers, such as bacteria and fungi, recycle the nutrients from dead organisms and make them available to other species in the ecosystem.

The living things in an ecosystem are referred to as **biotic** factors. The non-living things, such as soil and water, are known as **abiotic** factors.

A habitat contains living and non-living factors

A **habitat** is the natural environment where a species or an organism lives. A habitat must supply food and water, shelter and protection, and mates to reproduce with. Every species needs specific conditions in a habitat; a koala and a polar bear would not live in the same place, because they have different requirements.

Habitats can change over time. This might be a result of naturally occurring events, such as a volcanic eruption, or be due to human causes, such as climate change or land clearing. The introduction of a new, non-native species can also have a major effect on native habitats and the organisms living there. These will be explored in more depth later in the chapter.

Each organism has an ecological niche

Each organism has a particular role within the ecosystem; this is its ecological niche. An organism's niche includes the interactions it has with other organisms, the type of food it eats, where it lives and how it reproduces. Only one organism can fill a specific niche in an ecosystem. Many different species can co-exist within an ecosystem if no two species have the same niche.

Figure 5.27: Feeding zones of warblers. Each warbler species occupies a different niche. The birds feed in different parts of the spruce tree, so they are not competing with each other for food.

Introduced species cause problems if they take over a niche and out-compete the native animals. If an organism disappears from an ecosystem, then there will be flow-on effects throughout the entire community. For example, if a pollinator disappears, then plants cannot reproduce, which limits the amount of food available to the herbivores, in turn limiting the food for carnivores.

Learning Ladder

Living systems

1. Outline the general components of an ecosystem.
2. Copy and complete.
   A community is made up of __________ of all the different __________ that live in the ecosystem. All organisms within a community __________ on each other for __________, __________ and __________. Each organism maintains its ecological niche within its specific __________.
3. Explain the difference between abiotic and biotic factors and give two examples of each.
4. Discuss the similarities and differences between a habitat and a niche.
5. Summarise the natural and human-made changes that can affect habitats. Brainstorm as many as you can think of.

Communicating

see page 343

1. Classify the following terms as either biotic or abiotic: trees, wind, birds, sunlight, rain, insects, rocks, soil, flowers.
2. Construct a visual way (such as a chart) to communicate the abiotic and biotic factors listed in Question 1.
3. Construct a digital diagram of a habitat. Annotate the diagram by labelling and describing all its parts, including biotic and abiotic factors.
4. Construct an annotated flowchart to show the levels that make up an ecosystem and the associated factors.

In context

Consider your favourite wild animal. Describe its habitat and suggest some features of its habitat that are important to the animal's survival.

Success criteria

- I can define what an ecological community is.
- I can describe the difference between a habitat and a niche.

5·10 ▸ Interactions between organisms

Figure 5.28: Predators kill and feed on their prey.

Learning intention

At the end of this lesson, I will be able to discuss key interactions between organisms, including predators, prey, parasites, competitors and pollinators.

Key terms

adaptation: a feature that an organism is born with that helps it to survive in its environment

parasite: an organism that lives in or on another organism, causing it harm

pollinator: an organism that transfers pollen from the male part of a plant to the female part

predator: an organism that kills and feeds on prey

prey: an organism that is killed by a predator

Content group: Ecosystems

There is more to interactions in an ecosystem than just who eats whom. Interactions between organisms can be helpful, such as the interaction between bees and flowers, or harmful, such as the interaction between a parasite and its host. Table 5.4 describes some ecological relationships.

Table 5.4: Ecological relationships

Interaction	Effects on the population	Example
Mutualism	The interaction is beneficial to both species.	A cassowary eating the fruit of plants and distributing the seeds in its dung
Commensalism	One species benefits and the other species is unaffected.	A staghorn fern growing on the branch of a tree
Competition	Both species are negatively affected because they compete for resources.	Sugar gliders competing with superb parrots for nesting hollows
Predator–prey	The predator benefits and the prey is harmed or killed.	A bronze whaler shark eating sardines
Parasitism	The parasite benefits and the host is harmed.	Mistletoe taking nutrients from a host tree

Competitors compete for resources

Competition often happens when organisms need the same limited resource (such as food) in the same ecosystem. Competitors can be different species or members of the same species. Depending on the amount of food available, organisms may be harmed, may starve or may have to find a new food source. Some members of the same species will even compete for a mate; it is all part of trying to survive and thrive.

Plants also compete for resources, but not in the same way as animals. Plants compete for space, light, water and nutrients, which are the things they need to produce energy. Some plants will grow and survive, while others will die.

Predators and prey keep each other in balance

Predators are organisms that kill and feed on other organisms: their **prey**. For example, a fox is a predator and its prey is a rabbit. Predator–prey relationships might seem ruthless, but their interactions play a crucial role in an ecosystem: they stabilise the numbers of prey organisms. If the numbers of prey increase, then numbers of predators will increase, which keeps the predator–prey ratio balanced. Predators have some special **adaptations** that enable them to catch their prey, such as good vision or hearing, sharp claws or teeth, and strong jaws.

Parasites are a type of predator that live on (or in) another organism, the host. Unlike other predators, parasites are usually much smaller than their prey and harm the other organism without killing it. For example, the hookworm lives in the intestines of its human host where it feeds on nutrients from digested food.

Pollinators and plants help each other

Flowering plants have female and male sex organs. For fertilisation to occur, they need a **pollinator**. A pollinator transfers pollen from the male sex organ of plants to the female sex organ.

Bees are important pollinators: without bees, a lot of plant life would cease to exist. The interaction is beneficial for the plant, which is able to reproduce, and for the bee, which gets a meal of nectar to take back to its colony. Other organisms, including birds, some lizards and even monkeys, can also be pollinators, but bees are especially adapted to this very important job.

Figure 5.29: Bees as pollinators enable plants to reproduce.

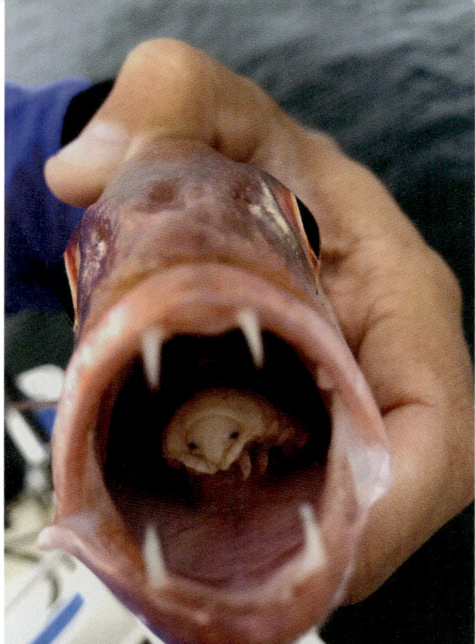

Figure 5.30: Parasites live on (or inside) their host organism. The parasitic isopod *Cymothoa exigua* attaches to a fish's tongue, eventually replacing it with its own body!

Learning Ladder

Living systems

1. Brainstorm and list as many different predator–prey relationships as you can think of.
2. Competition occurs in every ecosystem.
   - a Describe what competition is in your own words.
   - b Outline some resources that animals compete for.
   - c Outline some resources that plants compete for.
3. Explain how predators and prey keep each other in balance. Identify an example.
4. Discuss the example you provided in Question 3 in more depth. You may need to conduct an online search for information.

Processing data and information *see page 329*

1. Use Table 5.4 to classify the following interactions.
   - a Bandicoots feed on insects in your backyard.
   - b Dung beetles feed on kangaroo dung. The kangaroos are unaffected.
   - c A tick sucks blood from a wallaby.
   - d A cleaner shrimp removes parasites from a sea turtle.
2. Construct a table to organise the brainstorm you conducted for Question 1 in the previous set of questions.

In context

There has been a worldwide decline in the number of bees. Why do you think this is occurring? What could be done to protect and increase bee population numbers?

Success criteria

- I can describe the different ways organisms interact in an ecosystem.
- I can describe the role of pollinators in an ecosystem.

5·11 ▸ Abiotic factors in ecosystems

Learning intention

At the end of this lesson, I will be able to discuss key abiotic components of the environment and their impact on ecosystems.

Key term

zone of tolerance: the range of an abiotic factor that an organism can survive in

Investigation 5.11

Measuring abiotic factors, p.445

Content group: Ecosystems

Factors such as water availability, soil type, temperature and availability of nutrients affect what organisms will live in the ecosystem. If the abiotic factors change outside of their normal level, this affects the rest of the ecosystem.

Abiotic factors affect what can live in an ecosystem

Abiotic factors are chemical and physical factors within an ecosystem. Abiotic factors include:

- rainfall and water availability
- soil type
- temperature
- pH of soil or water
- sunlight
- nutrient availability in soil or water
- gas (oxygen and carbon dioxide) availability
- salinity (saltiness) of soil or water.

The abiotic factors in an ecosystem affect what can live there. The amount of sunlight, nutrients and water determines the types of plants that live in an ecosystem. This, in turn, affects what consumers can live there.

Organisms have adaptations to abiotic factors

Organisms have evolved adaptations to help them survive in their environment (see Figure 5.31). The adaptations can be:

- **structural**: a feature of an organism's body; for example, the sharp claws of a koala help it to climb trees to reach its food source
- **behavioural**: the way an organism responds to its environment; for example, the huddling of emperor penguins to stay warm
- **physiological**: a process inside an organism's body; for example, a camel's ability to retain water by producing low volumes of urine.

Organisms can still only tolerate specific levels of each abiotic factor. If the level of an abiotic factor moves out of this **zone of tolerance**, the organism will not survive.

Figure 5.31: Structural, behavioural and physiological adaptations

Coral bleaching is caused by abiotic factors

Coral bleaching is an example of what happens when abiotic factors move out of the zone of tolerance.

Corals live in a mutualistic relationship with an alga called zooxanthellae. The algae live inside the structure of the coral. The coral provides the algae with a home, and the algae photosynthesise and provide the coral with 90 per cent of its energy requirements. The algae are also responsible for the coral's colour.

If water temperature or pollution levels increase to outside the coral's zone of tolerance, the coral expels the algae and becomes bleached. If temperatures and pollution levels return to normal, the coral allows the algae to return, and it recovers. If the levels remain too high for too long, then the coral dies.

The optimum water temperature range for most reef-building corals is 20–32 °C, although different species have their own zones of tolerance. Increasing temperatures in Australian waters have led to coral bleaching all along the Great Barrier Reef.

Figure 5.32: A marine biologist measures bleached coral.

Learning Ladder

Living systems

1. Consider an ecosystem you visited recently. List five specific abiotic factors you observed.
2. Describe some abiotic factors that are important for a:
   a bird.
   b saltwater fish.
3. Explain the difference between abiotic and biotic factors. Provide examples to support your response.
4. Discuss, using a specific example, what is meant by an organism's tolerance level to an abiotic factor.
5. Predict ways that a kangaroo might adapt if:
   a pollution causes the salinity level of a freshwater pond to increase.
   b the summer months become drier and hotter.
   c grazing paddocks are cleared for development.

Questioning and predicting

see page 319

1. Predict what will happen to a plant if you remove the abiotic factors from its habitat.
2. Identify which of the following is the *most* suitable question to investigate a scientific concept.
   A How does the type of soil affect the growth of a plant?
   B How long will a plant survive without soil?
   C Will plants die if they get too much sunlight?
3. Construct a scientific question to investigate the difference in abiotic factors in a wetland ecosystem as compared to a bushland ecosystem.
4. Refer to Investigation 5.11 on page 445 and read the aim, materials and method. Predict what will be observed.
5. Construct a scientific question for Investigation 5.11 that considers relevant variables.

In context

Conduct research to compare the abiotic factors in the Daintree Rainforest (Queensland) and the Simpson Desert (Central Australia). Find out about rainfall (water availability), temperature range and nutrient availability in the soil. How do these factors influence what organisms live in these ecosystems?

Success criteria

- I can identify at least three examples of abiotic factors.
- I can explain how abiotic factors can determine what organisms live in an ecosystem.

5·12 ▸ Matter flows through ecosystems

Learning intention

At the end of this lesson, I will understand the process and importance of how matter such as nitrogen flows through ecosystems.

Key terms

atmospheric nitrogen: nitrogen in the atmosphere

mutualism: a relationship between two organisms in which both organisms benefit

Content group: Ecosystems

Matter cannot be created or destroyed. Matter cycles through different parts of ecosystems, including the atmosphere, soil and organisms. The atoms that make up your body were once part of the air you breathed and the food you ate. The atoms in a steak came from the plants that the cow ate, and before that from the carbon dioxide, water and nutrients in the plants.

Nitrogen is an important element for life

Nitrogen (N) is an important element that cycles through ecosystems. Nitrogen is in amino acids, the building blocks of proteins, which are essential for all life. Nitrogen is the most abundant element in the atmosphere, making up approximately 78 per cent of air, where it mostly exists as molecular nitrogen (N_2).

Plants take up inorganic nitrogen (from non-living sources) through their roots from the soil. In this way, nitrogen enters the biosphere and cycles through living things. Bacteria convert nitrogen from one form to another.

▲ **Figure 5.33:** Soy bean roots contain nodules of bacteria that take in nitrogen from the soil.

Nitrogen cycles from the atmosphere

Plants cannot use **atmospheric nitrogen**. Plants need to obtain nitrogen from the soil (through their roots) in the form of nitrate (NO_3^-). They use this nitrogen to make amino acids.

Most nitrogen is transformed by different species of nitrogen-fixing bacteria into forms that plants can use. This usually happens in several steps, each involving a different species of bacteria. Some of these bacteria live in the roots of legumes (for example, peas and beans). In this **mutualism** relationship, both the bacteria and the plant benefit. There are also denitrifying bacteria, which release nitrogen back into the atmosphere.

Herbivores (animals that only eat plants) produce proteins from the amino acids in the plants they eat. When a herbivore is consumed by a higher-order predator, the amino acids are then used by the predator. Decomposers such as bacteria and fungi break down the amino acids in dead matter, releasing nitrogen back into the soil so that it can be reused by plants.

▲ **Figure 5.34:** Mushrooms are a type of fungi that help other plants share nutrients, cycling matter through ecosystems.

Other elements also cycle through ecosystems

Carbon, oxygen, hydrogen and phosphorus also cycle through ecosystems. All these elements are important for the formation of molecules that are vital for life. Carbon, oxygen and hydrogen cycle through ecosystems through the processes of photosynthesis and cellular respiration. They move from inorganic molecules, such as carbon dioxide, molecular oxygen (O_2) and water, into organic molecules, such as glucose and other carbohydrates, which are then passed up through the food chain.

Phosphorus plays an important role in the formation of DNA. Plants take up phosphorus from soils and convert it into molecules that are then passed up through the food chain, returning to the inorganic forms through excretion and decomposition.

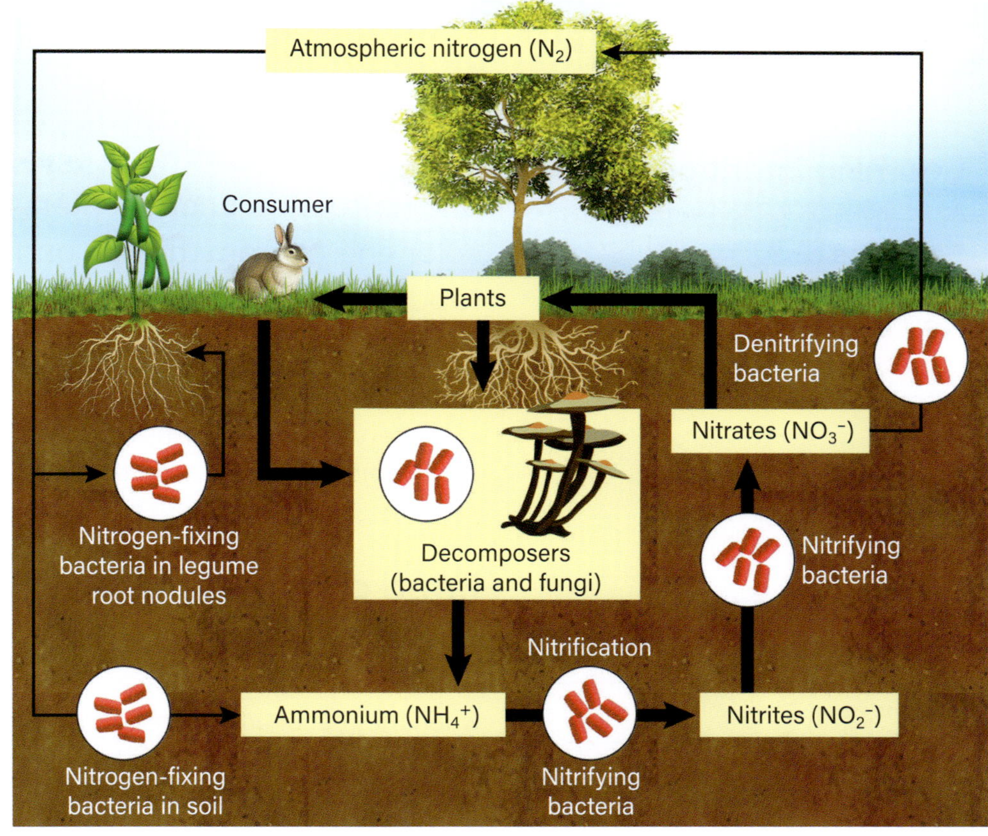

▲ **Figure 5.35:** The nitrogen cycle is crucial for plants to grow.

Learning Ladder

Living systems

1. Identify the two types of organisms that are important for the nitrogen cycle. What role do they play?
2. Summarise the stages of the nitrogen cycle, using Figure 5.35 to assist you.
3. Explain what is meant by a matter *cycle*. Use specific examples in your response.
4. Compare the nitrogen and carbon cycles. You may conduct an online search to find more information about each cycle. Discuss how are they similar and different.
5. Carnivorous plants catch insects *not* to gain energy, but to extract nutrients such as nitrogen. Discuss what this might indicate about the soils they live in.

Questioning and predicting

see page 319

1. Predict what would happen to the nitrogen cycle in an ecosystem:
   a where there were no decomposers.
   b that became too cold for nitrogen-fixing and nitrifying bacteria to survive.
2. Identify which of the following is the *most* suitable question to investigate a scientific concept.
   A Will plants die if they do not get enough nitrogen?
   B What happens if you remove nitrogen from the soil?
   C How do different nitrogen levels in the soil affect the growth of a plant?
3. Construct a scientific question to investigate the effect of various types of decomposers in soil on flowering plants.
4. Construct hypotheses for each of your predictions from Question 1. Use an 'If ... then' statement to structure your hypotheses.

In context

Identify and explain a process that allows carbon, oxygen and hydrogen to cycle through ecosystems. Use an annotated diagram to show the parts of the process and how they are related to cycles of matter.

Success criteria

- I can outline a specific example of how matter is cycled through ecosystems (for example, nitrogen).
- I can discuss the importance of why a variety of elements are cycled through ecosystems.

5.13 ▸ Energy flows through ecosystems

Learning intention

At the end of this lesson, I will be able to use food webs to understand how energy flows through ecosystems.

Key terms

food chain: a path of energy through an ecosystem

food web: a system of interlocking food chains

trophic level: the position of an organism in a food chain

Investigation 5.13

Algal balls, p.447

Content group: Ecosystems

All living organisms need energy to grow, repair damage and reproduce. Energy enters most ecosystems as solar energy from the Sun and is transferred through food chains and food webs.

Plants bring energy into ecosystems

Producers convert light energy from the Sun into chemical energy during photosynthesis. In photosynthesis, plants use light energy to produce glucose ($C_6H_{12}O_6$) from carbon dioxide (CO_2) and water (H_2O).

Food chains show the path of energy

Food chains are diagrams that show one possible pathway of energy through an ecosystem. Energy is transferred between organisms when they eat or are eaten. Food chains always begin with a producer: a plant or microorganism that can produce energy. Food chains have arrows between each organism to show the direction that energy moves. In an Australian grassland ecosystem, a grasshopper gains energy from grass, so there is an arrow from the grass to the grasshopper (see Figure 5.36).

Wedge-tailed eagle
Quaternary consumer

Eastern brown snake
Tertiary consumer

Blue-tongue lizard
Secondary consumer

Grasshopper
Primary consumer

Grass
Producer

▲ **Figure 5.36:** Each organism in a food chain provides energy for the next organism. Organisms can be described according to their position in the chain.

The other organisms in the food chain are consumers. Consumers can be described in terms of the organisms they feed on (see Table 5.5). The different types of consumers help maintain balance in ecosystems.

Table 5.5: Types of consumers and how they obtain their energy

Consumer	Obtains energy by ...	Examples
Herbivore	feeding on producers	Koala, grasshopper
Omnivore	feeding on producers and consumers	Sea star, most humans
Carnivore	feeding on other consumers	Kookaburra, brown snake
Decomposer	breaking down dead matter	Bacteria, fungi

Food chains interact to form food webs

Organisms are usually part of more than one food chain because they have more than one food source or get eaten by more than one consumer. A species needs multiple food sources in case one runs out.

Food webs show all the feeding relationships in an ecosystem. They show a much bigger picture of the flow of energy and how each organism is involved. Food webs also show how organisms are arranged into different **trophic levels**. Producers are at the bottom and the highest-order consumer is at the top. Energy passes from low trophic levels to the higher trophic levels through consumption.

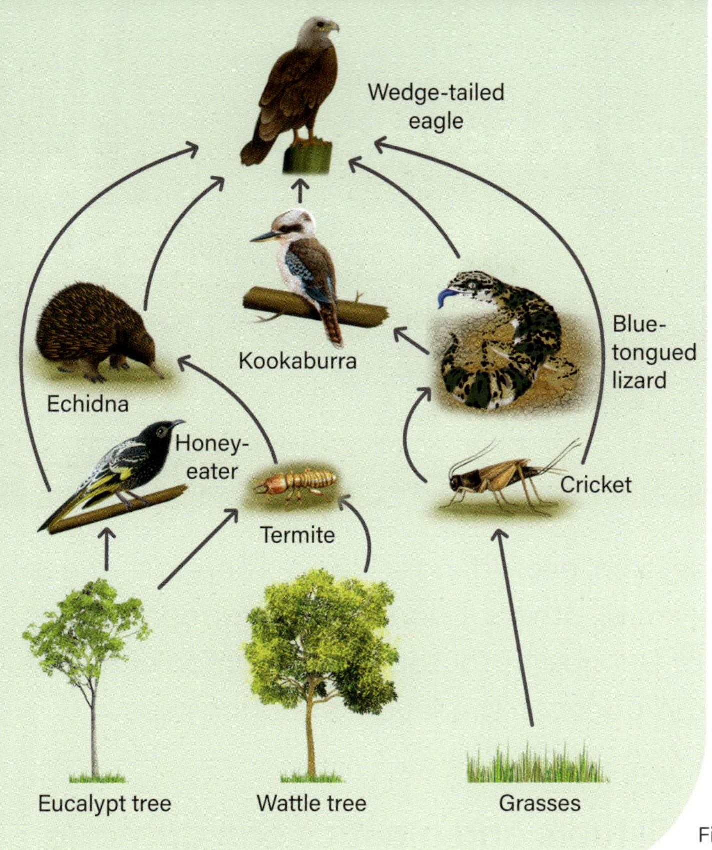

Figure 5.37: This bushland food web shows how different food chains can connect. An organism may be a tertiary (third-order) consumer in one food chain but a secondary (second-order) consumer in a different food chain.

Energy is lost at each trophic level

Only about 10 per cent of the energy in a trophic level is passed on to the level above. This is why many producers and lower-level consumers are required to support a small number of higher-order consumers. This relationship can be represented in an energy pyramid (see Figure 5.38). The energy pyramid shows how energy is lost to the surroundings between each trophic level of the food web.

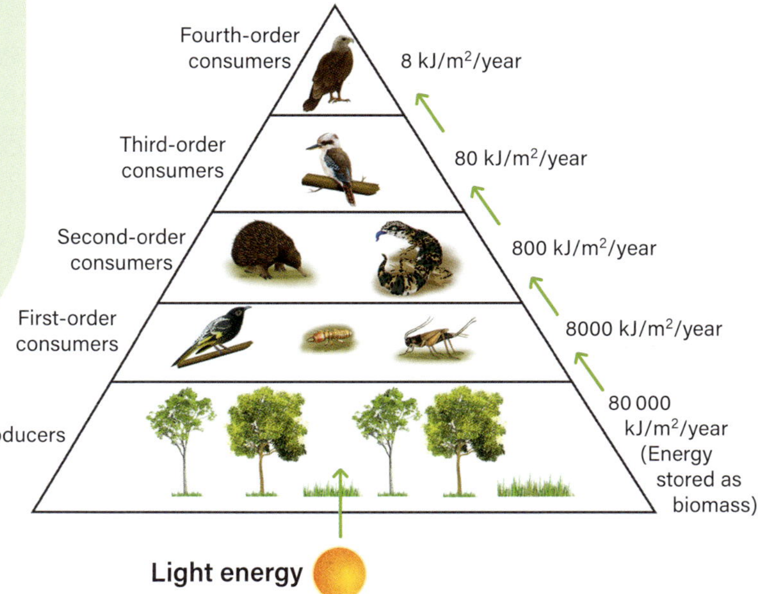

Figure 5.38: An energy pyramid shows the flow of energy from one trophic level to the next.

Learning Ladder

Living systems

1. Identify the main source of energy in any ecosystem. How does it enter the first level?
2. Describe what the arrows represent in a food chain or food web.
3. Explain how a food web can be used.
4. Select an organism from Figure 5.37. If it were removed, predict the impact on the other organisms in the food web. Be specific and provide reasons.
5. A large population of fourth-order consumers is released into an ecosystem. Predict the outcome.

Processing data and information

see page 329

1. Select an organism from the food web in Figure 5.37. Identify the food chain or chains to which it belongs.
2. **a** Construct a table to collect information about the ecosystems in your garden. Include consumers and producers, what they eat or where they get their energy, what level of consumer they may be, and any other facts about them.
   b Propose at least one food chain for your ecosystem.
3. Refer to the energy pyramid in Figure 5.38. Construct a column graph to represent the amount of energy transferred to each trophic level, assuming the producers begin with 80 000 kJ. (Hint: Put energy (kJ) on the *y*-axis.) See page 370 in the Science how-to section for help in constructing your column graph.
4. Construct a pie chart to show the proportion of different types of consumers in the food web in Figure 5.37. See page 371 in the Science how-to section for help in constructing your pie chart.

In context

Propose what would happen if an invasive species was introduced to a food web. Describe how the energy flow would be affected.

Success criteria

- I can identify where energy in a food web originates.
- I can describe the role of food webs in representing how energy flows through an ecosystem.

5·14 ► Human impacts on ecosystems

Figure 5.39: Algal blooms prevent light from penetrating to deeper water, which means producers cannot photosynthesise and support food webs.

Learning intention

At the end of this lesson, I will be able to examine secondary source data on the natural and human factors that change populations.

Key terms

biodiversity: the variety of organisms in an ecosystem

endangered: at risk of becoming extinct

extinct: having no living members of the species; died out

non-biodegradable: does not break down in the environment

pesticide: a chemical used to kill insects or other organisms that feed on plants

threatened: likely to become endangered in the near future

Content group: Ecosystems

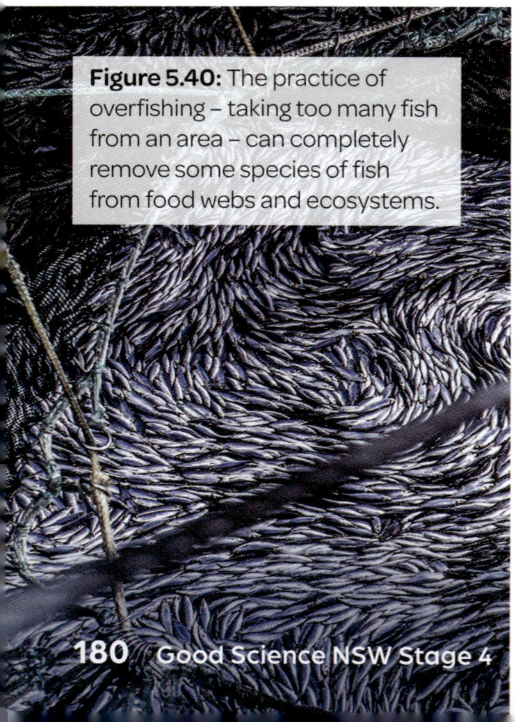

Within an ecosystem, everything is linked. Therefore, if one thing changes, so will others. Changes can happen to the biotic factors or the abiotic factors. Many human activities have dramatic impacts on the feeding relationships in an ecosystem.

Farming, fossil fuels and plastics can damage the environment

Pesticides are chemicals farmers use to kill unwanted plants and insects as a way to protect their crops. These chemicals may be passed on to animals that eat the pests. The chemicals can build up to toxic levels in the ecosystem and harm organisms (including humans) that consume the pesticides in their food.

Algal blooms happen when large amounts of nutrients from farms or industry run off into streams. The algae thrive on the extra nutrients and grow quickly, forming a bloom that covers the surface of the water. The bloom prevents light reaching the producers below, which die and so cannot support the rest of the food chain.

The burning of fossil fuels is a cause of global warming, which has a big impact on habitat and food sources in food webs. Fossil fuels are also used to make plastics, which are **non-biodegradable**. They can enter food webs when animals accidentally consume them.

Removing or introducing organisms affects biodiversity

Biodiversity is the variety of species within an ecosystem. In an ecosystem with high biodiversity, there are more species in a food web, so there are more food chains and a larger variety of food sources. Thus, species in a biodiverse ecosystem have more chance of survival if the environment changes.

Introduced species – such as cane toads and blackberries – often thrive and take over, reducing biodiversity. In some cases, this has led to species becoming **extinct**. The Australian Government is committed to supporting the recovery of any **threatened** species.

Figure 5.40: The practice of overfishing – taking too many fish from an area – can completely remove some species of fish from food webs and ecosystems.

Keystone species are essential to ecosystems

A keystone species has an irreplaceable role in its ecosystem. No other species in its ecosystem can fill its niche. If a keystone species is removed, the whole ecosystem can collapse.

Southern cassowaries (*Casuarius casuarius johnsonii*) are a keystone species in the North Queensland tropical rainforest ecosystem. They eat fruit and then deposit the seeds in their dung. Without cassowaries, some rainforest plants could become extinct.

Southern cassowaries are **endangered** because of habitat destruction and other human activities. Recent conservation efforts have helped to reverse the population decline, including habitat protection and restoration. However, cassowaries are still being killed by traffic, dogs and feral pigs.

Figure 5.41 shows the predicted changes to the population of cassowaries if different conservation strategies are used. In 2015, the Threatened Species Strategy was implemented. Although no clear improvements were seen in the following three years, conservationists are hopeful that more vigorous management efforts outlined in the revised Threatened Species Action Plan (2022–2032) will result in an increase in cassowary numbers. Increasing the area of rehabilitated rainforests is an obvious way to improve the cassowary population.

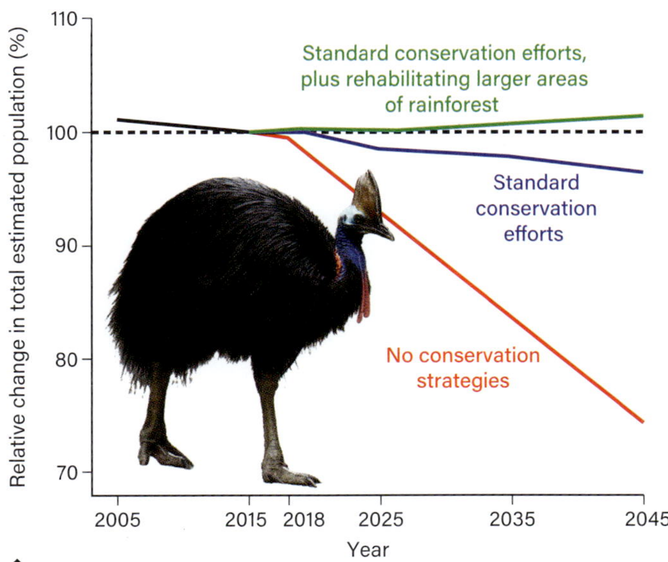

Figure 5.41: Predicted responses to management scenarios of the southern cassowary

Learning Ladder

Living systems

1. Outline at least four human activities that affect ecosystems.
2. Describe the effects of algal blooms on food webs.
3. Explain why organisms are more likely to survive in an ecosystem that has high biodiversity.
4. Discuss which has higher biodiversity: a tropical rainforest or a wheat farm. Provide a reason.
5. Evaluate the following statement: The cassowary population is meant to increase if we continue with standard conservation strategies.

Processing data and information *see page 329*

1. Refer to Figure 5.41.
   - **a** Identify which management strategy will have the most positive effect on cassowary population.
   - **b** What percentage decline in population will be seen in cassowary population from 2025 to 2045 if no conservation efforts are implemented?
   - **c** Compare the red and green. How many times more effective at conserving population numbers is the second option than no conservation, as projected for 2045. Use the *y*-axis for scale.
2. Imagine you are comparing the number of cassowary sightings by park rangers, residents and tourists in each month of the year. Construct a table to collect the data. Use clear headings and conventions.
3. **a** Refer to Figure 5.41. Construct a column graph to show relative change in population of cassowaries predicted from 2018 to 2025, based on which management strategies are taken.
   - **b** Represent the same information in a different format to clearly communicate the findings; for example, a chart or an infographic.
4. Select two or more extinct species. Conduct research to gather a variety of related statistics. Construct an infographic poster to display the data.

In context

Choose an aspect of agriculture that has a negative impact on ecosystems. Discuss ways to minimise its impact or compensate for the negative effects.

Success criteria

- I can suggest how human activity can affect interactions within food webs and chains.
- I can give an example of the effect of human activities in Australia on natural environments.

5·15 ▸ Environmental impacts on ecosystems

Learning intention

At the end of this lesson, I will be able to explain how ecosystems change as a result of environmental events such as bushfires.

Key terms

germination: the process in which previously dormant seeds put out shoots for new growth

intensity: how much heat is released from a fire front

understorey: ground-level vegetation

Content group: Ecosystems

▲ **Figure 5.42:** Lightning strikes can cause natural bushfires.

Australian ecosystems are often harmed by unpredictable natural events that reduce biodiversity, such as droughts, floods and bushfires. However, some natural events – such as low-intensity bushfires – can trigger new growth and cause other changes in an ecosystem.

Bushfires happen every year in Australia

Bushfires are common natural events in Australia. The hot, dry summers of many regions provide perfect conditions for bushfires. Lightning strikes are the most common natural cause of bushfires. When lightning hits a dry or dead tree, the heat of the strike makes the tree catch fire and break apart. Burning wood scatters and ignites other plants. Bushfires are also often started by humans.

The **intensity** and duration of a bushfire depend on the plants in the ecosystem, because they provide the fuel needed for the bushfire to burn.

Low-intensity bushfires are helpful in some ecosystems

Low-intensity bushfires usually do not last very long and do not destroy many plants. In fact, they can be helpful for some ecosystems because the fires remove older and dead plants, making more space for young plants to grow. Fire can even trigger **germination** in some plants. The ash from the fire returns some nutrients to the soil, and the ecosystem recovers relatively quickly. Controlled burns are part of fire safety, and lighting low-intensity bushfires has been a practice of First Nations Peoples for thousands of years.

Figure 5.43: The Waripiri People in the Northern Territory burn spinifex during winter, so that the grasses will not catch fire during summer.

High-intensity bushfires usually harm ecosystems

High-intensity bushfires damage ecosystems. Tall, thick and compact grasses and **understorey** plants provide a large fuel supply, so high-intensity bushfires are more likely to happen in areas where there has not been a fire for many years. In this case, there is a large amount of dry plant material, such as fallen branches and dead trees at the ground layer. This results in high-intensity fires that take a long time to burn and can destroy entire trees, right up to the top leaves. Millions of animals can be killed and the loss of plants and animals has a lasting negative impact on the entire ecosystem.

Figure 5.44: Dry eucalypt leaves are being used as a fuel source in this experiment at the Pyrotron.

Technology can help us understand bushfires

Technology can help scientists monitor bushfires and learn from previous fires.

The Commonwealth Scientific and Industrial Research Organisation (CSIRO) has a national bushfire research facility in Canberra called the Pyrotron. There, scientists safely carry out bushfire experiments.

The Pyrotron is a 25-metre-long wind tunnel with different sections and compartments. Fuel sources, such as leaves, are placed in the tunnel and set on fire while a fan blows air through the tunnel. Sensors and cameras measure how the fire spreads, while observers watch through fireproof windows.

The Pyrotron improves our understanding of how bushfires spread in different conditions. It also helps firefighters train, be better prepared and make better decisions during a fire.

Computer models can also be used to predict the movement, intensity and duration of a bushfire. The CSIRO's bushfire modelling system, Spark, simulates a bushfire. It uses the information it receives to predict the direction, speed and intensity of a bushfire.

Learning Ladder

Living systems

1. Identify at least three types of natural events that cause changes to living ecosystems.
2. Describe the benefits of a controlled burn.
3. Explain the different impacts low-intensity and high-intensity bushfires have on ecosystems.
4. Propose what plant characteristics you would see in an ecosystem that is prone to low-intensity bushfires. Discuss ways in which we can promote these characteristics.
5. Discuss reasons why First Nations land management practices include regular burning of areas of vegetation.

Questioning and predicting

see page 319

1. Predict how an ecosystem would change after a natural event such as a flood or a drought.
2. What question could you ask to determine if a bushfire was low intensity or high intensity?
3. Modify your question from Question 2 so that it can be investigated scientifically.
4. Propose a hypothesis for the scientific question you constructed in Question 3. Use an 'If ... then' statement to structure your hypothesis.

In context

Discuss how technology is helping to manage the impact of fires on Australian ecosystems. To extend yourself, explore technologies not included in the text.

Success criteria

- I can name at least three natural events that cause environmental change to Australian ecosystems.
- I can describe changes that could occur in an ecosystem after a natural event such as a bushfire.

5·16 ▸ Living systems in context

Learning intention

At the end of this lesson, I will be able to identify factors that lead to a species becoming endangered or extinct, specifically in the Australian context.

Key terms

deforestation: the removal of trees to make land available for other uses

endemic: only found in a certain location or community

fauna: animals

flora: plants and microorganisms

urbanisation: the creation of urban areas such as cities

Content group: Living systems in context

Australia has some of the most unique **flora** and **fauna** in the world. Almost half of all species in Australia are **endemic**. This is what makes them so distinct but also vulnerable to extinction. Australia faces one of the world's most pronounced declines in flora and fauna populations.

Australia has the highest mammal extinction rate

Small mammals are particularly vulnerable in Australia because for millions of years they had no major predators and little competition. This means they are not equipped to cope with major changes in their ecosystems.

After European settlement in the late 1700s, extinction rates increased (see Figure 5.46). Current extinction rates are higher than in any other country. The top contributing factors to the decline in biodiversity are introduced species and habitat destruction.

Figure 5.45: Introduced rabbits compete with the native wallaby for food and other resources because they reproduce much faster.

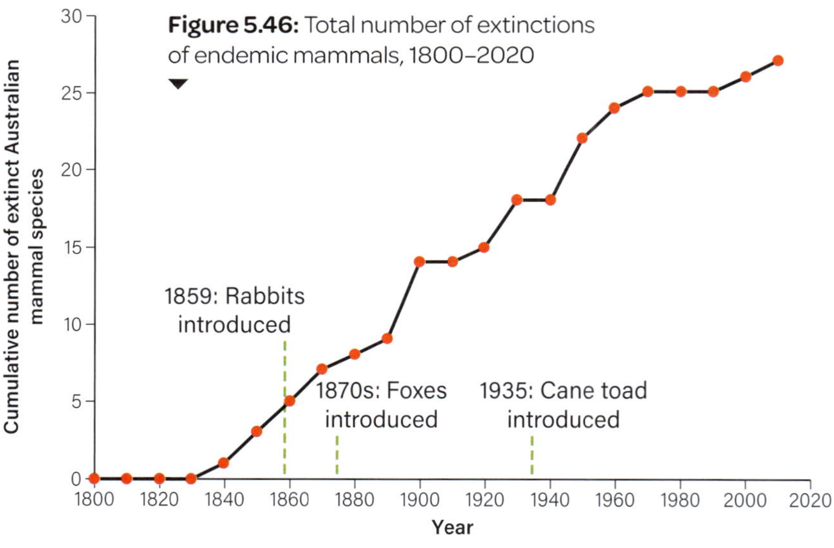

Figure 5.46: Total number of extinctions of endemic mammals, 1800–2020

Introduced species compete with native species

Introduced species compete with native species for the same resources, remove species through predation or toxins, and even damage soils and plant life through trampling, burrowing and digging.

In Australia, most ecosystems have been affected by introduced species. Foxes and cats are skilled predators and have hunted many species to extinction. Rabbits out-compete small marsupials. Cane toads (and their tadpoles) are toxic if eaten by native mammals. Brumbies (feral horses) and camels trample fragile soils and plants that have not evolved to cope with hard-hooved animals. Figure 5.46 shows the sharp increases in mammalian extinction rates after the introduction of rabbits, foxes and cane toads.

Each introduced species was brought to Australia for specific reasons, but the Europeans lacked knowledge about Australian ecosystems and did not foresee the damage the introductions would cause.

Urbanisation and deforestation damage food webs

Urbanisation is the replacement of natural ecosystems with cities and suburbs. As cities grow, the surrounding land is cleared and many of the organisms in the ecosystem die.

Deforestation is the removal of large trees to make space for towns and farms. All organisms within food webs are affected by deforestation because they all depend on the original energy source from these producers.

Removing trees can increase soil erosion and water pollution. This is because tree roots hold the soil together and filter the water flowing through the ecosystem. Without trees, erosion can increase as wind and rain carry away the topsoil needed for plants to grow.

The following animals have become extinct:

- 1960s: desert bandicoot (*Perameles eremiana*)
- 1970s: central hare-wallaby (*Lagorchestes asomatus*)
- 1980s: Christmas Island shrew (*Crocidura trichura*)
- 1990s: Kangaroo River Macquarie perch (*Macquaria* sp.)
- 2000s: Bramble Cay melomys (*Melomys rubicola*)
- 2010s: Lister's gecko (*Lepidodactylus listeri*) (in the wild).

To safeguard Australia's unique natural heritage for future generations, we need to address the causes of biodiversity loss.

▼ **Figure 5.47:** The land covered by this housing development had large trees and an ecosystem for a variety of organisms.

Learning Ladder

Living systems

1. Identify the main factors that contribute to species population decline in Australia.
2. Describe the terms 'urbanisation' and 'deforestation'.
3. Explain how urbanisation can affect interactions in food chains and food webs. Use specific examples in your response.
4. Compare the effects of urbanisation and introduced species. Suggest which factor you think will have the biggest impact on extinct species rates in Australia. Why?
5. **a** Discuss how an introduced species can affect other organisms in an ecosystem, with reference to biotic and abiotic interactions in food webs.
   b Refer to Figure 5.46. If the trend in extinction rates continues, predict the total number of extinct endemic mammals in 2040.

Communicating *see page 343*

1. Identify the quantitative statistic in the opening paragraph of this section.
2. **a** Imagine you are upset about the lack of conservation efforts in your community. Describe an appropriate way to communicate your frustrations to the local council.
   b To support your argument to the local council, you need to use relevant evidence and present it in a digital format.
      i Suggest websites that provide reliable statistics about animal populations in the area.
      ii Suggest digital platforms that could be used to communicate your findings to council.
3. Construct a presentation for your peers about a threatened or endangered species in Australia, using the tools highlighted in Question 2b.
4. Modify your presentation from Question 3 so it can be presented to local council officials.

Success criteria

- I can describe the link between urbanisation and deforestation, and how it contributes to the decline of species populations.
- I can explain the impact of introduced species on interactions in ecosystems.

▸ Living systems summary

Body systems

In organisms made of more than one cell, cells with similar functions group together to form tissues, organs and body systems.

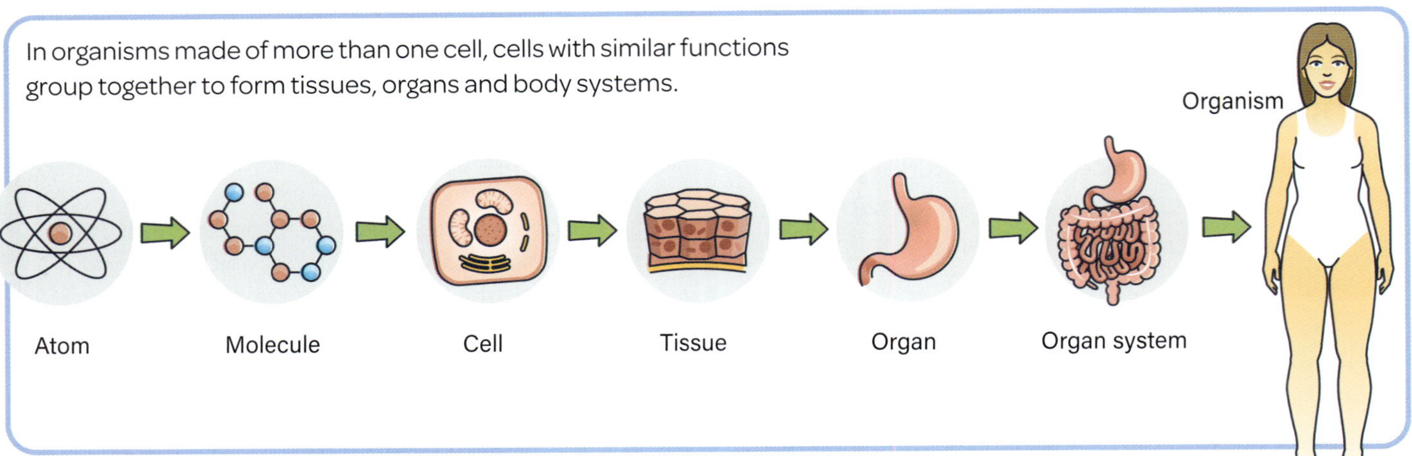

The role of the human **digestive system** is to obtain nutrients and energy from food.

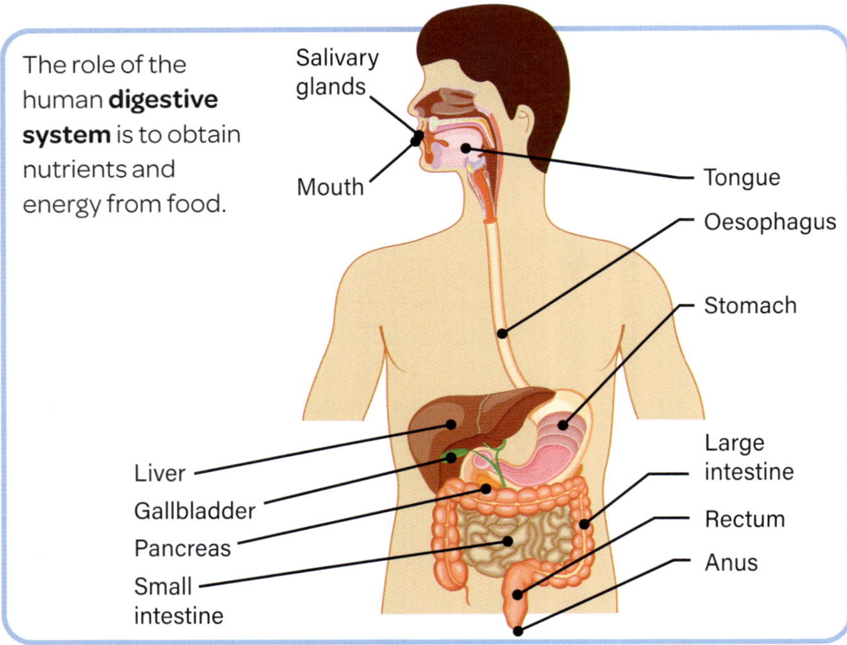

The role of the human **circulatory system** is to transport nutrients and oxygen throughout the body.

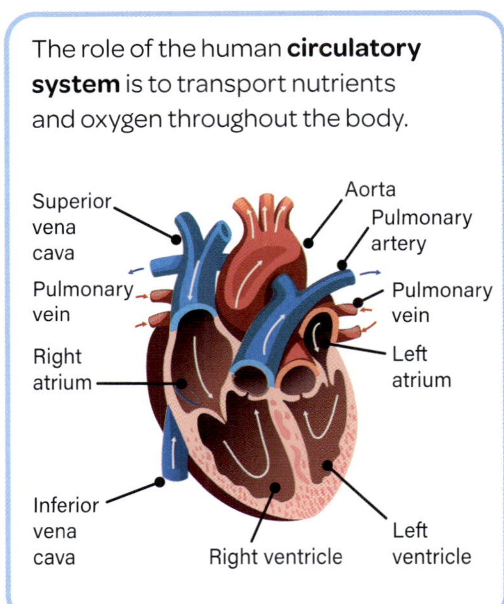

The role of the human **respiratory system** is to bring oxygen into the body and remove waste gases.

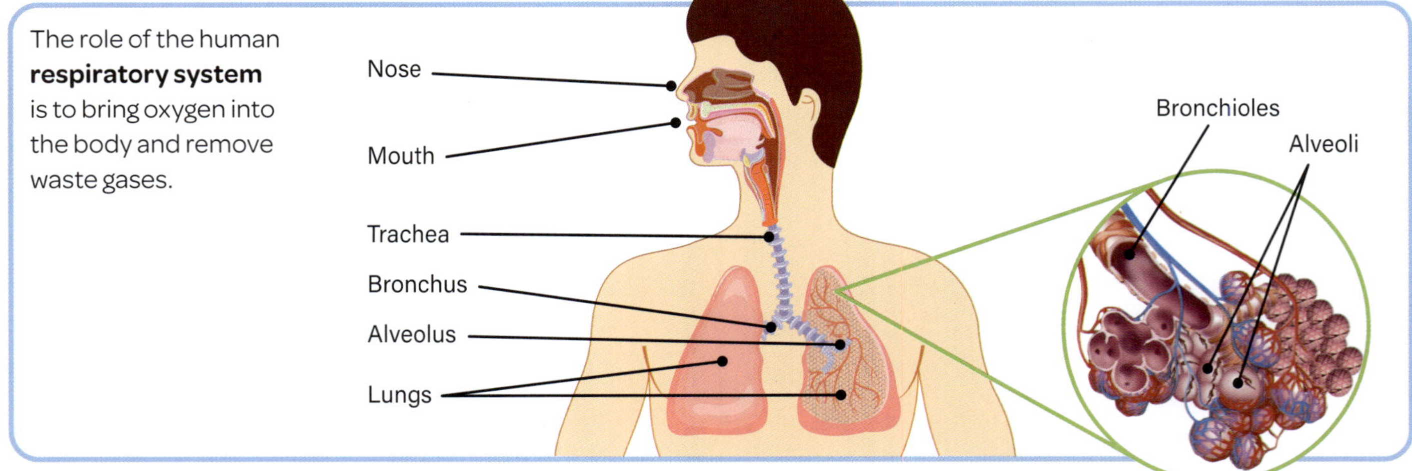

The role of the human **excretory system** is to eliminate cell waste from the body through urine.

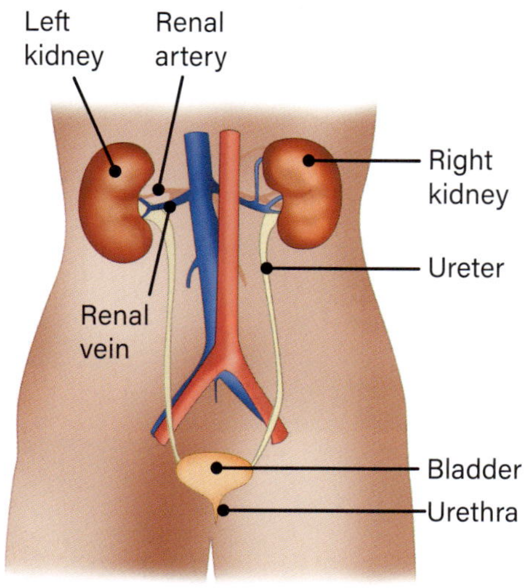

Body systems must work together to provide cells with everything they need to function and survive, such as gases, nutrients and water.

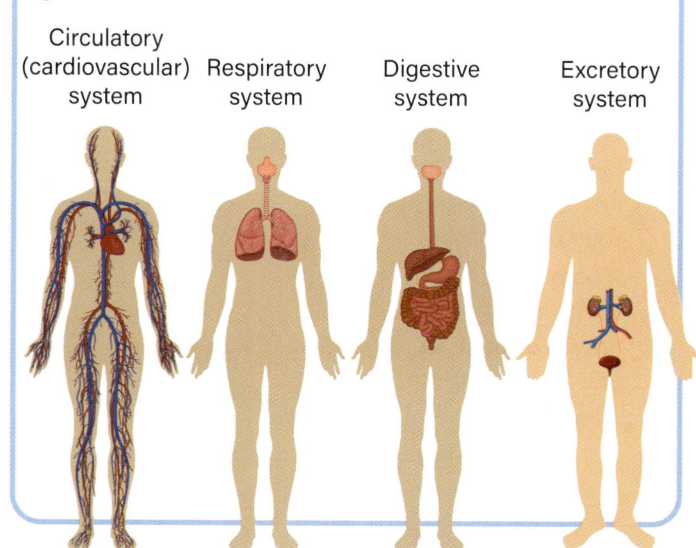

Plant systems

- Plants have two organ systems – the root system and the shoot system – which include the roots, stem, leaves, and reproductive organs like flowers.
- The four tissue systems of plants – the epidermis, vascular tissue, ground tissue and meristematic cells – are made up of specialised cells that perform important functions.
- The xylem and phloem are the two primary tissues that form tubular structures that transport water, nutrients and food throughout a plant:
  - The **xylem** only allows a flow of water and nutrients in one direction: up from the roots.
  - The **phloem** allows a two-way flow of substances up and down the shoot and root systems.

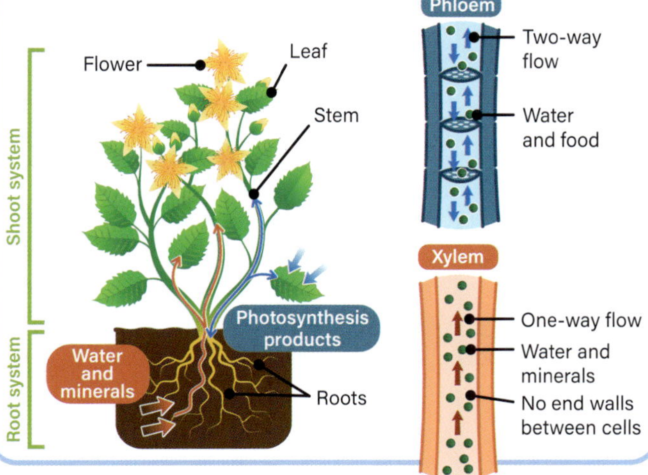

- The male reproductive part of a flower is called the **stamen**. The female reproductive part of a flower is called the **pistil**.
- **Pollination** is the movement of pollen from the male part of a flowers to the female part of a flower. Pollen is moved from the stamen to the pistil to fertilise the ovum (egg), which is required for plants to reproduce.

Parts of a flower

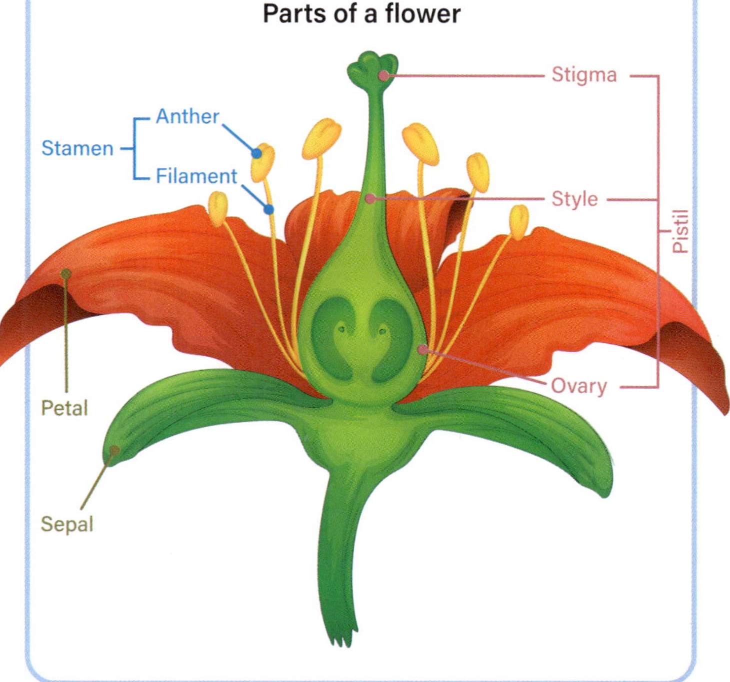

Ecosystems

- An **ecosystem** is a community of living things interacting with one another and the non-living things in their environment.
- A **community** is all the living things in an ecosystem that rely on each other to survive and interact in different ways.
- A **habitat** is the natural environment where a species or an organism lives that supplies food, water, shelter and protection, and enables reproduction.
- **Energy** enters most ecosystems as solar energy from the Sun.

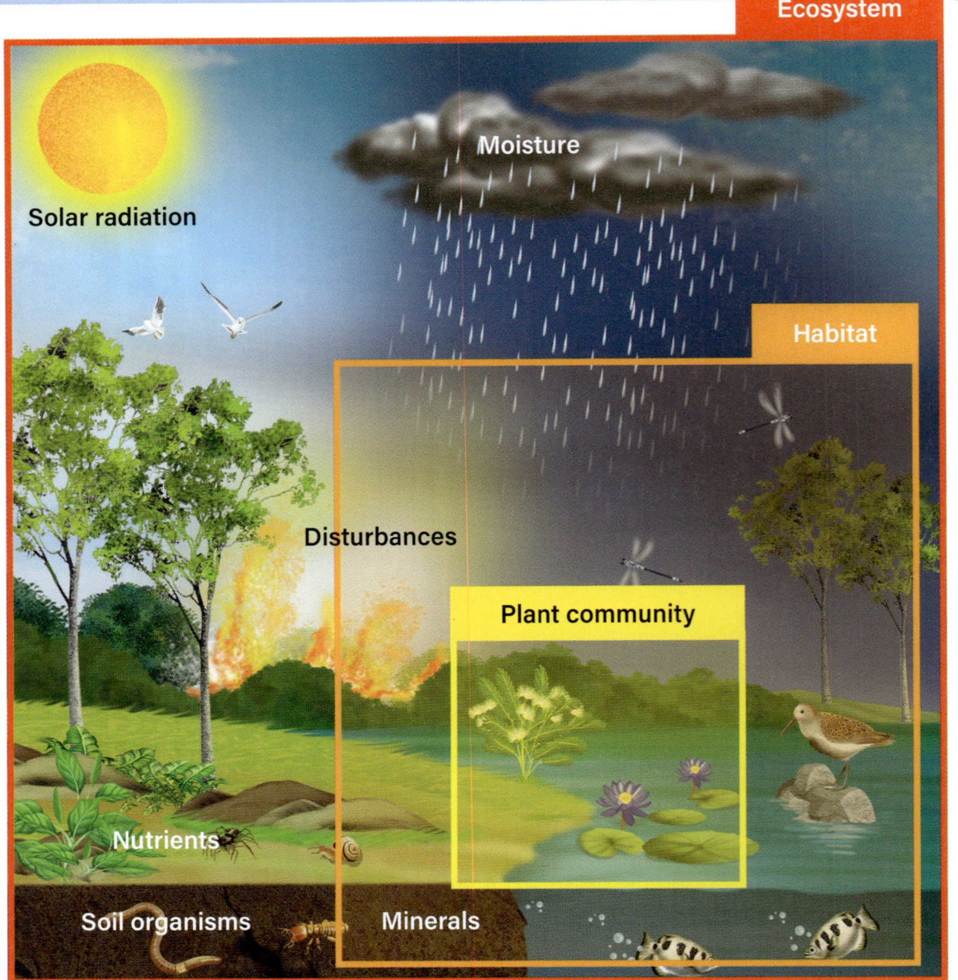

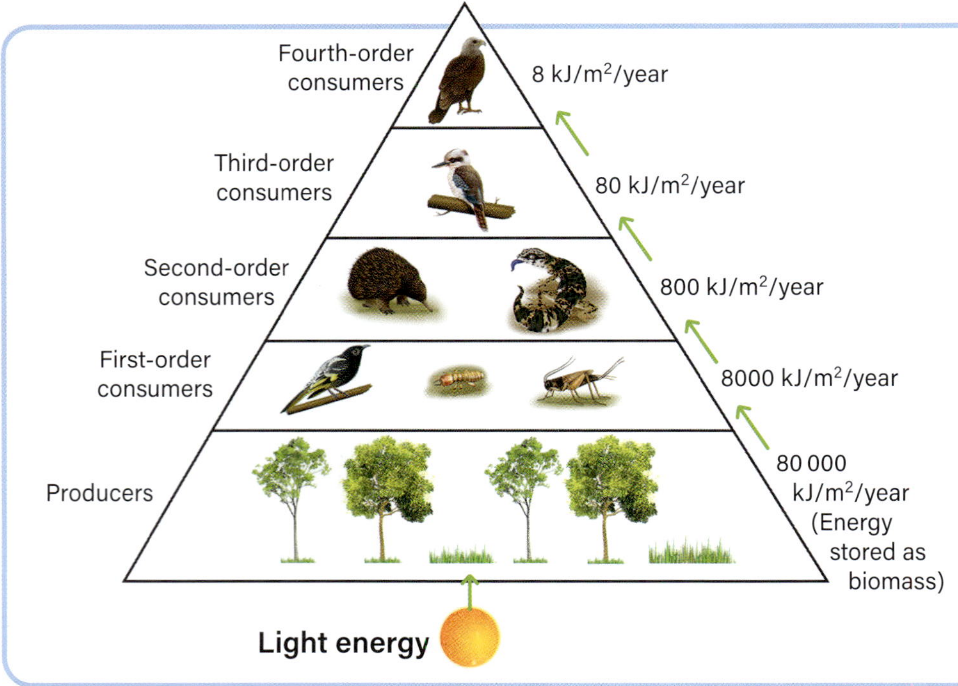

- The living things in an ecosystem are called the **biotic** factors.
- The non-living things – such as soil and water – are called the **abiotic** factors.
- Matter cycles through different parts of ecosystems, including the atmosphere, soil and different organisms.
- The feeding relationships between organisms, and the energy transfers in an ecosystem, can be shown using food chains, food webs and energy pyramids.

- Natural events – such as droughts, floods and high-intensity bushfires – can impact biodiversity and harm ecosystems.
- Low-intensity bushfires are helpful in some ecosystems. First Nations Peoples have used low-intensity bushfires to successfully manage Country for thousands of years.

- Human activities – such as farming, the use of fossil fuels, removing or introducing species – can dramatically impact the feeding relationships in an ecosystem.
- Australia has some of the world's highest rates of species population decline and extinction due to increased vulnerability to introduced species, urbanisation and deforestation.

Masterclass

Steps in progression

		1	2
Content	Living systems	Identify three land-dwelling apex predators.	**a** Name the body system that is made up of bones, muscles and joints. **b** Describe one of the purposes of this body system for apex predators.
Processes	Questioning and predicting	Suggest why: **a** saltwater crocodiles have no natural predators. **b** kangaroos are not apex predators.	Complete the following question so it investigates a scientific concept: How does a wedge-tailed eagle's _______ make it an effective _______?
Processes	Processing data and information	Categorise the apex predators in Table 5.6 into three categories based on their lifespans.	Construct a table to organise the data of the number of apex predators in their natural habitats, for every year over the past five years (see Table 5.6).
Processes	Communicating	Identify two pieces of scientific information about the orca and the grey wolf.	How would you choose to communicate information about the average lifespan of apex predators?

From body systems to ecosystems

What do body systems have to do with ecosystems? Is there a relationship between the organisms at each trophic level of ecosystems and how their organ systems operate? Consider that plants, the producers, are at the bottom of the food chain, and their organ systems are significantly simpler than animals' body systems. Let's now consider some of the animals that occupy the highest trophic level of ecosystems throughout the world.

Apex predators are predators at the top of the food chain. To be an effective predator, we assume these animals must be fast and strong. Bones, muscles and joints make up an animal's musculoskeletal system. Ligaments and tendons around joints enable muscles to control skeletal movement, which is responsible for providing animals with the strength and agility required to hunt, chase and capture prey. Could this be the body system that sets them apart from other animals in their ecosystems?

Or maybe the status of these animals as apex predators is unrelated to body systems. Perhaps their excellent predator skills come from structural, behavioural and/or physiological adaptations?

Demonstrate your understanding

3	4	5	
Explain how tendons and ligaments are involved in the musculoskeletal system, even though they are not muscle or bone.	Compare the adaptations of the apex predators in Table 5.6. Select the adaptation that you think is most likely to contribute to the animal's success as a predator.	Predict how producers in an ecosystem might be affected if an apex predator were removed from the ecosystem.	
Construct a question that investigates how one apex predator's adaptation relates to its trophic level.	Write a hypothesis to predict what would happen if all the apex predators in Table 5.6 were placed in one ecosystem.	Construct a series of clear questions that could be used to test your hypothesis from Step 4.	Science how-to 319
Construct a scatter plot to see if there is a relationship between the average lifespan of an apex predator and the maximum speed it can travel (see Table 5.6).	Construct an infographic to display the maximum speed at which each apex predator in Table 5.6 can travel.	**a** Discuss the links between the qualitative and quantitative information in Table 5.6. **b** Conduct research and find data that supports or refutes the information in Table 5.6.	Science how-to 329
Explain how you might present information about the adaptations of apex predators using a named digital technology.	Construct a slide presentation to communicate the information about the apex predators listed in Table 5.6. Also include information acquired from additional research.	Prepare speaker note cards to help you present the slides you prepared in Step 4 to a group of your peers.	Science how-to 343

Figure 5.48: Great white sharks were thought to be the apex predators of the ocean until it was discovered that orcas prey on sharks.

Table 5.6: Common apex predators

Apex predator	Lifespan (years)	Maximum speed (km/h)	Adaptations
Lion	15–30	74	Strong claws and paws
Siberian tiger	15–25	96	Nocturnal hunters and night vision
Saltwater crocodile	60–100+	29	Water-tight flaps over eyes, ears and throat
Orca	30–90	56	Hunt in groups
Bald eagle	20–50	40 (flying)	Sharp, pointed beak
Grey wolf	6–18	60	Long legs
Dingo	4–16	60	Hip and limb flexibility
Polar bear	15–45	40	White fur for low visibility
Wedge-tailed eagle	20–45	35 (flying)	Binocular vision

Focus area 6

6.0 Periodic table and atomic structure

Our physical world is made up of over 100 chemical elements. These can be organised into groups based on their physical and chemical properties. The periodic table is a tool created by scientists to classify elements.

Learning Ladder

The Learning Ladder contains the scientific content and processes you will learn in this chapter. Each area has five levels of progression. To move to higher levels, you need to practise activities at the earlier levels. This will help you develop the ability to complete tasks that are more complex.

Steps in progression	Content: Periodic table and atomic structure	Working scientifically processes: Processing data and information	Working scientifically processes: Analysing data and information	Steps in progression
5	I can investigate the use of elements and compounds based on their properties.	I can process and interpret quantitative and qualitative data from a range of sources.	I can evaluate data and information for accuracy, reliability and validity.	5
4	I can explain how different elements and compounds are used.	I can collect and present data in a range of appropriate formats.	I can draw conclusions based on patterns in data and information.	4
3	I can compare the properties of elements and compounds, and relate them to the periodic table.	I can construct appropriate graphs with headings and units for data.	I can explain relationships between datasets and information.	3
2	I can describe the structure and properties of an atom, element or a compound.	I can construct appropriate tables with headings and units for data.	I can describe trends from collected data and information.	2
1	I can identify that everything is made up of different types of matter.	I can identify data from graphs, tables and digital sources.	I can identify trends within given data and information.	1

Figure 6.1: The organisation of the periodic table is closely linked with the atomic structure of the elements.

6·1 ▸ Classifying elements

Figure 6.2: Aluminium is a lightweight and malleable metal, which means it is a good material for packaging food and drinks.

Learning intention

At the end of this lesson, I will be able to:

- distinguish between metals and non-metals based on their properties
- describe some of the uses of metals and non-metals.

Key terms

atom: a particle that makes up all matter; made up of protons, neutrons and electrons

brittle: not able to be bent; will break or shatter if stressed

ductile: can be drawn out into a wire

element: a pure substance made of one type of atom

malleable: able to be bent and shaped

metalloid: an element with properties of both metals and non-metals

property: a characteristic of a substance that affects what it looks like and how it behaves

substance: matter that has a fixed chemical make-up

Investigation 6.1

Comparing metals and non-metals, p.448

Content group: Classification of matter

Figure 6.3: Sulfur is a non-metal element; it is dull yellow in colour and is brittle.

An **element** is a pure **substance** made of a single type of **atom**. Every object in the world is made of either one type of element or a combination of elements.

Each type of element has different **properties**. This means that different elements are suited to be used for different purposes. Elements can generally be classified into two groups: metals and non-metals.

Non-metal elements share many properties

Non-metal elements share many of the same properties. The properties of non-metal elements include:

- low melting temperature
- poor electrical conductivity (electricity does not easily flow through the element)
- **brittle** (not **malleable**)
- low boiling temperature
- poor thermal (heat) conductivity (heat does not easily pass through the element)
- dull in appearance (not shiny).

Oxygen is a very common non-metal element. In its standard form, oxygen is a gas. Every animal on Earth needs oxygen to make energy for living cells. Burning things in combustion reactions is another example of something that requires oxygen; without it, we would not be able to use stoves, engines, rockets or gas heaters.

Carbon is another common non-metal element. Pure carbon has several different forms. Diamonds and graphite (the 'lead' in your writing pencil) are both pure carbon. The properties of pure carbon vary, depending on its form. For example, graphite is dull but can conduct electricity, while a diamond does not conduct electricity but is shiny. These are exceptions to the general properties of non-metals.

Metal elements have the opposite properties of non-metal elements

Metal elements generally have the opposite properties of non-metal elements. The properties of metal elements include:

- high melting temperature
- high boiling temperature
- good electrical conductivity
- good thermal (heat) conductivity
- malleable
- lustrous (shiny).

Aluminium is a common metal element. Aluminium is useful because it is very strong and relatively lightweight. This makes it an excellent material for making things such as aircraft and bridges. Aluminium is also used in food packaging and storage – such as in aluminium foil and soft-drink cans – and to protect products (such as some medicines) from air, light and moisture.

Copper is another common metal that is used for many purposes. It is used for electrical wiring because it is a good conductor of electricity and is not only malleable but also **ductile**. This means it can be stretched out into a long wire without breaking.

Figure 6.4: Copper is a metal that is a good conductor of electricity and is malleable, which means it is a good material to be used for electrical wiring.

Metalloids have properties of both metals and non-metals

There are several elements that cannot be easily classified as either a metal or a non-metal because they have properties of both. These elements are called **metalloids**. Metalloids are usually brittle solids. Metalloids can be used in many ways. For example, **silicon** is a semi-conductor of electricity, meaning it has insulator properties but can also conduct electricity under certain conditions. For this reason, it is used in a variety of electronics.

Figure 6.5: The properties of the metalloid silicon mean that it is a good material to use for a range of electronic products, including microchips.

Learning Ladder

Periodic table and atomic structure

1. Define the term 'element'.
2. Identify two pure elements. What are their properties?
3. **a** Compare the properties of the two elements you identified in Question 2 by identifying their similarities and differences.
   b Discuss whether each element's properties match those of metals or non-metals.
4. **a** Match each element with its common use:

oxygen	food storage
aluminium	lighting a fire
silicon	insulative electronics

   b Explain why oxygen, aluminium and silicon are used in these ways. Link your explanations to the properties of each element.
5. Graphite and diamonds are two forms of pure carbon.
   a Identify the different uses for graphite and diamonds.
   b How are these uses linked to their properties?
   c Investigate additional uses for graphite and diamonds.

Processing data and information *see page 329*

Read Investigation 6.1 on page 448 and then answer the following questions.

1. **a** Outline the substances that will be tested in this investigation.
   b What properties will be tested and therefore included in the results table?
2. Using the information from Question 1, construct a results table that could be used to record data collected during this investigation.
3. **a** Based on the properties of each substance, classify each substance as a metal or a non-metal.
   b Explain how you could represent these investigation results in a visual format.

In context

Identify a structure or object that is made entirely of metal. Suggest why it is made of metal by describing the properties of the metal that was used to make the structure or object.

Success criteria

- I can distinguish between metals and non-metals based on their general properties.
- I can describe two uses of metals and two uses of non-metals.

6·2 ► Elements, compounds and alloys in action

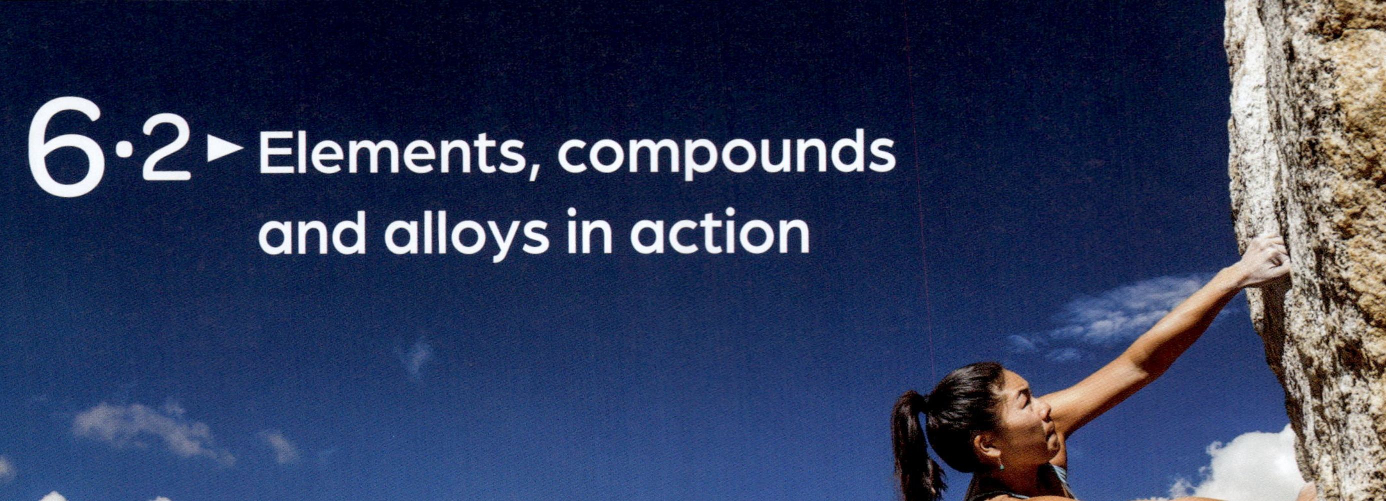

Figure 6.6: Oxygen is a versatile and important element, whether in elemental form in the air we breathe or in compound form like in the rocks we climb.

Learning intention

At the end of this lesson, I will be able to describe the properties and uses of some common elements, compounds and alloys.

Key terms

alloy: a metal mixed with other metals or non-metals

compound: a substance containing atoms of two or more elements; the atoms are chemically bonded together and are present in the compound in a fixed ratio

Investigation 6.2

Comparing the properties of elements, compounds and alloys, p.449

Content group: Classification of matter

Figure 6.7: A ruby is a compound called aluminium oxide.

Elements, compounds and alloys have different properties and are used in many ways in everyday life. Let's explore these substances in action.

Elements and compounds are pure substances

A **pure element** is a substance made of one type of atom. For example, carbon (C) and oxygen (O) are pure elements. A **compound** is a substance made up of atoms from two or more elements. For example, carbon dioxide (CO_2) is a compound made from the elements carbon and oxygen.

Compounds are often referred to by their chemical formulas. These formulas are expressed using chemical symbols and subscript numbers. For example, water has the chemical formula H_2O; this means that there are two hydrogen (H) atoms for every one oxygen (O) atom; this is a 2:1 ratio. This ratio will never change. Water will always have exactly two hydrogen atoms for every one oxygen atom.

If you change the ratio of atoms in a compound, you change the substance altogether. For example, if you add one oxygen atom to a water molecule (H_2O), it is not water anymore. It is now H_2O_2, which is hydrogen peroxide, a common antiseptic.

Pure elements have different properties to their compounds

Carbon forms many compounds with other elements. All living things contain large amounts of carbon. So does crude oil (oil that has not been separated into usable petroleum products) because it is a fossil fuel formed from the remains of once-living things. Crude oil is used for manufacturing petrol, engine oil, candle wax and plastics. Pure carbon can have very different properties to compounds of carbon. For example, diamonds are pure carbon, while the compound carbon monoxide (CO) is a gas.

Oxygen also forms many compounds with other elements. One of these compounds is water (H_2O), which contains hydrogen (H) and oxygen (O) (see Figure 6.8). When oxygen is part of a compound, its properties can be extremely different. For example, the oxygen we breathe is a gas at room temperature, while water is a liquid.

The oxygen we breathe has very different properties from the rocks we climb. The compound silicon dioxide (SiO_2) – made up of silicon (Si) and oxygen (O) – is found in all forms of rock. Silicon dioxide (SiO_2) forms a non-metallic, solid lattice (a three-dimensional shape made of a repeating pattern of atoms) with a hard, rough texture, making it an ideal type of rock for climbers (see Figure 6.9).

Figure 6.8: Water is a compound. The chemical formula for water is H_2O.

Another compound of oxygen is **aluminium oxide** (Al_2O_3), which is made up of the elements aluminium and oxygen. This compound forms a similar lattice as silicon dioxide, but with a different ratio of atoms that gives a crystalline appearance and shine. 'Crystalline' means having the structure and form of a crystal, with well-defined edges and faces. Aluminium oxide is known as ruby or sapphire, depending on whether it is red or blue, but the chemical and physical structure is the same in both gems.

Figure 6.9: A molecular model of the compound silicon dioxide

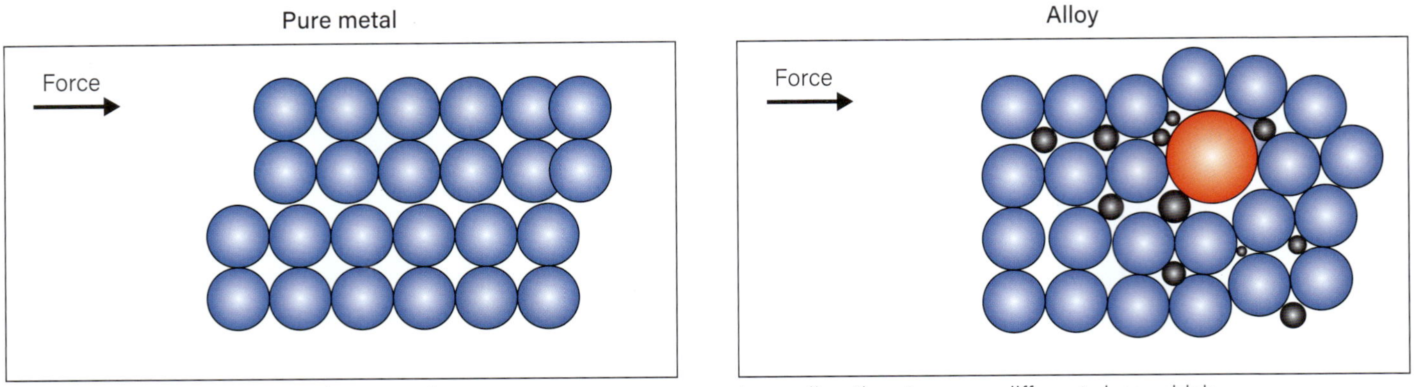

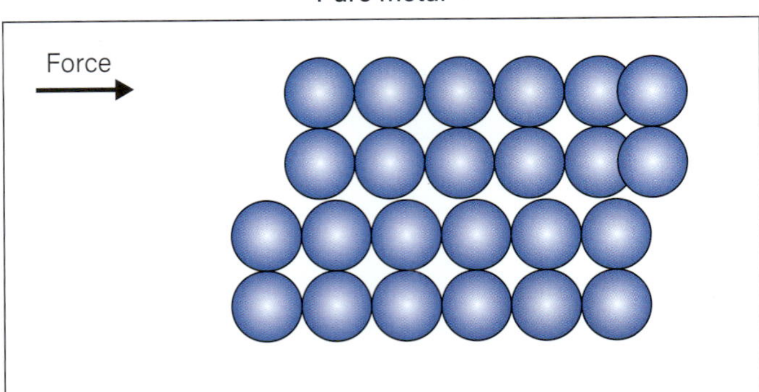

In a pure metal, the atoms are the same size and can slip past each other.

Alloy

Force

In an alloy, the atoms are different sizes, which prevents the atoms slipping past each other.

▲ **Figure 6.10:** Atoms in pure metals and alloys

Alloys are metals that have been chemically modified

An **alloy** is formed when a pure metal is mixed with another element (usually a different metal or carbon). For example, bronze is an alloy made from copper and tin; carbon steel is an alloy made from iron and carbon. In general, alloys are harder than pure metals.

For example, iron is a metal element. In its pure form, iron is soft and rusts easily. To create a stronger, rustproof substance, iron is mixed with carbon to make the alloy steel. Steel is very strong and more resistant to rust, so it is widely used for tools and construction.

Figure 6.11: Carbon steel is an alloy made from iron and carbon. This type of steel is used in many buildings and constructions, as it is very strong and rustproof.
▼

Mohs' scale compares the hardness of substances

In 1812, a German geologist, Friedrich Mohs, developed a scale for comparing the hardness of substances. Mohs' hardness scale is a chart of the hardness of various substances (see Figure 6.12). Hardness is ranked from 1 (very soft) to 10 (very hard). Hard minerals (those with larger numbers on the scale) can scratch softer minerals (those with a smaller number on the scale).

For example, diamonds, a pure form of carbon, are one of the hardest natural substances on Earth, so diamonds can scratch most other solid substances. This is why diamonds score a 10 on Mohs' scale. Interestingly, graphite, another pure form of carbon, is at the opposite end of the scale with a score of 1.

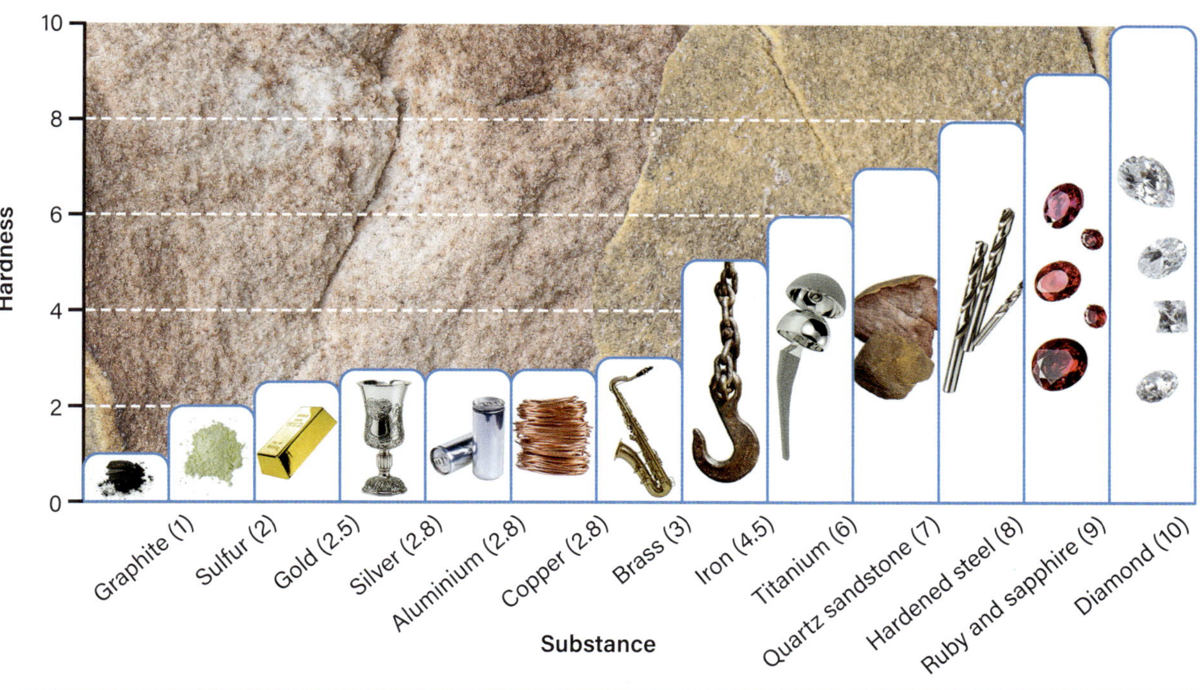

Figure 6.12: Mohs' hardness scale ranks the hardness of substances from 1 to 10.

Learning Ladder

Periodic table and atomic structure

1. Classify each of the following substances as either a compound or an element.
   - **a** Water (H_2O)
   - **b** Oxygen gas (O_2)
   - **c** Diamond (C)
   - **d** Petrol (C_8H_{18})
   - **e** Brass (copper alloy)
   - **f** Platinum (Pt)
2. Describe the general difference between a pure metal and an alloy.
3. Carbon steel is made up of at least 98 per cent iron. Explain why this type of steel is so much harder than pure iron.
4. **a** Match each element with the compound it is used to make.

Element	Compound
oxygen	steel
iron	rock
carbon	plastic

   b For each element and compound pair in Question 4a, explain how the properties of the pure element are different from the properties of the compound.

   c Give an example of how each compound in Question 4a is used. Explain why each compound is suitable for this use.

Analysing data and information

see page 334

1. For each of the following pairs of substances, identify which substance is harder:
   - **a** iron or steel.
   - **b** silver or sterling silver alloy.
   - **c** bronze or copper.
   - **d** solder alloy or lead.

Refer to Figure 6.12 to answer Questions 2, 3 and 4.

2. **a** Describe the general trend in hardness across metals, non-metals, compounds and alloys.

   b Identify any substances in the graph that do not fit the trend.
3. Consider copper, aluminium and silver. Based on their hardness values, what can you assume about their physical structures?
4. Imagine Figure 6.12 presents the results of an investigation that addressed the aim 'To investigate the difference in hardness between metals, non-metals, compounds and alloys'. Construct a conclusion based on patterns in the data.

In context

You have been asked to build a rocket for Australia's space program. Compare aluminium and carbon steel. Which material would you select to build the rocket? Explain your decision.

Success criteria

- I can describe the properties and uses of some common elements, compounds and alloys.

6·3 ▸ Matter is made of atoms

Figure 6.13: Nitrogen is an element.

Learning intention

At the end of this lesson, I will be able to:

- identify that all matter is made of atoms, which are made up of protons, neutrons and electrons
- describe the structure of atoms in terms of the nucleus, protons, neutrons and electrons.

Key terms

electrically neutral: having an equal number of protons (positively charged) and electrons (negatively charged)

electron: a subatomic particle that orbits the nucleus of an atom; it is negatively charged

electron configuration: the number of electrons in each shell of an atom, starting with the smallest; for example, 2, 8, 8, 2

electron shell: the space around an atom's nucleus, in which electrons circulate

matter: any substance that has mass and takes up space

molecule: a unit made up of two or more atoms chemically bonded to each other

neutron: a subatomic particle located in the nucleus of an atom; it is neutrally charged

nucleus: the centre of an atom, which contains protons and neutrons

proton: a subatomic particle located in the nucleus of an atom; it is positively charged

subatomic particles: particles that make up atoms: protons, neutrons and electrons

valence shell: the electron shell furthest away from the nucleus of an atom; occupied by electrons

Content group: Atomic structure

Everything you see around you is made of matter. The atom is the smallest particle of matter. Atoms are made up of tiny subatomic particles called protons, neutrons and electrons.

Matter can be classified

You will remember from Chapter 4 that **matter** can be classified as pure substances and impure substances (mixtures). In this section, we will focus on pure substances. Pure substances include elements and compounds (see Figures 6.14 and 6.15).

Figure 6.14: Salt is a compound.

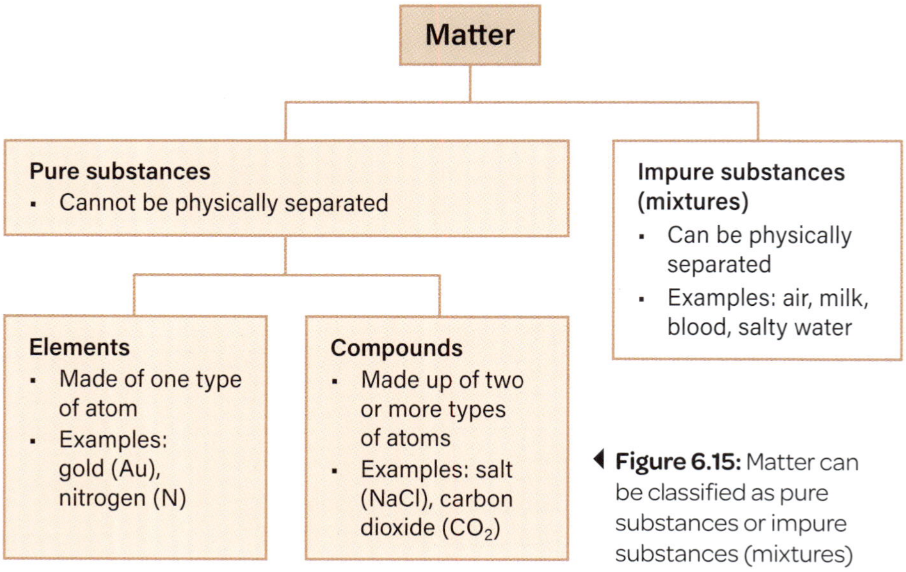

Matter

- **Pure substances**
  - Cannot be physically separated
  - **Elements**
    - Made of one type of atom
    - Examples: gold (Au), nitrogen (N)
  - **Compounds**
    - Made up of two or more types of atoms
    - Examples: salt (NaCl), carbon dioxide (CO_2)
- **Impure substances (mixtures)**
  - Can be physically separated
  - Examples: air, milk, blood, salty water

Figure 6.15: Matter can be classified as pure substances or impure substances (mixtures)

Molecules can be elements or compounds. Atoms of the same type of element can bond to form molecules, such as two oxygen atoms bonding to form an oxygen molecule. Glucose (sugar) is an example of a molecule that is also a compound because it contains atoms of more than one type of element.

Atoms are made up of particles

Atoms are the basic building blocks of matter. Atoms are made up of three types of **subatomic particles**:

- **protons**, which have a positive charge
- **neutrons**, which have no charge
- **electrons**, which have a negative charge.

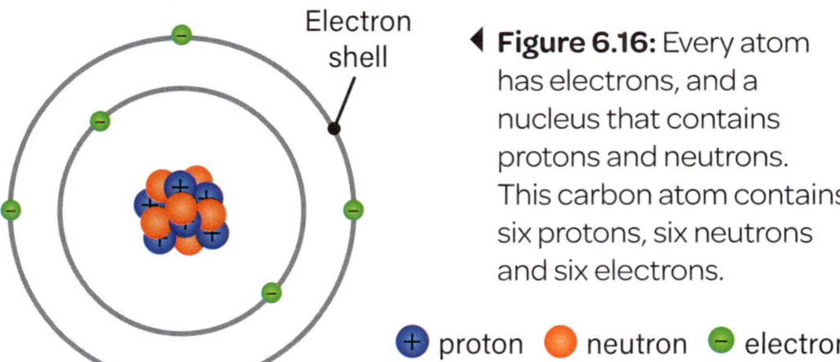

Figure 6.16: Every atom has electrons, and a nucleus that contains protons and neutrons. This carbon atom contains six protons, six neutrons and six electrons.

Electrons are much smaller than protons and neutrons. This means that almost all the mass of an atom is in the **nucleus**.

Atoms are held together by strong attractions between the protons and electrons, which are attracted to each other because they are oppositely charged, like opposite poles of a magnet. The electrical charge of a proton is exactly equal and opposite to the electrical charge of an electron. An atom is **electrically neutral**, which means it has no overall charge because it contains equal numbers of protons and electrons; the positive and negative particles cancel out.

Electrons circulate in shells

Electrons circulate around an atom's nucleus in **electron shells**. An atom's electrons are distributed across the electron shells using the 2, 8, 8, 2 **electron configuration**, starting with the smallest, innermost shell (that is, the shell that is closest to the nucleus) (see Figure 6.17).

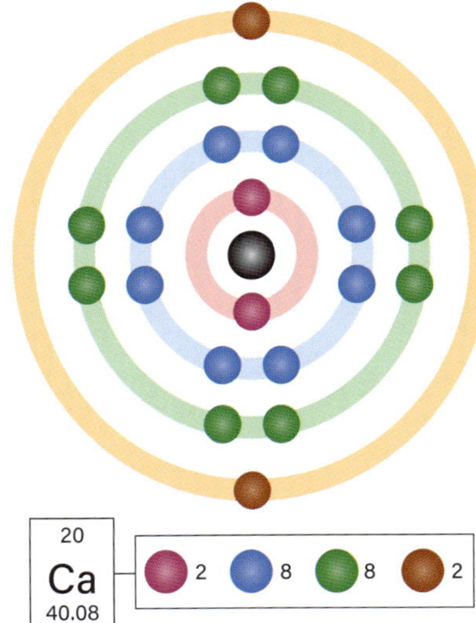

Figure 6.17: A calcium atom has 20 electrons, which are distributed across four electron shells. There are two electrons in its **valence shell**.

Table 6.1: Properties of the subatomic particles of electrons, protons and neutrons

	Electron	Proton	Neutron
Symbol	e^-	p^+	n^o
Charge	1^-	1^+	0
Relative mass	$\frac{1}{1840}$	1	1
Location	Electron cloud around the nucleus	Nucleus	Nucleus

Learning Ladder

Periodic table and atomic structure

1. Define matter with reference to atoms.
2. Describe the structure of a calcium atom (see Figure 6.17) using key terms from this section. Draw a labelled diagram to support your response.
3. Compare protons, neutrons and electrons.
4. Explain the difference between an element and a compound.
5. Discuss some of the uses of the compound glucose.

Processing data and information *see page 329*

1. Refer to Table 6.1. Identify:
   - **a** the symbol used to represent an electron.
   - **b** the charge of a proton.
   - **c** the relative mass of an electron.
   - **d** where neutrons are located in an atom.
2. Construct a table to present the information in Figure 6.15.
3. Construct a column graph (see page 370 in the Science how-to section) to represent the number of electrons in each shell of a calcium atom.

In context

A carbon atom usually has six protons and six neutrons. However, some carbon atoms have seven or eight neutrons. Research the significance of this and how it can be useful in forensic science.

Success criteria

- I can describe the structure of atoms, including the subatomic particles (protons, neutrons and electrons).

6.4 ▸ Changing ideas of the atom

Figure 6.18: Rutherford used a piece of gold foil in his famous experiment that tested Thomson's atomic model.

Learning intention

At the end of this lesson, I will be able to:

- outline historical developments of ideas of the atom
- explain how new technologies have led to a more detailed understanding of atomic structure.

Key terms

alpha particle: a positively charged particle that is four times larger than a proton

cathode ray: a stream of electrons observed in a high-vacuum tube

spectral line: a dark or bright line in an otherwise uniform spectrum

Investigation 6.4
Flame test, p.451

Content group: Atomic structure

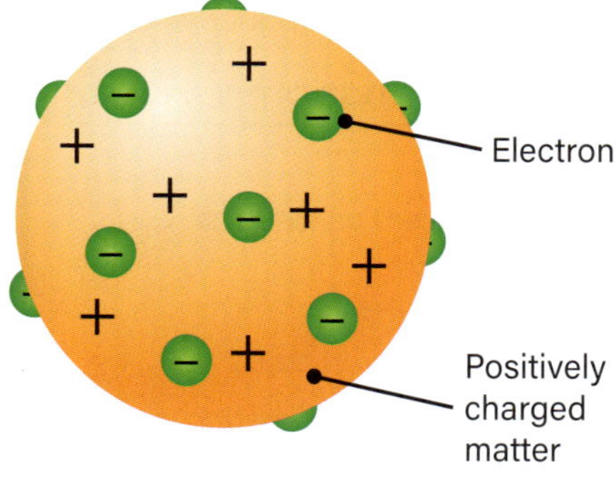

Figure 6.19: Thomson's plum pudding model describes the atom as a positively charged sphere (the 'pudding') with negatively charged electrons (the 'plums') embedded in it.

Early ideas developed by scientists described atoms as tiny spheres that could not be split into anything simpler. This idea was difficult to test because atoms are not visible to the naked eye. Over time, many scientists have collaborated and built on each other's work to develop the atomic model we use today. Some of the developments in our understanding of atomic structure (that is, the nucleus, protons, neutrons and electrons) have been made possible by new technology.

Democritus proposed that everything is made of atoms

Democritus was an ancient Greek philosopher (c. 460–370 BCE). His greatest contribution to science was to suggest that all substances are made of small particles that cannot be divided or destroyed. He called these particles *atomos*, which means 'cannot be cut or divided' in ancient Greek.

Dalton proposed that compounds are made of two or more different types of atoms

In 1803, English chemist John Dalton proposed his atomic theory. He suggested that all matter is made of tiny, indestructible particles that he called atoms. He also proposed that all the atoms that make up a particular element are identical in size, mass and other properties. For example, every carbon atom is exactly the same as every other carbon atom, in every way. Dalton concluded that each element has unique atoms and that atoms of different elements combine with each other in simple ratios to form new substances, called compounds.

Thomson discovered electrons

At the beginning of the 20th century, English physicist J.J. Thomson studied the effect of passing **cathode rays** through gases. A cathode ray tube is a glass tube that fires a beam of negatively charged particles. By observing how the cathode ray beam behaved, Thomson identified the particles as electrons.

Thomson proposed that if every atom contains negatively charged particles, then there must also be positively charged material in every atom to balance the electrical charge. His model was described as the plum pudding model in which the atom consists of a positively charged sphere with electrons embedded in it (see Figure 6.19).

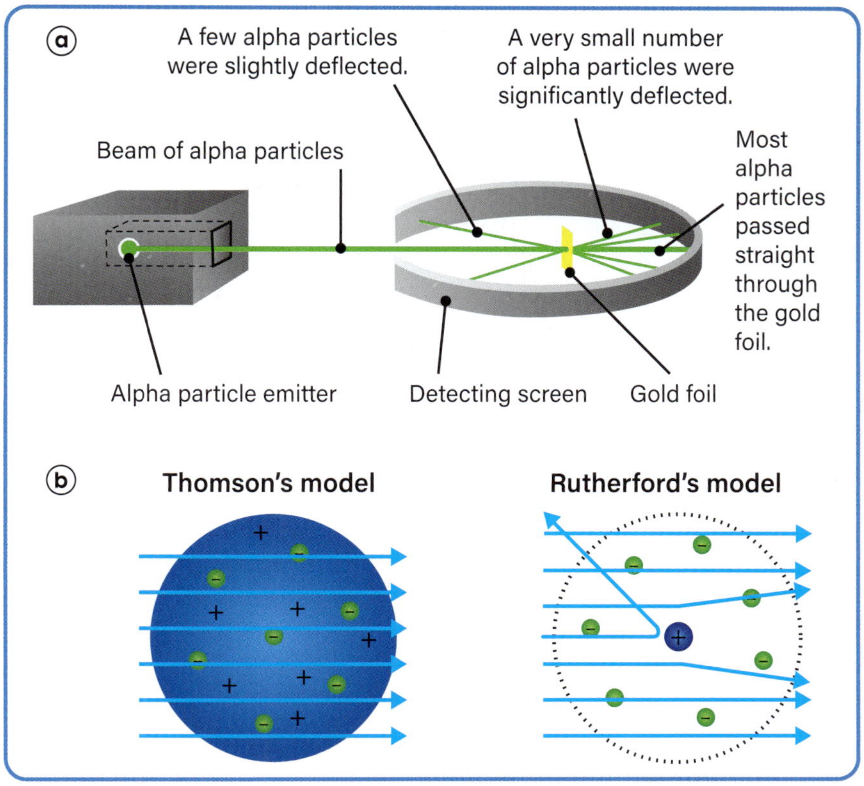

Figure 6.20: (a) Rutherford's famous experiment bombarded a sheet of gold foil with positively charged alpha particles. This experiment showed that the atom is mostly empty space, with a small, dense, positively charged nucleus. (b) According to Thomson's plum pudding model, there was nothing dense or heavy enough in the atom to deflect the alpha particles. However, Rutherford's observations could only be explained by the presence of a dense, positively charged nucleus.

Rutherford proposed the nuclear model of the atom

Ernest Rutherford was a New Zealand-born physicist. Working in England in around 1910, he tested Thomson's plum pudding model by bombarding a piece of gold foil with **alpha particles**, which are small and positively charged.

If the plum pudding model was correct, all the alpha particles would pass through the gold foil. This is because the plum pudding model assumed an atom's positive charge was spread out throughout the entire volume of the atom. This would mean that the atom's overall charge would be too weak to significantly affect the path of the alpha particles, which are relatively massive and fast moving.

However, although most of the alpha particles went straight through the gold foil, a small percentage were deflected at large angles (see Figure 6.20). This means that a small proportion of the particles changed direction after meeting the foil; their pathway was altered from passing directly through the foil in a straight line. This could be explained by the atom in the gold foil having a very small, yet dense, positively charged nucleus at its centre. This led Rutherford to conclude that the atom is mostly empty space with a tiny, central nucleus and electrons orbiting around it. Rutherford later identified that the positive components of the nucleus are protons.

Bohr proposed the concept of electron orbits

In 1913, Danish physicist Niels Bohr worked out mathematically that the electrons in Rutherford's model must move around an atom's nucleus in orbits of fixed sizes and energy levels, like planets orbiting the Sun.

This energy was explained by emission spectra. This is where atoms absorb and emit (send out) a fixed amount of radiation when the electrons move between the orbits or electron shells, which are at specific distances from the nucleus. This produces a series of regular **spectral lines** (see Figure 6.21 on the next page).

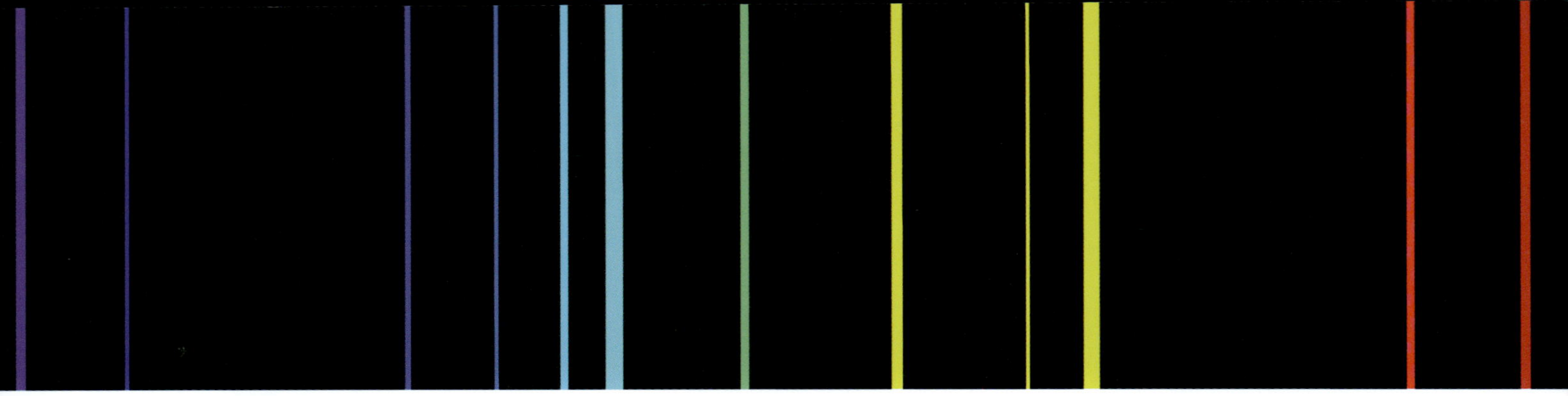

Figure 6.21: The emission spectrum of carbon provides evidence that electrons can occupy different shells, observed as a spectrum of specific colours.

Bohr proposed that each electron shell can only contain a certain number of electrons: a maximum of two electrons can occupy the first shell, and up to eight electrons can occupy the second shell (see Figure 6.22). Bohr also suggested that the configuration of electrons in different types of atoms explained how they can form bonds in chemical reactions.

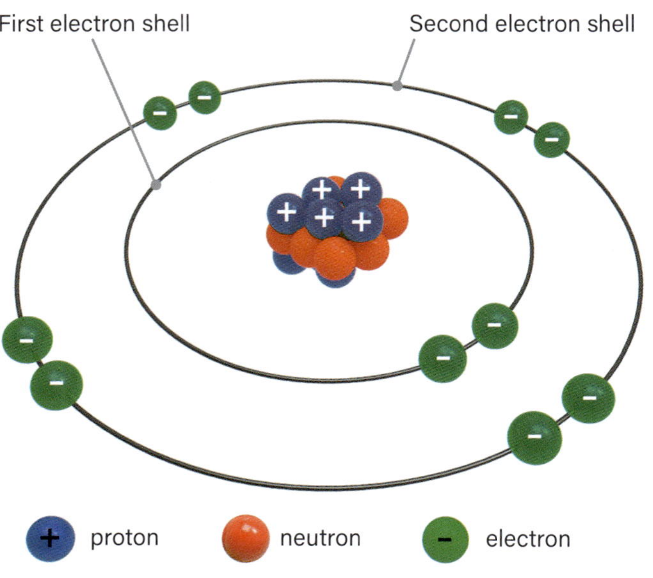

Figure 6.22: In Bohr's model of the atom, a small, dense nucleus is surrounded by electrons in orbits of a set size and energy.

Schrödinger developed Bohr's model of the atom

In 1926, the Austrian physicist Erwin Schrödinger used mathematical equations to describe the probable location of an electron. Schrödinger worked out that electrons do not move in set paths (orbits) around an atom's nucleus; instead, they move in clouds whose exact location is uncertain. Even though it is impossible to know the exact location of the electrons, Schrödinger's work shows that we can work out where they are most likely to be found.

Chadwick discovered the neutron

James Chadwick was an English physicist who worked with Ernest Rutherford. In 1932, Chadwick bombarded beryllium atoms with alpha particles. He found that a new particle was ejected that had almost the same mass as the proton but no charge: the neutron. His finding revolutionised understanding of atomic structure and gained him a Nobel Prize in Physics in 1935.

During World War II, Chadwick was extensively involved in the Manhattan Project, a research program that studied reactions to produce nuclear weapons.

Atoms are mostly empty space

Today, scientists know that every atom contains a nucleus that is made up of neutrally charged particles called neutrons and positively charged particles called protons, and that an atom's nucleus is orbited by negatively charged electrons.

We also know that atoms are mostly empty space. If the nucleus of an atom was the size of a pea and was placed in the middle of a stadium, the orbiting electrons would be outside of the stadium.

New technology allows scientists to learn more about atoms

New technology has enabled scientists to develop a more detailed understanding of matter, elements and atoms. For example, using the electron microscope, scientists have observed how atoms are arranged in different substances, which has led to a more detailed understanding of their properties.

Devices called particle accelerators are used to accelerate tiny particles to enormous speeds and smash them into one another. By analysing these collisions and what remains afterwards, scientists have developed their understanding of matter and the universe.

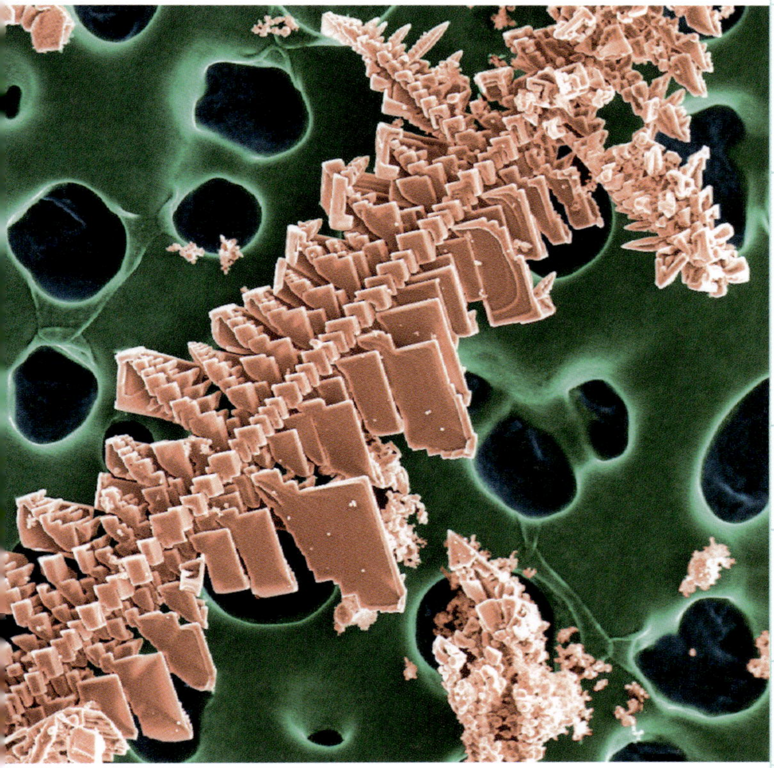

▲ **Figure 6.23:** This image from a microscope shows magnesium nanoparticles arranged like dominoes.

Table 6.2: A summary of theories and models about the atom from ancient times to the early 20th century

Date	Scientist	Discoveries or theories about the atom
c. 460–370 BCE	Democritus	• All substances are made of small particles that cannot be divided or destroyed.
1803	John Dalton	• Atoms of an element are identical. • Atoms of different elements combine to form new substances called compounds.
1904	J.J. Thomson	• Atoms contain negatively charged particles, now called electrons. • An atom is a sphere of positive charge (the 'pudding') with negatively charged particles (the 'plums') embedded through it.
1910	Ernest Rutherford	• Atoms are mostly empty space, with a tiny, dense, positively charged nucleus in the centre.
1913	Niels Bohr	• Electrons move around an atom's nucleus in orbits of fixed sizes and energies (like planets orbiting the Sun).
1926	Erwin Schrödinger	• Electrons do not move in set paths around an atom's nucleus, but in waves or clouds.
1932	James Chadwick	• The atom contains a neutral subatomic particle: the neutron.

Learning Ladder

Periodic table and atomic structure

1. Outline the main ideas of the atomic theory proposed by Dalton.
2. Draw a labelled diagram that shows our modern-day understanding of the structure of atoms.
3. Compare Chadwick and Thomson's ideas about the atom by identifying their similarities and differences.
4. Scientists' understanding of the atom has changed over time. Explain how this can help scientists in the future.
5. Discuss the evidence used by Bohr to support the theory that electrons orbit in electron shells.

Analysing data and information

see page 334

1. Identify the general trend in the development of scientists' understanding of the atom.
2. **a** Look up and draw diagrams of the atomic structures of boron, carbon, nitrogen and oxygen.
   b Describe the trend in the valence electrons.
3. Explain how Rutherford discovered the positively charged nucleus. Make sure you mention experimental method and evidence.

In context

Construct a timeline showing the development of atomic theory. Include the names of scientists and their contributions. Use a ruler to ensure the timeline is scaled correctly.

Success criteria

- I can outline historical developments of ideas of the atom.
- I can explain how new technologies have led to a more detailed understanding of atomic structure.

6.5 ▸ The periodic table

Learning intention

At the end of this lesson, I will be able to describe what the periodic table tells us about each element.

Key terms

atomic mass: the average mass of the atoms in a sample of an element

atomic number: the number of protons in an atom

chemical symbol: a symbol of one or two letters used to represent an element

periodic table: a table of all known elements and their chemical symbols

Content group: Periodic table

The periodic table includes every known element. The table provides a lot of information about each element, including its chemical symbol and information about its atoms.

▲ **Figure 6.24:** Silver has many names in different languages but its chemical symbol (Ag) does not change. This helps scientists to communicate with and understand each other and to collaborate.

Chemical symbols always stay the same

Every known element has a chemical symbol. For example, the chemical symbol for silver is Ag. Scientists throughout the world call this element Ag, regardless of which language they speak. So, while 'silver' is *argent* in French, *plata* in Spanish and *серебряный* in Russian, scientists from these countries can understand each other by referring to silver by its chemical symbol.

Using 'Ag' as the symbol for silver does not make much sense to English speakers, but it seems a perfect symbol if you are French and use the word *argent* for that element. See Table 6.3 for the origins of the names of some other elements.

Table 6.3: The symbols and origins of some element names

Element	Symbol	Origin of symbol
Sodium	Na	The Latin word for sodium is *natrium*.
Iron	Fe	The Latin word for iron is *ferrum*.
Strontium	Sr	Named after a Scottish village called *Sròn an t-Sìthein* in Scottish Gaelic.
Lead	Pb	The Latin word for lead is *plumbum*.
Potassium	K	The Latin word for potassium is *kalium*.

The periodic table tells us about the atoms of each element

In the **periodic table**, each element has its own square. This square contains an element's **atomic number**, **chemical symbol**, name and **atomic mass** (see Figure 6.25).

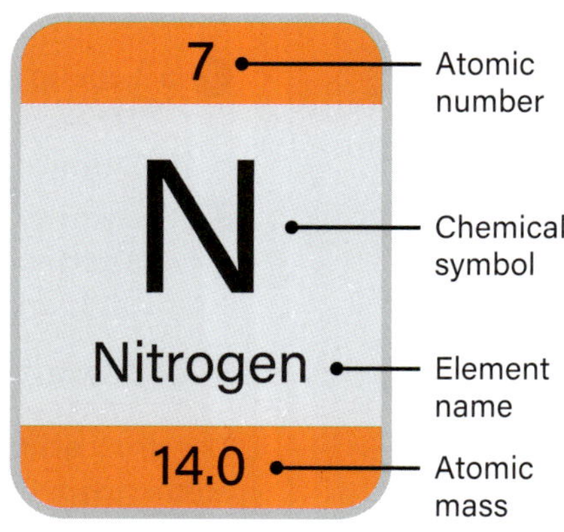

▲ **Figure 6.25:** The periodic table includes information for every known element.

An atomic number equals the number of protons

The atomic number states how many protons are in the nucleus of an atom of the element. For example, every atom of silicon (Si) has 14 protons, so the atomic number of silicon is 14. Similarly, every atom of barium (Ba) has 56 protons, so the atomic number of barium is 56.

An atom has the same number of electrons and protons

Atoms are assumed to be neutrally charged unless otherwise stated. Therefore, we can assume that an atom has the same number of protons and electrons. This is because the positive charges of the protons cancel out the negative charges of the electrons, resulting in an overall neutral charge. Look again at silicon (Si), which has an atomic number of 14: this means that silicon has 14 protons; therefore, a silicon atom has 14 electrons.

Atomic mass is an average

The atomic mass is the average mass of the atoms of a sample of the element. This is measured in atomic mass units (amu). The electrons that orbit an atom's nucleus do not weigh very much, so most of the mass of an atom is in the nucleus, where the neutrons and protons are located. The number of neutrons in an atom is often equal to the number of protons, but this can vary; you will learn more about this in Stage 5.

Learning Ladder

Refer to the periodic table on pages 208–9 to help you answer the questions below.

Periodic table and atomic structure

1 Copy the paragraph below into your notebook and fill in the blanks.

There are 118 elements in the periodic table, which means there are 118 different types of _______ that make up everything. Boron's atomic number is _______, which means it has _______ protons in its _______. It is next to _______ in the periodic table and _______ aluminium.

2 Locate lithium and beryllium in the periodic table. Describe the similarities and differences of the atomic structure of these two elements, based on their positions in the periodic table.

3 Consider these elements: potassium, magnesium, gold, nitrogen, helium, tin, barium and cobalt.

a Which two elements are likely to have similar properties? Explain your answer.

b Which two elements do you think have the least similar properties? Explain your answer.

4 Discuss the benefits and drawbacks of using chemical symbols instead of element names to identify elements.

Processing data and information

see page 329

1 Copy the table into your notebook and fill in the blanks.

Element	Symbol
Carbon	C
Oxygen	
	Na
Fluorine	
	Cu
Calcium	
Tungsten	

2 Construct a table that is similar to Table 6.3. Include 10 elements whose symbols do not match their English names. As well as the columns in Table 6.3, add a column for any interesting facts about the discovery of the elements.

3 **a** Classify the 118 known elements into two groups:

- Group A: The English name of the element matches its symbol.
- Group B: The English name of the element does not match its symbol.

b Calculate the percentage of elements in Group A and the percentage of elements in Group B (see 'Percentages' in the Science how-to section on page 364).

c Construct a pie chart to represent the percentages you worked out in Question 3b.

In context

Use the periodic table to find three elements named after scientists and three elements named after countries. Select one of the six elements and research the reasons it was named after a scientist or a country. Write a short story about how that element was named. Share your story with a classmate.

Success criteria

- I can describe what the periodic table tells us about each element.

Periodic table of elements

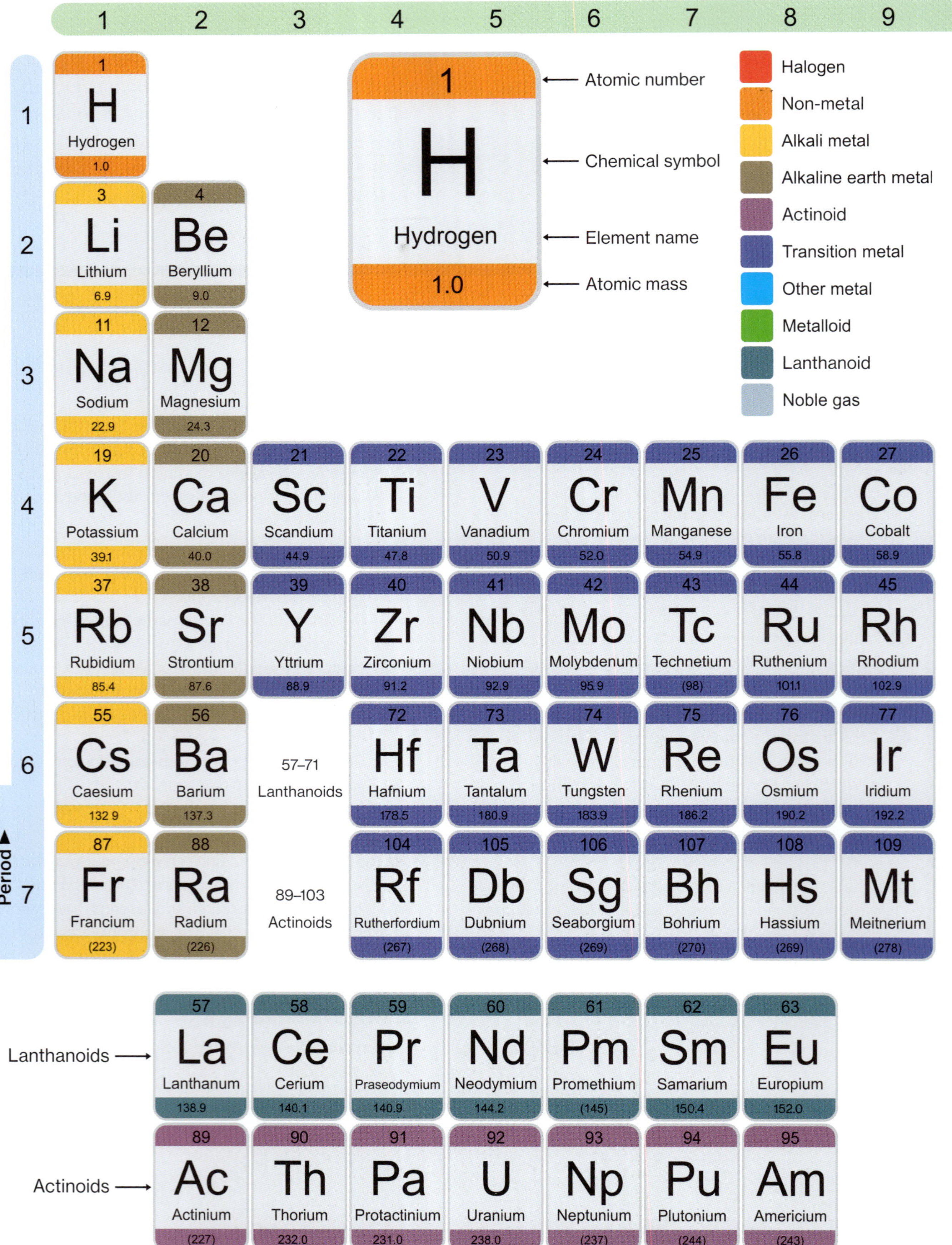

Figure 6.26: The periodic table shows all known elements and their chemical symbols.

◄ Group

10	11	12	13	14	15	16	17	18
								2 He Helium 4.0
			5 B Boron 10.8	6 C Carbon 12.0	7 N Nitrogen 14.0	8 O Oxygen 16.0	9 F Fluorine 19.0	10 Ne Neon 20.1
			13 Al Aluminium 26.9	14 Si Silicon 28.0	15 P Phosphorus 30.9	16 S Sulfur 32.0	17 Cl Chlorine 35.4	18 Ar Argon 39.9
28 Ni Nickel 58.6	29 Cu Copper 63.5	30 Zn Zinc 65.3	31 Ga Gallium 69.7	32 Ge Germanium 72.6	33 As Arsenic 74.9	34 Se Selenium 78.9	35 Br Bromine 79.9	36 Kr Krypton 83.8
46 Pd Palladium 106.4	47 Ag Silver 107.8	48 Cd Cadmium 112.4	49 In Indium 114.8	50 Sn Tin 118.7	51 Sb Antimony 121.7	52 Te Tellurium 127.6	53 I Iodine 126.9	54 Xe Xenon 131.2
78 Pt Platinum 195.1	79 Au Gold 197.0	80 Hg Mercury 200.6	81 Tl Thallium 204.4	82 Pb Lead 207.2	83 Bi Bismuth 209.0	84 Po Polonium (209)	85 At Astatine (210)	86 Rn Radon (222)
110 Ds Darmstadtium (281)	111 Rg Roentgenium (281)	112 Cn Copernicium (285)	113 Nh Nihonium (286)	114 Fl Flerovium (289)	115 Mc Moscovium (288)	116 Lv Livermorium (293)	117 Ts Tennessine (294)	118 Og Oganesson (294)

64 Gd Gadolinium 157.3	65 Tb Terbium 158.9	66 Dy Dysprosium 162.5	67 Ho Holmium 164.9	68 Er Erbium 167.2	69 Tm Thulium 168.9	70 Yb Ytterbium 173.1	71 Lu Lutetium 175.0
96 Cm Curium (247)	97 Bk Berkelium (247)	98 Cf Californium (251)	99 Es Einsteinium (252)	100 Fm Fermium (257)	101 Md Mendelevium (258)	102 No Nobelium (259)	103 Lr Lawrencium 281

6·6 ▸ Organisation of elements in the periodic table

Learning intention

At the end of this lesson, I will be able to:

- describe how the elements in the periodic table are organised
- predict the relative reactivity of elements based on their position in the periodic table.

Key terms

group: a column in the periodic table

period: a row in the periodic table

reactive: undergoes chemical reactions easily

valence electron: an electron in the valence shell

Investigation 6.6A

Modelling atomic structure alongside the periodic table, p.453

Investigation 6.6B

Investigating the reactivity of metals, p.455

Content group: Periodic table

In the periodic table, elements are organised into 18 columns called groups and seven rows called periods. Elements are listed in order by their atomic number from 1 (hydrogen) to 118 (oganesson). The table also groups together elements with similar properties.

Elements are organised into groups

In the periodic table, elements are arranged into 18 vertical columns (called **groups**). Each group has a number from 1 to 18. These groupings are based on the physical and chemical properties of the elements. For example, certain gases are in one group and certain metals are in another group.

Groups relate to valence electrons

In the periodic table, the number of each group relates to how many **valence electrons** are in an atom of each element in the group. The group number does not always match the number of electrons in the valence shell; however, a relationship can usually be seen (see Table 6.4). Elements with the same number of valence electrons have similar chemical properties.

Some of the groupings in the periodic table are:

- **Group 1 (alkali metals)**: The group 1 metals – lithium (Li), sodium (Na), potassium (K), rubidium (Rb), caesium (Cs) and francium (Fr) – are the alkali metals. These metals are very **reactive** (see Figure 6.28) and are soft, shiny and malleable, and conduct heat and electricity.

Group 1	Group 2	Groups 3–12	Group 13	Group 14	Group 15	Group 16	Group 17	Group 18
1 valence electron	2 valence electrons	Number of valence electrons varies	3 valence electrons	4 valence electrons	5 valence electrons	6 valence electrons	7 valence electrons	8 valence electrons

▲ **Table 6.4:** The relationship between group numbers in the periodic table and valence electrons

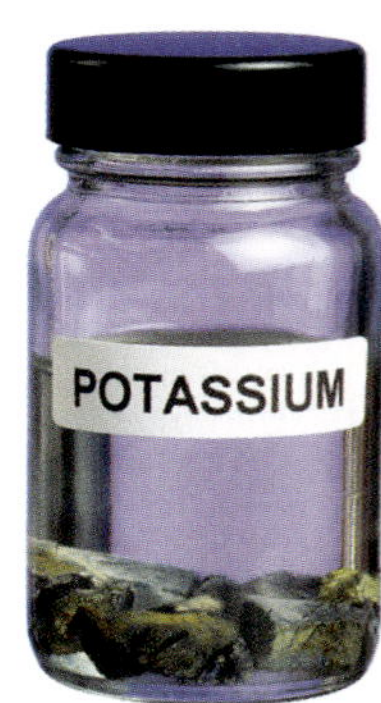

▲ **Figure 6.27:** The alkali metals in group 1 of the periodic table are so reactive they must be stored in oil to prevent the metals coming into contact with water or air.

Figure 6.28: When the alkali metals in group 1 come into contact with water or air, they react violently. In this image, potassium is reacting with water.

- **Group 2 (alkaline earth metals)**: The group 2 metals – beryllium (Be), magnesium (Mg), calcium (Ca), strontium (Sr), barium (Ba) and radium (Ra) – are the alkaline earth metals. These metals have similar properties to the metals in group 1, except they are less reactive.
- **Groups 3–12 (transition metals)**: The elements in groups 3–12 are the transition metals. These metals are all solid except for mercury (Hg), which is liquid. Transition metals can combine with other elements to form many different compounds, which are often brightly coloured (see Figure 6.30).

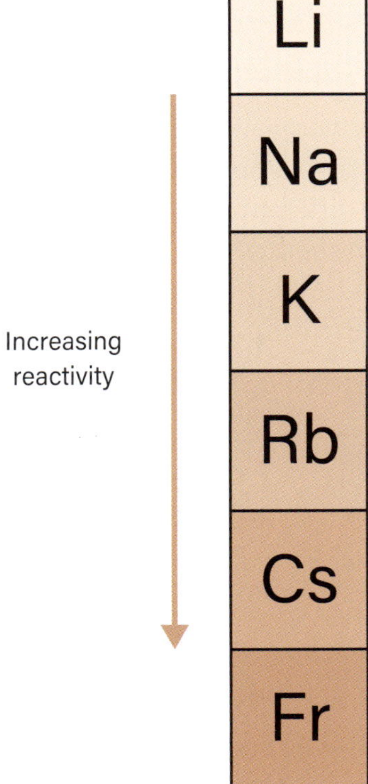

Figure 6.29: The reactivity of group 1 metals increases as you move down the group in the periodic table.

Figure 6.30: These brightly coloured salts are all compounds of transition metals.

▲ **Figure 6.31:** Neon lights are made from neon, one of the noble gases in Group 18 of the periodic table.

- **Groups 13–16 (metalloids)**: To the right of the transition metals in the periodic table are the semi-metals (these are called metalloids). The metalloids include boron (B), silicon (Si), germanium (Ge), arsenic (As), antimony (Sb) and tellurium (Te). These elements blend the properties of metals and non-metals.

The non-metals are found on the right side of the periodic table.

- **Group 17 (halogens)**: The six elements in this group – fluorine (F), chlorine (Cl), bromine (Br), iodine (I), astatine (At) and tennessine (Ts) – are the halogens. Halogens are non-metals; they are all gases except bromine (which is liquid at room temperature) and iodine and astatine (which are solids). The group 17 elements are very reactive; they are the most reactive non-metals.
- **Group 18 (noble gases)**: The seven elements in this group – helium (He), neon (Ne), argon (Ar), krypton (Kr), xenon (Xe), radon (Rn) and oganesson (Og) – are the noble gases. These elements are also known as 'inert gases' because they are mostly unreactive. This is because they have a full outer shell of eight electrons, which makes them extremely stable on their own.

Elements are organised into periods

In the periodic table, elements are also arranged into seven horizontal rows (called **periods**), which are numbered from 1 to 7. In the rows, the elements are placed in order of their atomic number.

Periods relate to electron shells

The elements in the same period have the same number of electron shells. For example, potassium (K) and bromine (Br) have very different properties, but they both have four electron shells and so are placed in period 4.

Period	Electron shells
Period 1	1 electron shell
Period 2	2 electron shells
Period 3	3 electron shells
Period 4	4 electron shells
Period 5	5 electron shells
Period 6	6 electron shells
Period 7	7 electron shells

◀ **Table 6.5:** The relationship between period numbers in the periodic table and electron shells

Two rows of the periodic table are separate from the other periods. These rows contain the lanthanoids (57–71) and the actinoids (89–103). These elements are rare and have very similar properties. These rows are separated so that they do not disrupt the arrangement of the periodic table.

Reactivity trends are observed in the periodic table

The elements in the periodic table are clearly divided into three main sections – metals, non-metals and metalloids – based on their physical and chemical properties.

The reactivity of metals generally increases as you move down a group and left across a period. In other words, the most reactive metals are in the bottom left corner of the periodic table.

The opposite is true for non-metals. They increase in reactivity as you go up a group and to the right; therefore, fluorine (F) is the most reactive non-metal. Noble gases are not included in this trend because they are generally not reactive at all.

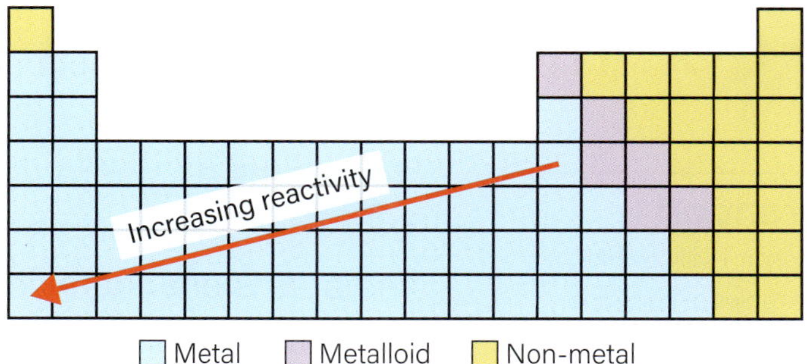

Figure 6.32: In the periodic table, the elements are grouped into metals, non-metals and metalloids.

Figure 6.33: Four metals reacting with hydrochloric acid; from left to right: calcium (Ca), magnesium (Mg), nickel (Ni) and copper (Cu).

Learning Ladder

Periodic table and atomic structure

1. a Identify which subatomic particle is linked to an element's atomic number.
   b Identify and describe groups 1 and 2 of the periodic table.
2. Distinguish between the groups and periods of the periodic table.
3. a Identify the type of element found in the far right side of the periodic table.
   b Describe the properties that these elements share.
4. Draw the periodic table.
   a Label the known groups (alkali metals, alkaline earth metals, transition metals, halogens, noble gases) and sections (metals, metalloids and non-metals).
   b Annotate one element from each group with information about how the element is used.

Processing data and information see page 329

1. Use the periodic table to identify the atomic number of:
   a zinc.
   b aluminium.
   c boron.
   d potassium.
2. a Identify three elements that are in the same:
      i group as potassium.
      ii period as calcium.
   b Construct a table to organise the information about the elements from Question 2a.
3. Figure 6.33 shows four metals reacting with hydrochloric acid. Based on the intensity of the bubbles in the photo, create a column graph to show the reactivity (zero, low, medium, high) of each of the metals. Follow the conventions for constructing a scientific graph (see 'Graphing' in the Science how-to section on page 368).

In context

Choose one metal, one non-metal and one metalloid. Research the three elements to:

a identify when and by whom each element was discovered.

b describe how each element was found in nature.

c explain the properties and uses of each element.

Share your findings with a classmate.

Success criteria

- I can describe how the periodic table is organised, using terms such as 'group' and 'period'.
- I can predict the relative reactivity of elements based on their position in the periodic table.

6·7 ▸ Common compounds

Learning intention

At the end of this lesson, I will be able to:

- identify some examples of common compounds and their uses
- recognise the ratio of atoms in a compound from the chemical formula.

Key term

chemical formula: an expression of the elements that make up a chemical compound, usually presented as a ratio using chemical symbols and numbers; for example, H_2O

Investigation 6.7

Investigating the different properties of a compound and one of its elements, p.456

Content group: Periodic table

You learnt in Chapter 4 that a compound is a substance that contains atoms of two or more different elements; the atoms are chemically bonded together in a fixed ratio. Compounds exist throughout the universe because many atoms are unstable on their own and so they bond with others. Compounds are expressed as **chemical formulas**.

Water is a compound vital for life

Every living thing on Earth needs water to survive. Each water molecule is made up of two hydrogen atoms bonded to one oxygen atom (H_2O). This is a two-to-one (2:1) ratio.

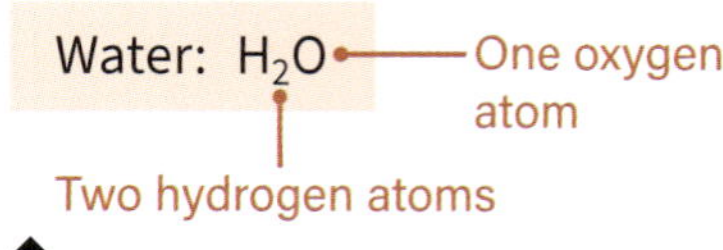

Figure 6.34: Water is a compound made up of hydrogen and oxygen in a 2:1 ratio indicated by the subscript values.

Carbon dioxide is a compound that is essential for plants

Carbon dioxide is another compound that is essential to life. Humans and other animals breathe in oxygen and breathe out carbon dioxide as a waste product. Plants use carbon dioxide to make energy to survive. Carbon dioxide is a gas at room temperature. Each carbon dioxide molecule contains one carbon atom bonded to two oxygen atoms. This is a 1:2 ratio.

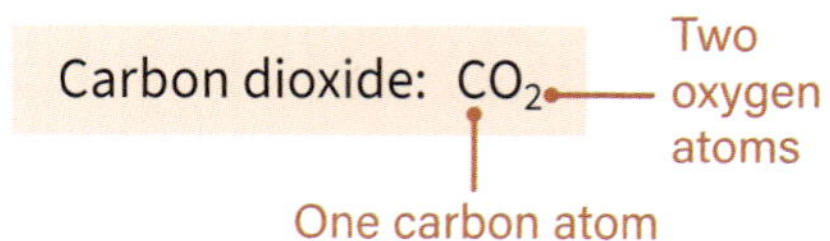

Figure 6.35: Carbon dioxide is a compound made up of carbon and oxygen in a 2:1 ratio indicated by the subscript values.

Water (H_2O)
Oxygen
Hydrogen
Chemical bond

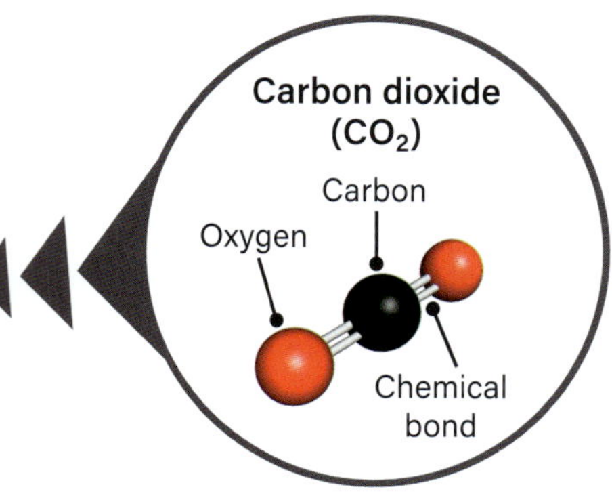

Figure 6.36: Plants need sunlight, carbon dioxide and water to survive and grow.

Ammonia is a compound used in farm and household products

Ammonia is used to make products such as fertilisers, which help plants grow, and cleaning products. At room temperature, ammonia is a poisonous gas. Ammonia is made safer in agriculture by combining it with other compounds to form ammonium salts, which are solid, white granules when they are at room temperature. Each ammonia molecule is made up of one nitrogen atom bonded to three hydrogen atoms. This is a 1:3 ratio.

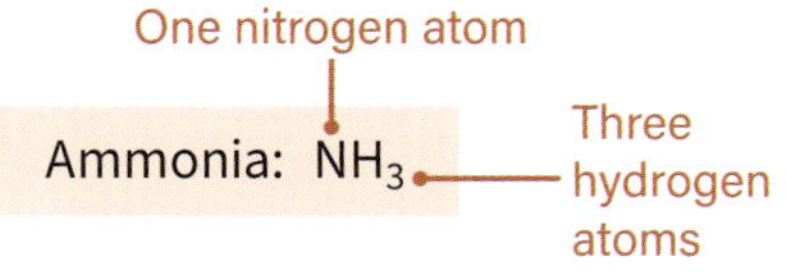

Figure 6.37: Ammonia is a compound made up of nitrogen and hydrogen in a 3:1 ratio indicated by the subscript values.

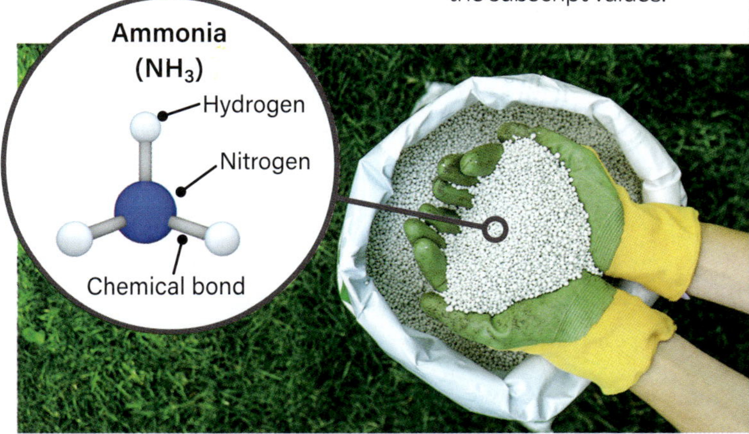

Figure 6.38: Ammonia is used to make fertilisers, which help crops to grow.

Sodium chloride is a compound used in many foods

Sodium chloride (NaCl) is better known as table salt. In sodium chloride, for every one sodium atom there is one chlorine atom. This is a 1:1 ratio. Sodium chloride is a compound that does not form molecules but has a three-dimensional lattice structure, with alternating sodium and chlorine atoms (see Figure 6.39).

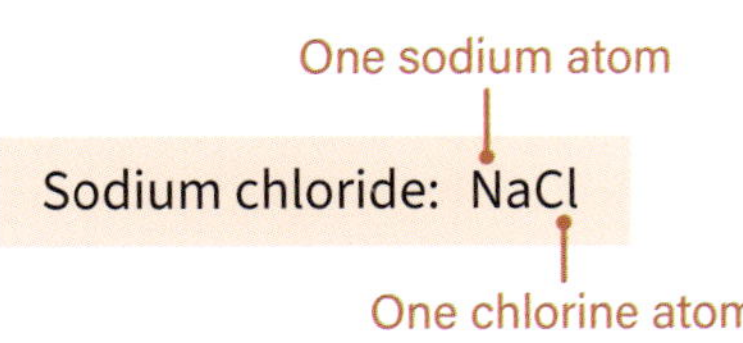

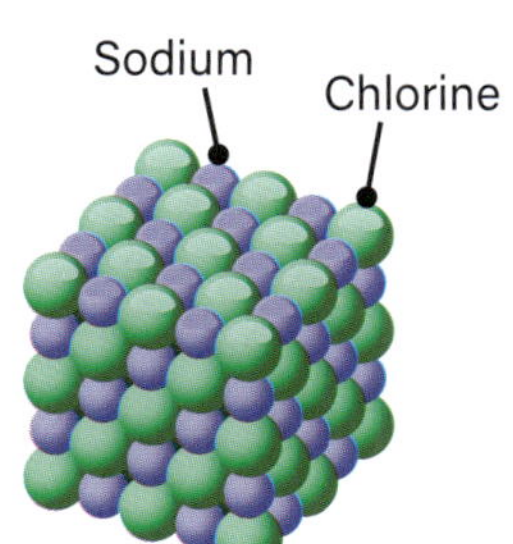

Figure 6.39: Sodium chloride is a compound made up of sodium and chlorine in a 1:1 ratio indicated by the subscript values.

Learning Ladder

Periodic table and atomic structure

1. Classify each of the following substances as either a compound or an element.
   - **a** Carbon monoxide (CO)
   - **b** Hydrogen gas (H_2)
   - **c** Gold metal (Au)
   - **d** Sulfuric acid (H_2SO_4).
2. **a** Identify the elements in ammonia.
   b Identify the ratio of the different atoms in ammonia.
3. Construct a Venn diagram to compare the:
   - element sodium (an alkali metal)
   - compound sodium chloride (salt)
   - element chlorine (a non-metal).
4. Explain why water and carbon dioxide are essential to life on Earth.

Analysing data and information *see page 334*

1. Hydrocarbons are a family of compounds. The first four members are methane (CH_4), ethane (C_2H_6), propane (C_3H_8) and butane (C_4H_{10}).
   - **a** Identify the pattern in the four chemical formulas.
   - **b** Predict the chemical formula of the fifth member of the hydrocarbon family: pentane.

For the following questions, refer to the graph below.

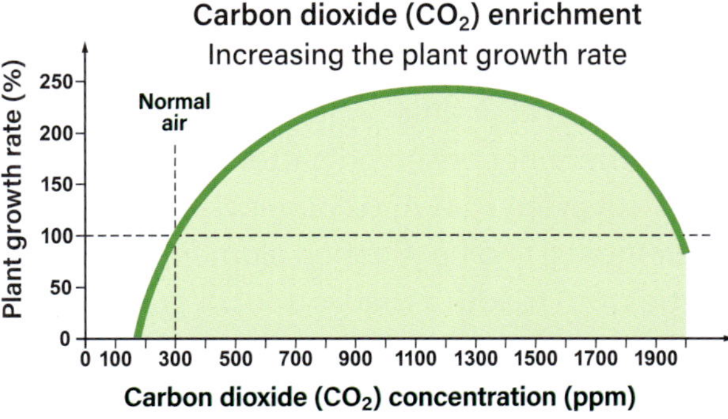

2. Describe the trend in the graph with reference to the independent and dependent variables.
3. **a** Identify how much CO_2 (ppm) is in normal air.
   b How much CO_2 maximises the growth rate of plants?
   c At what high concentration of CO_2 do growth rates decrease to below that of normal air?
4. Identify the variables that should have been controlled when this data was collected.

In context

Identify three compounds and explain how they are used.

Success criteria

- I can identify some common compounds and their uses.

6.8 ▸ Development of the periodic table

Learning intention

At the end of this lesson, I will be able to outline how the modern periodic table was developed over time by different scientists.

Key terms

chemical property: a property of a substance that is observed during a chemical reaction

hypothesis: a suggested explanation or prediction of a scientific problem that can be tested with an investigation

physical property: a property of a substance that can be observed and measured, such as colour, texture, melting and boiling points, density and hardness

Content group: Periodic table

The periodic table has been developed over thousands of years. It began with the ancient Greek philosopher Aristotle, who believed that matter consisted of four elements: water, fire, earth and air. The periodic table we have today is the work of many scientists from all over the world. Some of the key contributors are Dalton, Döbereiner, Mendeleev and Moseley.

Dalton produced an early periodic table of 20 elements

As you learnt in Section 6.4, in 1803, the English scientist John Dalton proposed his **hypothesis** that all matter is made of tiny particles that he called atoms. Dalton produced a table of 20 elements. Like in the modern-day periodic table, Dalton used a symbol to represent each element, although the symbols he used are not the ones used today. As in today's periodic table, Dalton listed the atomic mass of each element, though his calculations for these have been found to be incorrect.

Döbereiner grouped elements into triads

Scientists continued to expand Dalton's table as they discovered new elements. In 1829, German chemist Johann Döbereiner observed patterns or trends in the **physical properties** and **chemical properties** of some elements, and he grouped elements with similar properties into triads (a triad is a set or group of three related things) (see Figure 6.40). This grouping of elements that share similar properties laid the foundation for the structure of today's periodic table. To see the relationship between Döbereiner's triads and the modern-day periodic table, find each set of three elements in the periodic table on pages 208–9.

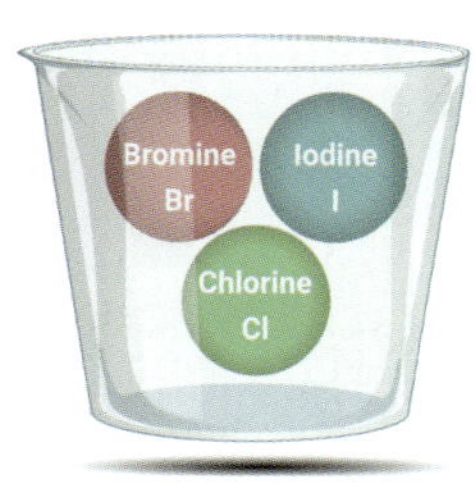

▲ **Figure 6.40:** Johann Döbereiner grouped some elements into triads.

Mendeleev arranged elements into groups and left gaps

In 1869, Russian chemist Dmitri Mendeleev arranged the elements known at the time in order of atomic mass number and put the known element 'families' into vertical columns.

Mendeleev left gaps in his periodic table for elements that were not yet discovered. He predicted that the unknown elements would have similar chemical properties to those of their 'family'. Later, when the elements were discovered, their properties closely matched Mendeleev's predictions.

Moseley arranged elements by their atomic number

In 1913, English physicist Henry Moseley suggested that it was an element's atomic number, not its atomic mass number, that was related to the element's physical and chemical properties. Moseley refined the previous periodic table and constructed a more accurate version with fewer elements missing.

1 H HYDROGEN																	2 He HELIUM
3 Li LITHIUM	4 Be BERYLIUM											5 B BORON	6 C CARBON	7 N NITROGEN	8 O OXYGEN	9 F FLUORINE	10 Ne NEON
11 Na SODIUM	12 Mg MAGNESIUM											13 Al ALUMINIUM	14 Si SILICON	15 P PHOSPHORUS	16 S SULFUR	17 Cl CHLORINE	18 Ar ARGON
19 K POTASSIUM	20 Ca CALCIUM	21 Sc SCANDIUM	22 Ti TITANIUM	23 V VANADIUM	24 Cr CHROMIUM	25 Mn MANGANESE	26 Fe IRON	27 Co COBALT	28 Ni NICKEL	29 Cu COPPER	30 Zn ZINC	31 Ga GALLIUM	32 Ge GERMANIUM	33 As ARSENIC	34 Se SELENIUM	35 Br BROMINE	36 Kr KRYPTON
37 Rb RUBIDIUM	38 Sr STRONTIUM	39 Y YTTRIUM	40 Zr ZIRCONIUM	41 Nb NIOBIUM	42 Mo MOLYBDENUM	43 Tc TECHNETIUM	44 Ru RUTHENIUM	45 Rh RHODIUM	46 Pd PALLADIUM	47 Ag SILVER	48 Cd CADMIUM	49 In INDIUM	50 Sn TIN	51 Sb ANTIMONY	52 Te TELLURIUM	53 I IODINE	54 Xe XENON
55 Cs CAESIUM	56 Ba BARIUM	57–71 LANTHANOIDS	72 Hf HAFNIUM	73 Ta TANTALUM	74 W TUNGSTEN	75 Re RHENIUM	76 Os OSMIUM	77 Ir IRIDIUM	78 Pt PLATINUM	79 Au GOLD	80 Hg MERCURY	81 Tl THALLIUM	82 Pb LEAD	83 Bi BISMUTH	84 Po POLONIUM	85 At ASTATINE	86 Rn RADON
87 Fr FRANCIUM	88 Ra RADIUM	89–103 ACTINOIDS	104 Rf RUTHERFORDIUM	105 Db DUBNIUM	106 Sg SEABORGIUM	107 Bh BOHRIUM	108 Hs HASSIUM	109 Mt MEITNERIUM	110 Ds DARMSTADTIUM	111 Rg ROENTGENIUM	112 Cp COPERNICIUM	113 Nh NIHONIUM	114 Fl FLEROVIUM	115 Mc MOSCOVIUM	116 Lv LIVERMORIUM	117 Ts TENNESSINE	118 Og OGANESEON

57 La LANTHANUM	58 Ce CERIUM	59 Pr PRAESODYMIUM	60 Nd NEODYMIUM	61 Pm PROMETHIUM	62 Sm SAMARIUM	63 Eu EUROPIUM	64 Gd GADOLINIUM	65 Tb TERBIUM	66 Dy DYSPROSIUM	67 Ho HOLMIUM	68 Er ERBIUM	69 Tm THULIUM	70 Yb YTTERBIUM	71 Lu LUTETIUM
89 Ac ACTINIUM	90 Th THORIUM	91 Pa PROTACTINIUM	92 U URANIUM	93 Np NEPTUNIUM	94 Pu PLUTONIUM	95 Am AMERICIUM	96 Cm CURIUM	97 Bk BERKELIUM	98 Cf CALIFORNIUM	99 Es EINSTEINIUM	100 Fm FERMIUM	101 Md MENDELEVIUM	102 No NOBELIUM	103 Lr LAWRENCIUM

Before CE
0–1749
1750–1799
1800–1849
1850–1899
1900–1949
1950 onward

Figure 6.41: This periodic table colour-codes elements according to when they were discovered.

Learning Ladder

Periodic table and atomic structure

1. Identify some similarities and differences between an early periodic table and the modern-day periodic table.
2. Describe how Moseley's suggestion contributed to the structure of the modern periodic table.
3. **a** Explain how Döbereiner grouped elements.
   b Select one of Döbereiner's triads and predict which properties they have in common.
   c Research the triad you selected in Question 3b to see if your prediction was accurate.
4. Over time, more and more elements have been discovered. Propose some reasons for this.

Processing data and information

see page 329

1. Use Figure 6.41 to construct a timeline that shows the number of elements discovered during each of the seven time periods.
2. Display the information from Question 1 in a graph.
3. **a** Construct a table with three columns headed 'Period of discovery', 'Elements in order of discovery' and 'Total number of elements discovered'. The periods of discovery could be ancients, 1600s, 1700s, 1800s, 1900s and 2000s.
   b Research all the elements and complete your table.
4. Display the information in the table you completed in Question 3a in a different format (that is, not a table) (for example, a pie chart or an infographic).

In context

Research Mendeleev's periodic table. Compare his periodic table with the modern-day periodic table. What are the similarities and differences?

Success criteria

- I can outline how the modern periodic table was developed over time by different scientists.

6·9 ▸ Periodic table and atomic structure in context

Learning intention

At the end of this lesson, I will be able to explain how the properties and availability of some materials – metals, alloys and compounds – influence their uses.

Key terms

bronze: an alloy made of copper and tin

steel: an alloy made of carbon and iron

Investigation 6.9

Modifying metals, p.458

Content group: Periodic table and atomic structure in context

Since ancient times, humans have used the materials available to them – including metals, alloys and compounds – in a variety of ways. The properties of the materials dictated what tasks they were suited to and how they were used by humans.

The properties of rocks make them useful tools

Rocks are compounds, and they are mostly made of the compound silicon dioxide, which contains the elements silicon and oxygen.

Humans have been using rocks or stones as tools for thousands of years. This was particularly the case during the Stone Age, as many metals and alloys had not yet been discovered or invented. The early humans who lived in the Stone Age discovered that different types of rocks suited different tasks. For example, obsidian is a rock that breaks into sharp pieces; these pieces were used as arrow tips and scrapers. Very hard rocks with coarse textures – like granite – were used to grind food, and hard rocks with smooth textures were used to make axes.

Figure 6.42: During the Stone Age, humans made and used stone tools.

The properties of bronze make it ideal for making strong weapons

Bronze is an alloy; it is made from two metals: copper and tin. Bronze is much stronger and harder than pure copper. The ancient Sumerians were the first people to make this alloy, in around 3500 BCE. After bronze was discovered, people began making tools and weapons from this material as it is much harder than other materials they had previously used to make these objects (such as pure copper), it is more resistant to corrosion (wearing away), and it is easier to melt and cast into shapes.

Figure 6.43: Ancient peoples used bronze to make tools and weapons, including swords.

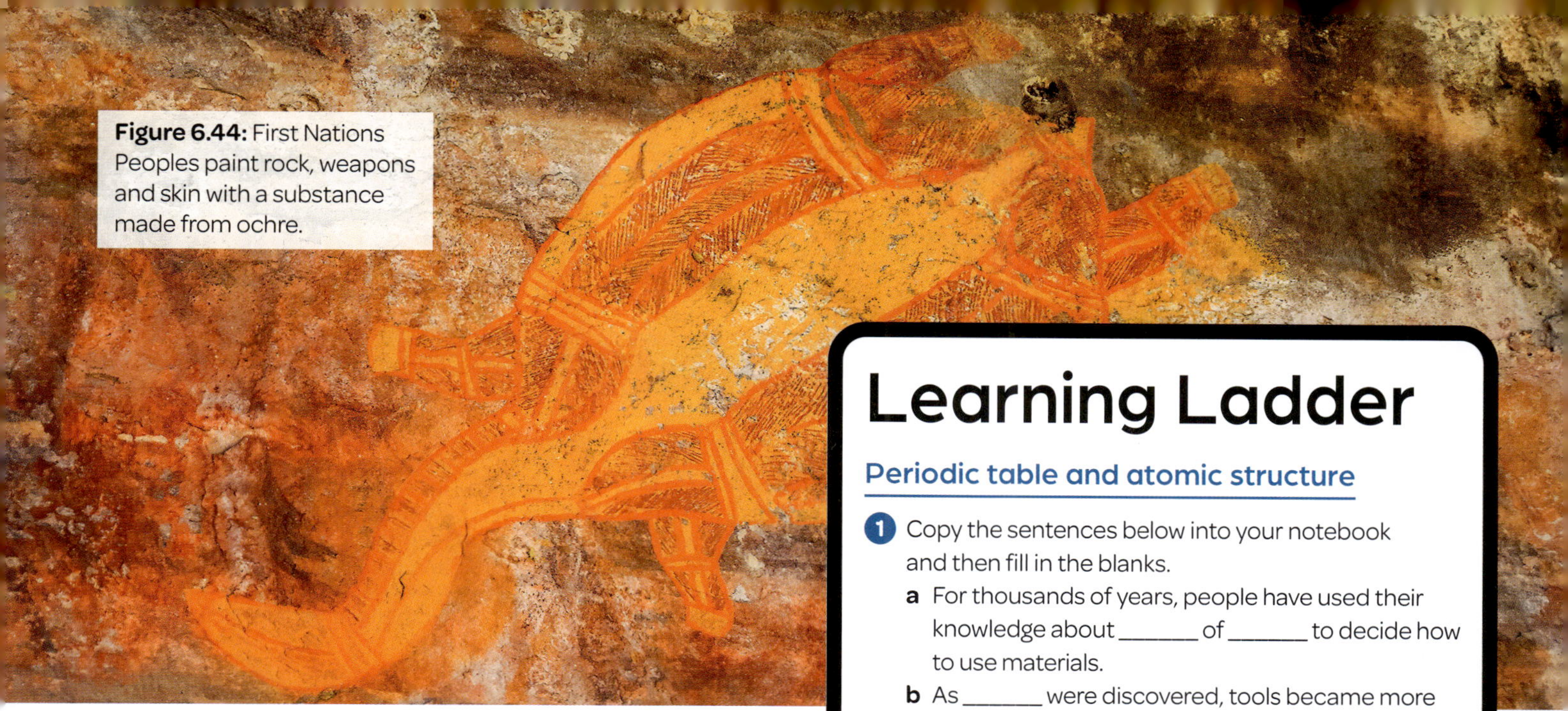

Figure 6.44: First Nations Peoples paint rock, weapons and skin with a substance made from ochre.

The properties of steel make it ideal for making medical instruments

Around 400 BCE, metalworkers discovered that adding carbon to iron created a stronger substance than pure iron. Today, this alloy is called **steel**. Steel is stronger than bronze, harder than iron and lighter than both these substances. Today, steel is used for a wide variety of purposes, including in construction (for example, the Sydney Harbour Bridge) and to make medical instruments.

Figure 6.45: Many medical instruments are made from steel.

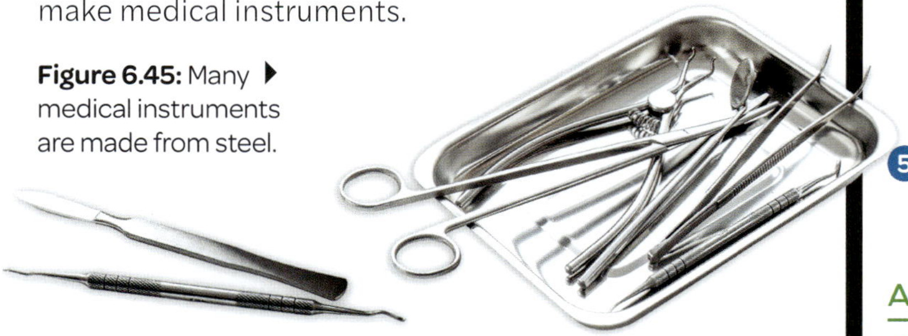

The properties of ochre make it ideal for making art

People from many cultures have developed and used dyes and paints from natural sources. A substance called ochre is widely used by Aboriginal and Torres Strait Islander Peoples for making art (see Figure 6.44) and for ceremonies. Ochre is a naturally occurring substance that is a mixture of clay, sand and a compound called iron oxide. The substance is mixed with water, animal fat, saliva or blood and then painted on rock, weapons, ceremonial objects and skin. Mixing different amounts of the ingredients results in different colours, such as reds, yellows and browns.

Learning Ladder

Periodic table and atomic structure

1. Copy the sentences below into your notebook and then fill in the blanks.
   - **a** For thousands of years, people have used their knowledge about ______ of ______ to decide how to use materials.
   - **b** As ______ were discovered, tools became more sophisticated.
   - **c** People of many different ______ have developed and used dyes and paints from natural sources.
2. **a** Describe the make-up of ochre.
   - **b** Explain how First Nations Peoples use the properties of ochre to create art.
3. **a** Construct a chart to compare the advantages and disadvantages of using stone versus copper to make tools and weapons.
   - **b** How did humans overcome the disadvantages of copper tools and weapons?
4. Steel is much stronger than bronze.
   - **a** Outline which two main elements form steel.
   - **b** Give examples of steel being used in everyday life.
   - **c** Suggest a circumstance where it is more appropriate to use bronze rather than steel.
5. Use the internet to analyse what people from other cultures – such as Chinese, Roman and Sumerian people – have used as paints.

Analysing data and information *see page 334*

1. List the following materials in order of strength (from the weakest to the strongest): bronze, copper, obsidian, ochre, stone, steel.
2. Justify your response to Question 1.
3. Look at the list of materials in Question 1.
   - **a** Explain the links between the elements that make up each material.
   - **b** Explain how the elements relate to the properties of the materials and how they are used.
4. Write a statement about the strength and use of stones, metals, alloys and metal oxide.

Success criteria

- I can explain how the properties and availability of some materials influence their uses.

► Periodic table and atomic structure summary

Classification of matter

An element is a pure substance made of a single type of atom. Elements can generally be classified into metals, non-metals and metalloids. Metalloids are elements that have properties of both metals and non-metals.

Classification	Element	Properties	Everyday uses
Non-metal	Oxygen	Forms a gas at room temperature	Reactant in combustion reactions like respiration and burning things
	Carbon	Varying appearance and conductivity, depending on structure	Graphite in pencils; gemstones
Metal	Aluminium	Lightweight and strong	Material for aircraft, bridges, foil, cans
	Copper	Conducts electricity, is malleable and ductile	Electrical wiring
Metalloid	Silicon	Semi-conductor, insulator	Microchips

A compound is a substance made up of atoms from two or more elements. Compounds have different properties to the individual elements they are made of. Common compounds include silicon dioxide (rocks), water, carbon dioxide, ammonia (used in fertilisers) and sodium chloride (table salt).

Hydrogen
Water: H_2O — Oxygen
Two hydrogen atoms

Compounds are represented using chemical formulas, which use chemical symbols and subscript numbers to show the ratio of atoms in a compound. For example, water has the chemical formula H_2O; this means that there are two hydrogen (H) atoms for every one oxygen (O) atom.

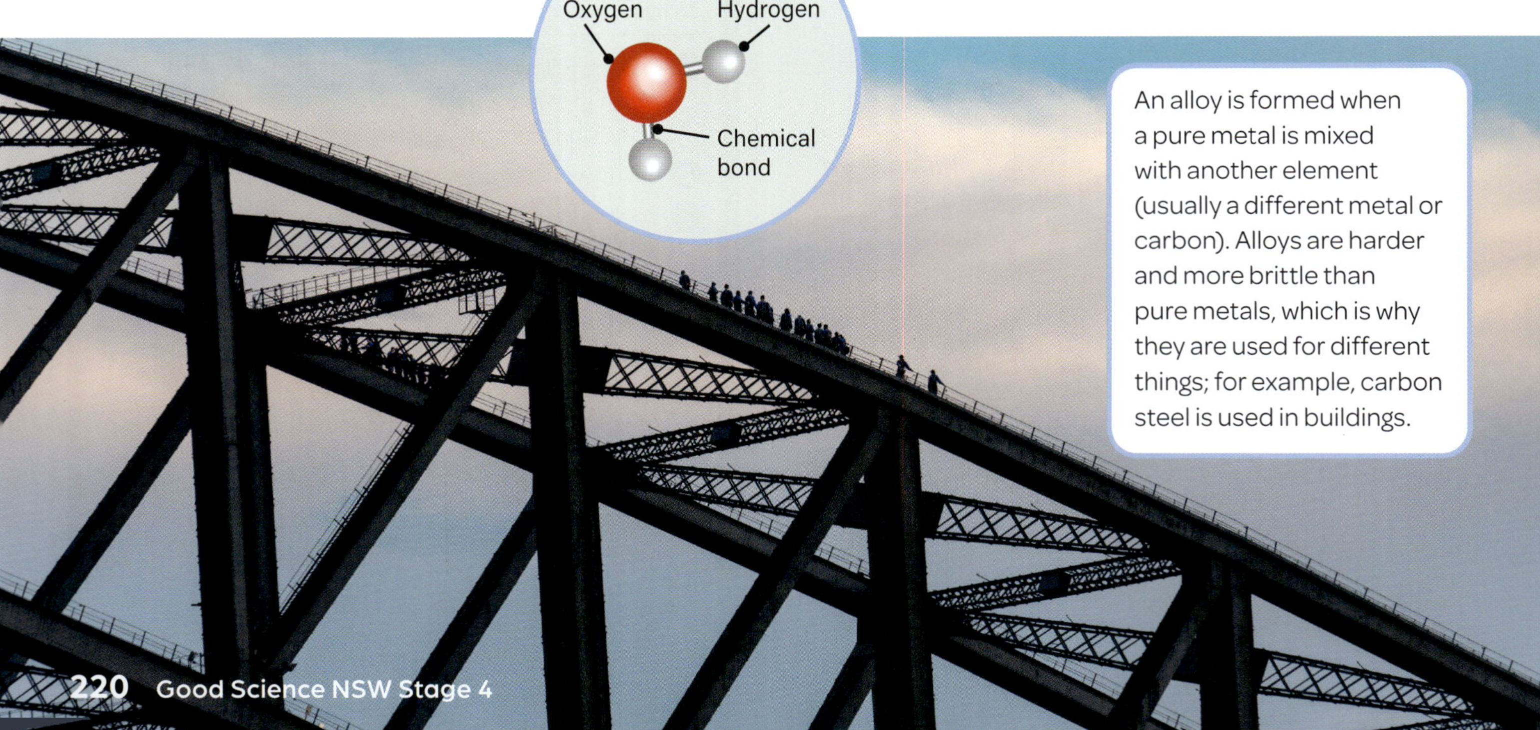

An alloy is formed when a pure metal is mixed with another element (usually a different metal or carbon). Alloys are harder and more brittle than pure metals, which is why they are used for different things; for example, carbon steel is used in buildings.

Atomic structure

- An atom is the smallest, stable unit of an element that retains the properties of that element; atoms are the smallest particles of matter.
- Atoms are made up of three types of subatomic particles: protons, neutrons and electrons.
- Models of atomic structure have changed over time due to observations made possible by evolving technologies.

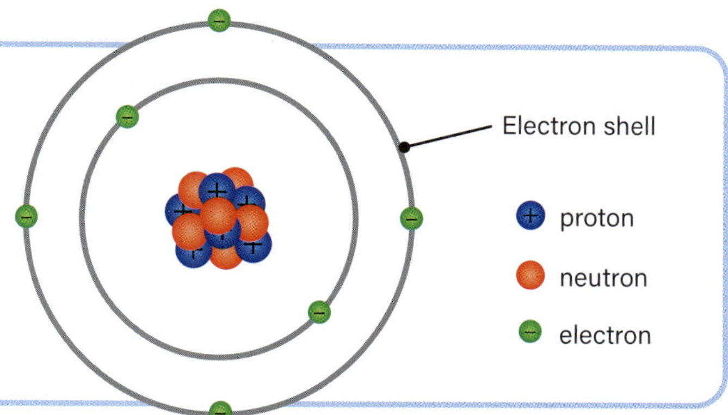

Periodic table

The periodic table provides information about every known element, including its chemical symbol and information about its atoms. Each square in the periodic table represents a specific element.

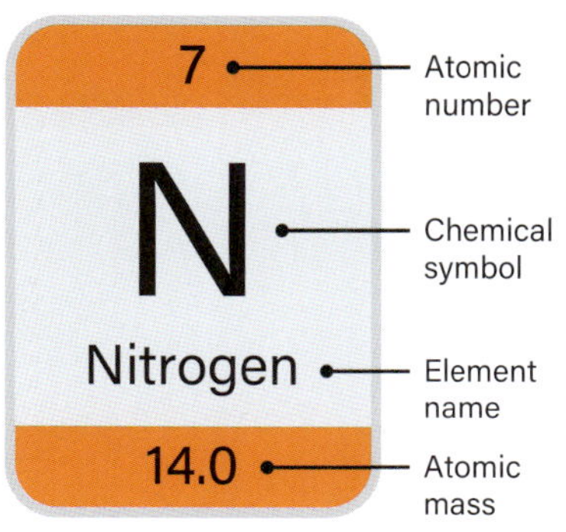

In the periodic table, the elements are organised into 18 columns called **groups** and seven rows called **periods**. Elements are listed in order by their atomic number from 1 (hydrogen) to 118 (oganesson).

Elements are also grouped based on their properties.

- the most reactive metals are in group 1 (alkali metals)
- the most reactive non-metals are in group 17 (halogens)
- the least reactive non-metals are in group 18 (noble gases)
- metals that make colourful compounds (transition metals) are in the middle.

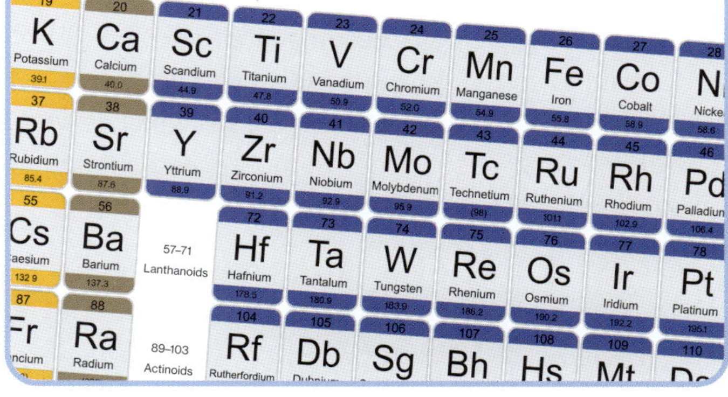

Döbereiner's triads

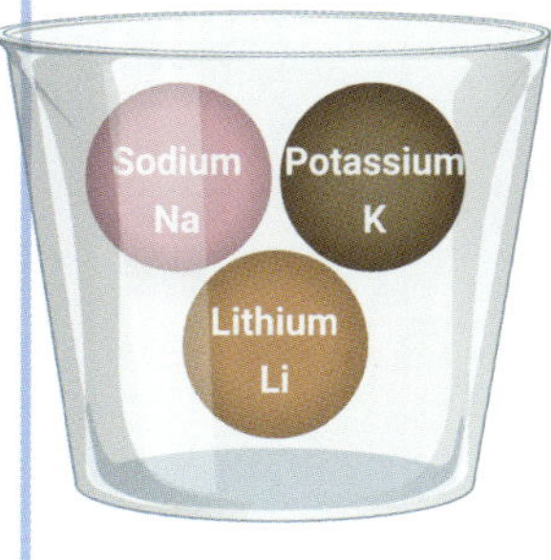

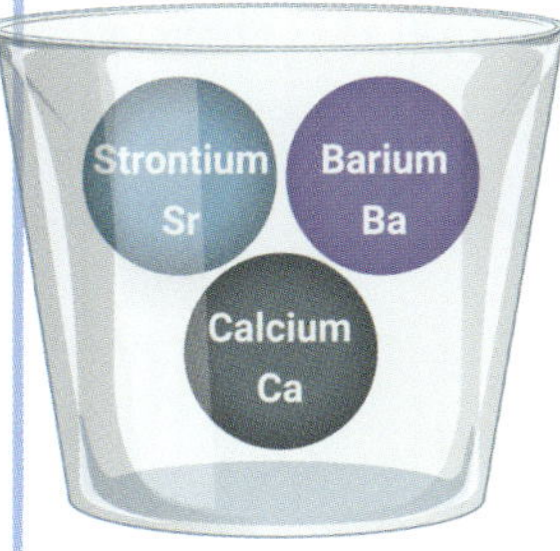

The historical development of the periodic table demonstrates the growing understanding of the chemical and physical properties of elements. The periodic table we use today is the work of many scientists from all over the world. Some of the key contributors are Dalton, Döbereiner, Mendeleev and Moseley.

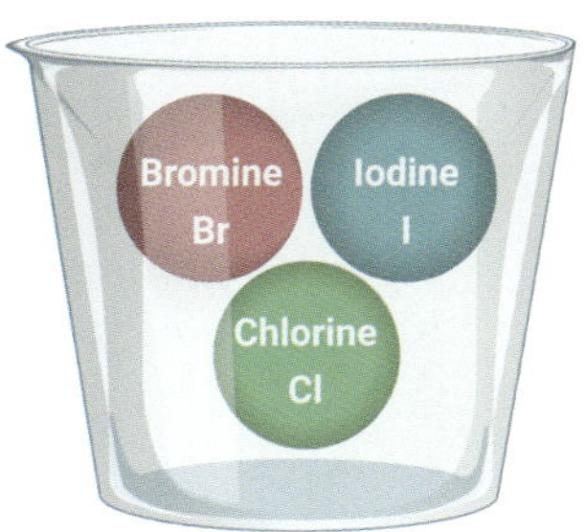

Since ancient times, humans have used the materials available to them – including metals, alloys and compounds – in a variety of ways. The properties of the materials dictated what tasks they were suited to and how they were used by humans.

Masterclass

Steps in progression → 1 → 2

		1	2
Content	Periodic table and atomic structure	How many protons does an atom of mercury contain?	Describe the property of mercury that makes it unique.
Processes	Processing data and information	Refer to Figure 6.47. Identify the percentage of mercury in amalgam dental fillings.	Construct a table to display the information in Figure 6.47.
	Analysing data and information	Identify whether the trend in Figure 6.48 has a positive or negative correlation or no correlation.	Describe the trend displayed in Figure 6.48.

When elements become deadly

Chemical elements are typically used for beneficial purposes such as medicine, technology and energy. However, some elements can be hazardous when misused or mishandled. For example, mercury (atomic number 80) is toxic if ingested and can be harmful to ecosystems. Unlike any other metal, mercury is a liquid at room temperature, which is why it was commonly used in thermometers and amalgam (alloy) dental fillings, before scientists understood the dangers of this element.

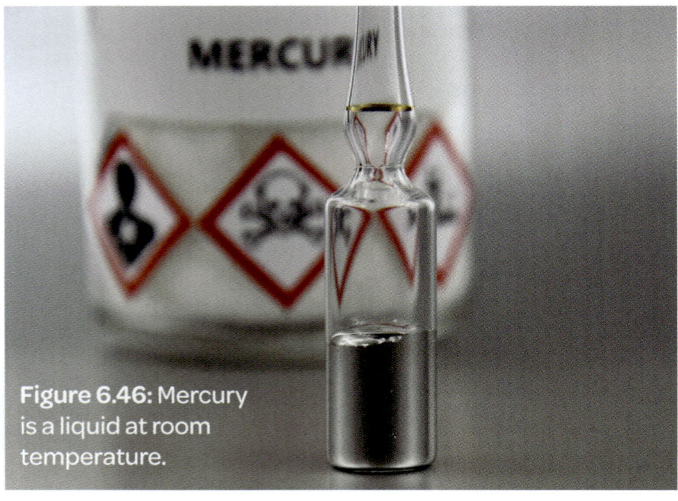

Figure 6.46: Mercury is a liquid at room temperature.

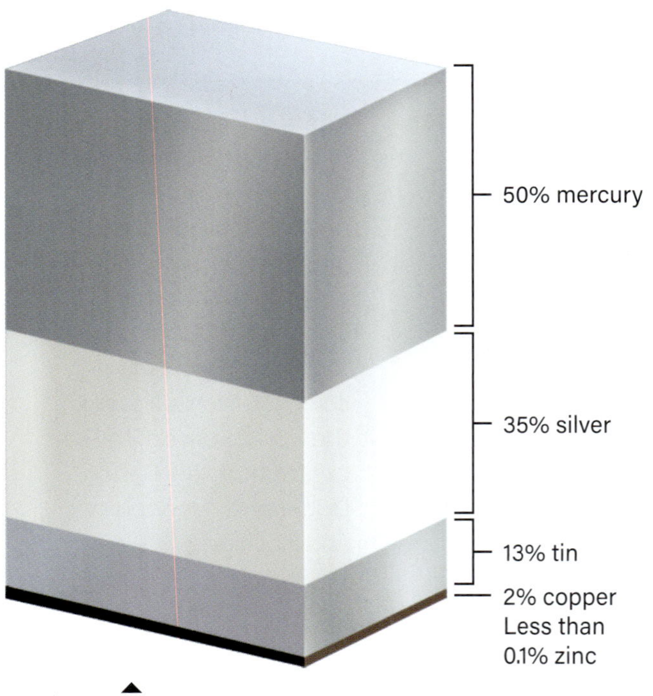

Figure 6.47: The composition of silver alloy amalgam dental fillings

Demonstrate your understanding

3	4	5	
Refer to Figure 6.47. **a** Compare the properties of the metals used in amalgam dental fillings. **b** Compare the positions on the periodic table of the metals used in amalgam dental fillings.	Discuss how mercury's unique properties enabled it to be used in thermometers to indicate temperature.	**a** Investigate how mercury interacts with other metals. **b** Discuss how this interaction made mercury ideal for amalgams.	
Construct a column graph to visually display the information in Figure 6.47.	Construct a pie chart to represent the information in Figure 6.47.	**a** Investigate each metal in amalgam alloy. **b** Construct a column graph to display the relative toxicities of the metals in amalgam alloy.	Science how-to p. 329
Explain the relationship between mercury levels and a pregnant woman's age (see Figure 6.48).	Suggest a conclusion you could draw from Figure 6.48.	Suggest questions you could ask about the data given in Figure 6.48 that would help you to evaluate the validity of this information.	Science how-to p. 334

Mercury can accumulate in our bodies slowly over time, particularly from eating fish that has high levels of mercury. Unborn babies are particularly sensitive to the effects of mercury. To educate expectant mothers about the risks of mercury, a study looked at the levels of mercury in pregnant women in the United States. The age of the women was one of the factors that was analysed (see Figure 6.48). The results of the study were used to educate the general population.

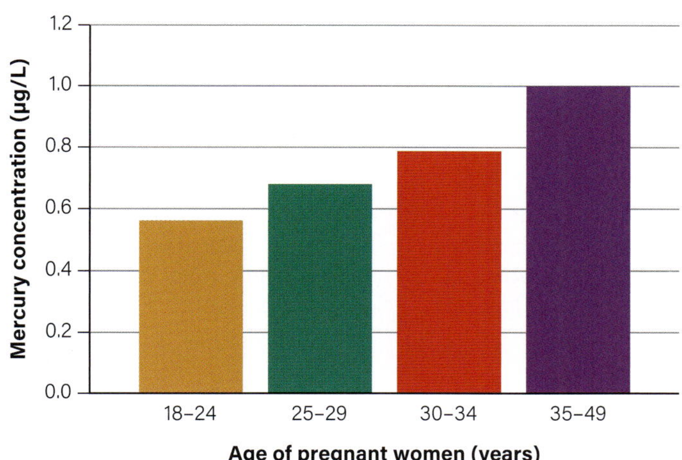

▲ **Figure 6.48:** The mean blood mercury levels in pregnant women aged 18–49 by age group in the United States, 2003–2008

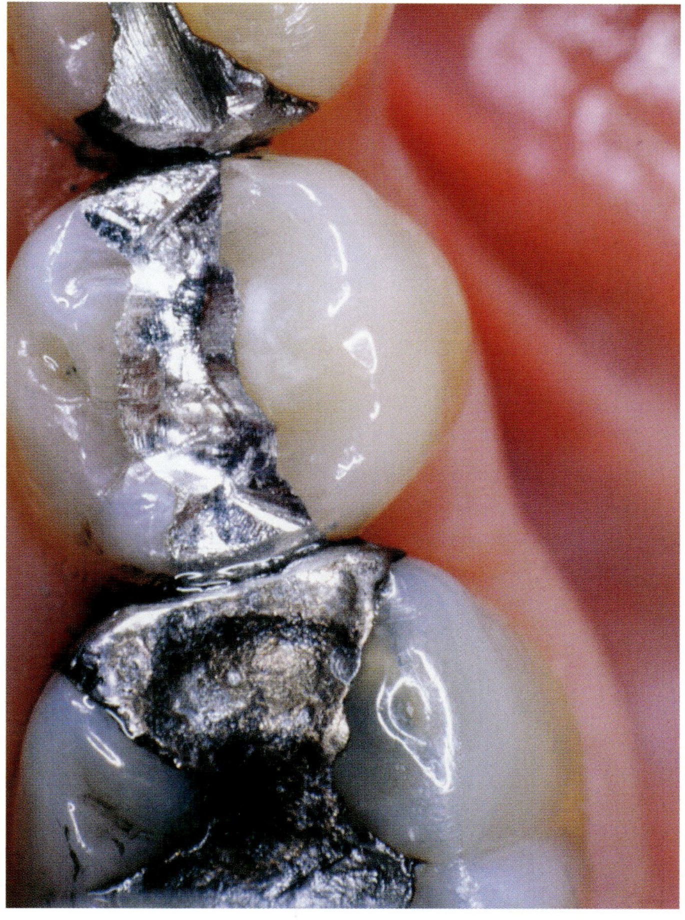

▲ **Figure 6.49:** Silver alloy amalgam dental fillings

Focus area 7

7.0 Change

Change is an essential part of life. Without change we would be unable to progress or even survive as a species. We can see the effects of energy change, chemical change and geological change all around us. From energy transformations powering life, to chemical reactions driving innovation, to our dynamic Earth-sculpting landscapes, this chapter delves into the science that explains change and how it affects our world.

Figure 7.1: Baking bread relies on a combination of chemical and physical changes.

Learning Ladder

The Learning Ladder contains the scientific content and processes you will learn in this chapter. Each area has five levels of progression. To move to higher levels, you need to practise activities at the earlier levels. This will help you develop the ability to complete tasks that are more complex.

Steps in progression	Change	Observing	Planning investigations	Conducting investigations
5	I can apply my knowledge of chemical changes to real-world contexts.	I can use observations and measurements to answer questions.	I can construct an appropriate plan before conducting a first-hand investigation.	I can conduct first-hand investigations and accurately record the collected data.
4	I can communicate the processes involved in a range of changes.	I can make inferences based on my observations.	I can develop appropriate scientific aims for a range of investigations.	I can collect accurate and reliable data from first-hand and second-hand investigations.
3	I can explain the role of energy in different types of changes.	I can record observations and measurements accurately.	I can distinguish between controlled, dependent and independent variables.	I can implement safe practices when using scientific equipment.
2	I can describe a scientific change.	I can use scientific tools to enhance observations.	I can describe ways to reduce risks for a range of investigations.	I can use scientific equipment to conduct investigations and gather first-hand data and information.
1	I can identify a scientific change.	I can make observations using my senses.	I can identify appropriate materials and technologies to conduct investigations.	I can identify correct processes for conducting investigations.
	Content	Working scientifically processes		

Figure 7.2: An explosion involves chemical change.

Figure 7.3: A rollercoaster ride involves many different types of energy.

7·1 ▸ Types of energy

Learning intention

At the end of this lesson, I will be able to:

- identify objects that have energy because of their motion (kinetic energy)
- identify objects that have energy because of other properties (potential energy).

Key terms

energy: the ability to do work

kinetic energy: the energy of movement

mass: the amount of matter that a physical body contains

motion: the change in position of an object over time

potential energy: stored energy; has the ability to cause movement

transfer: move from one place or object to another

transform: change from one type to another

work: when a force applied to an object causes the object to move

Investigation 7.1

Rolling balls, p.459

Content group: Energy transfers

Imagine you are riding a rollercoaster, climbing slowly up the track. You hear the clicking gears as you move further and further from the ground. By the time you reach the highest point, your heart is pumping. For a moment you are still, then suddenly you plunge down the other side, in the grip of gravity. Down and back up again, the wind in your face, over and over until finally the ride comes to a stop. You just experienced about a dozen different types of energy in that single rollercoaster ride. **Energy** is the ability to make things happen, to create change, to do work. There are many different types of energy, all of which can be classified into two main forms: kinetic energy and potential energy.

A moving object has kinetic energy

The **kinetic energy** of an object such as a rollercoaster is the energy that it has because of its **motion**. Many common types of energy, such as 'sound' and 'light', are classed as kinetic energy because they involve the movement of particles, even if the motion cannot be seen. Thermal energy is also a type of kinetic energy that you cannot see. Particles of objects vibrate at different rates based on how much 'thermal' energy the objects have.

The amount of kinetic energy a moving object has depends on its **mass** and its speed (see Figure 7.4).

Kinetic energy can be **transferred** from one object to another. It can also be **transformed** into other types of energy. This will be discussed further in Sections 7.2 and 7.3.

30

The bicycle and the truck are travelling at the same speed, but the truck is heavier, so it has more kinetic energy.

30

10

4

Both people weigh the same amount, but the one who is moving more quickly has more kinetic energy.

Figure 7.4: What type of energy do moving objects have?

Potential energy is stored until release

Potential energy is stored energy that can be used to do **work**. This work might be hammering a nail, shooting an arrow or cooking dinner. Matter naturally tends to exist in a form in which it contains the least amount of energy. There are different ways that energy can be stored and released to do work.

Gravitational potential energy is the energy an object has because of its position above the ground (see Figure 7.5). An object lifted up high has the potential to do work. When lifted up, a heavier object has more gravitational potential energy than a lighter object. When the object falls because of gravity, it moves and gains speed as it falls. Gravitational potential energy has been transformed into kinetic energy.

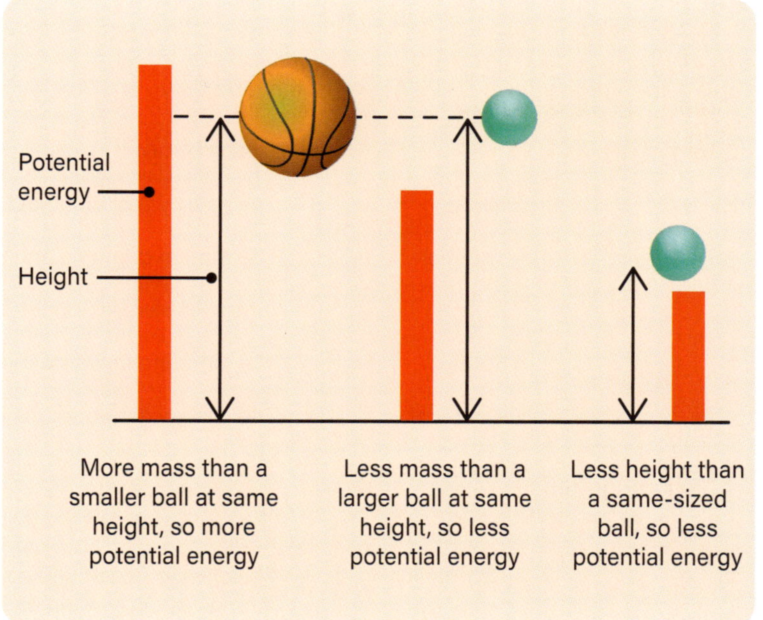

Figure 7.5: The higher above the ground an object is, and the greater its mass, the more gravitational potential energy it has.

Elastic potential energy is the energy transferred in an object such as a spring or an elastic band. If you pull back on an elastic band, you have transferred stored energy to the elastic material. When you let it go, the elastic band returns to its original shape. The stored elastic potential energy has been transformed into kinetic energy. The stronger the elastic band or spring, the greater its potential to store energy. Also, the more an elastic object is stretched or compressed, the more potential energy it stores.

▲ **Figure 7.6:** When the elastic band is let go, the elastic energy is converted into kinetic energy.

Chemical potential energy is the energy stored in the bonds between atoms in a substance. Chemical reactions break the bonds and release the energy that is holding them together. One very common chemical reaction is combustion, such as when a campfire burns. Heat and light energy are released when the bonds between the atoms that make up wood are broken and rearranged.

Nuclear potential energy is different from chemical energy because it involves a change to the nucleus of an atom rather than to the bonds holding atoms to other atoms. When a nuclear reaction takes place, an enormous amount of energy is released from the nucleus of an atom, causing a chain reaction of energy release. When nuclear reactions are not properly controlled, the results can be disastrous (see Figure 7.7).

There are different types of kinetic energy and potential energy

Table 7.1 shows some common types of kinetic and potential energy.

Table 7.1: Common types of kinetic and potential energy

Energy type		Description
Kinetic	Motion	The energy that an object has because of its movement (change in position over time)
	Thermal	The motion of atoms and molecules in a substance or object that is being heated
	Light	A form of electromagnetic radiation emitted by hot objects; it is usually visible light
	Sound	The vibrating movement of particles that can be detected by the ear
	Electrical	The energy generated by the moving flow of electrons
Potential	Gravitational	The potential of an object to do work because of its position above the ground
	Elastic	The energy stored in an object when it is stretched or deformed
	Chemical	The energy stored in the bonds between atoms, released as heat, sound or light during chemical reactions such as burning
	Nuclear	The energy stored in the nucleus of atoms; it may be released quickly as a nuclear explosion or controlled slowly as a nuclear reaction

Figure 7.7: The uncontrolled release of stored nuclear potential energy causes a devastating explosion.

Learning Ladder

Change

1. Identify whether these objects are displaying potential energy or kinetic energy.
   - **a** A cup that is about to fall off a table
   - **b** A bouncing ball
   - **c** A stretched elastic band
2. Describe the difference between energy *transfer* and energy *transformation* and provide an example of each.
3. Explain the energy transformation that takes place when:
   - **a** a car accelerates after being stopped at a red light.
   - **b** someone is jumping on a trampoline.
4. Outline a scenario where object A and object B are travelling at different speeds, but they have the same kinetic energy. (Hint: Start by deciding what objects A and B are in your scenario.)
5. A boy is using an elastic band to launch balls of paper towards a target. However, he cannot reach the target, no matter how far he pulls the elastic band back. Use your knowledge from this lesson to offer him some advice.

Observing

see page 316

1. Using the table below, identify some examples of kinetic and potential energy that you can observe right now with your own senses (for example, sight, hearing, touch). Include examples from inside and outside the classroom, and the total number of each type.
2. Propose at least three tools (for example, equipment) that would help you to observe more types of energy, beyond what you can detect using your own senses.
3. Compare your table from Question 1 with that of a classmate. Check their table as they check yours.
   - **a** Discuss and record any errors or differences in the lists, tallies and/or totals.
   - **b** Based on the discussion with your classmate, adjust your table to ensure it is accurate.
4. Using your adjusted table from Question 3, what inferences (scientific assumptions) can you make about the types of energy around you? Refer to your observations to support your response.

In context

Research and compare the amount of potential energy stored in a stretched elastic band (elastic potential) and in a litre of petrol (chemical potential).

Success criteria

- I can identify examples of energy as kinetic energy or potential energy.
- I can explain what potential energy is, including naming some types of potential energy.

	Inside the classroom		Outside the classroom		Total
	List	**Tally**	**List**	**Tally**	
Potential energy					
Kinetic energy					

7·2 ► Energy transfer

Learning intention

At the end of this lesson, I will be able to:

- identify conduction, convection and radiation as different ways that energy can be transferred
- distinguish between and explain examples of conduction, convection and radiation.

Key terms

conduction: the transfer of energy through a substance

conductor: a substance that allows the transfer of energy

convection: the transfer of energy by movement of a liquid or gas

heat: the flow of thermal energy between objects

insulator: a substance that resists the transfer of energy

matter: a substance that has mass and takes up space

radiation: the transfer of energy that does not require contact with matter

Investigation 7.2A

Conduction: Heat energy transfer in a solid, p.460

Investigation 7.2B

Convection: Heat energy transfer in a liquid, p.461

Content group: Energy transfers

Heat is a transfer of thermal energy. The terms 'heat' and 'thermal energy' are sometimes used interchangeably. Thermal energy can be transferred in three ways: by conduction, convection and radiation. Heat always moves towards the object with less energy. In other words, thermal energy is always transferred from hot to cold.

Conduction is how energy is transferred through materials

Energy moves through solids by **conduction**. All the particles in the solid are vibrating in place. When heated, the particles gain energy and vibrate even more. As the particles bump into neighbouring particles, energy is transferred along the object. Heat moves through the solid (see Figure 7.8).

Some solids transfer thermal energy better than others. Substances that transfer heat well are called **conductors**. Metals are among the best conductors. Non-metals are usually poor conductors of heat and are known as **insulators**. Plastic, wood and rubber are insulating materials. This is why rubber thongs protect your feet from hot sand.

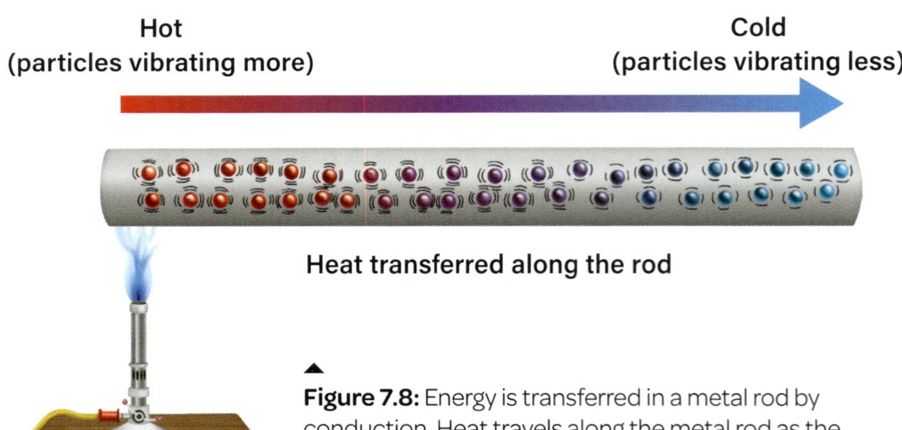

Figure 7.8: Energy is transferred in a metal rod by conduction. Heat travels along the metal rod as the metal particles vibrate and bump into each other.

Figure 7.9: A variety of materials can be used to insulate our homes. Why do you think spray-foam material is one of the most effective insulators for a roof space?

Figure 7.10: Convection currents can be seen in lava lamps, which contain wax in a transparent liquid. The bulb in the base heats the wax, which expands and becomes less dense. The warm wax rises, then cools and becomes denser as it moves away from the heat. It sinks to the bottom where it is warmed again.

Convection moves energy through moving particles

Thermal energy transfers through moving particles of matter by **convection**. Particles in liquids and gases can flow freely past one another, rather than just vibrating in place as they do in solids. When a liquid or gas is heated, particles move to take the place of particles with less heat. If a liquid is heated from underneath, particles move from the heat source at the bottom towards the top where the liquid is cooler. The cooler particles are therefore pushed downwards to where their thermal energy increases again. This continuing pattern of movement is called a convection current. The same pattern of movement occurs with gases.

Convection currents are due to density differences. When a liquid or gas is heated, the particles move further apart from each other, causing the substance to expand and decrease in density. This causes it to rise into cooler areas. The denser cold liquid or gas falls into the warm areas, triggering the convection currents that transfer energy from place to place.

Convection explains why hot air-balloons rise, how lava lamps work (see Figure 7.10), and why it is often hotter upstairs than downstairs in a two-storey house.

Radiation transfers heat through a vacuum

Heat energy can also be transferred from one place to another without **matter** by **radiation**, which travels by electromagnetic waves. This is how the heat of the Sun travels through the vacuum of space to reach Earth. Many types of energy can be transferred by electromagnetic radiation, such as visible light and microwaves. Heat energy, or infrared radiation, causes objects to gain thermal energy and heat up when placed near a fire or left outside on a sunny day. Infrared heaters transfer thermal energy throughout a room by radiation.

Learning Ladder

Change

1. True or false? Thermal energy is a type of potential energy because it is not an object in motion.
2. Describe how heat affects the thermal energy of:
   - **a** solid objects.
   - **b** liquid substances.
   - **c** gaseous substances.
3. Heat is applied to an object.
   - **a** Describe how this affects its thermal energy.
   - **b** Explain how you could measure the observed change.
4. Energy from the Sun changes several times before it reaches Earth. Explain how this energy can reach Earth, even though space has no matter.
5. Identify and explain the types of energy transfer (conduction, convection, radiation) in each of the following systems. (Hint: There could be more than one type for each.)
   - **a** When hot coffee is stirred with a spoon, the spoon gets hot.
   - **b** Warm air over hot sand rises, whereas cooler dense air blows in from the ocean.
   - **c** You try to sit on a rock to relax after a long, sunny day, but the rock is too hot to sit on.

Planning investigations

see page 323

Consider this investigation question: 'Which method of popping popcorn is the most effective based on the percentage of kernels left unpopped: in a microwave, in a pan on the stovetop, or in a popcorn air machine?'

1. Make a list of the equipment and materials required to carry out an investigation to address the question.
2. Identify four hazards associated with the investigation. Describe two ways to reduce the risk for each hazard.
3. Identify the independent variable, the dependent variable and at least three controlled variables.
4. Propose a suitable aim for this investigation.
5. Write a step-by-step procedure for a Year 8 student to address the investigation question.

In context

The cold water in your plastic bottle always warms up too quickly. Use your understanding of conductors and insulators to propose ways to solve this problem.

Success criteria

- I can interpret basic heat transfer diagrams of convection, conduction and radiation.
- I can explain the difference between convection and conduction with reference to states of matter and particle theory.

7.3 ► Energy changes

Learning intention

At the end of this lesson, I will be able to identify a system and the energy transfers and transformations within it that cause observable changes.

Key term

system: a set of simple things that work together to perform a function

Investigation 7.3

Energy transformations, p.462

Content group: Energy transfers

All energy on Earth originally comes from the Sun. Nuclear reactions in atoms power the Sun, and this energy is transformed from nuclear potential energy into light and heat. The Sun then radiates an enormous amount of light and heat energy into space in all directions. Only a small amount of this energy is transferred to Earth, where it is then stored, released, transferred and transformed over and over again.

Energy causes change in systems

When several simple things work together to perform a function, this is called a **system**. Energy transfers and transformations cause change in systems. These changes are happening all the time around us.

Energy *transfers* happen when the energy in one object is transferred to another object, which causes observable changes in those objects. An example is when you fry eggs: thermal energy from the stove is transferred to the frying pan, which heats up to cook the eggs. The transfer of energy involves one type of energy and more than one object. It is still thermal energy, but it has been transferred from the stove to the pan and to the eggs.

Energy *transformations* happen when a form of energy is changed to a different form of energy in the same object. This also causes change in the system. An example is when the elastic potential energy of a stretched rubber band changes into kinetic energy when it is released. The rubber band is a single object that experiences two types of energy: elastic potential energy transformed into kinetic energy. The rubber band flying across the room is the observable change in the system.

Devices can be used to transfer and transform energy

How did you get to school today? Maybe you came in a car or bus; maybe you rode a bicycle or walked. Your travel to school involved a transformation of energy. Chemical potential energy (the petrol in the car or the food you ate) was transformed into kinetic energy (causing the motion of the car or of your legs). We use many devices to transform energy. For example, televisions transform electrical energy into light and sound energy, and kettles transform electrical energy into thermal energy.

Imagine that you threw a book and it fell to the ground with a bang (see Figure 7.11). How many energy transfers and transformations would happen? Recall the many different forms of energy from Table 7.1 on page 228.

① Energy transformation in arm throwing

Chemical potential energy

↓

Kinetic energy

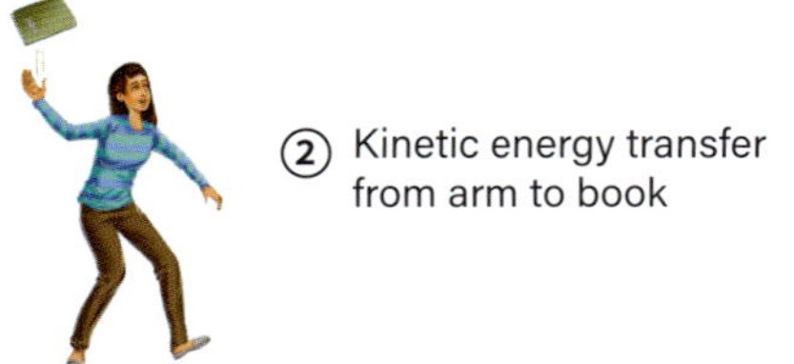

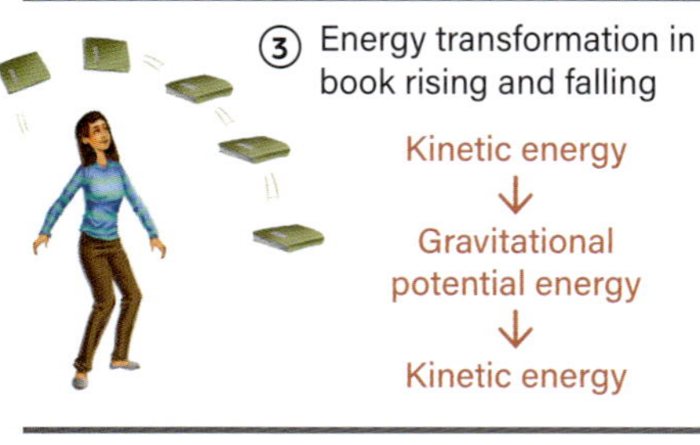

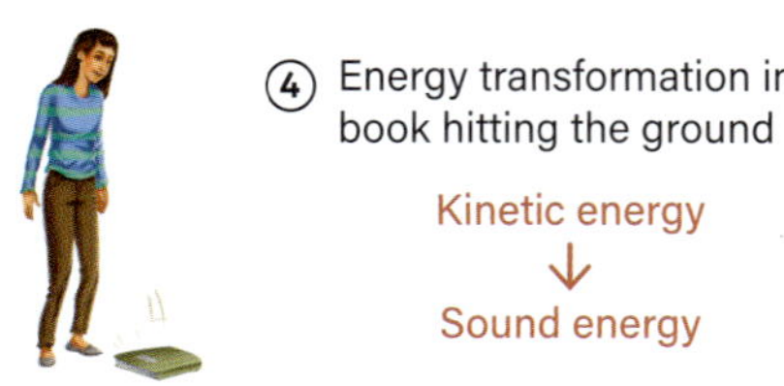

Figure 7.11: How many energy transformations are happening here?

In a suitable system, any one of these forms could be transformed into one of the other forms or the energy transferred from one object to another.

Flowcharts can be used to represent energy transformations

An apple can grow because of light and heat energy from the Sun, which it transforms and stores as chemical potential energy. You eat the apples and your body transforms the chemical potential energy in the apple into kinetic energy for movement such as walking.

These energy transformations can be displayed on a flowchart. Flowcharts clearly show where energy has come from, what it is doing and where it goes.

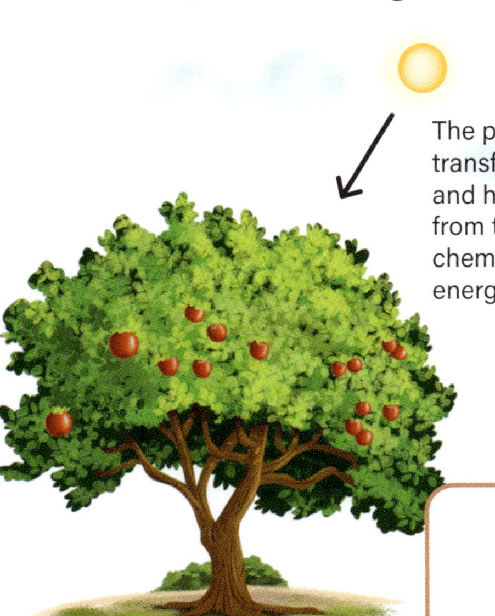

The plant transforms light and heat energy from the Sun into chemical potential energy.

Chemical potential energy is stored in the leaves, stems, fruits and roots of plants.

The chemical potential energy in the apple is transformed into kinetic energy through digestion and then respiration to create energy in muscles that move.

Figure 7.12: Energy transformations taking place from growing food to using energy from the food

Learning Ladder

Change

1. Propose the main form of energy related to each example and give a reason for your answer.
   - **a** Standing on a tightrope, 10 metres off the ground
   - **b** An arrow ready for release from a bow
   - **c** A fireworks display
2. Categorise the following as an energy transfer or transformation and give a reason for your answer.
   - **a** Kicking a ball
   - **b** Lighting a candle
   - **c** Hitting a drum
3. Propose a specific system or situation that could result in the following energy transformations.
   - **a** Chemical energy → light energy
   - **b** Electrical energy → heat energy
   - **c** Gravitational potential energy → sound energy
4. Consider electrical energy transforming into sound and motion (kinetic energy).
   - **a** Suggest a real-life example that would result in the energy transformations.
   - **b** Construct a flowchart to explain the energy transformations.

Conducting investigations

see page 326

Consider this investigation question: 'Which method of popping popcorn is the most effective based on the percentage of kernels left unpopped: in a microwave, in a pan on the stovetop, or in a popcorn air machine?'

1. Identify the most suitable approach to address the investigation question.
   - **A** Conduct a survey asking people's preferred method for popping popcorn.
   - **B** Trial one of the methods four times to see if the same amount of popcorn pops each time.
   - **C** Trial each method four times to determine the average proportion of popped to unpopped kernels for each method.
2. Construct a risk assessment for this investigation.
3. Propose a piece of equipment that could help to determine the percentage of kernels left unpopped. Would your suggestion provide accurate results? Explain why or why not.

In context

Most cars burn petrol as their source of energy. With reference to energy transfer and transformation, outline why more and more vehicle manufacturers are designing cars that are powered by alternative energy sources.

Success criteria

- I can describe the difference between energy transfer and energy transformation.
- I can give at least one example of an everyday energy transformation.

7·4 ▸ Systems in energy

Learning intention

At the end of this lesson, I will be able to:

- recognise that energy is transferred within systems, and into and out of open systems
- apply the law of conservation of energy to familiar energy systems.

Key terms

closed system: a system from which energy can flow in and out but matter cannot

conservation of energy: the scientific law that states that energy cannot be created or destroyed

isolated system: a system from which neither energy nor matter can flow in or out

open system: a system that allows energy and matter to transfer in and out

Content group: Energy transfers

The law of **conservation of energy** tells us that energy cannot be created or destroyed, only transferred from one object to another, or transformed into different types of energy. Therefore, we can assume that the total amount of energy in the universe cannot change. But what about energy systems? We know that a ball rolling on a flat surface will eventually stop, and that hot tea will eventually cool. Where has the energy gone? To answer this question, we must first discuss open and closed systems in relation to energy.

Open systems allow the flow of matter and energy in and out

In an **open system**, energy and matter can enter and leave. Imagine a classroom with open windows: air (matter) from the classroom can flow out, and air from outside can flow in (see Figure 7.13). Sun shining in (energy) warms the room, and heat can also leave through the open windows. The amount of energy and matter in the classroom depends on surrounding factors.

Earth is an open system. All of Earth's energy originates from the Sun. About half of it is reflected back and half is absorbed by the land and oceans (see Figure 7.14). The energy cycling through the land and oceans enables life on Earth. Photosynthesis transforms solar energy into chemical energy to feed plants and animals. Animals transfer the chemical potential energy into heat energy, which is transferred back through the atmosphere and radiated into space.

Energy enters and escapes Earth's atmosphere, but it is not created or destroyed. It cycles through Earth, and back and forth from space.

▼ **Figure 7.13:** This classroom is an example of an open system. Energy and matter can flow in and out.

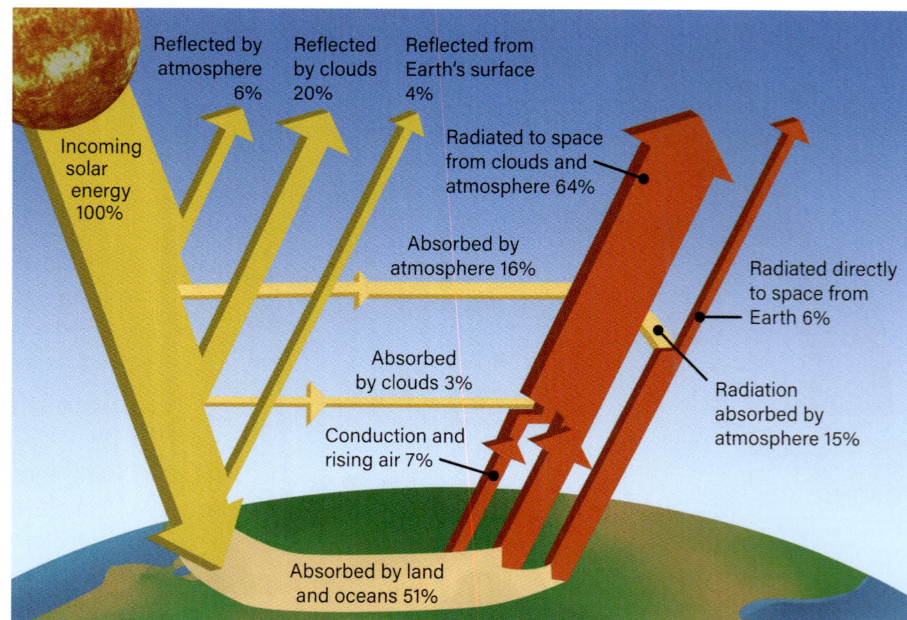

Figure 7.14: Earth's energy budget ▸

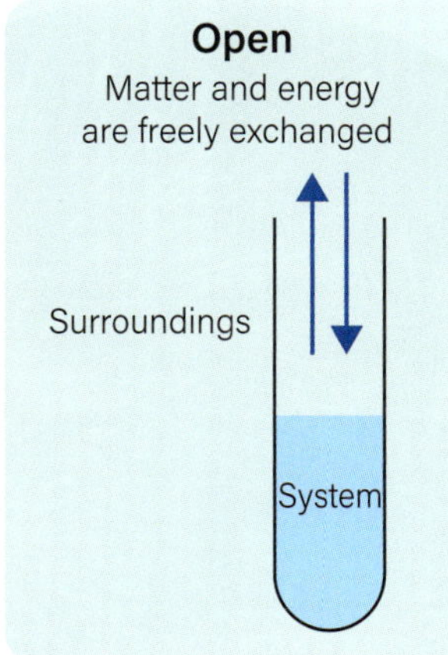

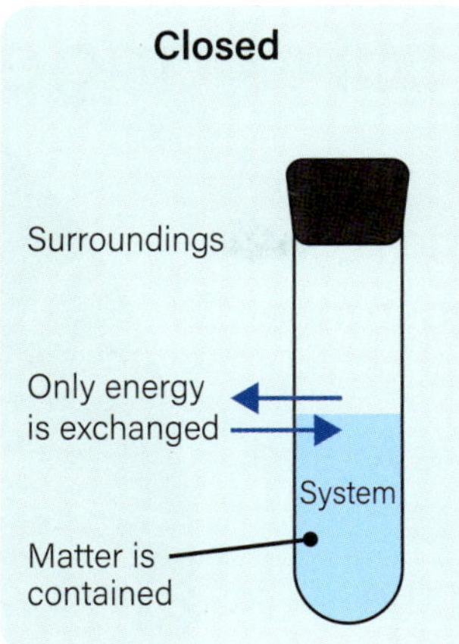

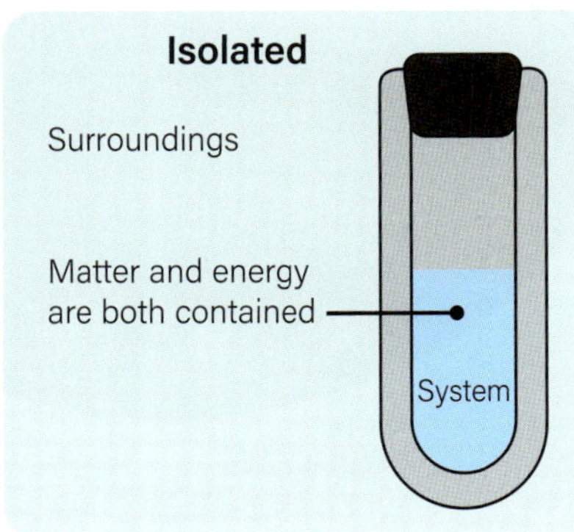

Figure 7.15: The flow of matter and energy in and out of an open, closed and isolated system

Closed systems restrict the flow of matter and energy in and out

Matter and energy are unable to flow freely in or out of a **closed system**. For a classroom to be a closed system, we would need to close the windows and doors, and shut the blinds and curtains. This would stop air flowing in and out, as well as prevent sunlight and other forms of energy getting in or escaping.

In a closed system, matter cannot enter or leave, but energy can usually flow in and out. For example, when you are cooking pasta on the stove, having the lid on the pan keeps the heat in so the water will boil faster, but the stove is still able to transfer energy into the system.

Isolated systems have no flow of matter and energy in or out

If a system is completely isolated from its surroundings, neither matter nor energy is able to flow in or out. This is an **isolated system**. Energy can cycle within the system, but the total amount of energy stays the same. However, a system that has no exchange of energy is not realistic, so isolated systems are only seen in scientific models and simulations.

Learning Ladder

Change

1. List the changes that occur in a classroom with open doors and windows.
2. Identify which of the following best represents a closed system of energy.
   a Turning off a pressure cooker but allowing it to continue cooking the food for several more hours
   b Having the fan on to prevent the shower steaming up the bathroom
3. Justify your response to Question 2 by referring to energy systems, transfer and transformation.
4. With reference to open and closed systems, compare the energy changes that occur in an outdoor herb garden and a herb garden in a greenhouse.
5. Elaborate on your response to Question 4 by discussing links between energy changes and the law of conservation of energy.

Observing

see page 316

1. Identify the senses you would use to support the idea that a cup of hot chocolate is an open system.
2. Identify the tool you could use to measure the difference in cooling rate between two cups of hot chocolate, one with a lid (closed) and one without a lid (open).
3. Construct a table to record observations on the cooling rate of open and closed hot chocolates over a 10-minute period.
4. With reference to first-hand evidence, justify the fact that hot chocolate with a lid is not a truly closed system.
5. Justify the law of conservation of energy with reference to Figure 7.14. Discuss how energy that originates from the Sun can cycle through Earth without being created or destroyed.

In context

With a classmate, brainstorm naturally occurring systems that cycle energy through Earth. Select a system and construct an annotated poster diagram to represent how energy cycles through the system. Be sure to explain scientific terms related to energy systems and include flowchart diagrams where appropriate. Present your poster to the class.

Success criteria

- I can distinguish between open and closed systems.
- I can use diagrams to interpret and represent energy transfer and transformations in systems.

7·5 ▸ Physical and chemical change

Learning intention

At the end of this lesson, I will be able to:

- define physical and chemical change and recognise examples of each
- distinguish between physical and chemical change by using senses to observe indicators of change.

Key terms

chemical change: a change in properties with a new substance formed

physical change: a change in appearance with no new substance formed

property: a characteristic of a substance that affects what it looks like and how it behaves

reversible: can be changed back to its previous state

Content group: Chemical change

A physical change is one in which no new substance is formed and where the process can be reversed. If a new substance is produced and the process is not reversible, then this is a chemical change.

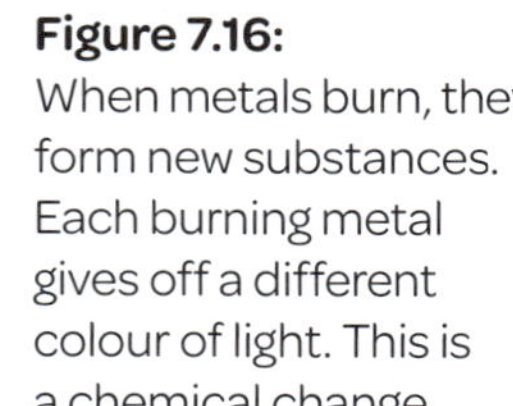

Figure 7.16: When metals burn, they form new substances. Each burning metal gives off a different colour of light. This is a chemical change.

Physical changes do not affect the properties of a substance

When a substance undergoes a physical change, its **properties** do not change. A change of state is a **physical change**; for example, water changing from solid (ice) to liquid (water) to gas (steam). All three states still consist of water molecules (H_2O).

Physical changes are **reversible**. Wax can be heated and melted, and then cooled and solidified; it looks the same as before it was melted. The wax has undergone physical changes.

Other examples of physical changes are salt and sand being mixed, paper being cut, evaporation, a ruler breaking, and sugar dissolving in water.

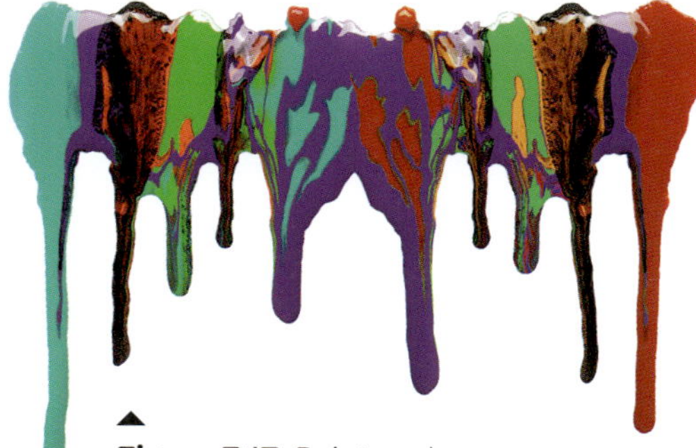

Figure 7.17: Paint undergoes a physical change when it dries.

Chemical changes can make new substances

A **chemical change** happens when a new substance is formed during a chemical reaction. Chemical reactions cause the atoms in the substance to rearrange and form new substances.

Evidence of a chemical change can be seen when:

- there is a change in colour
- gas evolves (bubbles)
- a new substance forms/appears
- there is a change in energy (heat, cooling or light is observed)
- a precipitate (solid particles within a liquid) forms.

Chemical changes are not easily reversible. You cannot uncook a pancake to make it batter again.

Some other examples of chemical changes are burning wood, a banana rotting, a battery in use, fireworks going off, iron rusting and cookies baking.

Identifying the type of change can be difficult

A reaction can sometimes include both physical and chemical changes; for example, baking a cake. While the cake is being cooked, water from the batter evaporates, which dries out the cake. The colour of the cake batter changes permanently, indicating a chemical change.

It is not always straightforward to distinguish a physical change from a chemical change. An apparent colour change can sometimes result from a physical change; for example, white snow melts to colourless water, and blue and red paint mixed together look purple. However, a single snowflake would be colourless, and the paint mixture up close would show both red and blue particles (not purple). No new substances have been formed.

Similarly, the production of a gas is not always a sign of a chemical change. When water is boiled, bubbles form, indicating a change of state. Several types of observations may be important in distinguishing between physical and chemical changes. Relying on only one observation might lead to incorrect or incomplete conclusions.

▼ **Figure 7.18:** The sugar in the top pan is melting, which is a physical change because no new substances are formed. The sugar in the bottom pan has been heated more, and some has chemically changed into a new substance: caramel.

Learning Ladder

Change

1. Identify the following as physical or chemical changes.
   - **a** Ice melting
   - **b** A steak cooking on a barbecue
   - **c** Paper burning
   - **d** Chocolate powder mixing into milk
2. Describe the difference between a physical and a chemical change.
3. When you bend an instant ice pack, it immediately becomes ice cold. With reference to energy, explain how this can be identified as a chemical change.
4. What processes are involved as water changes state? Is this a physical or a chemical change? Justify your decision.

Observing

see page 316

1. Identify the sense you would use to observe the following indicators of change.
   - **a** Two colourless solutions combined together turn pink.
   - **b** Inserting a flame into a reaction test tube makes a loud pop.
   - **c** An alcohol is mixed with an acid to release a fruity aroma.
2. Describe a scientific tool you could use to record and organise observations as your teacher demonstrates some physical and chemical changes.
3. Your observations from Question 2 seem to be different from a classmate's.
   - **a** Propose ways you could determine the accuracy of your or their observations.
   - **b** Identify observation tools you could use next time to avoid accuracy issues.

In context

Fireworks involve chemical reactions that produce coloured light. Use the internet to find out what chemical reactions take place and what produces the different colours.

Success criteria

- I can explain the difference between a physical change and a chemical change.
- I can identify evidence for a chemical change and make links to common examples.

7·6 ▸ Representing chemical change

Learning intention

At the end of this lesson, I will be able to:

- describe a chemical reaction in terms of reactants and products
- recognise the law of conservation of mass.

Key terms

chemical reaction: a process in which one or more substances are changed to form new substances

conservation of mass: the scientific law that states that mass cannot be created or destroyed

product: a substance formed in a chemical reaction

reactant: a substance that undergoes a chemical change

word equation: a representation of a chemical change in words

Investigation 7.6

Law of conservation of mass, p.463

Content group: Chemical change

Humans need several things to be able to survive, including oxygen to breathe, water to drink and food to eat. All of these substances are involved in chemical reactions that give us the energy we need to grow, move and survive.

A chemical reaction is a rearrangement of atoms

A **chemical reaction** occurs when two substances combine in such a way that the atoms rearrange themselves. Consider water: the formula for water is H_2O because each molecule of water contains two hydrogen atoms and one oxygen atom. We can form water by combining hydrogen and oxygen gas and adding some energy to get the reaction going. This will cause the hydrogen and oxygen molecules to split apart into atoms and recombine into water molecules.

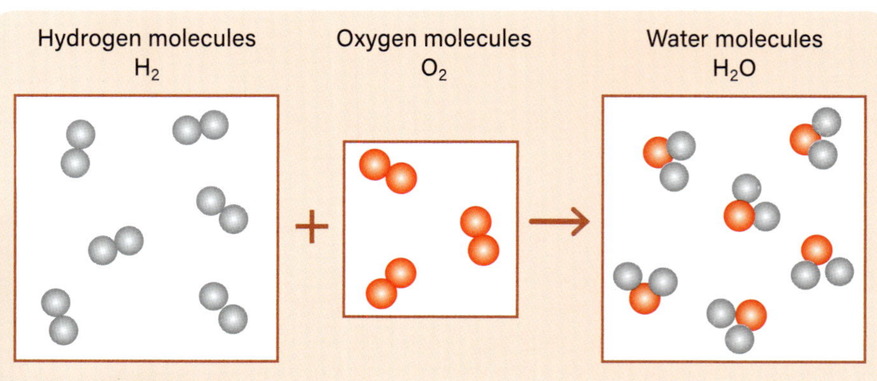

▲ **Figure 7.19:** Hydrogen (H_2) and oxygen (O_2) molecules split apart and recombine to form water molecules (H_2O).

Reactants — Product

oxygen gas + hydrogen gas → water

Figure 7.20: The word equation for the reaction between oxygen gas and hydrogen gas to form water

Word equations can describe chemical reactions

In the reaction in Figure 7.20, the oxygen gas and hydrogen gas are the initial substances before the reaction: the **reactants**. Water is the substance created in the reaction, known as the **product**. We can represent this process as a **word equation**.

The arrow (→) represents the reaction. The reactants are always on the left of the arrow with plus (+) signs between each reactant to indicate that they are 'added' together. The products are always on the right of (after) the arrow, indicating that the reaction has taken place.

Conservation of mass

During a chemical reaction, the atoms of the reactants recombine to form new substances; no new atoms are created, and no atoms are destroyed. This means that the same number of atoms in the reactants are in the products. The atoms are just arranged differently. The total mass of products must be the same as the mass of reactants. This is known as the law of **conservation of mass**.

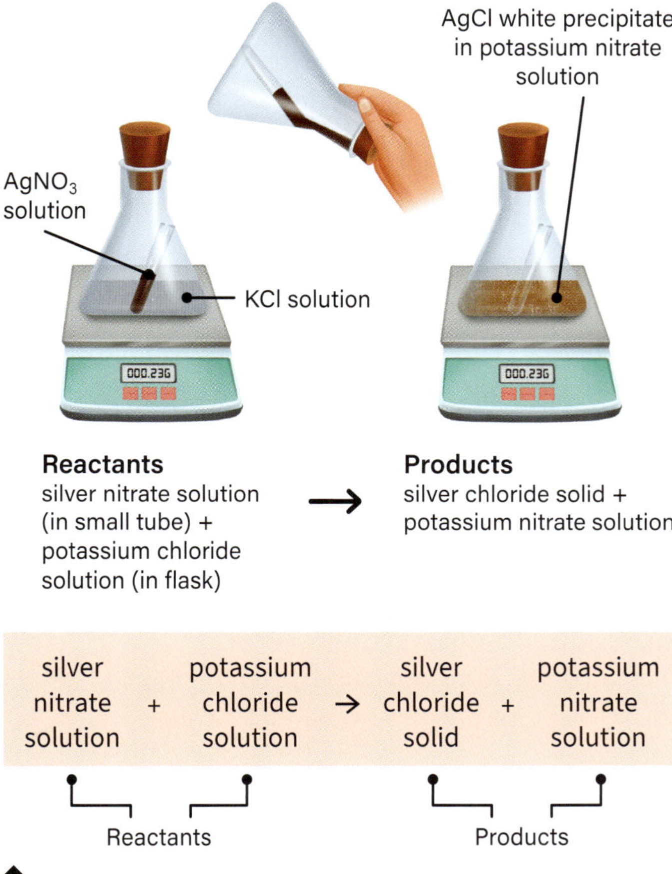

Figure 7.21: If a reaction occurs in a sealed container, the mass will not change, even if new products are formed.

Sometimes, the law of conservation of mass might seem wrong. Think about a piece of wood burning in a fireplace. The wood is undergoing a chemical reaction known as combustion. After the reaction has finished, the ash left over is much lighter than the original wood. This is because the ash is not the only product of the reaction: carbon dioxide and water are formed, but they have drifted away as gases. If we were to weigh all of the ash, carbon dioxide and water, they would be equal to the mass of the original block of wood and oxygen that reacted.

Learning Ladder

Change

1. Identify the reactants and products in the following word equations.
   - **a** zinc + hydrochloric acid → zinc chloride + hydrogen gas
   - **b** oxygen gas + magnesium → magnesium oxide
   - **c** hydrogen peroxide → oxygen gas + hydrogen gas
2. Write a word equation for photosynthesis, where carbon dioxide and water combine to form glucose and oxygen gas.
3. Suggest why an arrow (→) is used in chemical equations, and not an equals (=) sign.
4. Dissolving salt in water is not an example of a chemical reaction. Explain why.
5. A piece of wood is burned in air. The wood forms ash, which is lighter than the original wood. Outline what happens to the rest of the mass of the wood. Refer to the law of conservation of mass.

Conducting investigations *see page 326*

You are conducting an experiment to support the law of conservation of mass by weighing substances before and after a chemical change.

1. Identify the purpose of having a stopper on the reaction flask.
2. Propose how the word equation of the chemical reaction in the experiment could be used to gather materials to carry out the investigation.
3. Propose three ways to minimise the risk associated with glassware used in the investigation. Describe what you would do if you broke the flask during the investigation, and what you would do if you cut your finger on a piece of the broken glass.
4. Explain how you would ensure that the data collected from your investigation was accurate.

In context

Discuss the importance of being able to represent chemical change in an equation. Use the internet to research how representations of chemical reactions developed over time. How have these developments benefited society?

Success criteria

- I can identify the reactants and products in a chemical reaction.
- I can use word equations to represent chemical change.

7.7 ▸ Chemical changes in everyday life

Learning intention

At the end of this lesson, I will be able to identify and describe some chemical changes that occur in everyday life.

Key terms

combustion: the reaction of a substance with oxygen to give off heat

corrosion: the reaction of a metal with oxygen

Investigation 7.7A
Modelling respiration and photosynthesis, p.464

Investigation 7.7B
Corrosion of iron, p.466

Investigation 7.7C
Preventing corrosion, p.467

Content group: Chemical change

A chemical reaction produces a chemical change. In a chemical reaction, one or more reactants are changed into one or more products. The atoms rearrange to form new substances as products.

Photosynthesis is a chemical reaction in plants

In photosynthesis, plants use energy from sunlight to transform water and carbon dioxide into glucose and oxygen. The reactants (water and carbon dioxide) are substances that change during a reaction, while the products (glucose and oxygen) are substances that form as a result of the reaction.

The equation in Figure 7.22 shows that atoms in carbon dioxide and water react and rearrange to form glucose and oxygen. The products are very different from the reactants; they are completely new substances.

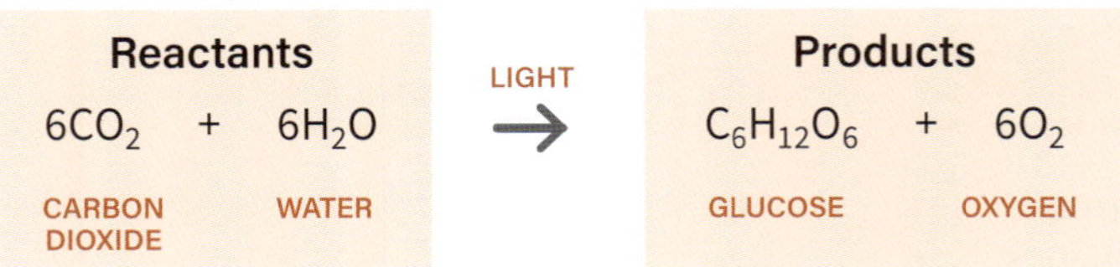

▲ **Figure 7.22:** The chemical equation for photosynthesis

▲ **Figure 7.23:** This iron boat has rusted; this is a chemical change.

We can use a word equation or a chemical equation to represent a chemical reaction. In a chemical equation, the chemical formulas are used instead of the words, or names, of the substances involved. For now, you are only expected to write and interpret word equations.

Cellular respiration combines oxygen and glucose

Cellular respiration is the way that cells of all living organisms produce energy. During cellular respiration, oxygen (from the air) and glucose (from food) combine. This chemical reaction produces carbon dioxide, water and energy.

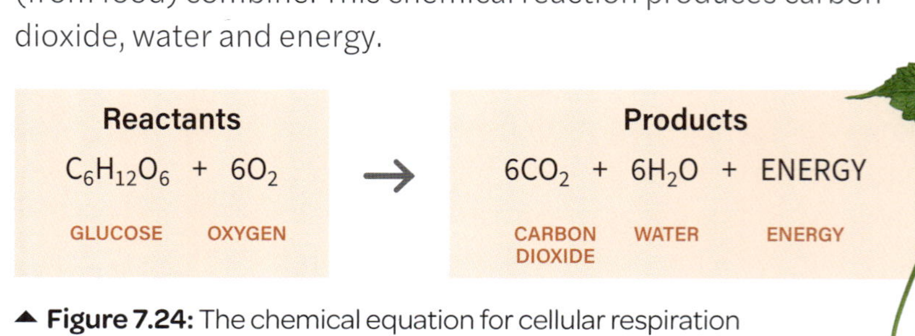

▲ **Figure 7.24:** The chemical equation for cellular respiration

Cellular respiration is the opposite reaction to photosynthesis.

Rust is a chemical change

When some metals come into contact with oxygen in the air, they combine in a chemical reaction.

Reactants
$4Fe + 3O_2$
IRON OXYGEN GAS

Products
$2Fe_2O_3$
IRON OXIDE

◀ **Figure 7.25:** Iron reacts with oxygen to form rust.

This type of reaction is known as **corrosion**. The reactants are the metal and oxygen. The product is a new substance called a metal oxide. When objects made from iron react with oxygen, this produces iron oxide. This is commonly known as rust. Corrosion can cause metal objects to become weaker and break down. Some metals are less likely to corrode than others. Galvanised iron is made by coating iron with a thin layer of zinc. The zinc is less likely to react with oxygen, so this coating protects the iron.

Combustion is a reaction with oxygen gas

Many substances react with oxygen gas (O_2) to produce heat and other chemical products in **combustion** reactions. Combustion reactions can be fast and explosive, such as the reaction of hydrogen gas (H_2) with O_2. They can also be slow, such as corrosion and respiration.

When H_2 gas reacts with oxygen, it produces water vapour.

hydrogen + oxygen gas → water vapour

Reactive metals combust easily in oxygen gas

A metal's reactivity is determined by how readily it combusts with oxygen gas. For example, magnesium undergoes combustion to release heat and light, which makes it ideal for use in sparklers and fireworks.

magnesium + oxygen gas → magnesium oxide + heat/light

The combustion of hydrocarbon compounds (for example, fuels such as petrol and natural gas) produces carbon dioxide and water.

methane (natural gas) + oxygen gas → carbon dioxide + water vapour + heat

Combustion reactions are used to power engines in cars and aeroplanes, and to generate electricity.

Learning Ladder

Change

1. Identify three common chemical changes.
2. Use the word equation for photosynthesis to write a sentence to describe the chemical reaction that occurs within plants. Use the terms 'reactant' and 'product'.
3. Identify and name the reactants and products in each of the following chemical changes.
   - **a** Photosynthesis in plants
   - **b** Cellular respiration in living things
   - **c** Rusting of iron
   - **d** Combustion of hydrocarbons
4. Discuss how chemical change is linked to energy for one of the chemical changes in Question 3.
5. Compare photosynthesis and respiration.

Planning investigations *see page 323*

Imagine you are investigating which of three types of wood burns the hottest.

1. List all of the materials you would require.
2. Identify three hazards that might be present in your investigation. Suggest a minimisation strategy for each hazard.
3. Identify three things that would need to be controlled in your investigation.
4. Summarise the aim of the investigation in a context beyond the classroom.
5. Write a step-by-step investigation plan that a Year 8 student could follow to carry out the investigation. Include a title or investigation question, aim, materials, method and safety considerations.

In context

Research and explain the difference between breathing and cellular respiration. Refer to chemical change in your response.

This lesson includes some important and well-known chemical reactions. Working with a partner, think of as many things as you can that work because of chemical reactions.

Success criteria

- I can identify and describe some chemical changes that occur in everyday life.

7·8 ▶ Energy changes in chemical reactions

Learning intention

At the end of this lesson, I will be able to:

- identify that chemical reactions involve a transfer of energy
- distinguish between exothermic and endothermic chemical reactions.

Key terms

endothermic: a reaction that absorbs energy in the form of heat

exothermic: a reaction that releases energy in the form of heat

Investigation 7.8

Exothermic and endothermic reactions, p.468

Content group: Chemical change

Chemical reactions involve more than just elements and compounds; they also involve energy, usually heat. Some reactions release heat. Others require heat to be added. Without energy being transferred, chemical reactions would not happen.

Exothermic reactions release heat, and endothermic reactions absorb heat

In **exothermic** reactions, the reactants have more energy than the products. This additional energy is released as heat. Exothermic reactions include respiration, combustion reactions, corrosion and burning magnesium.

In **endothermic** reactions, the products have more energy than the reactants. To happen, these reactions require energy from their surroundings, usually in the form of heat. Endothermic reactions include photosynthesis, bread-baking, and decomposition of water (H_2O) into oxygen and hydrogen gas.

Figure 7.26: Endothermic reactions absorb energy, whereas exothermic reactions release energy.

Figure 7.27: Fuel being burnt in a rocket launch is an example of an exothermic reaction.

Bonds break and form during chemical reactions

During a chemical reaction, bonds in the reactants break, and new bonds form to make products. Energy is needed to break the bonds of the reactants, and energy is released when new bonds form in the products.

Figure 7.28: Total energy levels of reactants and products in an (a) exothermic reaction and (b) endothermic reaction

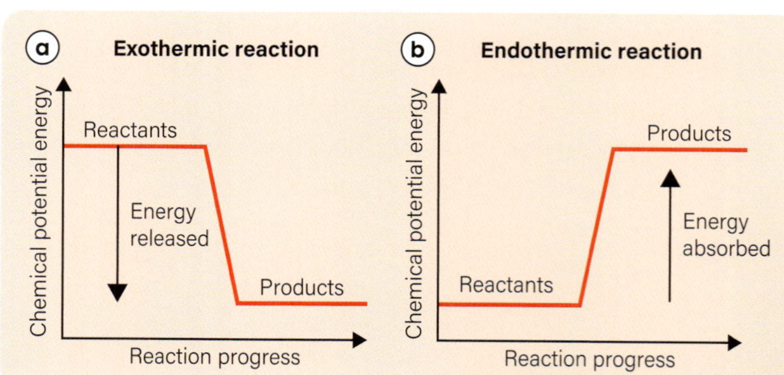

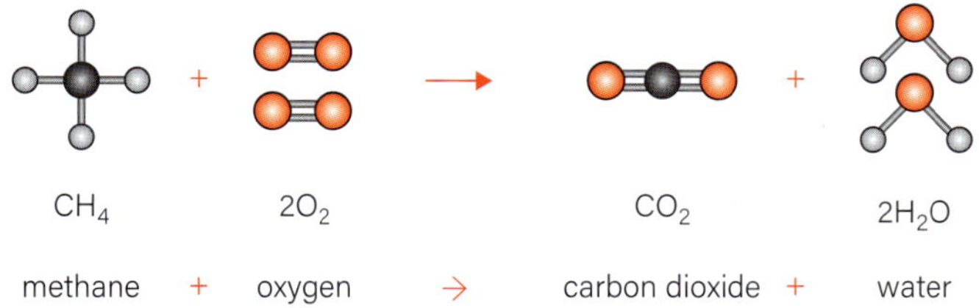

Figure 7.29: The reaction between methane and oxygen

When methane reacts with oxygen, bonds within methane and oxygen break, and new bonds form between carbon, hydrogen and oxygen to form carbon dioxide and water (see Figure 7.29). Depending on the atoms involved, different amounts of energy are required to break and form bonds.

Chemical reactions produce or absorb heat

To identify whether a reaction is exothermic or endothermic, we need to look at the overall energy of the reaction: whether energy was released, or whether it was required for the reaction to proceed.

An exothermic reaction occurs when the energy used to break the bonds in the reactants is less than the energy released when new bonds are made in the products. This additional energy is released as heat and the temperature of the surroundings increases. Overall, these reactions release energy.

Chemical equations for exothermic reactions can be written like this:

reactants → products + ENERGY

In an endothermic reaction, the energy required to break the bonds in the reactants is greater than the energy released when the products are formed. This means that overall the reaction requires energy from its surroundings, so the temperature decreases around the reaction container or surroundings. For example, instant ice packs work by an endothermic reaction between ammonium nitrate and water, which absorbs energy from the surroundings.

Chemical equations for endothermic reactions can be written like this:

reactants + ENERGY → products

Learning Ladder

Change

1. Identify how the energy of the reactants in an exothermic reaction changes (increases or decreases) when the products are formed.
2. Describe why energy is required for chemical reactions.
3. Identify the following changes as exothermic or endothermic. Give a reason for each choice.
   - **a** Ice melting
   - **b** Burning wood in a campfire
   - **c** Frying bacon and eggs
   - **d** Using a cold pack on an injured leg
4. Compare the energy required to break and form bonds in both endothermic and exothermic reactions.
5. Search the internet for common endothermic and exothermic reactions. Identify and explain at least two examples of each.

Conducting investigations *see page 326*

Refer to Investigation 7.8 on page 468 to answer the following questions. Read the aim, materials and method.

1. Explain whether the method provides a suitable way to address the investigation aim.
2. If you do not have access to a thermometer, outline a way you could observe energy changes during each chemical change.
3. Identify four hazards associated with the investigation, including your suggestion for Question 2. Describe two ways to reduce the risk for each hazard identified.
4. Propose three things you could do to ensure consistent measurements are collected using the thermometer.

In context

A sample of sodium hydroxide was dissolved in a test tube of water at 19 °C. The temperature increased to 28 °C. A sample of potassium nitrate was dissolved in a test tube of water at 20 °C. The final temperature was 11 °C. Determine the temperature change for each and state which reaction was endothermic and which was exothermic.

Success criteria

- I can describe exothermic and endothermic reactions.
- I can explain why chemical reactions involve energy transfer.

7·9 ▸ Earth's structure

Learning intention

At the end of this lesson, I will be able to describe Earth's structure in terms of the core, mantle, crust and lithosphere.

Key terms

asthenosphere: the portion of Earth's mantle underneath the lithosphere that can flow

atmosphere: the layer of gas that surrounds Earth

core: Earth's central layer, made up of a liquid outer core and a solid inner core

crust: Earth's thin outer layer, made up of continental crust and oceanic crust

density: how heavy something is for its size; mass divided by volume

lithosphere: Earth's rigid outer zone (crust and most of the upper mantle), made up of tectonic plates

mantle: Earth's middle layer, made up of an upper mantle and a lower mantle

tectonic plate: a section of Earth's lithosphere

Investigation 7.9

Modelling Earth's structure, p.470

Content group: Geological change

When Earth formed approximately 4.5 billion years ago, it was a ball of molten (melted) rock. As it gradually cooled, this molten rock separated into three main layers. The most dense metallic elements moved to the centre to form the core. Elements of medium density formed the mantle, and the least dense elements moved to the surface to form its crust and atmosphere.

Earth's core is solid on the inside, liquid on the outside

Earth's centre where elements are of the greatest **density**, is the **core**. It is made up of a very hot liquid outer core and a solid inner core. The core is mostly iron (85%), with a small amount of nickel (5%) and some lighter elements, including oxygen and silicon. The pressure in the inner core is so immense that the atoms are forced together to form a solid.

The mantle sits between Earth's core and the crust

Earth's thickest layer is the **mantle**. Even though the mantle is made of solid rock, the very high temperatures and pressures enable the rock to flow very slowly. The process is similar to what you might see when you put a blob of silly putty on the edge of a desk. The rocks in the mantle are mostly made up of oxygen (48.9%), silicon (21.5%) and magnesium (22.8%). Other elements include iron, aluminium, calcium, sodium and potassium. The processes in the mantle cause a lot of the change and movement on Earth's surface. The mantle can be thought of as two parts: the upper mantle and the lower mantle.

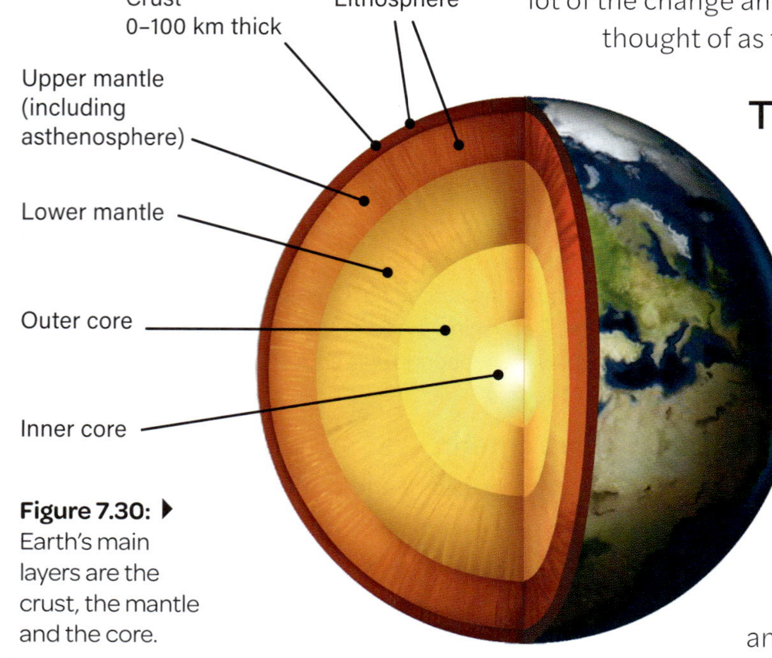

Figure 7.30: ▸ Earth's main layers are the crust, the mantle and the core.

The crust is Earth's thin outer layer

The **crust** is Earth's hard outer layer, and its thinnest layer. The crust consists of a variety of rocks that are mostly made of the elements oxygen (46.1%), silicon (28.2%) and aluminium (8.23%). Other elements include iron, calcium and sodium. There are two types of crust: continental crust and oceanic crust. The continental crust forms the continents and the shallow seas around the continents. It covers about 40 per cent of Earth's surface and is 10–100 kilometres thick. The oceanic crust is formed in Earth's ocean basins. It covers about 60 per cent of Earth's surface and is 5–7 kilometres thick.

Tectonic plates move around on the asthenosphere

The **lithosphere** is Earth's rigid, rocky outer zone. It includes the crust and most of the upper mantle. The lithosphere is made up of **tectonic plates** that 'float' and move around on a zone called the **asthenosphere**. There are 15 major tectonic plates and some smaller ones.

The asthenosphere is a thin zone of the upper mantle that sits just beneath the lithosphere. The rocks here are at very high temperatures and pressures, so they almost melt. This makes them very viscous and ductile compared to other parts of the mantle. This allows the tectonic plates of the lithosphere to move and act on each other, which can cause earthquakes and volcanoes at Earth's surface.

The atmosphere is a layer of gas

The **atmosphere** is the 600-kilometre thick layer of gas that surrounds Earth. It protects life on Earth from harmful radiation by either absorbing the radiation or reflecting it back into space. The atmosphere is made up of nitrogen (78%) and oxygen (21%), along with tiny quantities of argon, carbon, neon and helium.

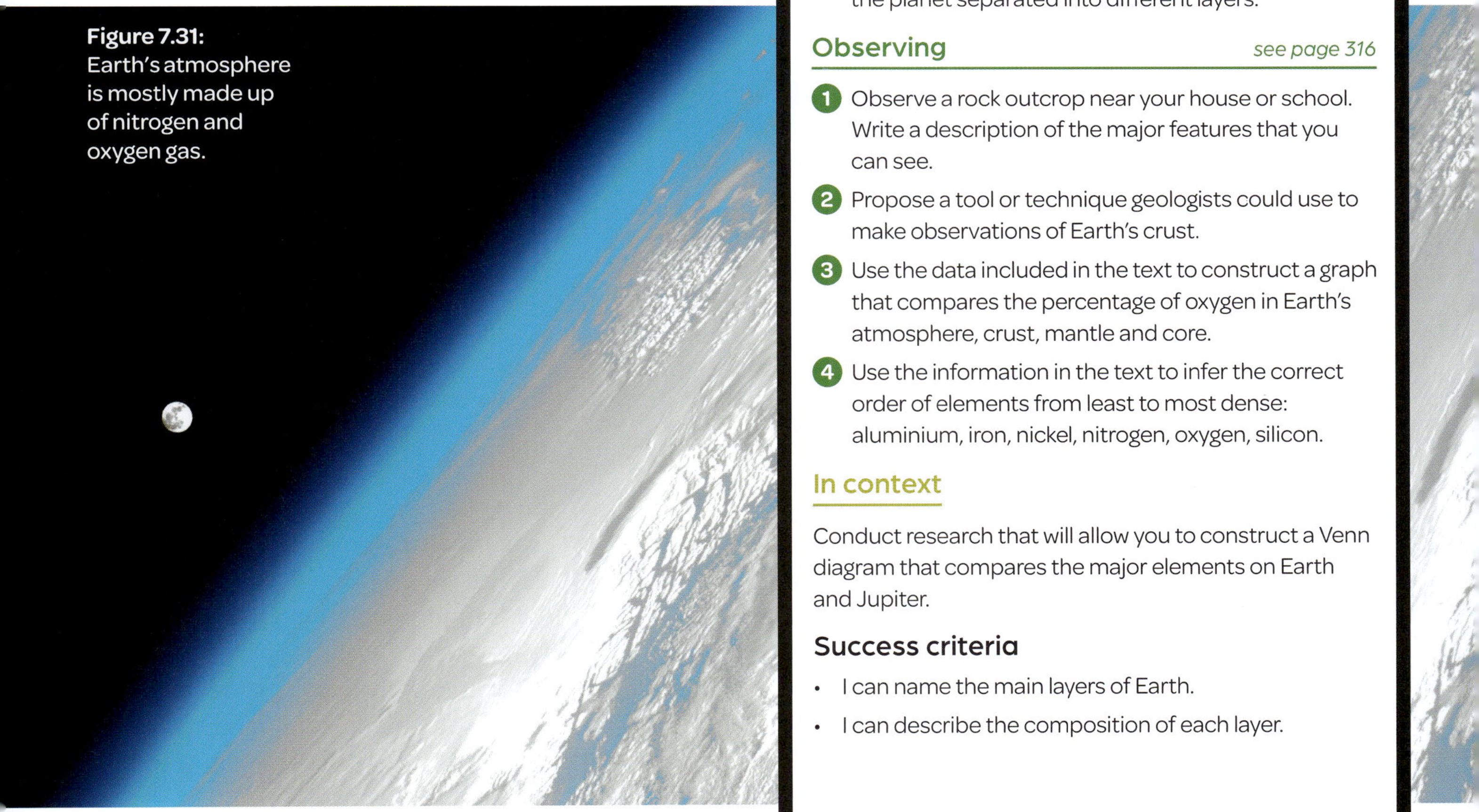

Figure 7.31: Earth's atmosphere is mostly made up of nitrogen and oxygen gas.

Other planets have different compositions

The rocky planets are all thought to be made of the same elements found in the rocks on Earth; however, the gas and ice giants in the outer parts of the solar system have a much higher percentage of hydrogen and helium. Jupiter – the largest planet in the solar system – is mostly hydrogen and helium but is thought to have an incredibly hot and dense core of iron, silicon and oxygen.

Learning Ladder

Change

1. Identify Earth's thickest and thinnest layers.
2. Describe how the rock moves in the mantle.
3. Explain how heat energy in the asthenosphere helps the tectonic plates to move.
4. Construct a flow-chart that shows the transfers and transformations of energy from the mantle to move the tectonic plates.
5. Propose what the structure and elemental composition of the early Earth was like before the planet separated into different layers.

Observing

see page 316

1. Observe a rock outcrop near your house or school. Write a description of the major features that you can see.
2. Propose a tool or technique geologists could use to make observations of Earth's crust.
3. Use the data included in the text to construct a graph that compares the percentage of oxygen in Earth's atmosphere, crust, mantle and core.
4. Use the information in the text to infer the correct order of elements from least to most dense: aluminium, iron, nickel, nitrogen, oxygen, silicon.

In context

Conduct research that will allow you to construct a Venn diagram that compares the major elements on Earth and Jupiter.

Success criteria

- I can name the main layers of Earth.
- I can describe the composition of each layer.

7·10 ▶ Evidence for plate tectonics

Learning intention

At the end of this lesson, I will be able to outline the evidence for the theory of plate tectonics.

Key terms

continental drift: the theory that the continents have moved over time

mid-ocean ridge: a long chain of mountains under the ocean formed by plate tectonics

rift valley: a valley formed when a continent was pulled apart

subduction: where one tectonic plate moves underneath another

Investigation 7.10

Modelling seafloor spreading, p.472

Content group: Geological change

The theory of plate tectonics states that Earth's outer layer – the lithosphere – is divided into more than 15 tectonic plates. These plates move around on the asthenosphere (in the upper mantle), interacting at their boundaries, shifting the continents and producing new landforms. For a long time, people thought that the continents were in the same place as when Earth first formed. However, evidence gathered in the first half of the 20th century indicated that the continents are moving.

The continents move over time

In 1912, German meteorologist Alfred Wegener published his theory on **continental drift**. He proposed that the continents were once joined in one large landmass that he called Pangaea. Over time, this landmass split apart and the continents moved to their current positions.

Wegener found evidence for past glacial climates in equatorial Africa and for tropical climates in north-western Europe. The only way to explain this was that the continents had moved.

Further evidence to support Wegener's theory includes:

- how the continental shelves of continents fit together like pieces of a jigsaw
- identical rock formations on either side of the Atlantic Ocean
- identical plant and animal fossils on different continents separated by oceans (see Figure 7.32).

Wegener reasoned that it was unlikely that identical rocks would have formed and identical species would have evolved so far apart. Therefore, the continents must have once been joined together. He could not suggest how the continents moved, and so his theory was not supported by most of the scientific community. Modern-day geologists used Wegener's work to help them develop the theory of plate tectonics as new evidence was discovered showing how the continents move.

▼ **Figure 7.32:** Identical fossil records across the southern continents (Gondwana) support the theory of continental drift.

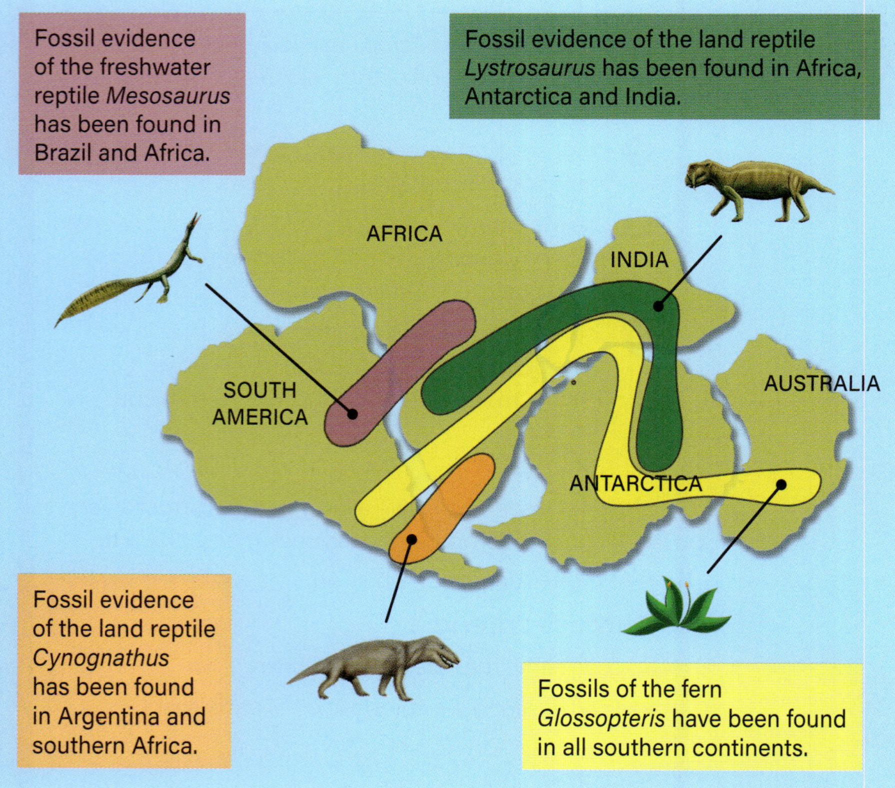

The sea floor is spreading apart

Seafloor spreading happens when molten rock rises up from the mantle at the **mid-ocean ridges** and solidifies to form new oceanic lithosphere.

There are four major pieces of evidence to show that the sea floor is spreading:

- **Rift valleys along mid-ocean ridges**: In 1952, US cartographer Marie Tharp found that there was a V-shaped valley running along the bottom of the Atlantic Ocean. This rift valley is where new lithosphere is formed.
- **Magnetic striping**: As lava cools, the magnetic minerals in it align with Earth's magnetic poles, just like a compass needle does. The positions of the magnetic poles have changed over time, even reversed, and so, in rocks formed at different times, the minerals are aligned differently.
- **Depth of sediments**: Sediments on the oceanic crust are deeper closer to the continents. This implies that those rocks are older because there has been more time for the sediment to accumulate.
- **Age of the sea floor**: Radiometric dating shows that the oceanic crust closer to the continents is much older than the rock closer to the mid-ocean ridges.

Old lithosphere is subducted

If new lithosphere is being formed at mid-ocean ridges, why is Earth not getting larger? Scientists discovered that this is because of **subduction**, which is where the edge of one tectonic plate is pulled under the one next to it.

Geologists use three main pieces of evidence to show that the lithosphere is being subducted:

- **Ocean trenches**: Seafloor mapping shows the existence of deep ocean trenches. We now know this is where two tectonic plates are colliding.
- **Earthquake locations**: Earthquakes get deeper further away from ocean trenches. This is evidence that one plate is moving down under the other.
- **Volcano chains above subduction zones**: The formation of a chain of volcanoes above a subduction zone shows that one plate is moving under another. When a plate subducts, the subducting lithosphere starts to melt and the molten rock rises to the surface to form volcanoes.

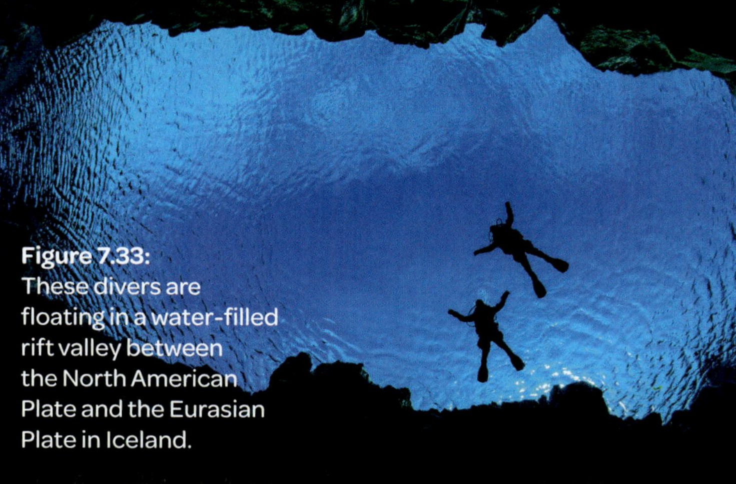

Figure 7.33: These divers are floating in a water-filled rift valley between the North American Plate and the Eurasian Plate in Iceland.

Learning Ladder

Change

1. Outline how the position of the continents changed over time.
2. Describe the evidence Wegener used to explain why continents must have moved over time.
3. Explain one piece of evidence for the theory of plate tectonics that is observed at a:
   - **a** mid-ocean ridge.
   - **b** subduction zone.
4. Conduct some research into Marie Tharp's work and contribution to the theory of plate tectonics. Use this to write a social media post that she could have made about her discoveries.
5. Outline why the discovery of subduction was important for the development of the theory of plate tectonics.

Observing

see page 316

1. Identify four different observations made by scientists that support the theory of plate tectonics.
2. Refer to Investigation 7.10 on page 472.
   - **a** Propose a way you could use scientific tools to record observations of your model.
   - **b** Construct a table that you could use to measure the thickness of each 'stripe' in your model.
3. If plate tectonics did not exist and the continents had not moved, what would Wegener have observed in the geological record instead? How would this have been different from what he actually observed?

In context

Explain why Wegener's theory of continental drift was not supported by the scientific community at the time.

Success criteria

- I can explain the theory of plate tectonics.
- I can outline some of the evidence that was used to support the theory of plate tectonics.

7.11 ▸ What causes tectonic plates to move?

Learning intention

At the end of this lesson, I will be able to describe how tectonic plates interact with each other and describe some of the forces that move them.

Key terms

convergent boundary: where two tectonic plates are moving towards each other and colliding

divergent boundary: where two tectonic plates are moving away from one another

transform boundary: where two tectonic plates slide past each other

Investigation 7.11A

Modelling slab pull, p.474

Investigation 7.11B

Observing convection currents, p.475

Content group: Geological change

As tectonic plates move on the asthenosphere, they slowly transform Earth's surface. We can see particular types of geological features where the plates interact.

Three factors cause the tectonic plates to move. Gravity pulls cold, dense subducting plates down into the mantle, pulling newly formed lithosphere away at mid-ocean ridges. The push of new crust being formed at mid-ocean ridges also moves the plates. Convection currents in the asthenosphere bring hot, low-density rock up towards the crust.

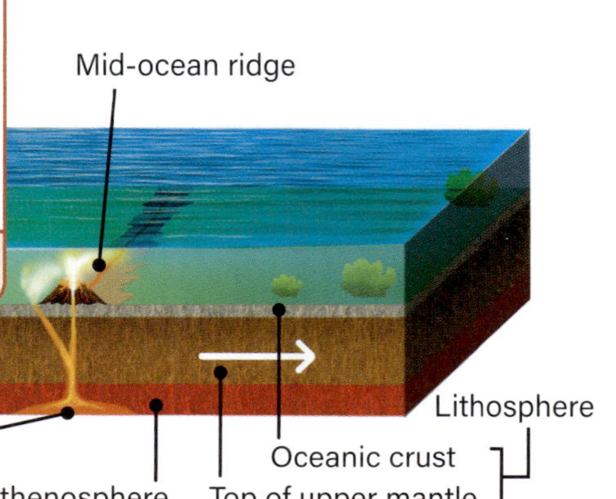

Figure 7.34a: ▸ Divergent plate boundary

Plates separate at divergent boundaries

Divergent boundaries (or constructive boundaries) are where two plates move apart (see Figure 7.34a). Magma rises up in the gap between the plates and solidifies to form new lithosphere. This forms a mid-ocean ridge. When this happens on a continent, it forms a rift valley and volcanoes.

Plates collide at convergent boundaries

Convergent boundaries (or destructive boundaries) are where two plates move towards each other and collide (see Figure 7.34b). The plate that is denser, usually older oceanic crust, subducts underneath the other. If two continental plates collide, the crust buckles and pushes together to form fold mountains.

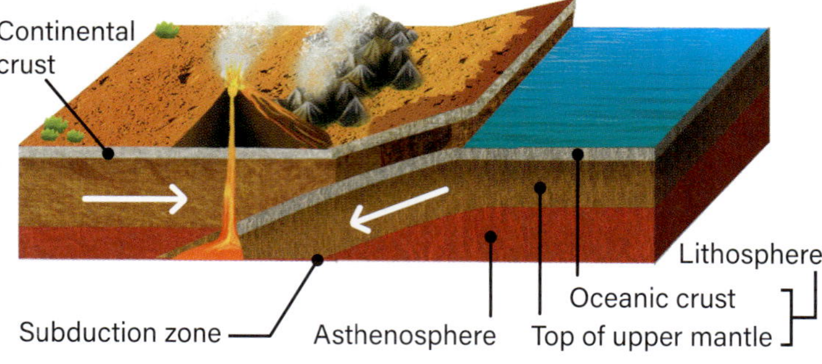

▲ **Figure 7.34b:** Convergent plate boundary

Plates slide past at transform boundaries

Transform boundaries (or conservative boundaries) are where two plates slide past each other (see Figure 7.34c). This causes a break in Earth's surface called a fault. As the plates slowly move, they cause earthquakes; however, usually lithosphere is neither created nor destroyed.

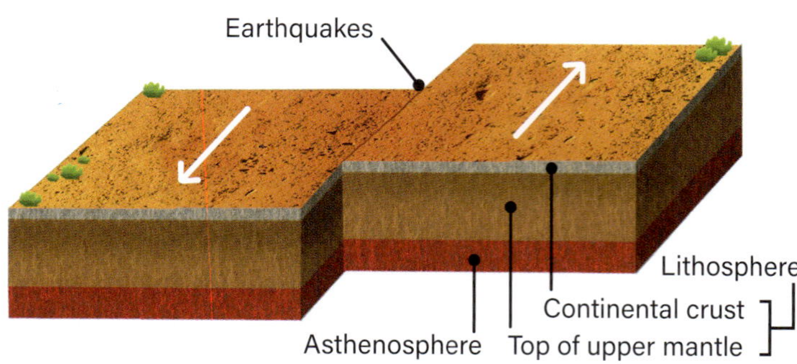

▲ **Figure 7.34c:** Transform plate boundary

Big plates can pull each other down

When a denser plate moves under another plate (subduction) at a convergent plate boundary, it will start to pull the rest of the plate along with it. This is known as slab pull. Scientists think slab pull is the major cause of tectonic plate movement. Because a subducting plate is cooler and denser than the warmer mantle, gravity causes it to sink towards Earth's core, pulling the rest of the plate along behind it. Plates with long subduction zones often move faster than plates with shorter subduction zones.

Ridge push is caused by gravity

Ridge push happens at mid-ocean ridges. When magma rises up at a mid-ocean ridge, it forms new lithosphere. This new lithosphere sits higher than the old one and so gravity causes it to slide downhill, pushing the old lithosphere in front of it. This push helps to move the tectonic plate along, away from the mid-ocean ridge towards the subduction zone.

Convection currents help to drag tectonic plates along

Convection is a way of transferring heat. Rock in the asthenosphere moves slowly in convection currents formed by hot rock near the core rising up, cooling and falling back down again. This brings molten material up to the mid-ocean ridges. As the rock in the asthenosphere slowly moves, it drags the plates along, moving them out and away from the mid-ocean ridges towards the subduction zones (see Figure 7.35).

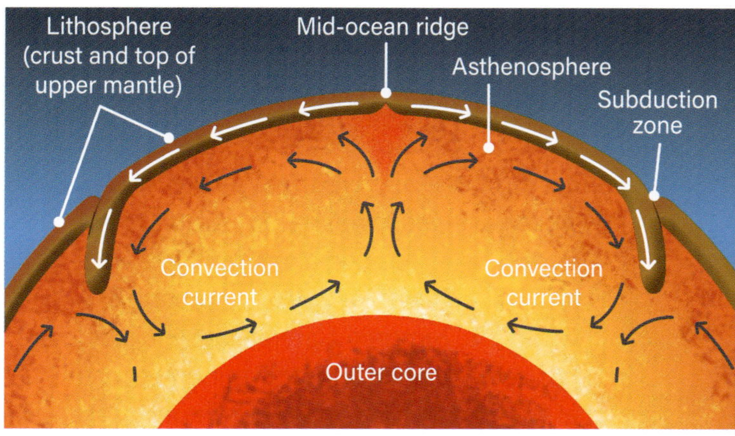

Figure 7.35: Convection currents bring molten rock up to mid-ocean ridges. This forms new lithosphere, which slides down due to gravity towards subduction zones.

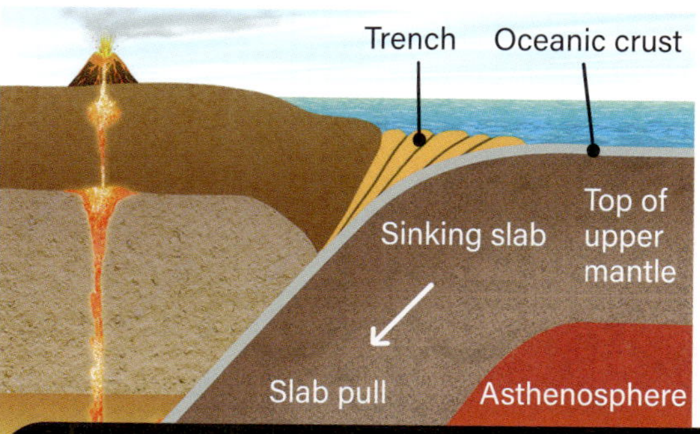

Figure 7.36: Slab pull

Learning Ladder

Change

1. Identify a change that could be observed in the landscape at a convergent, a divergent and a transform boundary.
2. Describe the change in the landscape that happens when a dense oceanic plate subducts under a continental plate.
3. Explain the factors that can cause tectonic plates to move and, for each factor, identify the type of energy involved.
4. Illustrate the energy transfers and transformations involved in convection currents moving tectonic plates apart.
5. Australia is moving northward at a rate of 10 centimetres per year. Predict what the landforms to the north of the continent would look like in 250 million years time. Justify your response. How would the distance between Australia and Antarctica change in 250 million years time? Justify your response.

Conducting investigations

see page 326

Refer to Investigation 7.11B on page 475.

1. Summarise the method of the investigation in a short paragraph.
2. Draw a labelled scientific diagram of the equipment you need for this investigation.
3. Identify two risks that you will need to mitigate during this investigation. Describe the safety practices you will use to mitigate them.
4. Outline how you might record your observations during this investigation.

In context

The Australian and Pacific plates interact in New Zealand. Suggest why there are active volcanoes on the North Island and a long mountain range on the South Island.

Success criteria

- I can describe the three types of movement that can happen at tectonic plate boundaries.
- I can describe how tectonic plates move.
- I can describe some forces that move tectonic plates.

7.12 ▸ Earthquakes

Learning intention

At the end of this lesson, I will be able to:

- outline how earthquakes can be explained by the theory of plate tectonics
- provide evidence for geological change.

Key terms

epicentre: the point on Earth's surface directly above the focus of an earthquake

fault: a break or an area of breaks between two blocks of rock

focus: the origin of an earthquake

intraplate earthquake: an earthquake that takes place in the middle of tectonic plates

seismic wave: a wave of energy that passes through Earth's layers that is caused by an earthquake

seismometer: a scientific instrument that detects seismic waves

Investigation 7.12

Slinky waves, p.476

Content group: Geological change

Earthquakes are caused by a build-up of pressure and the release of energy in Earth's crust. Most earthquakes happen along the boundaries of the tectonic plates, but they can also happen within a plate. A fault is part of the Earth's surface where blocks of rock slide past each other. Faults can be as large as a tectonic plate boundary or much smaller.

Earthquakes happen when tectonic plates move

Blocks of rock do not slide smoothly past each other. They catch and lock together, almost like Velcro. As they catch, pressure builds up until the rock is forced to move, releasing energy. The energy passes through Earth as **seismic waves** that move and shake the crust. This is an earthquake.

The point where an earthquake starts is called the **focus**. This is where the pressure has built up and been released, often causing rocks to rupture (break) and move along the **fault**. The point on the surface directly above the focus is called the **epicentre** (see Figure 7.37).

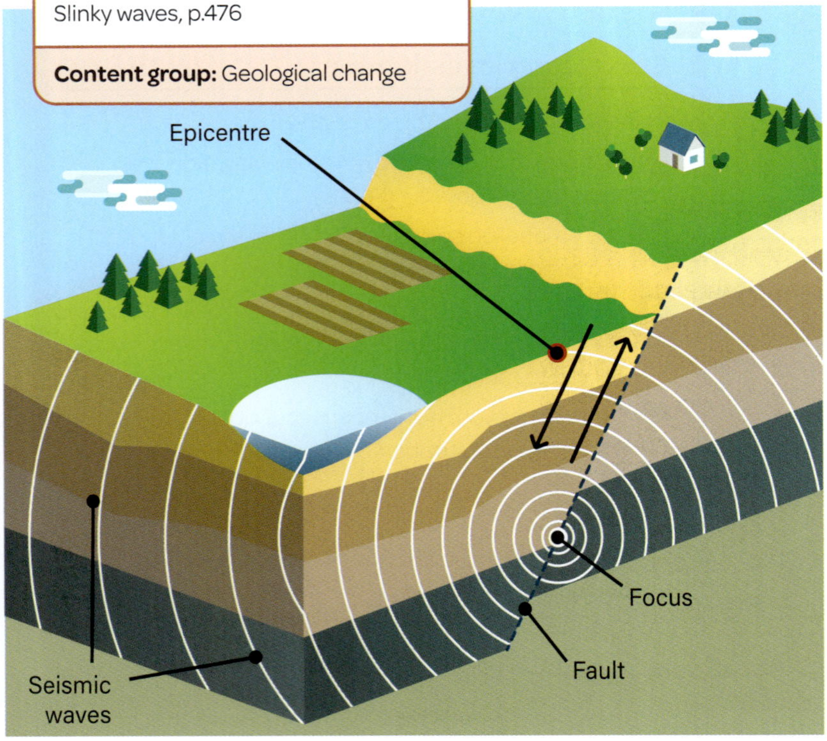

Epicentres are usually located on plate boundaries. This is because the boundaries are where most movement happens. Shallow earthquakes happen at divergent boundaries as the plates move apart. Shallow earthquakes also happen at transform boundaries because the plates are sliding past each other. The deepest earthquakes happen at subduction zones because the subducting plate is moving down into the asthenosphere.

Earthquakes can still happen in the middle of tectonic plates. These are called **intraplate earthquakes** and are the type of earthquake we experience in Australia. Intraplate earthquakes happen along fault lines and are caused by the build-up of pressure within the tectonic plate because it is being pushed and pulled in different directions at the boundaries. These earthquakes are usually shallow and release less energy than those that happen at the boundaries of a tectonic plate.

▲
Figure 7.37: The focus is the point where an earthquake starts, and the epicentre is the point on the surface above it.

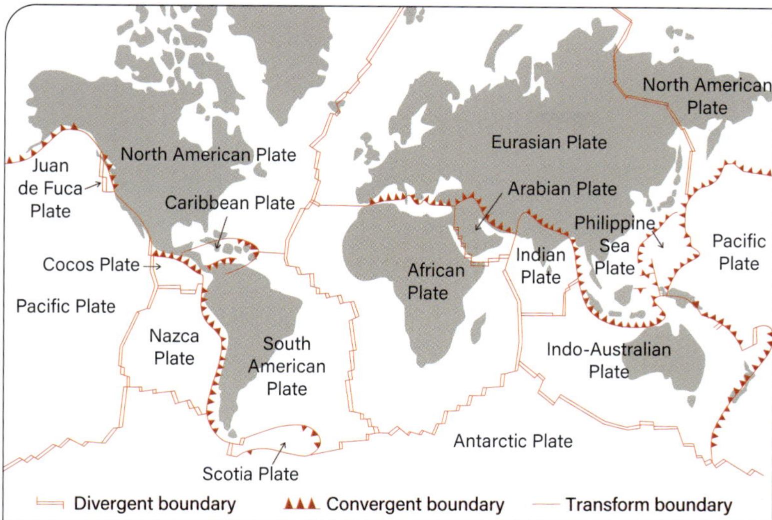

Figure 7.38: Tectonic plate boundaries and movement. Most earthquake epicentres are located on plate boundaries and are caused by the movement of the plates.

Earthquakes produce seismic waves

Seismologists are scientists who study earthquakes. They use sensitive equipment called **seismometers** to measure how much and how often Earth's outer layers move as a result of a seismic wave.

There are two main types of seismic waves: body waves travel through Earth; and surface waves travel around Earth's surface.

There are two main types of body waves: primary (P) waves and secondary (S) waves. Primary waves travel faster and are longitudinal waves. Secondary waves are transverse waves.

Figure 7.39: This road near Christchurch in New Zealand was damaged by a large earthquake.

There are two types of surface waves: Rayleigh and Love waves. They can move across the surface in all directions, like a rolling ocean wave. These waves travel more slowly than P and S waves, but their surface motion causes more destruction.

Learning Ladder

Change

1. Identify where most earthquakes happen.
2. Describe how an earthquake happens.
3. Explain how energy is released in an earthquake.
4. Draw a flowchart showing the energy transfers and transformations involved in a building collapsing during an earthquake.
5. Research the five largest earthquakes that have been recorded. Plot their locations on a map that shows tectonic plate boundaries. What do their locations have in common?

Planning investigations

see page 323

Imagine you want to conduct an investigation to model how construction material affects how long a building can withstand an earthquake.

1. Identify some materials and equipment you might need to conduct this investigation.
2. Identify two risks that might be involved in using the materials and equipment in Question 1. Describe a way to reduce these risks.
3. Identify:
   - **a** the dependent variable.
   - **b** the independent variable.
   - **c** at least two controlled variables.
4. Use the variables in Question 3 to propose an aim for your investigation.
5. Write a suitable method for your investigation.

In context

Explain why Australia has fewer earthquakes than New Zealand.

Success criteria

- I can outline what causes earthquakes and how this is linked to plate tectonics.
- I can identify how earthquakes provide evidence of geological change.

7·13 ▸ Volcanoes

Learning intention

At the end of this lesson, I will be able to:

- outline how volcanoes can be explained by the theory of plate tectonics
- provide evidence for geological change.

Key terms

hotspot (shield) volcano: a volcano formed by magma upwelling underneath a tectonic plate

igneous rock: rock that has solidified from lava or magma

lava: molten (melted) rock above Earth's surface

magma: molten (melted) rock under Earth's surface

strato volcano: a volcano formed at a subduction zone

volcano: a point in Earth's crust where lava erupts

Investigation 7.13A

Viscosity of lava, p.477

Investigation 7.13B

Wax volcano, p.478

Content group: Geological change

A **volcano** occurs where molten rock erupts at Earth's surface. Most volcanoes occur along the boundaries of tectonic plates, which helps to explain how they are formed and their behaviour. Volcanoes along diverging boundaries have runny lava that spreads out over large areas. Volcanoes formed along subduction zones have thicker lava and tend to be more explosive. Hotspot (shield) volcanoes are formed in the middle of a plate, rather than at a boundary.

Some volcanoes form at divergent plate boundaries

At divergent plate boundaries, volcanoes form when **magma** rises to fill the gap between the two diverging plates.

Most volcanoes on divergent plate boundaries form as fissure volcanoes – long fractures in the crust from which the **lava** erupts – and can be many kilometres long. Because the lava has come from the asthenosphere, it contains a lot of dark minerals and is very hot and runny, so it spreads out over large areas. When it cools and solidifies, the lava forms an **igneous rock** called basalt (see Figure 7.40).

▲ **Figure 7.40:** Basalt is an igneous rock that is formed from the fast cooling of lava from volcanoes at divergent boundaries. It makes up oceanic crust.

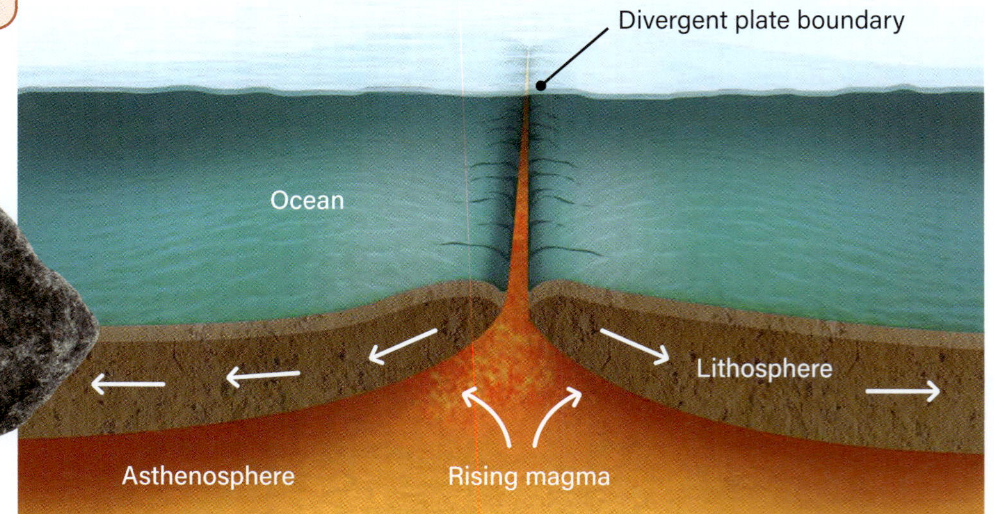

▲ **Figure 7.41:** Formation of a volcano on a divergent plate boundary under the ocean. As magma rises up in the asthenosphere, it forms a fissure volcano many kilometres long.

Volcano arcs form at subduction zones

Volcanoes also form in chains known as arcs along the length of subduction zones.

Subduction zone volcanoes erupt violently because the pressure of gases moving up to the surface can build. Because the lava is sticky, it does not spread very far. Instead, volcanoes build up with steep sides.

Volcanoes along subduction zones are known as **strato volcanoes**. When it cools quickly, magma from this type of volcano can form rhyolite (see Figure 7.42). Rhyolite is a fine-grained igneous rock that has a high silica content.

Figure 7.42: Rhyolite is an igneous rock formed by the rapid cooling of lava from a strato volcano.

Figure 7.43: A fissure eruption at Iceland's Eyjafjallajökull volcano. This volcano has formed as the North American and European plates move apart.

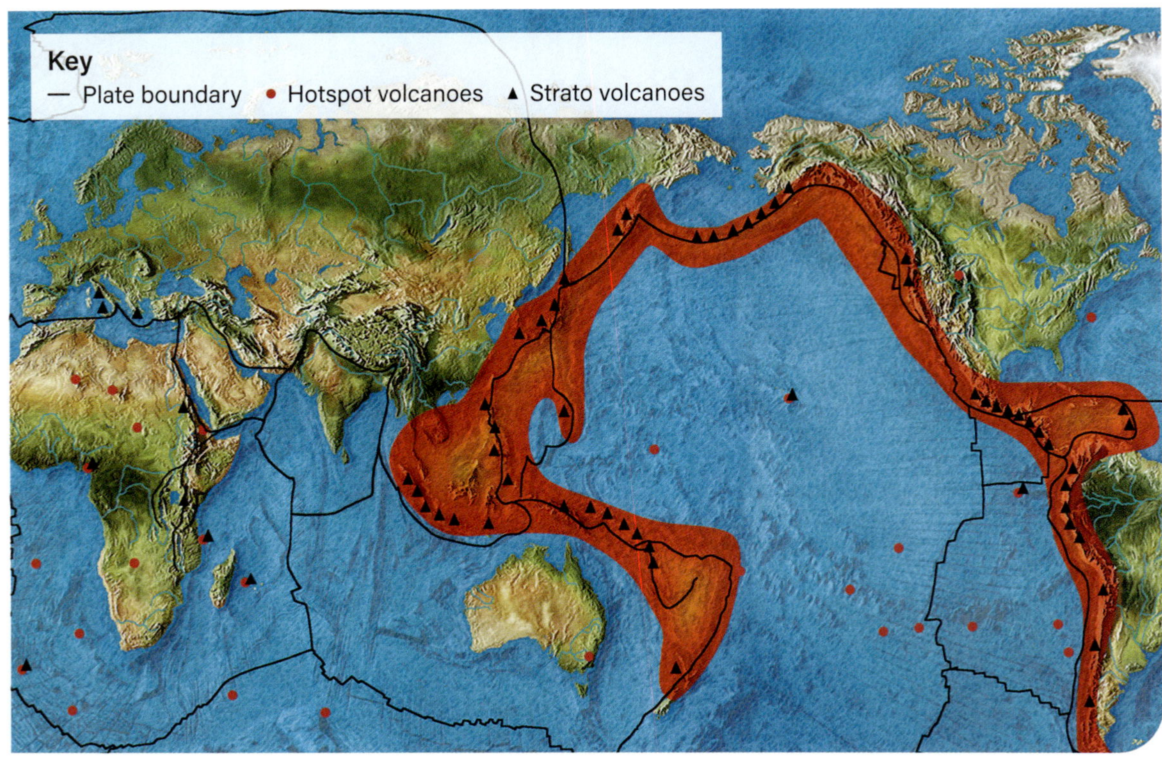

Figure 7.44: The Ring of Fire is a chain of volcanoes around the north, east and western edges of the Pacific Ocean. It cooled quickly so has very small crystals.

Some volcanoes form in the middle of plates

Some volcanoes are not on plate boundaries but are in the middle of the tectonic plates. These volcanoes are formed when there is an upwelling of magma (a hot spot) in a single location, so they are known as **hotspot (shield) volcanoes**. The lava that erupts out of hotspot volcanoes is similar to that of volcanoes at divergent boundaries because it comes from the asthenosphere. It is made of dark minerals, is very runny and forms the igneous rock basalt when it solidifies. The lava spreads out over large areas and over time forms volcanoes that are wide at the base compared to their height.

The hot spot in the mantle does not move, but the tectonic plates do. As the plates move, new volcanoes form in a chain (see Figure 7.45). Unlike volcanoes that form along a subduction zone, only the volcano over the hot spot is active and erupts. The volcanoes that have moved away from the hot spot no longer have a magma source and so are said to be extinct.

1 A hot spot breaks through the lithosphere and a volcanic island forms.

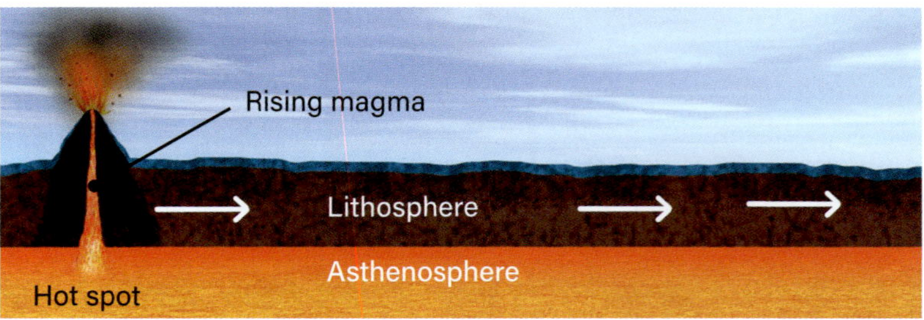

2 The crustal plate is constantly moving, so the island eventually moves off the hot spot. This makes room for a second volcanic island.

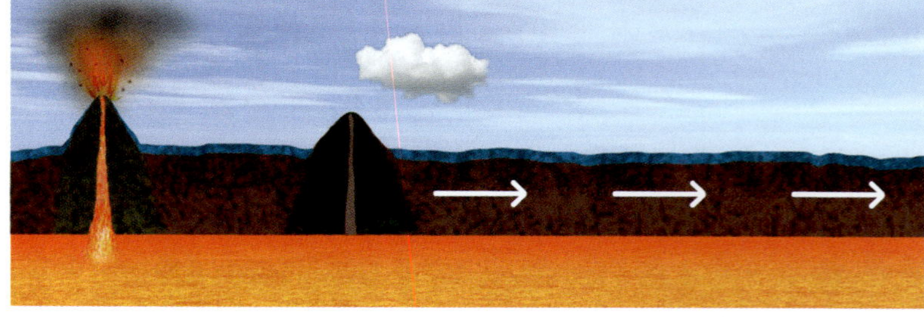

3 More islands form as the crustal plate keeps moving over the hot spot.

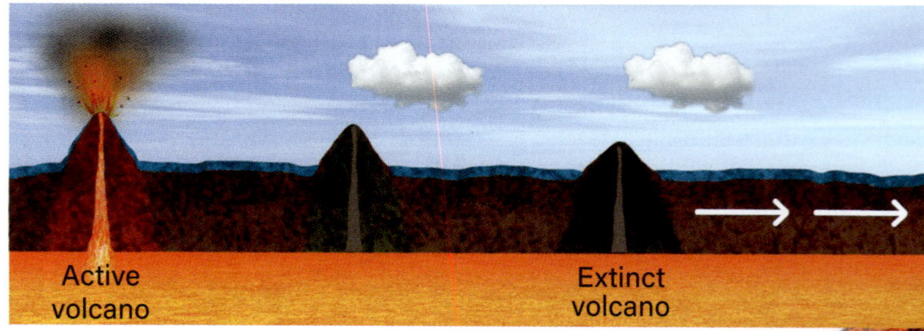

Figure 7.45: Steps in the formation of a volcanic island chain as a result of hotspot volcanism

Volcanoes make new landforms

We often think of volcanoes as destructive, but they can result in the formation of major landforms such as islands.

The Hawaiian Islands and Galapagos Islands have been formed by hotspot volcanism, and the positions of the islands in the chain can provide evidence for the direction and speed at which the plate is moving. Volcanoes in eastern Australia are part of a chain of hotspot volcanoes that formed as the continent moved north away from Antarctica. The oldest volcanoes are in Queensland, and the youngest is Mount Gambier in South Australia, which was last active about 10 000 years ago.

Figure 7.46: Volcanoes form at the boundaries between Earth's tectonic plates.

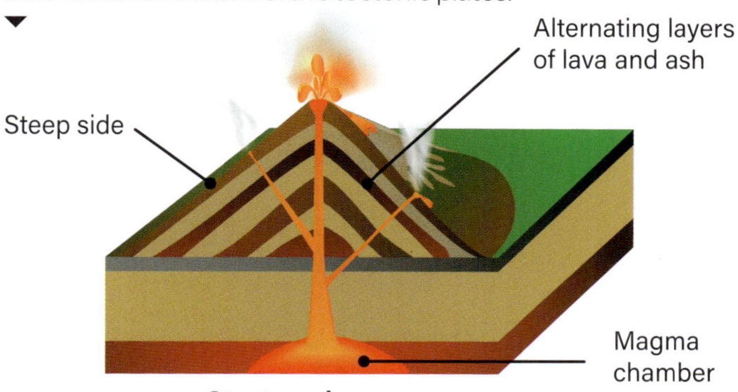

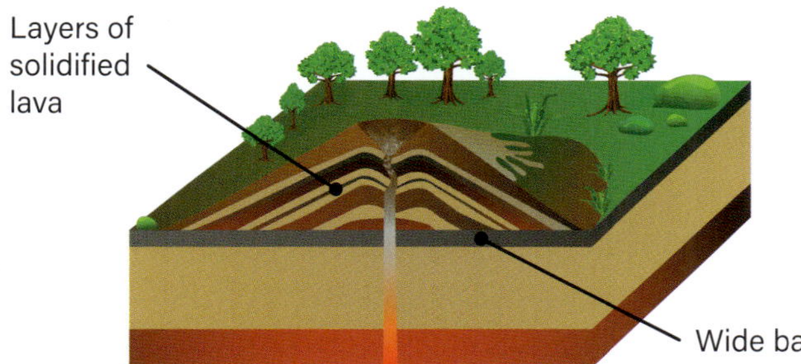

Figure 7.47: Strato volcanoes are made of alternating layers of lava and ash. Hotspot volcanoes are made of layers of solidified lava.

Learning Ladder

Change

1. Identify the best definition of a volcano.
   - **A** A cone-shaped mountain
   - **B** Where molten rock erupts at Earth's surface
   - **C** Where lava reaches the surface with an explosion
   - **D** A long fissure under the ocean
2. Describe where most volcanoes occur on Earth.
3. Explain the role of energy in a volcanic eruption.
4. Create a Venn diagram to compare and contrast a chain of islands formed by a hot spot with a chain of islands formed by a subduction zone.
5. You observe some volcanoes on a continent. What evidence do you need to look for to determine how they were formed? Justify your reasoning.

Conducting investigations

see page 326

Refer to Investigation 7.13A on page 477.

1. Summarise the method for the investigation in a short paragraph.
2. Draw a labelled scientific diagram of the equipment you need to use in this investigation.
3. Identify two risks that you will need to mitigate during this investigation. Describe the safety practices you will use to mitigate them.

In context

There are no active volcanoes on the Australian continent. However, Big Ben is an active strato volcano on Heard Island (an Australian territory), 4100 kilometres south-west of Perth. Identify the type of plate boundary that is nearby for Big Ben to have formed.

Success criteria

- I can outline what causes volcanoes and how this is linked to plate tectonics.
- I can identify how volcanoes provide evidence of geological change.

7·14 ▶ Aboriginal and Torres Strait Islander cultural accounts of earthquakes and volcanoes

Figure 7.48: An Awabakal Dreaming story describes the tail of an enormous kangaroo thudding into the earth, making it tremor.

Learning intention

At the end of this lesson, I will be able to understand how First Nations stories can inform Western scientists about the geology of Australia and how they can be accurate accounts of past events.

Key term

Newer Volcanics Province: an area covering roughly 15 000 square kilometres in south-east Australia; the most recent active volcanic area in Australia

Content group: Geological change

In geological terms, the continent of Australia is one of the most stable places on Earth. The continent is located at the centre of the Indo-Australian Plate, which means there are relatively few earthquakes, and no active and few dormant volcanoes, on the Australian mainland. Australia does have two active volcanoes in the Australian Antarctic Territory at Heard Island and the nearby McDonald Islands, located 4000 kilometres south-west of Perth.

Australia has earthquakes and volcanoes

The most recent volcanic eruptions on the Australian continent were around 5000 years ago, in what is now South Australia and Victoria. This area of more than 15 000 square kilometres, known as **Newer Volcanics Province**, has more than 400 dormant volcanoes.

The Newer Volcanics Province is thought to consist of hotspot volcanoes, formed when heat was transferred from the upper mantle through a thin spot in the crust. Evidence of this is that the volcanic rock is significantly younger in the southern parts of the Newer Volcanics Province. This is in line with the northern movement of the Indo-Australian Plate at approximately 6.9 centimetres per year.

Although Australia is not on a tectonic plate boundary, it does experience about 100 small earthquakes every year and a large one every few years. In 1989, a large earthquake near Newcastle in New South Wales damaged more than 50 000 buildings (see Figure 7.49) and killed 13 people.

Figure 7.49: Newcastle in the aftermath of the 1989 earthquake

Two feuding volcanoes are part of our history

All the volcanoes on the Australian continent are dormant, so there are few First Nations stories of volcanoes. One story is from Dja Dja Wurrung Country in the Central Goldfields region of Victoria. This story explores the relationship between two volcanoes, Tarrengower and Lalgambook. Lalgambook, a young and cheeky volcano, attempted to challenge Tarrengower's power and authority by throwing rocks at Tarrengower. Tarrengower did not respond but told Lalgambook that he was too weak to hit him with the rocks. Lalgambook got angrier and angrier and threw rocks high into the sky at Tarrengower but was unable to reach him. This caused Lalgambook to completely blow his core.

Lalgambook was named Mt Franklin by the Europeans and is classed as an extinct volcano. The last eruption of Lalgambook was around 5000 years ago. It is a prominent, conical scoria cone with a deep crater.

The story of the two feuding volcanoes is important for two main reasons. The first is that it details a relatively rare phenomenon of two volcanoes that share a magma reserve erupting simultaneously. Second, it is a lesson for young Dja Dja Wurrung People on how to speak and engage with their Elders ; that is, not to be too quick to get angry, and to be patient.

Awabakal Dreaming relates to earthquake activity

The Awabakal Dreaming story relates to earthquake activity on Awabakal Country in the Central Coast region of New South Wales, close to Newcastle. This story describes a giant kangaroo that was chased to Awabakal Country by a group of wallabies after it attacked one of them. The giant kangaroo hid there, and the Dreaming story explains that the tremors that occur are caused by the giant kangaroo's tail hitting the Earth.

This story is important because it shows how the Awabakal People have known about earthquakes and have experienced them for many thousands of years. When the Europeans colonised this part of the Australian continent, they had no knowledge of the long history of earthquakes, and so were not prepared for the 1989 Newcastle earthquake. If Western scientists had better understood the First Nations perspective, buildings could have been designed to be more earthquake resistant, and some of the deaths in Newcastle in 1989 might have been prevented.

Figure 7.50: Lalgambook (Mt Franklin). In this image, the cone of the extinct volcano is clearly visible.

Learning Ladder

Change

1. Outline what change caused the Newer Volcanics Province to be formed.
2. Describe the phenomena occurring in the story of the two feuding volcanoes.
3. Explain how the geological changes occurring in the Awabakal Dreaming story are linked to knowledge in Western science.
4. Summarise processes that are involved in the formation of volcanoes. Propose why this may be linked to a small number of Dreaming stories indicating the existence of volcanic activity.
5. Consider the two feuding volcanoes and answer the following questions.
   - **a** What are the names of the two feuding volcanoes?
   - **b** Why is this story an important geological resource?
   - **c** How does this story describe changes in the two volcanoes?

Observing

see page 316

1. Where is the Newer Volcanics Province located in Australia?
2. Examine the picture of Lalgambook (Mt Franklin) and list the features that mark it as being a volcano.
3. Describe how you could record your observations of Lalgambook (Mt Franklin) in a scientific format.
4. Examine the image of Newcastle in the aftermath of the 1989 earthquake. If scientists had been more familiar with the Awabakal Dreaming story, how could the damage to the buildings and the deaths have been prevented?
5. Find a map of the Newer Volcanics Province. Identify how far this area extends from east to west and from north to south (approximately).

In context

Explain how understandings of seismic activity contained in First Nations stories improve town planning, engineering and building designs in Australia.

Success criteria

- I can understand how First Nations Dreaming stories contain important scientific information.
- I can see how knowledge of First Nations Dreaming stories could enhance Western science in Australia.

7.15 ▸ Minerals

Learning intention

At the end of this lesson, I will be able to:

- identify that rocks contain different types of minerals
- classify minerals into groups using their observable properties.

Key terms

crystal: a solid substance made up of very ordered microscopic parts

mineral: a naturally occurring inorganic (non-living) substance

physical property: a characteristic or attribute of a substance that can be observed and measured, such as colour, texture, melting and boiling points, density and hardness

Investigation 7.15A

Observing minerals, p.479

Investigation 7.15B

Extracting copper, p.480

Content group: Geological change

Minerals are the building blocks of rocks. The types and amounts of minerals in a rock can suggest how the rock was formed and help geologists to classify it. Also, identifying rocks that contain useful mineral resources means that the minerals can be mined and used.

Minerals are made up of elements

Minerals are inorganic (non-living) substances that are found in nature. Each mineral contains one or more of the 98 naturally occurring elements. Elements are pure substances, made of only one type of atom. Most minerals are compounds, made up of two or more elements. For example, the mineral quartz is made up of silicon and oxygen (SiO_2). Some minerals are pure substances, made of only one element. These are called native elements.

Minerals have specific physical properties

Geologists identify different minerals by observing and measuring their different **physical properties**. Because each mineral has a different chemical make-up, it has a unique set of properties, including lustre, hardness, streak and **crystal** shape.

Lustre

'Lustre' refers to how light reflects off the surface of a mineral. Some minerals have a metallic lustre, which means they look shiny, like polished metal. Others have a non-metallic lustre, which means they look dull and earthy.

▲ **Figure 7.51:** Some minerals, such as pyrite, have a metallic lustre.

Hardness

Talc is a mineral so soft that you can scratch it with your fingernail. Diamond is also a mineral but is one of the hardest substances known. In 1812, German geologist Friedrich Mohs developed a scale to compare the hardness of minerals. His scale ranks minerals from 1 (very soft) to 10 (very hard). Hard minerals (those with larger numbers on the scale) can scratch softer minerals (those with a smaller number on the scale) (see Table 7.2).

Colour and streak

Colour is not usually a reliable way to identify a mineral because the mineral may be many different colours, or the same colour as another mineral.

Geologists use the streak of the mineral to help identify it. The streak is the powder made by scratching the mineral on an unglazed white tile. For example, gold has a gold streak colour, whereas pyrite, also known as fool's gold, has a black streak colour.

▲ **Figure 7.52:** Some minerals, such as jade, have a non-metallic lustre.

Table 7.2: Mohs' scale of relative hardness

Index mineral	Hardness	Everyday material
Talc	1	
Gypsum	2	Fingernail (2.5)
Calcite	3	Copper coin (3.5)
Fluorite	4	Wire nail (4.5)
Apatite	5	Glass and knife blade (5.5)
Orthoclase	6	Streak plate (6.5)
Quartz	7	
Topaz	8	
Corundum	9	
Diamond	10	

▲ **Figure 7.53:** Black haematite (left) has a red-brown streak. Malachite (right) has a green streak, which is similar to its body colour.

Crystal shape

Minerals usually have a crystal structure. Crystals are solid substances that have a regular shape because their atoms are bonded together in a regular, repeating pattern. Different minerals have different chemical make-ups, which can be seen in the shapes of their crystals.

Figure 7.54: Crystal ▶ shape can be used to identify minerals.

Learning Ladder

Change

1. Copy and complete the sentences with the correct terms.
   Minerals are the _______ _______ of rocks. They are _______ occurring substances. A native _______ is a mineral that contains only one type of _______.
2. Describe the main properties used to identify minerals.
3. Explain how a native element is different from other minerals.
4. Minerals in rocks can be thought of as being like Lego bricks in a tower. This comparison is called a simile. Create another simile for the same concept.
5. In the mining industry, diamonds are used in saw blades to cut through rock. Use your knowledge of mineral properties to propose why this tool would be useful.

Observing

see page 316

1. Describe the pyrite mineral pictured in Figure 7.51 in terms of:
   a lustre.
   b colour.
   c crystal shape.
2. Use Mohs' scale to identify which mineral(s) can:
   a be scratched by apatite.
   b be scratched by topaz, but can scratch orthoclase.
   c scratch corundum.
3. Construct a table that could be used to classify minerals as having metallic lustre or non-metallic lustre.
4. Propose three inferences you could make about the green mineral in Figure 7.54. What information might this tell you about that mineral?

In context

Minerals are everywhere. Brainstorm where you might find minerals in your everyday life.

Success criteria

- I can explain what minerals are and how they are different from rocks.
- I can describe some common physical properties used to classify minerals.

7·16 ▸ The rock cycle

Learning intention

At the end of this lesson, I will be able to:

- describe how sedimentary, igneous and metamorphic rocks are related through a variety of naturally occurring processes
- classify rocks into groups according to their observable properties.

Key terms

deposition: a process in which sediment is left in a new place

erosion: a process in which sediments are moved from one place to another

metamorphic rock: rock formed from another rock that has been changed by heat or pressure, or both

metamorphism: the process of change that happens to a rock because of heat or pressure, or both

sediment: small particles of rock, such as clay, sand and pebbles

sedimentary rock: rock formed by sediments that have been pressed together

weathering: a process in which rocks are worn down into smaller particles

Investigation 7.16

Modelling the rock cycle, p.481

Content group: Geological change

▲ **Figure 7.55:** There are three types of rock: igneous, sedimentary and metamorphic.

Rocks are constantly changing and being recycled. As the tectonic plates of the lithosphere move and act on each other, rocks can be pulled under Earth's surface or forced upwards. Rocks beneath the surface can change due to extreme heat and pressure. Rocks at the surface can change due to natural processes such as weathering (rocks breaking down into smaller particles) and erosion (rocks moved to new locations).

There are three main types of rock

Geologists classify rocks into three main types based on how they were formed:

- Igneous rocks are made by the cooling of molten (melted) rock, either on the surface or within Earth's crust. 'Igneous' comes from the Latin word *ignis*, which means 'of fire'. The minerals in the rock can be seen as individual crystals.
- **Sedimentary rocks** are made up of **sediments** such as clay, sand, pebbles, shells and other pieces of material. The sediment is usually deposited in layers by wind, water or gravity. Over time the sediment is buried and squashed, and the particles in it are stuck together to form rock. The minerals in the rock are within the sediments and may not be easily seen.
- **Metamorphic rocks** form when other rocks change (metamorphose) because of high temperatures and pressure within Earth's crust. The minerals in the rock can be seen as individual crystals.

Rocks can change over time

The three main rock types – igneous, sedimentary and metamorphic – are related because they can form from one another. Forces within Earth bring rocks to the surface or cause them to sink deep within the crust. The rock cycle is a model used to show how different processes can form the three different types of rock (see Figure 7.55). Because some of these processes take millions, even billions, of years and often occur only in certain environments, it is very likely that a rock will not undergo all possible processes in the life of the planet.

Figure 7.56: The rock cycle explains that formations such as the Three Sisters in the Blue Mountains were formed from the weathering and erosion of sandstones. These were formed from sediments from the weathering and erosion of much older igneous and metamorphic rocks.

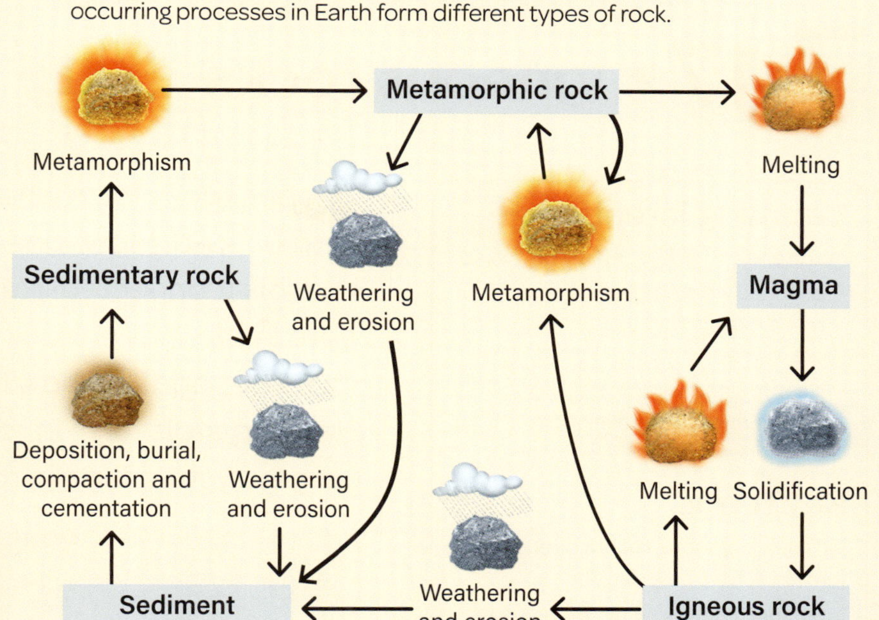

Figure 7.57: The rock cycle shows how different, naturally occurring processes in Earth form different types of rock.

These are some common processes that act on rocks:

- **Weathering**: Rocks break down into smaller pieces called sediment.
- **Erosion**: Sediment moves from one place to another.
- **Deposition**: Sediment settles in one place.
- **Burial and compaction**: As more sediment is deposited, the sediments below it are buried and squashed together.
- **Cementation**: Sediments are chemically glued together into rock.
- **Metamorphism**: Rocks change due to heat and pressure.
- **Melting**: Rock melts to magma due to high temperatures.
- **Solidification**: Molten rock cools and hardens.

Learning Ladder

Change

1. Identify three ways that rocks can change.
2. Describe the rock cycle in your own words.
3. Explain the role that forces and energy play in the formation of:
   - a igneous rocks.
   - b metamorphic rocks.
   - c sedimentary rocks.
4. Write a short story from the point of view of a quartz crystal in the Three Sisters. Describe the changes you have experienced, beginning as a quartz crystal in magma that was solidifying into an igneous rock.

Conducting investigations

see page 326

Refer to Investigation 7.16 on page 481.

1. Read the method, then describe the major steps to a partner in 60 seconds.
2. Describe how you would record and gather data for this investigation. Identify the equipment you would use.
3. Identify two risks involved in conducting this investigation. Suggest safe practices you would implement to mitigate these risks.
4. Propose how you could make sure that the data you collect from this investigation is accurate.

In context

Explain how the meanings of the words 'igneous', 'sedimentary' and 'metamorphic' give clues to how each of these rocks is formed.

Success criteria

- I can identify the three types of rock.
- I can describe how each type of rock is formed.
- I can explain what the rock cycle is.
- I can classify rocks into groups according to their observable properties.

7·17 ▸ Igneous rocks

Learning intention

At the end of this lesson, I will be able to:

- describe what an igneous rock is and explain how these rocks are formed
- classify igneous rocks into groups according to their observable properties.

Key terms

extrusive igneous rock: igneous rock formed at Earth's surface

intrusive igneous rock: igneous rock formed under Earth's surface

solidify: become a solid

Investigation 7.17A

Cooling rate and crystal size, p.483

Investigation 7.17B

Observing igneous rocks, p.484

Content group: Geological change

Igneous rocks form when molten rock cools and becomes hard or solid. Most igneous rocks contain interlocking crystals (connected like a jigsaw) of the minerals that were in the molten rock. The size of the crystals in igneous rocks depends on how long the molten rock took to cool. Small crystals form in rocks that have cooled quickly. Large crystals form in rocks that have cooled slowly.

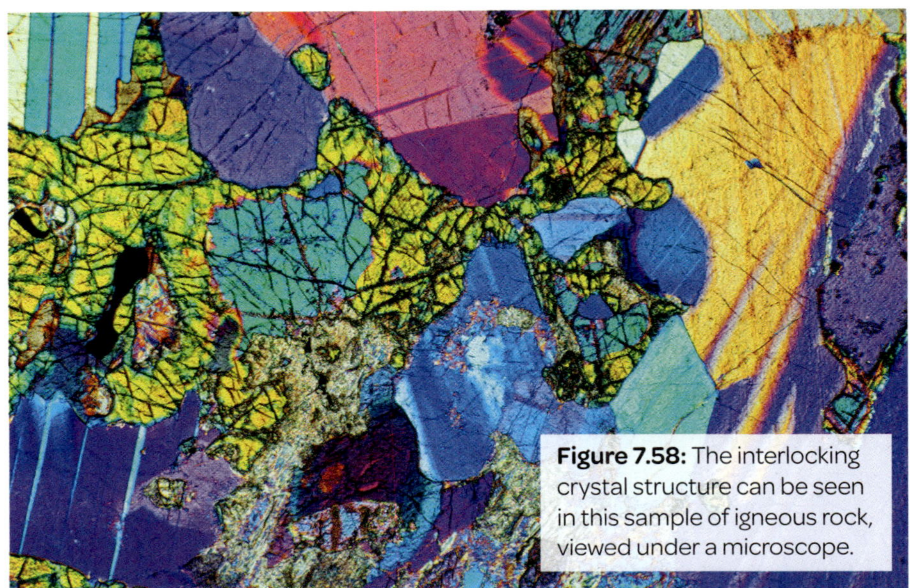

Figure 7.58: The interlocking crystal structure can be seen in this sample of igneous rock, viewed under a microscope.

Rocks can melt to form magma

What happens to chocolate when it is heated? It melts! The same thing happens to rocks. As rocks are pulled deep below Earth's surface towards the base of the lithosphere, the temperature increases. Rocks melt at temperatures of between 700 °C and 1300 °C, producing magma.

Magma is a liquid, so it is less dense than the surrounding rock and will rise to Earth's surface through faults, cracks and other weak spots, also melting the rock that it passes through. If it reaches the surface and erupts through a volcano, it is called lava.

Molten rock cools to form igneous rock

What happens to melted chocolate when it cools down? It **solidifies**! When molten rock cools and solidifies, the minerals form crystals that interlock like a jigsaw to form igneous rocks.

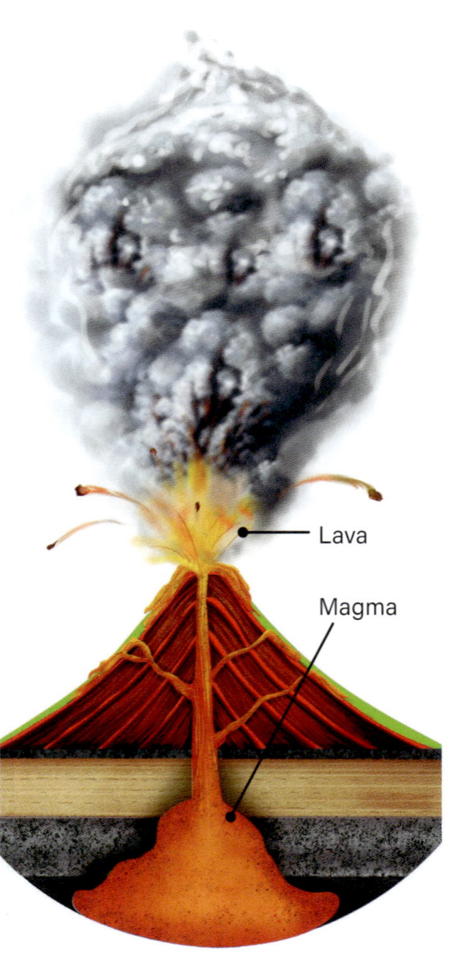

Figure 7.59: Magma collects in a chamber of a volcano before making its way to the surface to erupt as lava.

Figure 7.60: Sawn Rocks in Mt Kaputar National Park near Coonabarabran in New South Wales are made of columns of basalt, an extrusive igneous rock. As the lava flow cooled, it shrank, creating long hexagonal 'pipes'.

Rocks that form when lava cools on or near Earth's surface are **extrusive igneous rocks**. The magma has been extruded from (pushed out of) the crust. When magma reaches Earth's surface as lava, it cools very quickly, and even faster if it is under water. Extrusive igneous rocks have very small crystals, so small that they are often not visible with the naked eye. This is because they did not have very long to form before the lava solidified.

When magma cools under the surface, it forms **intrusive igneous rocks**. The magma has intruded on the rocks that were originally there. Magma under the surface cools very slowly, which means the crystals have more time to grow. The crystals in intrusive igneous rocks, such as granite, are much larger than those in extrusive igneous rocks and can be seen with the naked eye.

Crystal sizes tell us where an igneous rock was formed

The size of crystals (known as rock texture) in igneous rocks can tell geologists whether those crystals formed on the surface due to an erupting volcano, or deep underground. The bigger the crystals, the deeper the region where they formed. Geologists also look at the types of minerals that make up a rock to work out how and where the magma was formed.

Learning ladder

Change

1. Copy and complete these sentences with the correct terms.
   - **a** Molten rock below Earth's surface is known as _______ [lava/magma]. Molten rock at the surface is known as _______[lava/magma].
   - **b** Igneous rocks that have cooled quickly have _______ [large/small] crystals. Igneous rocks that have cooled slowly have _______ [large/small] crystals.
2. Describe how igneous rocks form.
3. Explain how the time taken to cool affects the observable characteristics of an igneous rock.
4. Draw a diagram that represents how heat energy from Earth's mantle is transferred and transformed during the formation of igneous rock.
5. Outline two key characteristics that could be used to classify a rock as an igneous rock.

Conducting investigations *see page 326*

Refer to Investigation 7.17A on page 483.

1. Summarise the major steps of the investigation into no more than five dot points.
2. Draw a labelled scientific diagram of the set-up used for Steps 3–6.
3. Explain two laboratory safety rules you will follow while conducting this investigation.
4. Propose a way that you could record your data to ensure that it was accurate.

In context

When magma loses gas as it moves to Earth's surface, gas bubbles may be preserved as the lava solidifies. The resulting igneous rocks contain many air bubbles. Draw a diagram to illustrate this process.

Success criteria

- I can describe an igneous rock.
- I can explain how igneous rocks are formed.
- I can classify rocks into groups according to their observable properties.

7·18 ▸ Metamorphic rocks

Figure 7.61: These heavily folded and metamorphosed rocks at Narooma in New South Wales formed as a result of intense pressure when the Pacific tectonic plate first started colliding with the east coast of Australia.

Learning intention

At the end of this lesson, I will be able to:

- describe what a metamorphic rock is and explain how these rocks are formed
- classify metamorphic rocks into groups according to their observable properties.

Key terms

contact metamorphism: the process of change that happens to rock over small areas, often near volcanoes

regional metamorphism: the process of change that happens to rock over large areas

Investigation 7.18

Modelling contact metamorphism, p.485

Content group: Geological change

Metamorphosis means 'change'. Like a caterpillar that has changed into a butterfly, a metamorphic rock has changed its form. Metamorphic rocks form from other rocks that have been changed by heat, pressure or both.

Rocks are changed by heat and pressure

The process in which rocks are changed by extreme heat, pressure or both is metamorphism. It happens when Earth's tectonic plates push together, move apart or slide past each other, burying rocks and making them very hot, and putting them under pressure. As rocks are heated and put under pressure, the minerals inside them rearrange. Some may even form new minerals by chemically reacting with each other or with fluids passing through the rocks. Metamorphosis results in rocks that have crystals. Sometimes these crystals have rearranged into layers.

Rocks can be squashed and folded, or 'cooked'

Rocks can be metamorphosed in two major processes.

Regional metamorphism happens over large areas, often when two tectonic plates push together. This puts rocks under much higher temperatures and pressures, and they may squash and fold. As they are pushed deeper and deeper, they keep changing as the temperature and pressure increase. Rocks that have been formed by regional metamorphism have crystals arranged in layers.

Contact metamorphism happens when a body of rising magma meets rock, increasing the temperature of the rock and 'cooking' it, forming new crystals. It happens in a small local area, often around volcanoes. The crystals in rocks formed by contact metamorphism are not in layers.

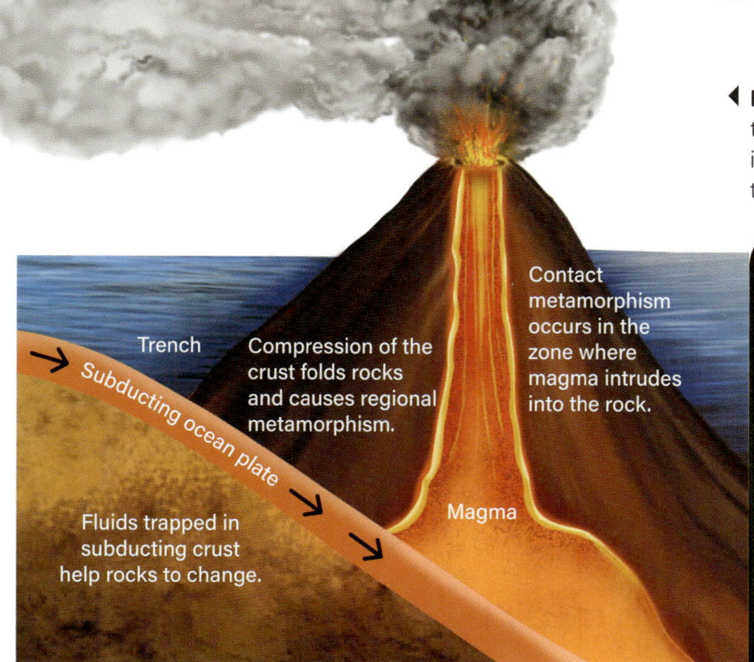

◀ **Figure 7.62:** Regional metamorphism happens when two tectonic plates push together. As the plates collide, the increase in pressure and temperature will cause the rocks to change over a very large area.

Metamorphism can be an ongoing process

A rock can undergo different amounts of change, depending on what happens to it. Even metamorphic rocks can undergo more metamorphism. Think about what happens when you bake chocolate chip cookies. A cool oven and a short cooking time result in soft, doughy cookies, and the chocolate chips keep their shape. A hot oven and a longer cooking time result in hard cookies with chocolate that melts, then resets.

Geologists compare the types of minerals in metamorphic rocks to work out where and how the rocks formed. This can help them to understand more about how Earth's tectonic plates move and act on each other.

▲ **Figure 7.63:** Garnets are gemstones formed when sedimentary rocks such as shale are subjected to heat and pressure at a convergent plate boundary (regional metamorphism).

Learning Ladder

Change

1. Identify one way a rock can be changed (metamorphosed).
2. Copy and complete this sentence with the correct terms.
   Metamorphic rocks are formed when _______ have been _______ due to _______, _______ or both.
3. Explain how heat and pressure cause the changes in metamorphic rocks.
4. Compare the features and formation of metamorphic and igneous rocks. In what ways are they similar and different?
5. Is it possible for metamorphic rocks to form on Earth's surface? Justify your response.

Planning investigations

see page 323

You have been asked to design an investigation that uses cookie dough to model how rocks change when they are subjected to different amounts of metamorphism. Consider this investigation when answering the following questions.

1. Identify the materials and equipment you would need to conduct this investigation.
2. Identify the risks that could be involved in using the materials and equipment in Question 1. Describe procedures you would use to mitigate these risks.
3. Identify the following for your investigation.
   a Dependent variable
   b Independent variable
   c At least two controlled variables
4. Use your response to Question 3 to write an aim for your investigation.
5. Write a step-by-step method for your investigation.

In context

Identify two locations on Earth where metamorphism could occur, and use evidence from the text to justify your answer.

Success criteria

- I can describe what a metamorphic rock is and how these rocks form.
- I can describe the difference between regional and contact metamorphism.
- I can classify rocks into groups according to their observable properties.

7·19 ▶ Sedimentary rocks

Learning intention

At the end of this lesson I will be able to:

- describe what a sedimentary rock is and explain how these rocks are formed
- classify sedimentary rocks into groups according to their observable properties.

Key terms

chemical sedimentary rock: sedimentary rock formed from layers of mineral crystals that have crystallised from water

clastic sedimentary rock: sedimentary rock made of sediments cemented together

organic sedimentary rock: sedimentary rock formed from the remains of plants or animals

Investigation 7.19A

Modelling the formation of sandstone, p.486

Investigation 7.19B

Observing sedimentary rocks, p.487

Content group: Geological change

Igneous and metamorphic rocks are formed by changes in Earth's crust. Rocks on Earth's surface also change. Wind, water, vegetation and changes in temperature can act on these rocks, wearing them down into smaller parts called sediments and depositing them in new locations. When these sediments are buried, compacted and cemented together over time, they form sedimentary rock.

Rocks are worn away by weathering

Weathering is the process in which rocks are worn away into sediments such as pebbles, sand and clay. There are two main ways this can happen.

Physical weathering breaks rocks into smaller particles through processes that do not change the chemical make-up of the minerals. This can happen when rocks expand, shrink and crack due to temperature changes, or if they are worn down by wind or water.

Chemical weathering breaks rocks into smaller particles through chemical reactions that change the minerals in the rocks. This can happen when rocks contact chemicals in the air or water. The new minerals may separate from the original rock to form sediments, or they may even dissolve in the water and be transported away.

Erosion moves sediment

Erosion is the movement of sediment from one place to another. The sediments can be transported by wind, water, ice or gravity; these are called the agents of erosion.

Deposition drops sediment in new areas

Deposition happens when erosion deposits (drops) sediments in a new place. Often this happens when there is no longer enough energy in the wind or water to carry them any further.

Figure 7.64: Sedimentary rocks are formed when sediments are eroded, deposited, compacted and cemented together. ▶

Clastic sedimentary rocks are made up of layers 'glued' together

If the sediments are not eroded after they are deposited, they may eventually be compacted and cemented together to form **clastic sedimentary rock**. As the layers of sediments build up, pressure on the sediments increases and they are compacted together. Chemicals in the groundwater that moves between the sediments will then cement the layers together, forming a hard rock. Clastic sedimentary rocks are classified by the size of their sediments.

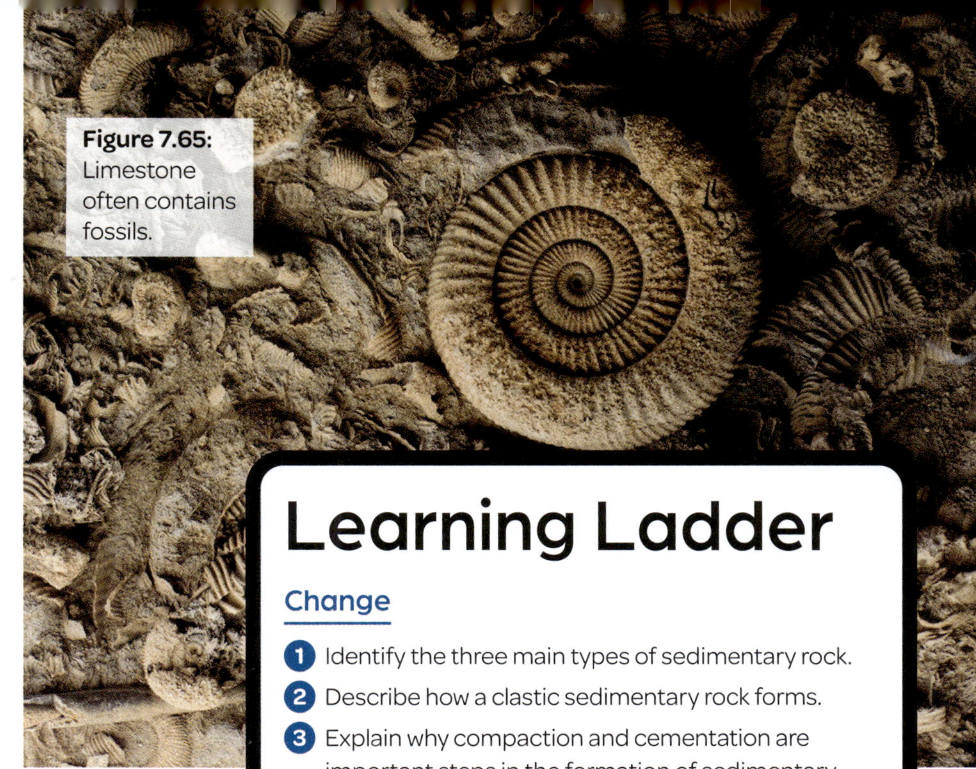

Figure 7.65: Limestone often contains fossils.

Pressure can make rocks from organic remains

Organic sedimentary rocks are formed from the remains of plants and animals. These remains have been deposited together, buried and compacted. Coal, limestone and chalk are examples of organic sedimentary rocks.

Coal is formed from the remains of plants and animals that died and became buried in ancient swamps. The remains did not decompose. Instead, when they were buried, the increasing pressure compacted and cemented the remains and changed them into coal.

Limestone is made of the remains of ancient coral reefs. The shells of the corals, shellfish and other marine animals made of a mineral called calcium carbonate were buried, compacted and cemented together. Limestone often contains fossils (see Figure 7.65)

Crystals can form rocks after evaporation

Chemical sedimentary rocks are formed when water that contains dissolved minerals evaporates, allowing the mineral crystals to grow. Halite (rock salt) is formed in this way. Some limestones can also be formed in this way: water evaporates and leaves behind crystals of calcium carbonate.

Learning Ladder

Change

1. Identify the three main types of sedimentary rock.
2. Describe how a clastic sedimentary rock forms.
3. Explain why compaction and cementation are important steps in the formation of sedimentary rocks. What could happen if they did not occur?
4. Limestone can form either as an organic or a chemical sedimentary rock. Compare and contrast the processes involved in the formation of each type of limestone.
5. Which of the three main types of sedimentary rock would be best at preserving fossils? Justify your answer with evidence.

Observing

see page 316

Use a specimen of a sedimentary rock to answer Questions 1–4.

1. Spend a few minutes observing the specimen using sight and touch only.
2. Describe how using a magnifying glass or microscope might add to your observations of this specimen.
3. Write down at least five observations you can make of the specimen.
4. Use your observations to infer what environment the sedimentary rock was formed in.

In context

Conglomerate, mudstone and sandstone are types of sedimentary rock. Research what sediments they are made from and the environments they are formed in.

Success criteria

- I can describe what a sedimentary rock is and how these rocks form.
- I can explain the difference between clastic, chemical and organic sedimentary rocks.
- I can classify rocks into groups according to their observable properties.

7·20 ▸ Fossils

Learning intention

At the end of this lesson, I will be able to describe how fossils form.

Key terms

cast: an object created when sediment or minerals fill a mould

fossil: preserved remains or traces of once-living things

fossil record: the history of life on Earth as documented by fossils

fossilisation: the process of the formation of a fossil

mould: a hollow impression formed by an imprint of an organism, or when the original bone or shell has dissolved

Investigation 7.20

Making 'fossils', p.488

Content group: Geological change

Fossils are the preserved remains or traces of once-living things. Fossils can be complete or part skeletons, shells and tree trunks, or traces that an animal existed, such as footprints, burrows and even poo. Fossils and the rocks that they are in can help palaeontologists, who study fossils, to reconstruct information about past living things and the environments they lived in.

Fossils form in sediment

The process that results in a fossil is called **fossilisation**, and it is rare. Usually when an organism dies, it decomposes or is eaten. Soft parts decay or are eaten quickly, and even hard parts such as bones and shells are eventually eaten or weather away if they are left exposed.

The organism dies and its remains are in an environment, such as under water, where they will be quickly buried by sediment.

The remains continue to be buried under more layers of sediment. These layers compact, and minerals in the groundwater cement them together into sedimentary rock.

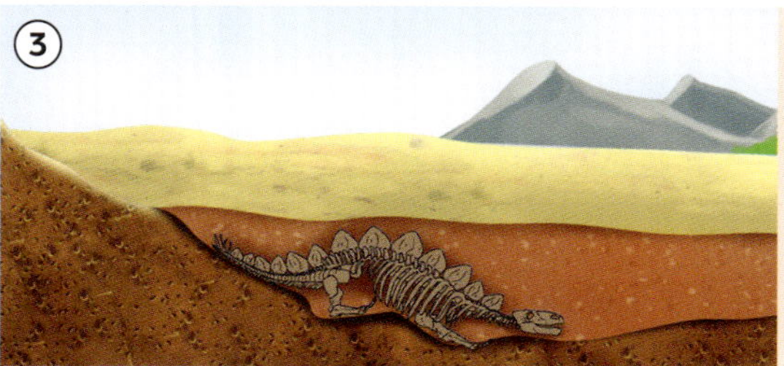

Groundwater minerals can also dissolve and replace the minerals in the bones of the organism, often making them harder than the original bone.

Movement of tectonic plates brings the fossil layer closer to the surface, and weathering and erosion or digging can expose them.

▲ **Figure 7.67:** For a fossil to form, a series of events that preserves an organism needs to happen.

▲ **Figure 7.66:** The fossils found in the rocks at Ediacara in South Australia are a very important part of the fossil record. They show the imprint of soft-bodied organisms in sandstone and are some of the oldest fossils of complex life in the world at more than 550 million years old.

The fossil record shows what life used to be like

Fossils provide evidence that different organisms existed at different times in the geological history of the Earth (the **fossil record**). Palaeontologists use different techniques to estimate how old a fossil of an organism is and to compare it to other fossils. This can provide them with information about how groups of organisms have changed over time, along with major events on Earth that caused many species to become extinct (mass extinctions). Some of the most important fossils in the fossil record are those that are rare or that show the evolution of new features in particular life forms.

▲ **Figure 7.68:** A skeleton of a *Muttaburrasaurus langdoni*, found at Muttaburra in Queensland. This dinosaur has been chosen as Queensland's fossil emblem.

Fossils can be found in other places

Most fossils are usually found in sedimentary rocks, although sometimes plant and animal remains can be found preserved in amber (fossilised tree sap), tar and permafrost (ground that is always frozen). These things often preserve the soft parts of organisms that would not usually be preserved in rock.

Footprints and burrows can also be preserved by fossilisation. These are called trace fossils. They can show palaeontologists how animals moved and lived, in a way that skeletons cannot.

◀ **Figure 7.69:** This 100-million-year-old mosquito has been preserved in amber.

Impressions of dead organisms and trace fossils can be moulds or casts. A **mould** is a hollow impression formed by an imprint of an organism, or when the original bone or shell has dissolved. A **cast** is created when sediment or minerals fill the mould.

Learning Ladder

Change

1. Define the term 'fossil'.
2. Identify an example of a fossil.
3. Explain the role of sedimentary processes of sedimentation and compaction in fossilisation.
4. Imagine you are the *Stegosaurus* in Figure 7.67. Write a social media update for each of the four steps, describing what is happening to you.
5. Propose why fossils that include feathers and other soft body parts are rare.

Planning investigations

see page 323

Design an investigation to identify which of these environments a fossil would most likely form in: sandy desert, mountain range, swamp. Use this to answer all of the following questions.

1. Identify materials and equipment you would use to model the different environments and the formation of a fossil.
2. Describe the safety precautions you would take to reduce the risks of using the materials and equipment you have listed.
3. For your investigation, identify:
   a the dependent variable.
   b the independent variable.
   c at least two controlled variables.
4. Use your answers for Question 3 to write an aim for your investigation.
5. Write a detailed method for your investigation.

In context

Explain why the Ediacaran fossils in Figure 7.66 are rare, and why they are important for understanding how life on Earth has evolved.

Success criteria

- I can describe how fossils form.
- I can explain how fossils show evidence that different organisms existed at different times.

7·21 ▸ History in the rocks

Learning intention

At the end of this lesson, I will be able to explain how geological history can be interpreted from a sequence of sedimentary layers.

Key terms

conglomerate rock: sedimentary rock made of large, rounded pebbles and fragments cemented together

geological history: how Earth has changed over time

law of superposition: older rocks are found beneath younger rocks in a sequence

relative age: the approximate age of a rock determined by comparing it to another rock

sequence: the order of something

Content group: Geological change

Rock layers are deposited from the bottom up ...

... so the deeper we dig, the further back in time we see.

▲ **Figure 7.70:** Geologists can work out the geological history of an area by comparing the layers of sediment.

It is possible that where you are right now was under water or near active volcanoes millions of years ago. Geologists can piece together the **geological history** of different locations by investigating the **sequence** of rocks in the area.

The youngest rocks are at the top

When sedimentary layers are deposited, younger ones are deposited on top of older ones. Unless something unusual has happened, the oldest rocks in a cliff face are at the bottom and the youngest rocks are at the top. This is known as the **law of superposition** and geologists use it to work out the **relative age** of the layers. Finding the same layers in different areas means the layers must have been deposited at the same time.

Rocks tell the story of geological history

If you are looking at a cliff face and observe different types of sedimentary rock, you can work out its geological history. This is because different rocks are deposited in different environments. Fossils in the rocks also provide information and can be used to work out how old the rocks are. The presence of igneous rocks might mean that a volcanic eruption happened.

Table 7.3: Sedimentary rock types and environments of formation

Rock type	Sediment size	Environment when formed
Limestone	Very small Often fossils of reef organisms	Warm, shallow seas
Siltstone and mudstone	Extremely small	Deep, calm water off the continental shelf
Sandstone	Small	Beaches and just offshore from land Deserts
Conglomerate	Large and varied	Rivers

Figure 7.71: Uluru is ▸ made of layers of very hard sandstone.

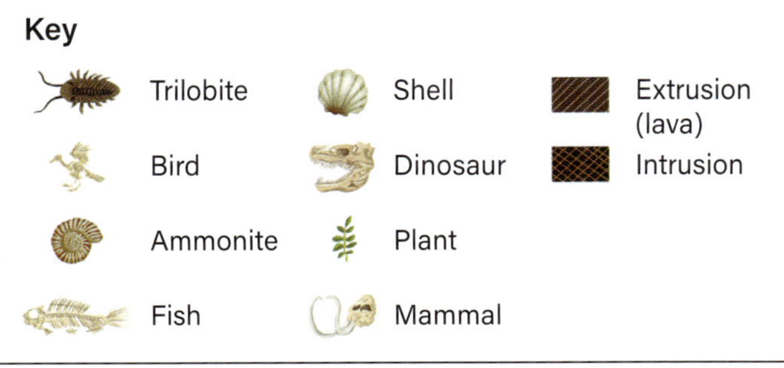

Figure 7.72: The law of superposition along with knowledge of how different types of rocks form can help us to interpret the geological history of an area. We can also use this information to link two different sites together if they contain the same rocks and fossils.

The origins of Uluru date back more than 500 million years

Uluru is made of layers of very hard sandstone. By studying these layers, geologists have been able to determine how the rock was formed. The sandstone was formed when a nearby ancient mountain range, similar to the modern Himalayas, was weathered, eroded and deposited in a valley more than 500 million years ago. Over time, this sediment was buried, compacted and cemented together, forming very hard rock. As tectonic plates pushed together, this rock was folded and pushed up to the surface again. The softer rock weathered and was eroded, leaving only the hard rock of Uluru at the surface. It is just the tip of a large section of rock that may reach as far as five kilometres below the surface.

Learning Ladder

Change

1. Copy and complete this sentence with the correct terms.

   Generally, the youngest layers of rocks are found on the __________ [top/bottom] and the oldest rock layers are found on the __________ [top/bottom].
2. Describe the law of superposition.
3. Explain why layers of different types of rock in a cliff face can tell you about the geological history of the location.
4. Uluru is a rock formation in Australia that formed over a long time. Describe the changes that occurred within the rock as it was formed.
5. **a** Identify the oldest and youngest rock layers in Figure 7.72.

   b Can you determine these from observing just the one location? Support your response with evidence from the figure.

   c Propose a reason why there are differences in the rock layers at the two locations.

Planning investigations

see page 323

Imagine how you could use food to model the deposition of different types of rocks and fossils over a period of time.

1. Identify the ingredients and equipment you would use.
2. Describe the safety precautions you would take to reduce the risks for the ingredients and equipment you have listed.
3. What material could you deliberately change in this investigation to see if it affected the rate of the rock or fossil deposition? How would you measure this?
4. Based on the variables you suggested in Question 3, construct a suitable aim you could use for this investigation.
5. Write a step-by-step method explaining how to create your model, including what each of the ingredients represents.

In context

Select an Australian rock formation other than Uluru and research how it was formed. What does this tell you about the history of these rocks?

Success criteria

- I can outline the law of superposition.
- I can identify in a sequence of sedimentary rocks where the youngest and oldest rocks can be found.

7.22 ▸ Change in context

Learning intention

At the end of this lesson, I will be able to recognise the role of change in a chain reaction machine. I will also be able to observe and discuss energy stores and energy transfers in a chain reaction machine.

Key term

Rube Goldberg machine: a machine or contraption that uses a chain reaction to carry out a simple task

Content group: Change in context

A chain reaction machine includes a series of deliberately complex and interconnected steps. Each small part triggers the next until eventually the end result is achieved. These machines are often creative and entertaining, but are not practical in terms of completing a task efficiently. Machines like this mainly exist for fun in games and cartoons, but they also provide a wonderful display of energy transfer and transformations.

A Rube Goldberg machine performs a simple task in a complicated way

Rube Goldberg machines, such as the one in Figure 7.74, are made up of a series of simple objects arranged to transfer and transform energy in a sequence to send something like a marble from the start point to the finish. Rube Goldberg competitions are quite popular and attract participants and audiences of all ages.

▲ **Figure 7.73:** High school students compete to create a Rube Goldberg contraption that can ice a cake in 20 steps or more.

Figure 7.74: A Rube Goldberg machine performs a simple task in a complicated way, using many transfers and transformations of energy. ▶

Rube Goldberg machines can be used to analyse energy changes

In Figure 7.75, we can see that the purpose of the contraption is to squeeze some orange juice, but there are 15 steps instead of just one. Each step is a display of an energy change. In some cases, the energy is changing from one type to another; for example, gravitational potential energy is transformed into kinetic energy. In other cases, the energy is simply transferred from one object to another. Can you identify each step in the series that contributes to the chain reaction, and the energy change it displays?

Because these machines were originally drawn for fun and entertainment, not only are they not practical in getting a particular task done, but they might not even be realistic. Can you identify any weaknesses in the machine? Are there any factors that are unpredictable and could interrupt the steps?

PROFESSOR BUTTS STEPS INTO AN OPEN ELEVATOR SHAFT AND WHEN HE LANDS AT THE BOTTOM HE FINDS A SIMPLE ORANGE SQUEEZING MACHINE. MILK MAN TAKES EMPTY MILK BOTTLE (A) PULLING STRING (B) WHICH CAUSES SWORD (C) TO SEVER CORD (D) AND ALLOW GUILLOTINE BLADE (E) TO DROP AND CUT ROPE (F) WHICH RELEASES BATTERING RAM (G). RAM BUMPS AGAINST OPEN DOOR (H) CAUSING IT TO CLOSE. GRASS SICKLE (I) CUTS A SLICE OFF END OF ORANGE (J) AT THE SAME TIME SPIKE (K) STABS "PRUNE HAWK" (L) HE OPENS HIS MOUTH TO YELL IN AGONY, THEREBY RELEASING PRUNE AND ALLOWING DIVER'S BOOT (M) TO DROP AND STEP ON SLEEPING OCTOPUS (N). OCTOPUS AWAKENS IN A RAGE AND SEEING DIVER'S FACE WHICH IS PAINTED ON ORANGE, ATTACKS IT AND CRUSHES IT WITH TENTACLES, THEREBY CAUSING ALL THE JUICE IN THE ORANGE TO RUN INTO GLASS (O).

LATER ON YOU CAN USE THE LOG TO BUILD A LOG CABIN WHERE YOU CAN RAISE YOUR SON TO BE PRESIDENT LIKE ABRAHAM LINCOLN.

Figure 7.75: An elaborate Rube Goldberg machine invented by cartoon character Professor Lucifer Butts

You can create your own Rube Goldberg machine

You do not need anything fancy to make your own version of a chain reaction machine. Things like empty paper towel rolls, string, paper cups, small toys and marbles may be all you need to build a contraption that displays a variety of energy transformations. What will you create?

Learning Ladder

Change

Consider the chain reaction machine shown in Figure 7.75.

1. Identify a type of stored energy that is commonly used in chain reaction machines.
2. Define the type of energy identified in Question 1.
3. Select an energy transformation included in the cartoon that you think is particularly clever. Explain the role of the objects and energy involved that leads to the intended change.
4. Construct an energy flowchart to explain the energy transformations in the machine shown in Figure 7.74.
5. A perpetual motion machine can perform a function forever without an external energy source. Justify why this type of machine is only hypothetical.

Observing

see page 316

1. Identify the main senses you would use to observe energy changes in a chain reaction machine.
2. Make links between your response to Question 1 and examples from the machines shown on this page.
3. Identify and describe the quantitative data you could collect by video-recording a Rube Goldberg machine from start to finish.
4. Describe how recording video footage of a Rube Goldberg machine in action might enhance your ability to make accurate observations of energy changes.
5. Is a Rube Goldberg machine an open or closed system? Justify your answer with evidence.

Success criteria

- I can describe the difference between the transfer and transformation of energy.
- I can give at least one example of an everyday energy transformation.

Change summary

Energy transfers

Types of energy

There are two main forms of energy:

- **kinetic energy**: the energy of movement
- **potential energy**: stored energy.

The common types of **kinetic energy** are motion, thermal, light, sound and electrical.

The common types of **potential energy** are:

- **gravitational energy** (stored in objects based on gravity; the potential to fall down)
- **elastic energy** (stored in elastic material like rubber bands; the potential to bounce back)
- **chemical energy** (stored in chemical bonds; the potential to undergo a chemical reaction)
- **nuclear energy** (stored in the nucleus of atoms; the potential to undergo a nuclear reaction).

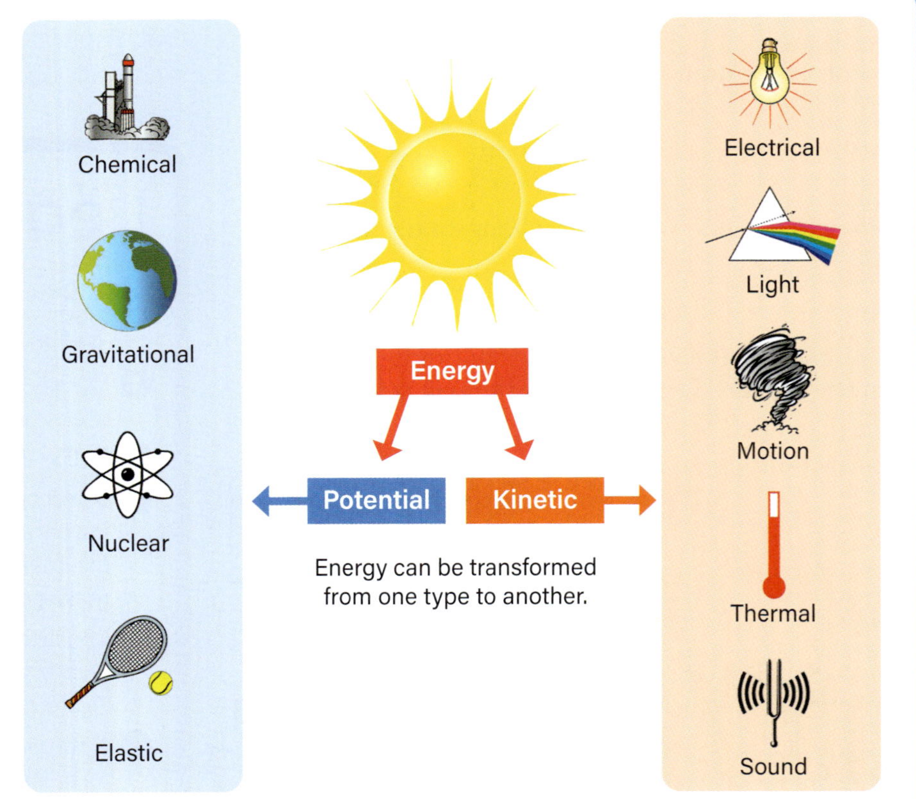

Transfers

Convection:
The movement of fluid molecules from higher-temperature regions to lower-temperature regions

Conduction:
Transmission of energy from one particle of the medium to another with the particles being in direct contact with each other

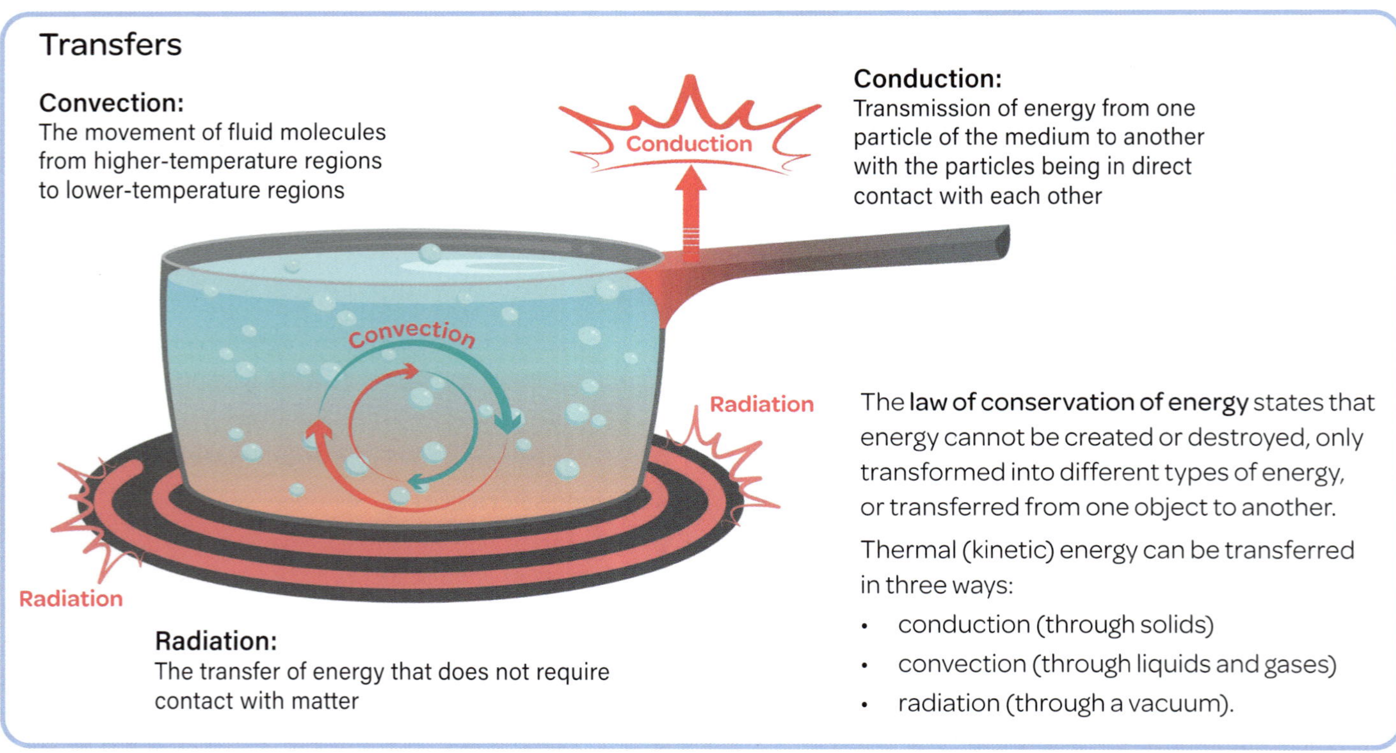

Radiation:
The transfer of energy that does not require contact with matter

The **law of conservation of energy** states that energy cannot be created or destroyed, only transformed into different types of energy, or transferred from one object to another.

Thermal (kinetic) energy can be transferred in three ways:

- conduction (through solids)
- convection (through liquids and gases)
- radiation (through a vacuum).

Systems can be described in many ways including open, closed and isolated:

- Open systems allow the flow of matter and energy.
- Closed systems allow the flow of energy but not matter.
- Isolated systems have no flow of matter or energy.

Earth is an open energy system. Energy and matter can flow into Earth, where it is cycled within Earth's systems. Energy and matter can also escape Earth and re-enter space.

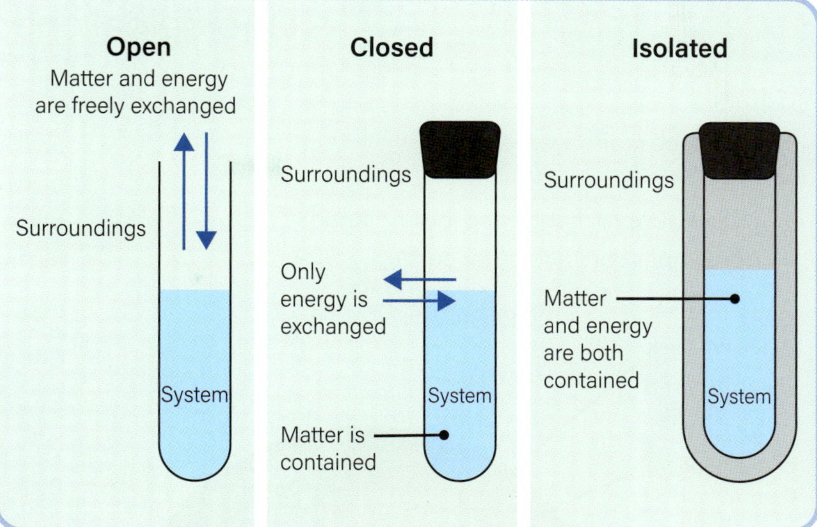

Chemical change

A **physical change** is a change in appearance, not properties; no new substance is formed, and the process is reversible. For example, paint undergoes a physical change when it dries.

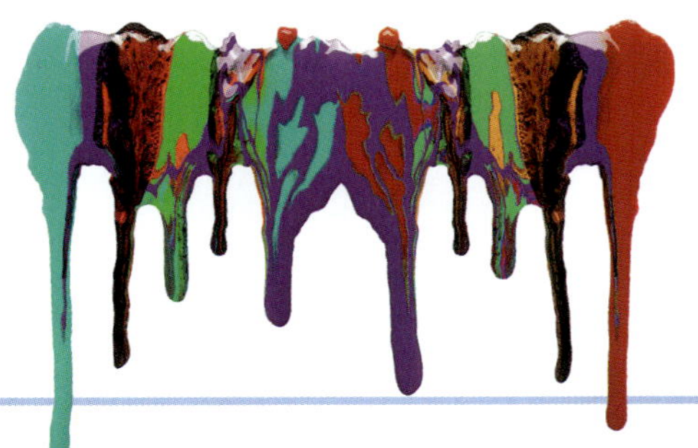

A chemical change can be represented by a word equation, with reactants on the left of the reaction arrow and products on the right.

The word equation for photosynthesis

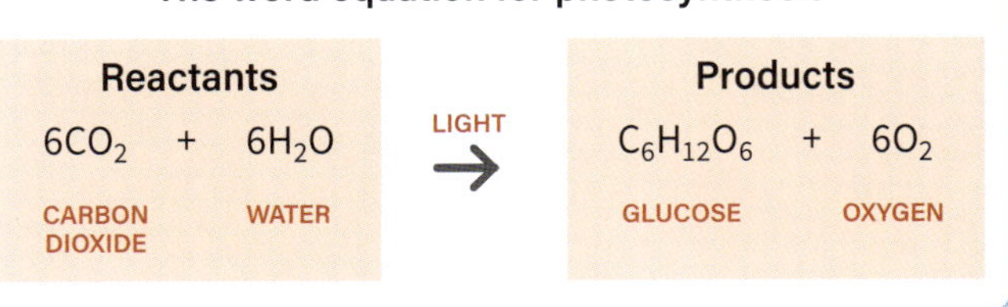

Exothermic chemical reactions release energy, while endothermic chemical reactions absorb energy.

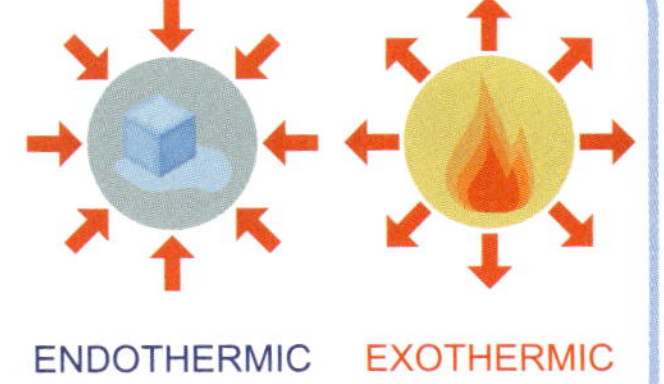

A chemical change forms new substances, and new properties are observed. For example, when metals burn, they form compounds with oxygen as they give off different colours of light.

Geological change

Earth's three main layers are the:

- crust
- mantle (upper mantle and lower mantle)
- core (inner core and outer core).

Crust
0–100 km thick
Lithosphere
Upper mantle
(including
asthenosphere)
Lower mantle
Outer core
Inner core

Earth's atmosphere is a 600-kilometre-thick layer of gas that surrounds the planet.

Earthquakes occur when pressure builds up between tectonic plates and is released as seismic waves. Earthquakes usually occur along the boundaries of tectonic plates.

Epicentre
Focus
Fault
Seismic
waves

The **lithosphere** is Earth's rigid, rocky outer zone (the crust and most of the upper mantle).

The **theory of plate tectonics** states that the lithosphere is divided into more than 15 tectonic plates. Evidence for this theory includes:

- identical rock and fossil formations on different continents
- mid-ocean ridges where the sea floor is spreading apart to create new lithosphere
- subduction zones where old lithosphere is destroyed
- volcanoes and earthquakes along subduction zones and mid-ocean ridges.

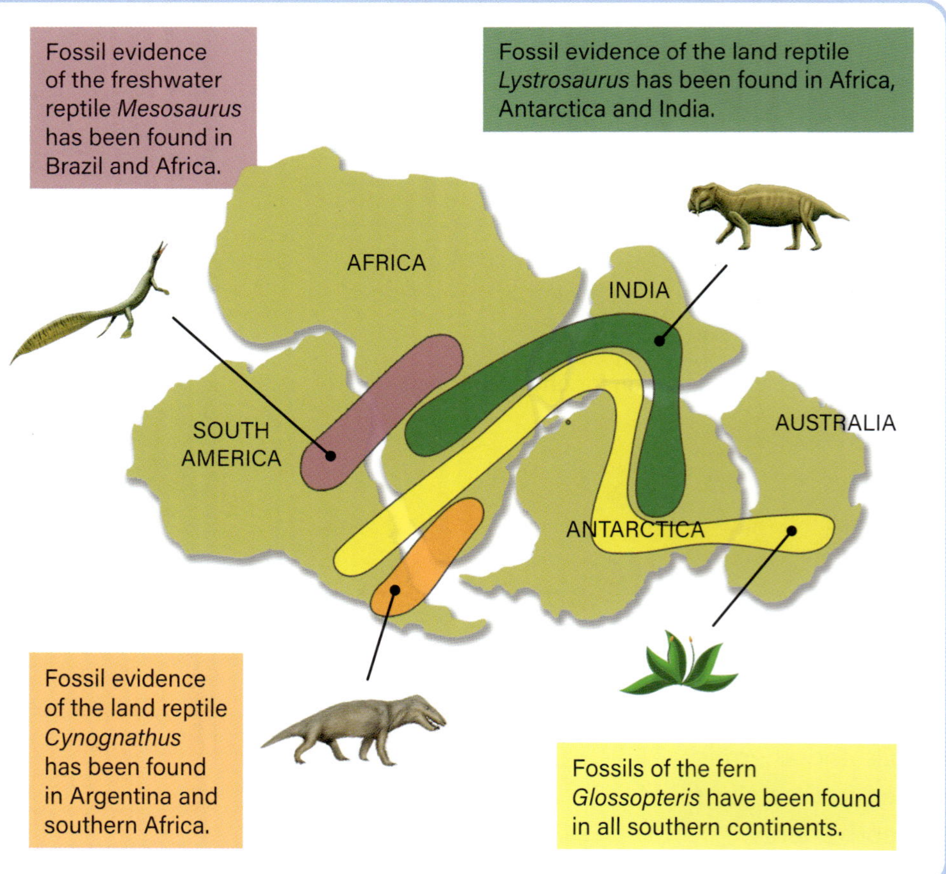

A volcano occurs where molten rock erupts at Earth's surface. Most volcanoes occur along the boundaries of tectonic plates.

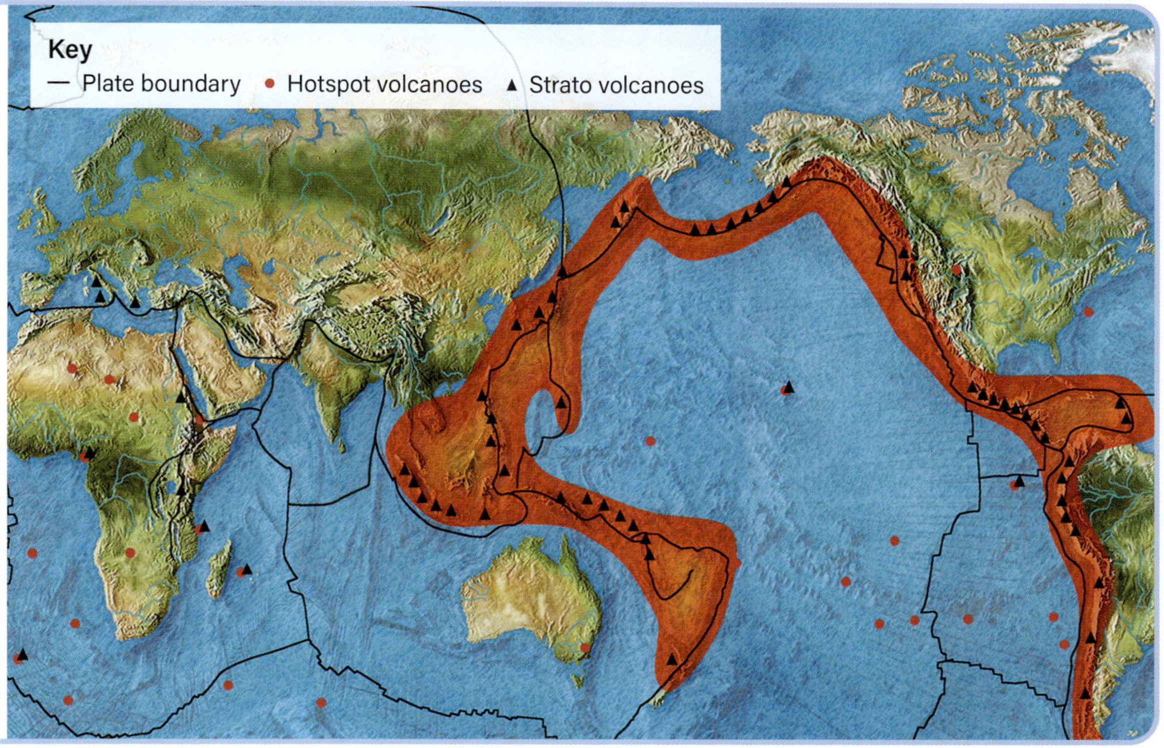

The rock cycle shows how different, naturally occurring processes in Earth form different types of rock.

There are three main types of rock:

- **igneous rocks**, which are made by the cooling of molten rock
- **sedimentary rocks**, which are made of sediments weathered and eroded from older rocks
- **metamorphic rocks**, which are made from older rocks that have been changed.

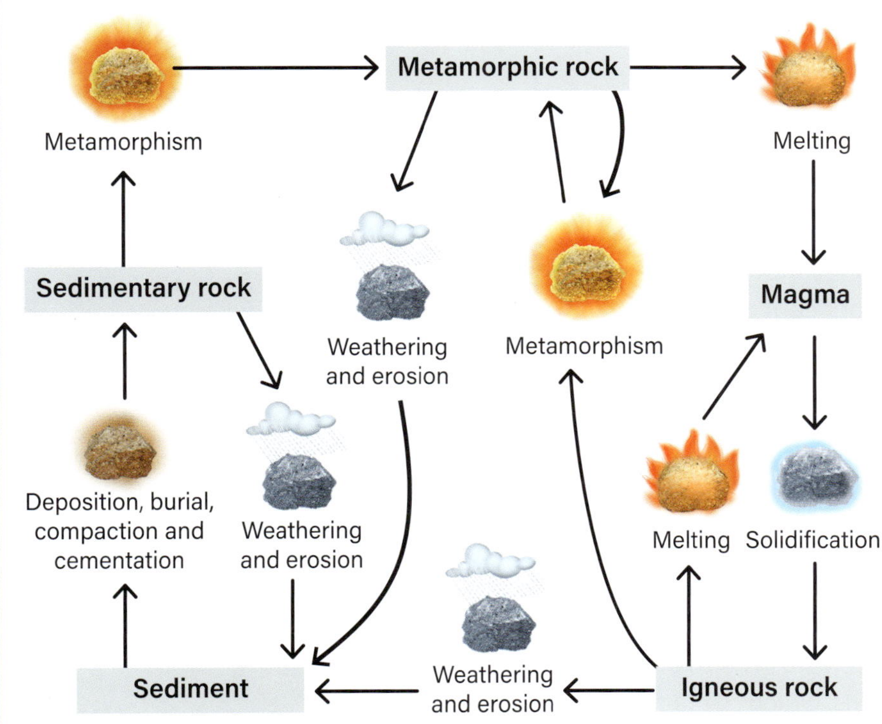

The **law of superposition** is used to work out the relative age of rock layers so geologists can determine how life and Earth have changed over time.

Rock layers are deposited from the bottom up ...

... so the deeper we dig, the further back in time we see.

Masterclass

Steps in progression		1	2
Content	Change	Identify the difference between the epicentre and the focus of an earthquake.	Describe how the tectonic plates interact along the South Island of New Zealand. (You may need to conduct research to answer this question.)
Processes	Observing	Identify a change to the landscape that may be observed after an earthquake.	Propose some equipment that geologists could use to make observations of how Earth has changed after an earthquake.
Processes	Planning investigations	Consider Table 7.4 below. Identify the most suitable approach (A, B or C) to investigate this question: How does soil type affect building movement due to liquefaction?	Construct a risk assessment for the investigation you identified in Step 1.
Processes	Conducting investigations	Identify the data you would need to gather to address the aim you wrote in Step 4 of the 'Planning investigations' row.	Describe how you could use video to gather data for this investigation.

Christchurch earthquake, February 2011

At 12.51 pm on Tuesday 22 February 2011, a large earthquake struck the city of Christchurch in New Zealand. The earthquake occurred along the boundary between the Indo-Australian Plate and the Pacific Plate. The epicentre of this earthquake was 10 kilometres west of Christchurch and the focus was five kilometres deep.

Although the shaking only lasted 10 seconds, the earthquake caused considerable damage to the city and resulted in 185 people losing their lives. More than 50 per cent of the buildings in the central business district were severely damaged. In Christchurch's suburbs, many houses, roads and other infrastructure were damaged due to a phenomenon known as liquefaction. Liquefaction occurs when the intense shaking of an earthquake causes soils to act like a liquid rather than a solid.

Wet soil that is made up of fine particles of sand and silt are prone to liquefaction. When an earthquake occurs, the pressure builds up and the liquefied soil moves underneath the surface until it escapes through cracks on the surface. The ground above the liquefied soil moves, tilts, cracks and sinks, which damages infrastructure.

Figure 7.76: This building in central Christchurch was damaged in the 2011 earthquake.

Demonstrate your understanding

3	4	5	
Explain how seismic energy built up and was released in the 2011 Christchurch earthquake.	Draw a diagram that represents what you explained in Step 3.	Explain why New Zealand is a geologically active country.	
Describe what information a seismometer provides about an earthquake.	Discuss the inferences you could make about the magnitude of an earthquake by observing the damage it caused.	Describe the data you would need to gather to determine the relationship between the magnitude of an earthquake and the damage it caused.	Science how-to p. 316
Identify the independent, dependent and controlled variables for the investigation you identified in Step 1.	Use the variables you described in Step 3 to construct an aim for this investigation.	Construct a step-by-step method and a list of materials for the investigation you identified in Step 1.	Science how-to p. 323
Explain the laboratory safety rules you would need to follow while conducting this investigation.	Discuss how you could make sure the data you gather in this investigation is accurate and reliable.	Construct a table to collect the data in this investigation.	Science how-to p. 326

Table 7.4: Different approaches to investigating the question: How does soil type affect building movement due to liquefaction?

Option	Approach
A	Place different types of soil in a container. Place a model building on top of the soil. Shake the container to replicate an earthquake.
B	Mix different types of soil with a fixed amount of water in a container. Place a model building on top of the soil. Shake the container to replicate an earthquake.
C	Mix soil and water in a container and shake it to replicate an earthquake.

Focus area 8

8.0 Data science 1

To understand the world around us, we are constantly looking for information to explain what we see and how we feel. To help us to do this, scientists collect and use data. Data allows us to record and measure observations using categories and numbers. From the streaming services that recommend the shows you watch, to computer models that predict weather patterns, so much of what we do relies on understanding data.

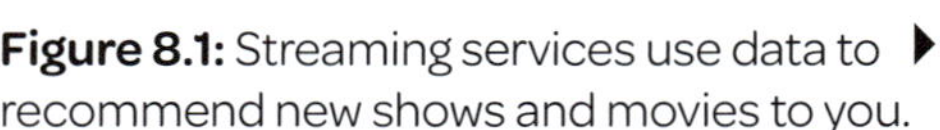

Figure 8.1: Streaming services use data to recommend new shows and movies to you.

Learning Ladder

The Learning Ladder contains the scientific content and processes you will learn in this chapter. Each area has five levels of progression. To move to higher levels, you need to practise activities at the earlier levels. This will help you develop the ability to complete tasks that are more complex.

Steps in progression	Content: Data science 1	Working scientifically processes: Analysing data and information	Working scientifically processes: Problem-solving	Steps in progression
5	I can create and evaluate data-based models that explain scientific phenomena.	I can evaluate data and information for accuracy, reliability and validity.	I can evaluate problem-solving strategies used to solve an identified problem.	5
4	I can explain how data is used to make predictions.	I can draw conclusions based on patterns in data and information.	I can use given criteria to find solutions to scientific problems.	4
3	I can recognise the relationship between data and scientific phenomena.	I can explain relationships between datasets and information.	I can explain scientific problems and phenomena using cause-and-effect relationships.	3
2	I can describe the relevance of different types of data.	I can describe trends from collected data and information.	I can suggest solutions to familiar scientific problems.	2
1	I can differentiate between types of data sources.	I can identify trends within given data and information.	I can identify scientific problems.	1

Figure 8.2: Scientists use data to model and predict scientific phenomena.

8.1 ► Data sources

Figure 8.3: A survey that asks students to describe why they like a flavour of ice-cream collects qualitative data.

Learning intention

At the end of this lesson, I will be able to describe a range of data sources and their applications.

Key terms

big data: large, diverse sets of information that grow at increasing rates

continuous data: data that can take any value along a continuum or between fixed values

data: a collection of information gathered through observations, measurements, study or analysis

discrete data: data that can only take certain values; fixed values; for example, a quantity count

experimental data: information collected by conducting an investigation and recording the results

first-hand investigation: an investigation where the data is collected by the investigator

qualitative data: data with qualities or characteristics that can be observed and described

quantitative data: information based on numbers that can be counted, measured or represented by numbers

second-hand investigation: an investigation that uses data previously collected in a separate investigation

trend: when data moves in a general direction, usually up or down

Investigation 8.1

Pin drop, p.489

Content group: Data

Figure 8.4: ► The number of apps on a phone is a type of discrete, numerical data.

Every day, all around us, there are people asking questions about how the world works. One way scientists can help to answer these questions is by conducting investigations to collect data about how the world works. But what is data, and how does it help us to understand things?

Data is information people collect

Data refers to information that has been collected. We can use this information to present ideas, or we can analyse it to see if any **trends** or patterns exist. Data can be collected by conducting investigations and recording numbers or other information, such as qualities or characteristics.

There are different types of data

When conducting investigations, scientists have to decide what type of data they are going to collect. The two types are qualitative data and quantitative data.

- **Qualitative data** is categorical. This data is usually based on observations or descriptions. For example, categorising items by colour or asking people to describe why they like their favourite food would give you qualitative information.
- **Quantitative data** is numerical and can be discrete or continuous. **Discrete data** includes fixed values, such as the number of students that attend class each day, while **continuous data** has many different measures that fall between fixed points, such as weather temperatures.

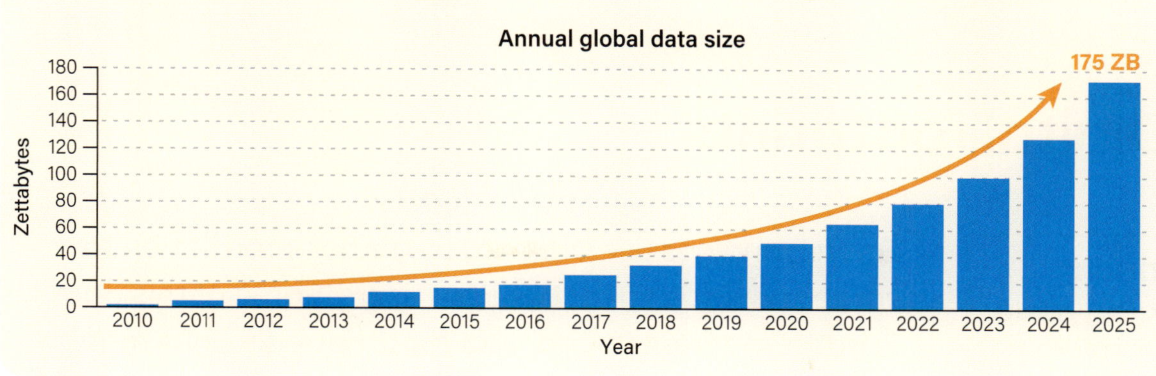

Figure 8.5: Search engines are used to generate big data, which continues to add to the amount of collected data across the world. (A zettabyte equals a trillion gigabytes.)

Data can come from different sources

The main way to collect data is by conducting investigations. This is called **experimental data**. To obtain this data, a **first-hand investigation** is conducted, and the information collected is recorded. For example, you could test the effect of rock music on the growth of plants. You expose one set of plants to rock music all day, every day, while another set grows in silence. The information you collect on the height of the plants throughout the investigation is experimental data. Once you have your data, you can draw a conclusion. **Second-hand investigations** are similar, but instead of the investigator collecting the data directly, data from previous investigations is gathered and can be considered in new ways.

Another type of data is **big data**. This is when large and complex sets of data are generated, often by search engines such as Google. As people use these to search for different things, the data is stored and added to datasets.

For data to be considered 'big data', there needs to be very large amounts of data that are continually collected and analysed, from multiple sources. For example, Netflix tracks and records the viewing habits of over 250 million subscribers and uses this information to make decisions about which shows to continue or cancel, and what new shows to develop.

Data can be used in different ways

There are many ways in which data can be applied, depending on the type of data that has been collected, how it has been sourced, and the objectives of the person investigating.

Data drives informed decisions across industries, shaping strategies, improving efficiency and personalising experiences. For example, it can be used to:

- inform businesses on customer preferences to increase sales
- help medical professionals to evaluate patients more accurately
- help governments form policies to address problems
- optimise how resources are allocated after natural disasters.

From targeted marketing to scientific breakthroughs, the application of data accelerates progress and innovation, underpinning growth and advancement.

Learning Ladder

Data science 1

1. Identify the data type collected when:
   a. asking people what is their favourite football team.
   b. counting the number of people who attended a concert.
   c. recording the heights of a number of people.
2. Describe why it might be important to collect big data.
3. Explain how the collection of data relates to ads you see on your phone.

Analysing data and information see page 334

Refer to Figure 8.5.

1. Identify a trend in the data.
2. Describe what that trend tells us about data collection.
3. Explain what this graph shows us about big data collectors such as search engines.
4. Make a conclusion relating to the amount of data used to inform companies that advertise online.

In context

Qualitative data can be collected in small or large amounts. Conduct a qualitative data survey on your classmates, on a topic of your choosing, and record the data you collect as a pie chart.

Success criteria

- I can differentiate between different types of data.
- I can describe at least three data sources.

8·2 ▸ Digital footprints

Learning intention

At the end of this lesson, I will be able to:

- describe the digital footprint created by different online activities
- recognise ways to safely engage with online platforms.

Key terms

digital footprint: information about a particular person that exists on the internet as a result of their online activity

paid promotion: when someone is paid to say they like a product (even if they do not)

personal data: potentially sensitive, private information connected to a particular individual

social media: the means of interacting and sharing information with other people via virtual networks

Investigation 8.2

Have you been influenced?, p.490

Content group: Data

Search engines use big data to gather information to target advertising for different users. Some of this data is gathered from your **digital footprint** – the information you provide about yourself on different online platforms.

You leave a digital footprint

Many apps and online sites require you to provide **personal data** to use them. While most sites and apps state that they will not share your personal data without your permission, you have often provided this permission without realising. For example, when you see a long list of terms and conditions you must agree to for access, do you read all of it? Most of us just click 'agree'. This is where you have provided permission for a site or app to use your personal data. Sites use this information to create a digital footprint that can follow you on the internet.

Facebook is the biggest **social media** platform in the world and about 80 per cent of Australians use it. When you sign up, you must provide your name, email address, mobile number, date of birth and gender. Facebook's data policy states that it will not share your personal information with other advertising, analytical or measurement partners without your permission. Consider if you have given permission. Did you have to agree to terms and conditions before making an account? If so, you have probably given permission.

You can be influenced online

An influencer is someone who can change or affect the way that other people behave. Some online influencers persuade people to buy a product or service by recommending it on social media, often because they are paid to do so. Some influencers blur the lines between genuine recommendations of products they like and **paid promotions**.

An example of this is the recent craze for the Stanley cup. The company used influencers on social media to spread the word about the new Quencher cup, offering an array of colour options. Once the cups took off on social media, Stanley's annual sales reportedly increased 10 times in 2023 alone. While similar cups are available for much less money, people want to pay a higher price for the influencer-recommended product.

▲
Figure 8.6: The Stanley company increased profits by more than 900 per cent after influencers advertised Stanley cups on social media.

You can stay safe on online platforms

To stay safe online, carefully consider what you share on the internet. Once information goes online, it is published publicly and permanently! Here are a few things you can do to limit the personal data you share.

1 Avoid posting identifying photos

Ensure your photos do not include details that would make you easy to identify. For example, a post of a photo of you in your sports uniform with the location of the venue in the background could provide enough information for someone to locate you.

Avoid tagging images and posts with people's names or your location, as this could provide strangers with information they could use to convince you to trust them.

2 Only make online purchases through secure websites

When using a credit card to pay for an online purchase, you need to enter your name, address, email, phone number and financial details such as credit or debit card information.

▲ **Figure 8.7:** Team colours or logos can allow strangers on the internet to identify your location.

This gives the seller all the details needed to fraudulently use your card. This is why you should never provide payment details to unverified individuals or organisations online. A secure website is usually labelled with a padlock icon next to the URL as shown in Figure 8.8. This is one way to tell whether a website is verified as a safe place to use payment details.

3 Do not include personal details in online reviews

Companies that provide products and services often ask the consumer (you) to leave online reviews. You will often leave your name alongside these reviews. When you complete an online review, consider what you have reviewed, whether your review was fair and who the target audience is. Who is going to see what you have written? Avoid using identifying or personal details in your reviews. Otherwise, you could be vulnerable to online attacks, such as fraud or identity theft.

▲ **Figure 8.8:** A padlock next to the URL of a website can indicate that your payment details are secure.

Learning Ladder

Data science 1

1. Identify the number of apps on your phone that require data.
2. Describe a digital application that asked you to provide more than one piece of personal data.
3. Explain how the data you provide to Facebook can be used.
4. Based on the personal data Facebook requires when you sign up, explain what data might be used if you signed up for a platform such as Instagram or WhatsApp.

Problem-solving

see page 339

1. Identify one problem with providing data to an insecure online source.
2. Describe a piece of data you are happy to provide to Facebook. Would this change if you were signing up for a different site? Why?
3. Explain a problem that can occur when using a credit card to pay for an online purchase.

In context

Many websites require personal data for you to use them. Construct a checklist that you could use to ensure that you only provide data to safe websites and platforms.

Success criteria

- I can identify platforms where I leave a digital data footprint.
- I can describe ways to safely use online platforms that collect data.

8.3 ▸ Scientific approaches

Learning intention

At the end of this lesson, I will be able to:

- recognise the ways that science uses data to gain insights
- compare scientific and non-scientific approaches to inquiries.

Key terms

algorithm: a rule or set of rules that specifies how to solve a problem, often used in mathematics or information technology

census: a complete count of everything

extrapolate: to estimate or guess what could happen by using information that is already known

hypothesis: a suggested explanation or prediction of a scientific problem that can be tested with an investigation

inference: a conclusion that is based on evidence and observations

insight: applying inferences to real-world scenarios

interdisciplinary: combining more than one branch of knowledge

non-scientific approach: an approach relying on tradition, intuition and personal belief to arrive at a conclusion

pseudoscience: beliefs mistakenly regarded as being based on scientific method

scientific approach: the process of using testing and experiments to objectively establish facts

statistics: the analysis of large collections of data to make inferences

unsubstantiated: not supported with any evidence

Investigation 8.3

Can you believe your eyes?, p.491

Content group: Scientific models

Scientists use data collected from investigations to make inferences that allow them to show evidence for scientific theories. They use the data to draw conclusions, and use insights to apply these findings to the real world. Decisions made by scientists must be based on evidence, not just opinion.

Data science is interdisciplinary

The collection and use of data by scientists is **interdisciplinary**. By collecting data from across different branches of science, and using **statistics** to analyse it, scientists are able to draw conclusions about the world around them and apply these conclusions to the real world.

Statistics is the collection and analysis of data to make **inferences**. Data may be collected either as a **census** or by taking a sample. When researchers conduct a census, they collect data from an entire population and can use **algorithms** to analyse the data. An example of some simple algorithms is calculating the mean, median, mode and range (see page 363 of the Science how-to section).

A sample uses smaller collections of data and **extrapolates** those findings to represent the entire population. For example, imagine your school principal wants to know the favourite sport in Year 8, but they do not have time to ask everyone. They could instead ask a sample of students from each Year 8 class. Using the data gained from these students, the principal can extrapolate to make an educated guess about the favourite sport of the whole of Year 8. When taking samples, it is important to make the sample representative. For example, if the principal wanted to know the favourite sport of the whole school, they would need to take a sample from all of the school years, not just Year 8.

Scientists use many methods to gain insights from data

Scientists use the scientific method to draw conclusions, gain **insights** and explain aspects of our world. The scientific method usually follows five steps. The scientist will:

1. make an observation
2. form a question to investigate the observation
3. form a **hypothesis** (the idea that is to be tested)
4. carry out investigations, collect data and analyse the results
5. draw conclusions about whether the hypothesis was valid or not.

For example, a hypothesis could be that tall people all have a large shoe size. The hypothesis can be tested by measuring people's height and recording their shoe size. An analysis of the data collected would show that while many tall people wear large shoes, some tall people wear small shoes and some short people wear large shoes. The conclusion is that there is no direct connection between height and shoe size, so the hypothesis is not supported by the scientific evidence.

Inquiries can be scientific or non-scientific

When you are presented with a scientific statement, it is important to confirm that it is supported by evidence. Otherwise, the statement is simply an opinion, which may or may not be correct.

When a statement is not supported by scientific evidence, we say it is **unsubstantiated**. Some statements are made in the belief that 'we have always done it this way'. But this is a **non-scientific approach**, rather than a **scientific approach**. For example, phrenology – which claims to reveal a person's personality traits by measuring the bumps on their head – was popular in the late 1800s, but has no scientific basis.

While phrenology is **pseudoscience**, branches of study such as neuroscience have collected evidence to support findings about areas of the human brain. As neurology develops, scientists have isolated areas of the brain that are primarily responsible for speech, memory, coordination and some personality traits. Now data collected on neuroscience can be used to help people who have suffered brain injuries.

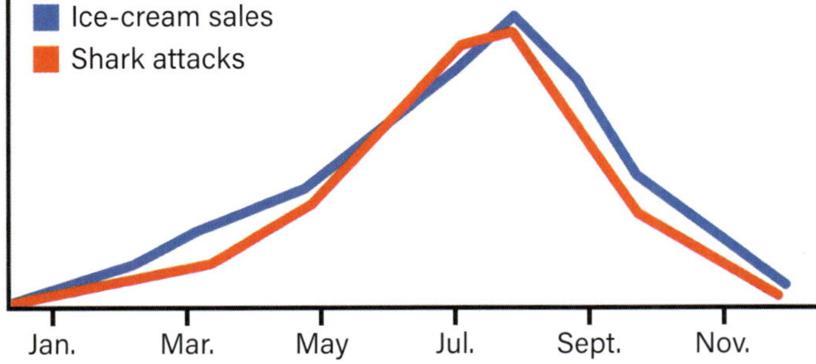

▲ **Figure 8.10:** Scientists analyse data and use it to draw conclusions about whether or not one thing causes another. For example, both ice-cream sales and shark attacks increase when the weather is warm. But they are not caused by each other – they are each caused by good weather.

▲ **Figure 8.9:** Making predictions without scientific evidence, as a fortune teller does, is pseudoscience.

Learning Ladder

Data science 1

1. Outline the differences between a scientific approach and a non-scientific approach.
2. Describe the reason that neurology is more relevant to our knowledge than phrenology.
3. Explain how data has been used to help us understand the human brain.
4. You observe that people who wear white running shoes tend to look fit.
   a Describe how you would turn your observation into a scientific hypothesis.
   b How would you collect data?

Problem-solving

see page 339

1. Identify and outline a problem with the data in the graph in Figure 8.10.
2. Describe how scientists solve problems using the scientific method.
3. Explain the link that was found between height and shoe size from the example given.
4. Imagine that you wanted to find a way to grow the largest pumpkin possible for an upcoming agricultural show. You know that pumpkin growth is based on temperature, water intake and sunlight exposure. Discuss how you could use your understanding to test how to grow the largest pumpkin.

In context

There are many forms of pseudoscience. Research and describe another form of pseudoscience, and compare it to a branch of science that collects similar evidence. What are the conclusions these methods have reached?

Success criteria

- I can describe some ways that scientists use data to gain insights.
- I can compare and contrast a named science and a pseudoscience.

8·4 ▸ Scientific models

Learning intention

At the end of this lesson, I will be able to identify types of scientific models and the reasons they are used.

Key terms

geocentric model: a model of the solar system with Earth at the centre

heliocentric model: a model of the solar system with the Sun at the centre

model: a physical or mathematical representation of a scientific idea or concept

phenomenon (plural **phenomena**): an observable fact or event

Investigation 8.4

Modelling bridges, p.492

Content group: Scientific models

Scientific models are based on data and observations of phenomena in the real world. Scientists develop models as a way of showing scientific concepts that may otherwise not be able to be seen or understood. As technology improves and more data becomes available, scientists can refine their models further to advance our understanding.

Scientists develop and use many models

A scientific **model** can be used for many purposes. Scientists use them to communicate understandings about how the world works and to explain concepts in an accessible way, particularly for **phenomena** that we cannot observe easily. Table 8.1 shows some of the types of models scientists can use to present a scientific concept or idea.

Table 8.1: Scientific models

Model type	Description	Example
Diagram	A two-dimensional labelled diagram that shows the components of an item or a concept	A diagram showing the components of the atom (see Figure 8.13)
Three-dimensional representation	A three-dimensional structure that shows the structure of a particular scientific concept	A three-dimensional model of the solar system (see Figure 8.12)
Computer simulation	A program on a computer that uses large amounts of data to construct representations and predictions about a scientific concept	A weather simulation on a computer (more on this in the next spread)
Mathematical formula	A formula that can be used to represent a scientific concept – often from a scientific law	$F = ma$ (An object's force in newtons is equal to its mass in kilograms times its acceleration in metres per second)

Figure 8.11: Computer simulations can make models of structures like water molecule arrangement in solids.

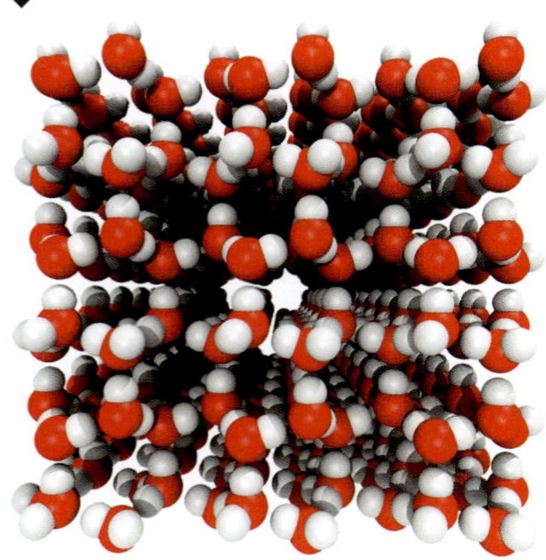

Models tell us about different concepts

Scientists have used models to communicate for hundreds of years. One example is the **heliocentric model** of the solar system, with the Sun at the centre, proposed by Copernicus and later supported with evidence collected by Galileo. Before this model, many people believed that Earth was the centre of the universe (the **geocentric model**). However, observations about 'retrograde motion', or the seemingly backwards movement of planets across the sky, contradicted this idea. By putting the Sun at the centre of their model, scientists could explain this movement as Earth being in different positions relative to the other planets as they revolved around the Sun in their orbits.

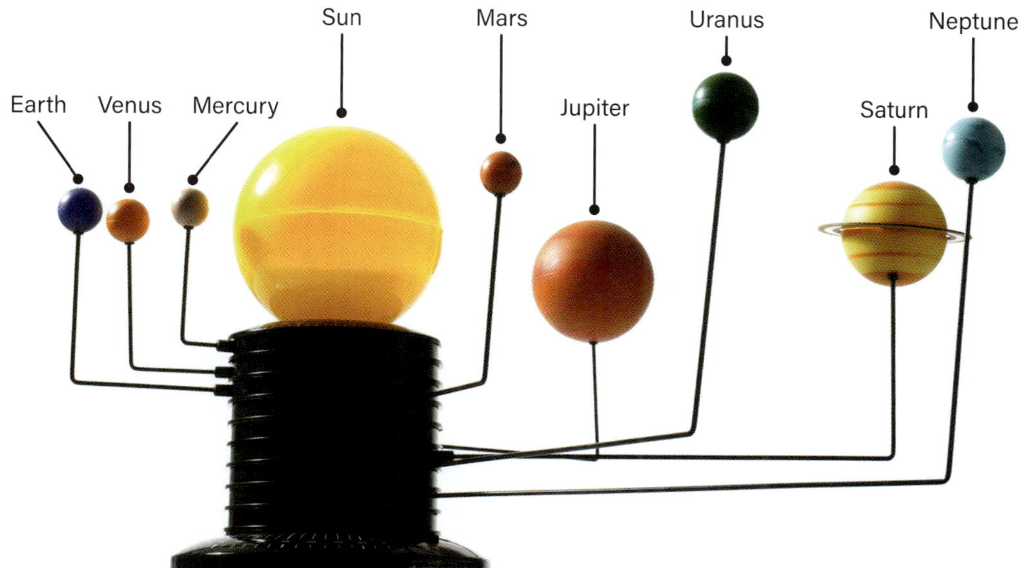

Figure 8.12: This three-dimensional model helps us understand the solar system.

Another well-known scientific model is the components and structure of an atom. Scientists often represent this using a two-dimensional model. Our current model of the atom is the electron cloud model, which shows how negatively charged electrons exist in a 'cloud' around the outside of a nucleus containing protons and neutrons. A common three-dimensional model is the Bohr model, which shows the correct location of the particles in the atom, but does not correctly represent how electrons orbit the nucleus.

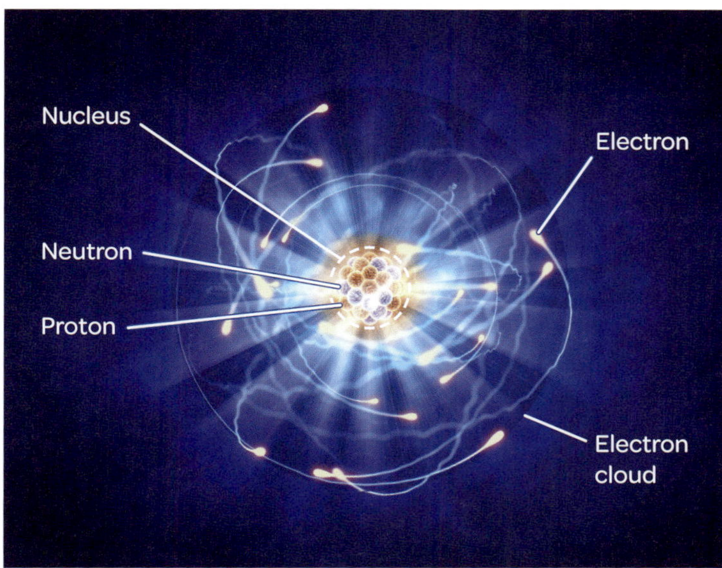

Figure 8.13: Some concepts can be represented with two-dimensional or three-dimensional models, but some representations are more accurate than others!

Learning Ladder

Data science 1

1. Outline a difference between a physical model and a computer model.
2. Describe why it is important to be able to use different types of models.
3. Explain how a model of the solar system shows us information about the arrangement of the planets.
4. Explain one way that using mathematical formulas allows us to make predictions about data trends.

Problem-solving

see page 339

1. Outline one problem with scientific models.
2. Describe how the electron cloud model of atoms solved a problem.
3. Retrograde motion is when planets appear to move backwards compared to others in the solar system. This was a problem for the geocentric model of the solar system, which placed Earth at the centre. Explain how the heliocentric model solved this problem.

In context

Many concepts use models to represent information. What models have you seen? Select a scientific concept and construct a two-dimensional or three-dimensional model to represent it scientifically.

Success criteria

- I can identify different types of scientific models.
- I can describe why scientific models are used.

8·5 ► Using digital models

Figure 8.14: Computer models help scientists to make weather forecasts, by generating images that show the formation and movement of weather patterns.

Computers can generate scientific models to help us represent and analyse large sets of data. By using a computer model, scientists can create data representations that they can manipulate and change, to help us make predictions about the future.

Learning intention

At the end of this lesson, I will be able to identify and analyse how computer models are used to represent data and make predictions.

Key terms

computer model: a model generated using computer software to show a scientific concept

simulations: experimental results generated by a computer model

trend: when data moves in a general direction, usually up or down

variable: a factor in an investigation or a model that can be changed, measured or controlled

Investigation 8.5

Trends in the weather, p.493

Content group: Scientific models

Computer models use data to create simulations

To create a **computer model**, scientists input a set of **variables** into a computer. The computer is able to link large amounts of numerical data to different factors based on complex algorithms it runs. This data is processed by the computer, which can create **simulations** of different phenomena.

Changing variables can alter model predictions

Scientists can alter the variables they enter into a computer model to predict different outcomes. They do this by editing different bits of data or information that the computer processes when it makes the model. For example, when constructing weather models using a computer, scientists input data based on variables such as humidity, temperature, wind speed and rainfall. The computer uses this data to predict upcoming weather patterns.

By changing the data they input, scientists are able to test which collections of variables can result in specific weather patterns – sunny or rainy days, hot or cold weather – and even when events like tropical cyclones are likely to occur.

Weather models can make predictions about the future

By writing mathematical equations into computer code, scientists can use computers to make predictions about future weather patterns, beyond just a 7- or 14-day forecast. By adding in previous weather **trends**, along with current weather patterns, computers are able to generate predictions about weather trends in the future.

Scientists use these models to ask questions about the future, such as:

- Are human factors going to cause a change in global temperature averages?
- Will certain factors affect rainfall trends in future years?

Previous and existing patterns in the weather help us to predict future weather patterns. By changing variables such as the amount of pollution we release into the atmosphere, scientists can use models to predict how global temperatures may change.

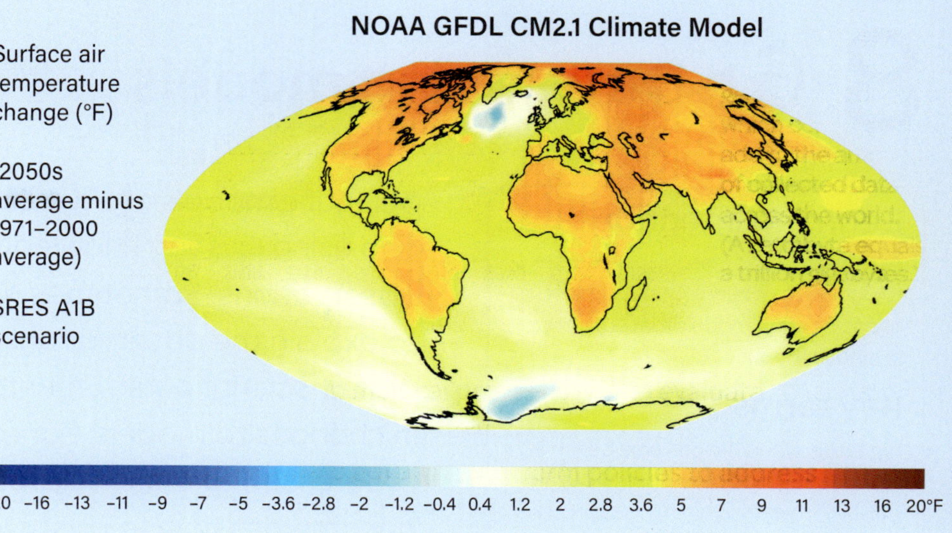

▲ **Figure 8.15:** Computer models can be used to predict future global trends, such as average temperature changes.

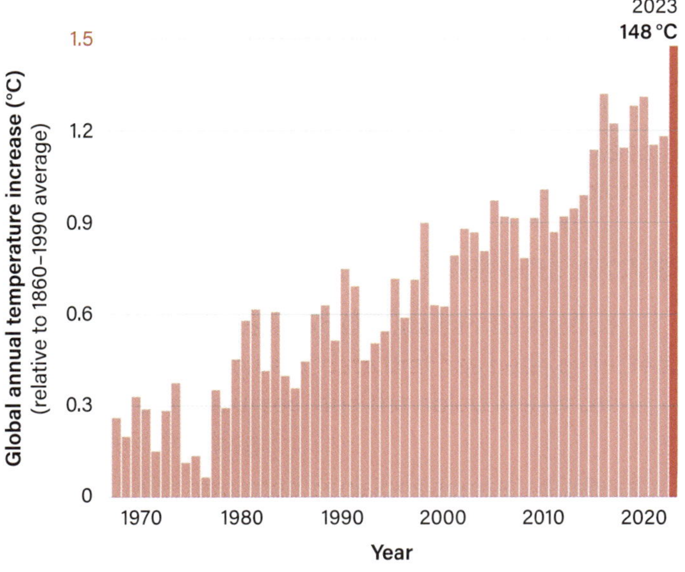

▲ **Figure 8.16:** Computer models can be used to graph patterns in temperature change, and the visual representations can help when predicting trends.

Learning Ladder

Data science 1

1. Outline how computer models can be different from two- and three-dimensional models.
2. Describe the importance of pollution as a variable in weather models.
3. Describe the relationship between the data that is put into a digital weather model and the predictions that can be generated.
4. Explain how a weather model can be used to make predictions about future weather patterns.

Analysing data and information *see page 334*

1. Identify a trend from the graph in Figure 8.16.
2. Describe how this trend could be used in a weather model.
3. Explain what a weather model that included the data in Figure 8.16 may predict about future global temperatures.
4. Draw a conclusion about how human activity may be influencing global temperature patterns.

In context

Many models are used to show global trends. Find an example of another model, and explain what predictions it is able to make about future global trends.

Success criteria

- I can identify a range of computer models and how they can be changed.
- I can analyse how a computer model can be used to make predictions.

8·6 ▸ Developing models

Learning intention

At the end of this lesson, I will be able to outline the observations scientists use to develop models that support scientific theories.

Key terms

big bang: when the universe began as an incredibly hot, dense point roughly 13.8 billion years ago

dark energy: a hypothetical form of energy that expands the fabric of the universe

dark matter: particles that do not absorb, reflect or emit light, so they cannot be detected by observing electromagnetic radiation

scientific theory: an explanation of a natural phenomenon that is supported by evidence and the results of repeated tests

Investigation 8.6

Modelling the big bang, p.494

Content group: Applications of models

Scientific models are developed by making observations and using existing knowledge to explain those observations. Sometimes new knowledge is required to explain the observations. At times, the development of new scientific instruments or technology will result in changes to models. For example, as the technology progressed to make telescopes more powerful, it became possible to observe more of the universe.

Scientists use data to develop models about the universe

As our knowledge of the solar system expanded, scientists wanted to expand their knowledge to understand the universe as a whole. Using different types of telescopes, they were able to make and record more observations. For example, scientists were able to use special microwave detectors to observe cosmic radiation, which contains light and radiation from the **big bang**.

In the 1920s, the scientist Edwin Hubble used telescopes to measure the distance between objects in space, and discovered that they were slowly moving further apart. This data allowed scientists to develop a model showing the formation of the universe. The model showed all matter rapidly expanding from a single point, and continuing to expand to this day. Over time, stars formed from matter created during this initial big bang.

Figure 8.17: The big bang theory is based on models made from data that scientists have collected about the universe.

Models helped develop the big bang theory

Using this model as a basis, scientists were then able to develop the big bang theory. This **scientific theory** states that the universe started at a single point, and exploded outwards from that tiny point around 13.8 billion years ago, and is still expanding today. Without the data from their observations, the formation of their models, and supporting evidence, this theory would not have been possible!

Models continue to develop as new evidence is found

While scientists have found evidence that strongly supports the big bang theory, they continue to collect data on, and build models of, the universe to try to increase our understanding.

In 1990, the North American and European space agencies launched the Hubble Space Telescope (named after Edwin Hubble). In 1998, it provided images to scientists that showed the speed at which the universe is expanding. New telescopes now show that this expansion is speeding up, which is the opposite of what scientists expected!

Needing a model to understand what was happening, scientists created the concepts of **dark matter** and **dark energy** to attempt to explain this departure from the accepted model. Some scientists think that about 80 per cent of the universe is made of dark matter or dark energy. However, their theories are unproven. Dark matter and dark energy are now two of the biggest mysteries in the observed universe.

Figure 8.18: The Hubble Space Telescope helped to provide images of the universe that could be used to make models.

▲ **Figure 8.19:** As technology advances, scientists can look for additional data to support their theories, such as evidence to support the existence of dark matter and dark energy.

Learning Ladder

Data science 1

1. Outline the difference between an observation and a model.
2. Describe why models are important in helping us to understand scientific theories.
3. Summarise ways in which data collected by scientists helps us to explain the universe.
4. Discuss how new data about dark energy could allow us to make more predictions about the universe.

Problem-solving *see page 339*

1. Identify one problem with the big bang model.
2. Outline the data that scientists are looking for to try to solve this problem.
3. Explain the link between the distances between objects in the universe and the big bang theory.

In context

Many scientific theories are based on models. Identify another scientific theory, and discuss a model that provides supporting evidence for it. What data was used to create this model?

Success criteria

- I can describe how observations can be used to develop models.
- I can outline how scientists use models to develop theories.

8·7 ▸ Gathering data

Learning intention

At the end of this lesson, I will be able to:

- construct testable scientific questions
- understand how to collect and analyse scientific data.

Key terms

accuracy: how close a measured value is to the true, exact value; how closely a recorded value matches the expected outcome of an investigation

dependent variable: the thing that is measured in a first-hand investigation

experimental data: information collected by conducting an investigation and recording the results

formulate: to invent something, usually in scientific or mathematical terms

independent variable: the thing that is deliberately changed in an investigation

mean: a measure of centre (an average) calculated by adding all the numbers together and dividing by how many numbers there are

range: in a set of numbers, a measure of spread between the highest number and the lowest number

reliability: the extent to which the results or measurements can be reproduced when the research is repeated

Investigation 8.7A

Reaching great heights, p.495

Investigation 8.7B

Height and arm span, p.496

Content group: Collecting, using and analysing datasets

Scientists use data to understand the world around them. However, first they must design investigations to collect useful data. Scientific investigations can be designed by asking specific questions.

Scientific questions form the basis of investigations

Scientists use questions to study our world. They **formulate** scientific questions that they can then investigate. Sometimes scientific questions can be investigated by making observations, while other times an experiment is conducted. A scientific question that can be investigated is written using a specific structure.

They usually ask something like, 'Does changing one factor affect how another factor can be measured?' In this question, the changed factor becomes the **independent variable** of the investigation, and the measured factor becomes the **dependent variable**. You can construct your own scientific questions by following this general format. For more information on constructing scientific questions, see page 321 of the Science how-to section.

For example, a scientific question could be: 'Does the number of times a coin is tossed increase the number of times it lands on tails?'

Data can be collected from investigations

Once a question has been asked, it can be tested through investigation. This allows scientists to collect and record **experimental data**, both qualitative and quantitative. Once multiple pieces of the same data have been collected, scientists are able to calculate values like the **mean** and **range** of the data. This data would be recorded in a scientific table.

From the question above, a coin might be tossed 5 times, then 10 times and finally 15 times. This process could then be repeated 5 times for each number of coin tosses. Table 8.2 shows some of the data that could be collected during this investigation.

Table 8.2: Coin toss investigation results table

Number of coin tosses	Number of times the coin landed on tails					
	Trial 1	Trial 2	Trial 3	Trial 4	Trial 5	Mean
5	2	3	2	2	4	2.6
10	3	6	7	5	4	?
15	11	7	5	10	6	?

We can then calculate the mean and range of this data. For example, in the five-toss investigation, the:

- range is: $4 - 2 = 2$
- mean is: $\frac{2+3+2+2+4}{5} = 2.6$

For more information on tabulating data and calculating the mean and range, see page 363 of the Science how-to section.

Data must be accurate and reliable

Experiments must be accurate and reliable. **Accuracy** means how close the results of an investigation are to the expected outcome. When tossing a coin, there are only two options that can occur: heads or tails. So the expected outcome (the number of tails) is half of the number of tosses. For five tosses, this would be 2 or 3 tails per trial. Accuracy is achieved by using precise measurements and specific equipment. For more information on accuracy, see page 338 of the Science how-to section.

An investigation is reliable if you conduct an experiment multiple times and achieve the same or very similar results each time. In the case of the coin toss results shown in Table 8.2, conducting five trials and then calculating the mean allows us to see if the data collected is reliable. Accuracy and **reliability** are discussed in more depth in the Science how-to section (see page 338).

Figure 8.20: Questioning forms the basis of all scientific investigation.

Figure 8.21: A coin toss

Learning Ladder

Data science 1

1. Identify two types of data that can be collected after asking a scientific question.
2. Describe how collecting data from investigations can help us to answer scientific questions.
3. Outline how reliability may affect data collection in an investigation.
4. Explain how the data from a coin toss investigation would help you to predict how many times tails would come up, on average, if a coin was tossed 20 times.

Problem-solving

see page 339

1. When conducting an investigation, what is one problem that would affect its reliability?
2. Propose one way you could make sure that the coin toss investigation is accurate.
3. Explain the relationship between first-hand investigations, accuracy and reliability.
4. You are going to conduct an investigation to look at the link between plant growth and exposure to sunlight. Propose one way you could make sure your investigation was accurate, and one way you could make sure it was reliable, based on the information on them in this section.

In context

Being able to calculate the mean and range of data is important. Using the data in Table 8.2, calculate the mean and range for the 10 and 15 coin tosses, and predict how many times, on average, you would expect tails to occur with 50 tosses. Give a reason for your prediction.

Success criteria

- I can construct a scientific question.
- I can see that the more times an experiment is conducted, the greater is its reliability.

8.8 ▸ Analysing data

Learning intention

At the end of this lesson, I will be able to analyse collected data.

Key terms

causation: a change in one factor being the reason for a change in another factor

correlation: a relationship between two factors or variables that shows them changing together, in either the same or opposite way

negative correlation: as one variable increases, the other decreases

positive correlation: as one variable increases, so does the other

validity: whether an investigation measured what it intended to measure

Investigation 8.8

Analysing rainfall, p.497

Content group: Collecting, using and analysing datasets

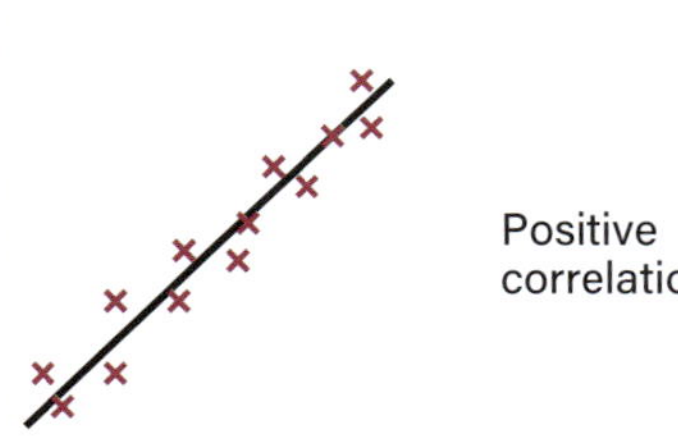

Positive correlation

Negative correlation

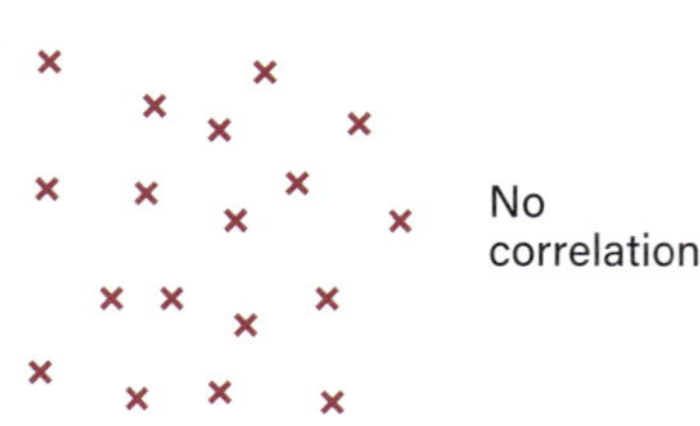

No correlation

▲ **Figure 8.22:** Scatter plots can be used to show relationships, such as a correlation between two factors.

Data can be collected from multiple sources. Scientists obtain data by conducting experiments first-hand, by observing a phenomenon, or by using physical or computer-generated models. Once data is collected, scientists analyse it to understand their findings.

Data can be graphed to look for patterns

When conducting an investigation, we can collect and record the data in a table. We can then graph the data to help us see if there are any trends.

In a scatter plot graph, we do this by placing points on a graph and then looking for a trend. For example, if the points form a straight line on the scatter plot, then there could be a **correlation**. There are three main possibilities:

- A **positive correlation**: as one variable increases, the other variable also increases; for example, age and weight in babies: as they get older, their weight increases.
- A **negative correlation**: as one variable increases, the other decreases; for example, rainfall levels and water usage: the more it rains, the less people water their gardens.
- No correlation: there is no link between variables; for example, your height and your favourite football team: there is no connection between how tall you are and who you support.

It is important to note that correlation does not always mean **causation**. Two factors can appear to be related, but that does not mean that one factor changing is the reason for the change in the other. We showed this with Figure 8.10 on page 287, where shark attacks and ice-cream sales have a positive correlation, but they do not cause each other. This is why it is also important to consider the investigation method.

Figure 8.23: There is a negative correlation between rainfall levels and water usage – the higher the rainfall, the less we water our gardens.

Results and methods can be analysed

In Section 8.7, we discussed the accuracy and reliability of an investigation as important factors to consider when looking at the quality of data. Another factor to consider is **validity**.

For an investigation to be considered valid, it must be a fair test. This means that all variables of the investigation must be controlled, except for the *one* independent variable that is changed. By doing this, scientists are able to show that any changes to the dependent variable must be *caused* by the independent variable, not just related to it.

When analysing the results of your investigation, look at whether your method was valid. Did you conduct a fair test? You can plot your results on a graph to analyse your results generally and to identify any patterns or trends. For more information on validity and graphing your results, see pages 338 and 368 of the Science how-to section.

Collected data can be compared to predictions

Finally, once your data and investigation have been analysed, it can be used to make an inference and draw a conclusion for the investigation. In the conclusion, scientists look at whether data is consistent with the initial predictions made at the start of the investigation. Data that aligns is said to support an experimental hypothesis.

Learning Ladder

Data science 1

1. Outline the difference between data with a positive correlation and data with a negative correlation.
2. Give an example of two variables with:
   a positive correlation. **b** negative correlation. **c** no correlation.
3. There is a positive correlation between hours of study and performance in examinations. Explain what this statement means.
4. Explain how data with negative correlation can be used to make predictions.

Analysing data and information

see page 334

1. Identify a trend in the rainfall data in the table that you can observe.
2. Describe any correlations you can see based on the trend you identified.
3. Explain the relationship between the number of rainy days and the amount of rainfall.
4. Based on the information in the table and in this section, draw a conclusion about which month people would spend the most time watering their gardens.
5. Based on the information from the table, can you assess the reliability of the data? Justify your answer.

In context

Once data has been collected and analysed, conclusions can be drawn. Use a previous investigation you have conducted in class, and analyse the data for its accuracy, reliability and validity. For each factor, suggest one way that it could be improved.

Success criteria

- I can analyse data from scientific investigations.
- I can determine whether data supports or does not support a hypothesis.

Weather in Penrith, 2023

Month	Total rainfall (mm)	Number of rainy days
January	97.4	17
February	56.4	9
March	46.8	13
April	88.0	15
May	4.8	3
June	14.2	15
July	8.2	12
August	40.8	11
September	7.0	6
October	15.6	7
November	70.2	14
December	114.8	11
Averages		
Mean	47.0	
Median	43.8	
Mode	none	
Range	110	

8·9 ▸ Data science 1 in context

Learning intention

At the end of this lesson, I will be able to create a model to explain an observable phenomenon.

Key terms

model: a physical or mathematical representation of a scientific idea or concept

phenomenon (plural **phenomena**): an observable fact or event

Investigation 8.9

Modelling the yowie, p.498

Content group: Data science 1 in context

Across this chapter, we have looked at how scientists gather, record, analyse and use data. One of the most important uses of data is in constructing models that can be used to explain scientific phenomena.

Data is collected from observations and measurements

In Chapter 1, we looked at how scientists make and record observations using a variety of scientific equipment. This is the first step towards making a **model** to explain a particular **phenomenon**. By making and recording observations, scientists put together data that forms the basis of their models.

For example, since early human history, people have observed that it is easier to lift and move heavy objects with assistance from machines. Early humans made observations that using a lever (a simple machine) made it easier to move heavy objects.

Models are constructed from data

Scientists use the observations they have made and the quantitative and qualitative data they have collected to build models. They can create two-dimensional and three-dimensional models of different concepts. Sometimes, when they have collected large amounts of data, scientists use computers to create digital models.

For example, using observations and data collected about levers, humans were able to construct two-dimensional sketches of a machine that could be used to throw heavy rocks and objects. These sketches were then built into machines that could be tested. We know these machines as catapults.

Observable phenomena are explained using models

The models scientists build can then be used to explain observable phenomena. These phenomena can be from many different branches of science. The models allow scientists to present ideas about phenomena so that many different audiences can understand these complex concepts.

Figure 8.24: Catapults are levers that can move heavy objects large distances.

For example, scientists continue to construct and test different models of catapults, to look at designs that can increase the distance an object can travel. They construct labelled models of these designs, to explain *why* they are able to move objects across these distances.

The model in Figure 8.25 shows a catapult. When the restraining rope is released, the arm moves upwards until it hits the frame. This force causes the object in the bucket to keep moving through the air until gravity pulls it back down to the ground.

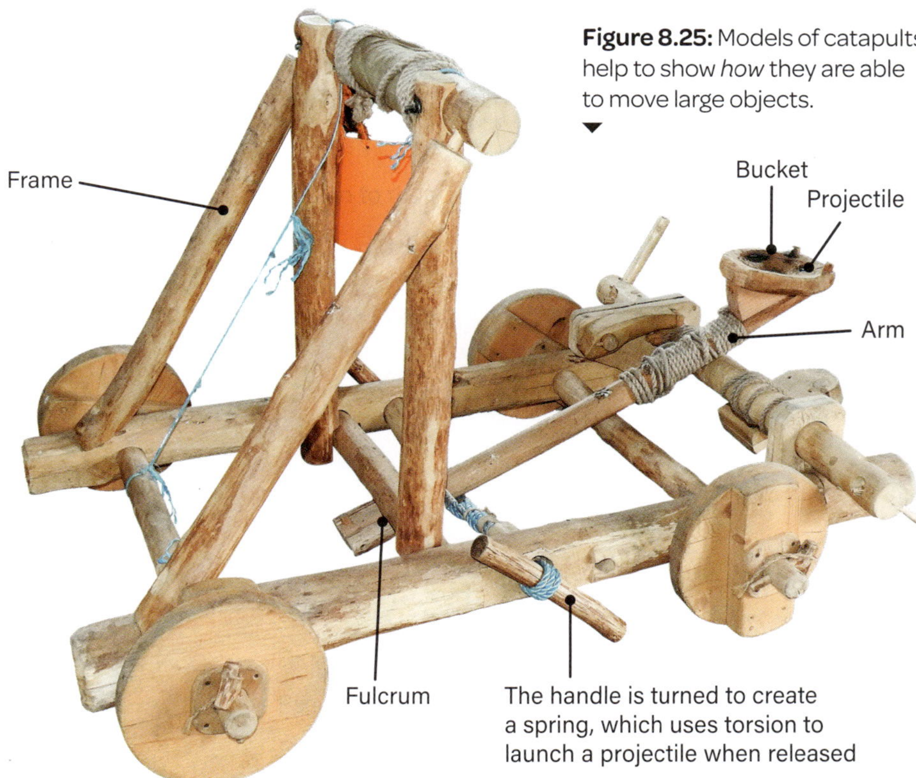

Figure 8.25: Models of catapults help to show *how* they are able to move large objects.

Learning Ladder

Data science 1

1. Identify the two types of data that can be found in models.
2. Describe why observational data is important when constructing models.
3. Consider the model of the catapult in Figure 8.25. Explain the link between the device and the movement of the object it throws.
4. Explain how an understanding of observations relating to catapults could allow someone to predict the movement of an object that is placed in a catapult's bucket.
5. Research catapult designs, and construct a labelled model that shows how another type of catapult is able to move heavy objects.

Problem-solving

see page 339

1. What is one problem with scientific models?
2. You have created a two-dimensional model of a catapult. Suggest a problem that you could have encountered in making the model. How would you resolve it?
3. Explain how a model you identify helps you to understand a specific scientific phenomenon.

Success criteria

- I can explain how scientific models are created.
- I can create a model that explains a scientific concept or phenomenon.

Data science 1 summary

Data

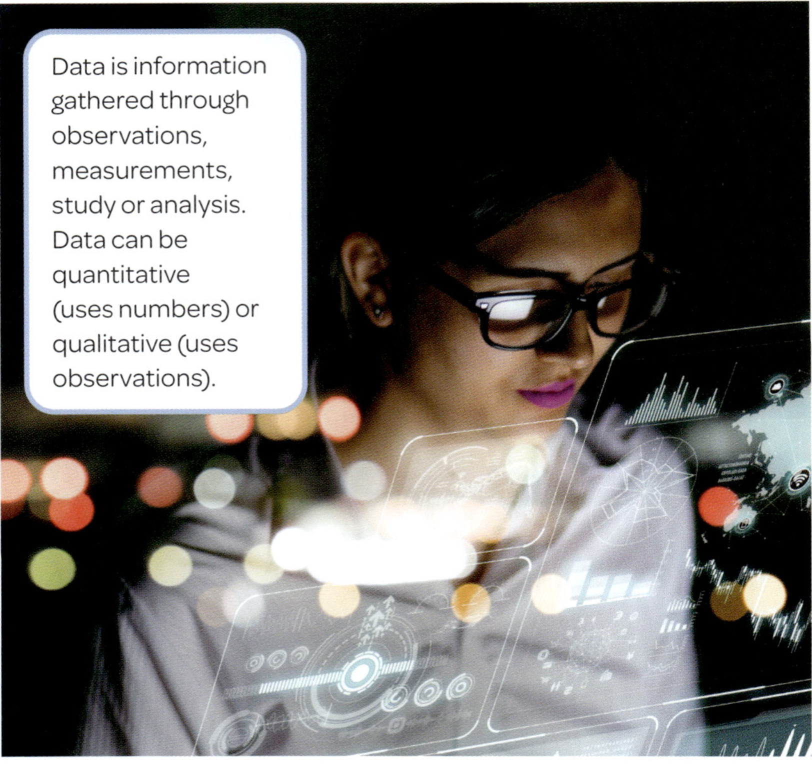

Data is information gathered through observations, measurements, study or analysis. Data can be quantitative (uses numbers) or qualitative (uses observations).

- Data can come from a range of sources, including investigations, websites and large datasets.
- When you go online, companies can access data about you, creating your digital footprint. You need to be careful about the data you share online, to prevent your information being misused.

Data science is interdisciplinary and combines multiple branches of knowledge. Scientists can use data to extract insights about phenomena and draw conclusions.

Scientific models

- Inquiries can be scientific or non-scientific. For an inquiry to be scientific, data must be collected using a scientific approach and used as supporting evidence for a particular phenomenon.
- Scientists use models to display data in many ways, including posters, three-dimensional representations, computer simulations and mathematical formulas.

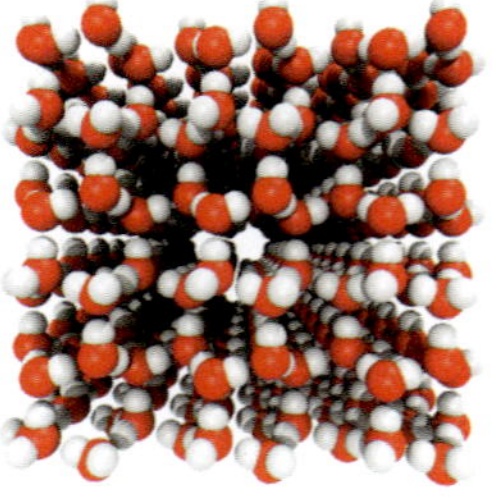

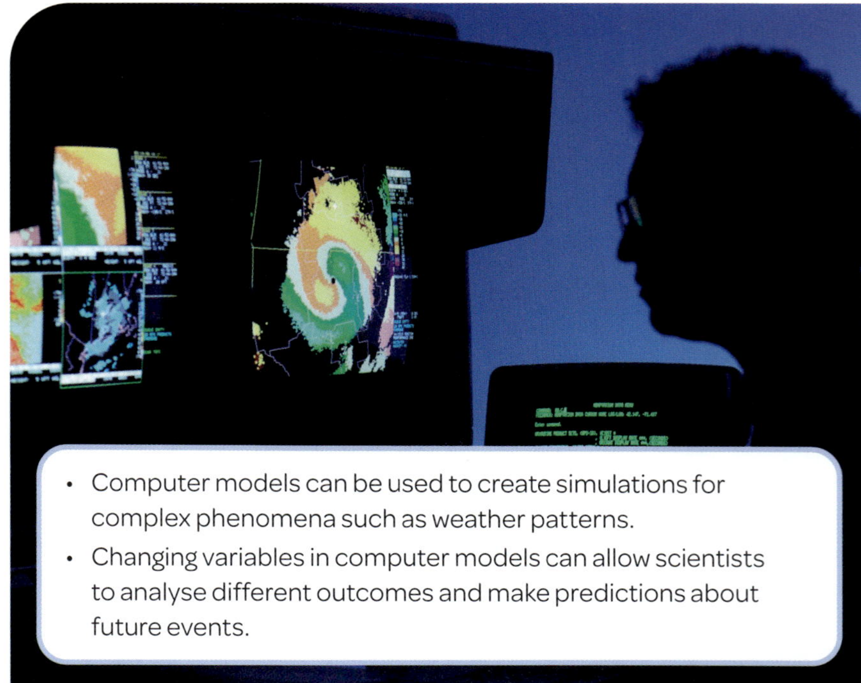

- Computer models can be used to create simulations for complex phenomena such as weather patterns.
- Changing variables in computer models can allow scientists to analyse different outcomes and make predictions about future events.

Applications of models

Scientists collect and use quantitative and qualitative data to create models.

Annual global data size

175 ZB

Zettabytes: 0, 20, 40, 60, 80, 100, 120, 140, 160, 180

Year: 2020, 2021, 2022, 2023, 2024, 2025

Models developed from data are then used to create theories, such as the big bang theory.

Collecting, using and analysing datasets

Scientific questions must be developed to conduct investigations and gather data.

- Looking at the range of data from an investigation and comparing it with data collected from other similar investigations can make data more accurate.
- Data from small groups can be compared to create a larger dataset, which helps to show if consistent patterns have been found that support a hypothesis.

Repeated trials in an investigation allow us to generate an average from our data, making it more reliable.

Masterclass

Steps in progression → 1 → 2

		1	2
Content	**Data science 1**	Identify some sources of data that could be used to train AI tools like Meta AI.	Meta AI interacts with us as users. Describe two reasons why data from Australia would be relevant to Meta AI.
Processes	**Analysing data and information**	Identify two relevant trends from the graphs in Figure 8.27.	Consider the data on the phone in Figure 8.26 that shows Facebook usage time. Describe any trends you can see
	Problem-solving	What is one problem with agreeing to the terms and conditions of social media platforms without reading them?	Outline one way to improve your digital safety when using social media platforms such as Facebook and Instagram.

Figure 8.26: Our devices can now show us how much time we are spending online and on different social media apps.

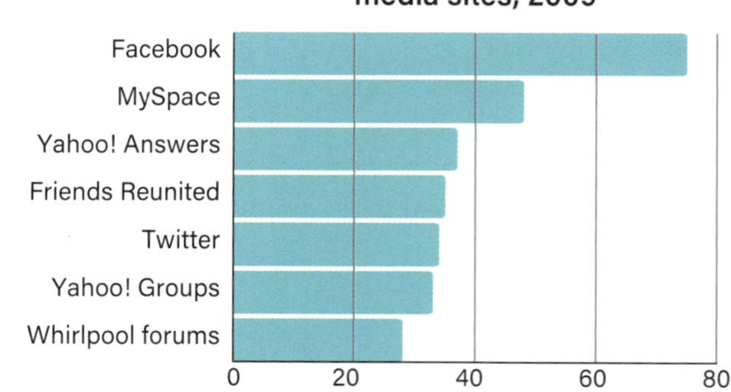

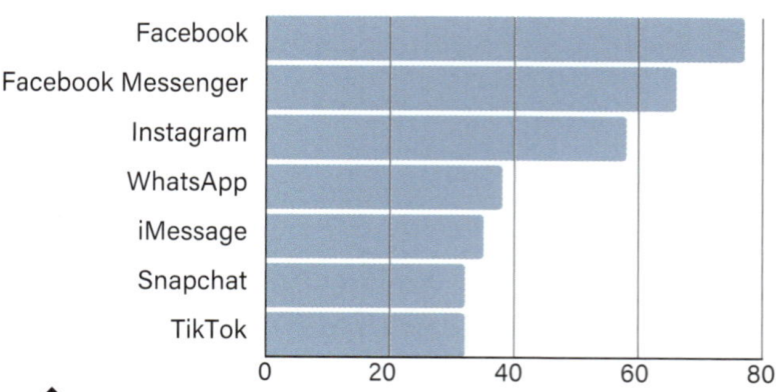

Figure 8.27: These graphs show the most popular social media platforms in Australia in 2009 and 2022.

Demonstrate your understanding

3	4	5	
What changes can we see to Australian social media usage over time? Consider both the written and graphical information in your response.	How could you use the data from the two graphs in Figure 8.27 to predict what Australian social media trends might look like over the next 20 years?	Use further online data on Facebook usage to construct a graph of user trends over the last 10 years. How effectively could you use this to predict what future usage might look like? Evaluate this method of predicting social media usage.	
Suggest a reason for the changes to the data in the graphs in Figure 8.27.	What conclusion could you draw from the data on the phone in Figure 8.26? How did you come to this conclusion?	Consider the data in the graphs in Figure 8.27. Evaluate the data for its accuracy, reliability and validity, based on what you see on the page, and what you can find online.	Science how-to 334
Suggest a link between the shift in time and the shift in social media platform usage. Were there any specific problems that may have caused this shift to occur?	'Teenagers should be unable to access any social media before the age of 16 to prevent them from revealing their personal information online.' Consider the above statement, and propose two solutions to solve the problem of teenagers' personal data being accessed, without increasing platform age limits.	Evaluate each of your identified strategies in Question 4, based on how widely they could be distributed and how easily they could be implemented by all teenagers.	Science how-to 339

Digital footprints used to train Meta AI

In 2021, the social media company Facebook was rebranded as Meta. Owners of the company stated that this shift was to 'bring the metaverse to life and help people connect, find communities and grow businesses'.

As part of this shift, Meta released a generative AI product, Meta AI, in 2023. This product has since been upgraded several times, and is now running on third-generation software. But how has this AI program been trained?

It turns out that our digital footprints are helping to train it! Meta has been using data from Facebook and Instagram accounts to help train the device on how to interact with users and search for information. People's posts can even help the software to learn skills such as image generation, by looking at uploaded photos and paintings from users. The software can look at any publicly available posts from Australian users and add them to its data bank. The reason for this is the user agreement all of us have to accept when signing up to the platforms. When you agreed to the terms and conditions of use, you agreed to let Meta use your data to train its AI tools.

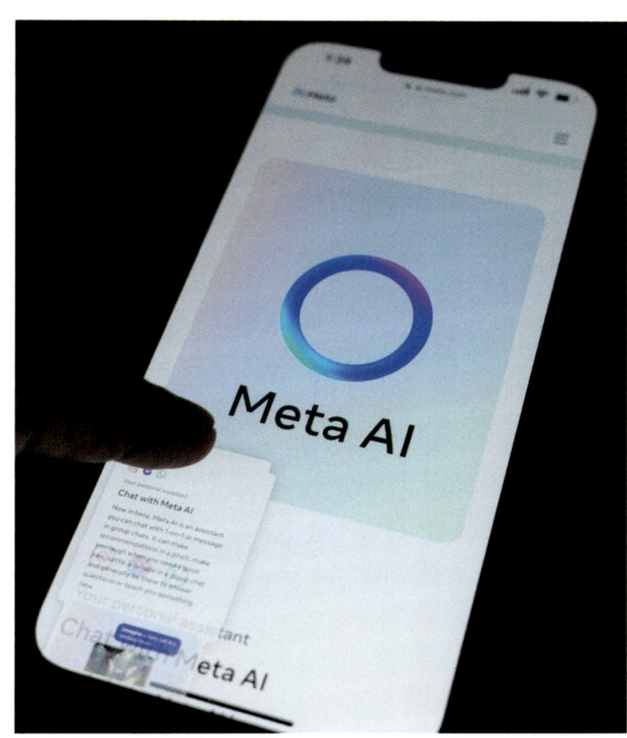

Figure 8.28: According to Meta's marketing, users of social media can use Meta AI to help them with finding information and grow their businesses.

Science how-to

Science skills are the tools you can apply when investigating the world around you. Learning how to use these skills will help you to understand scientific processes, ask relevant questions and communicate what you discover.

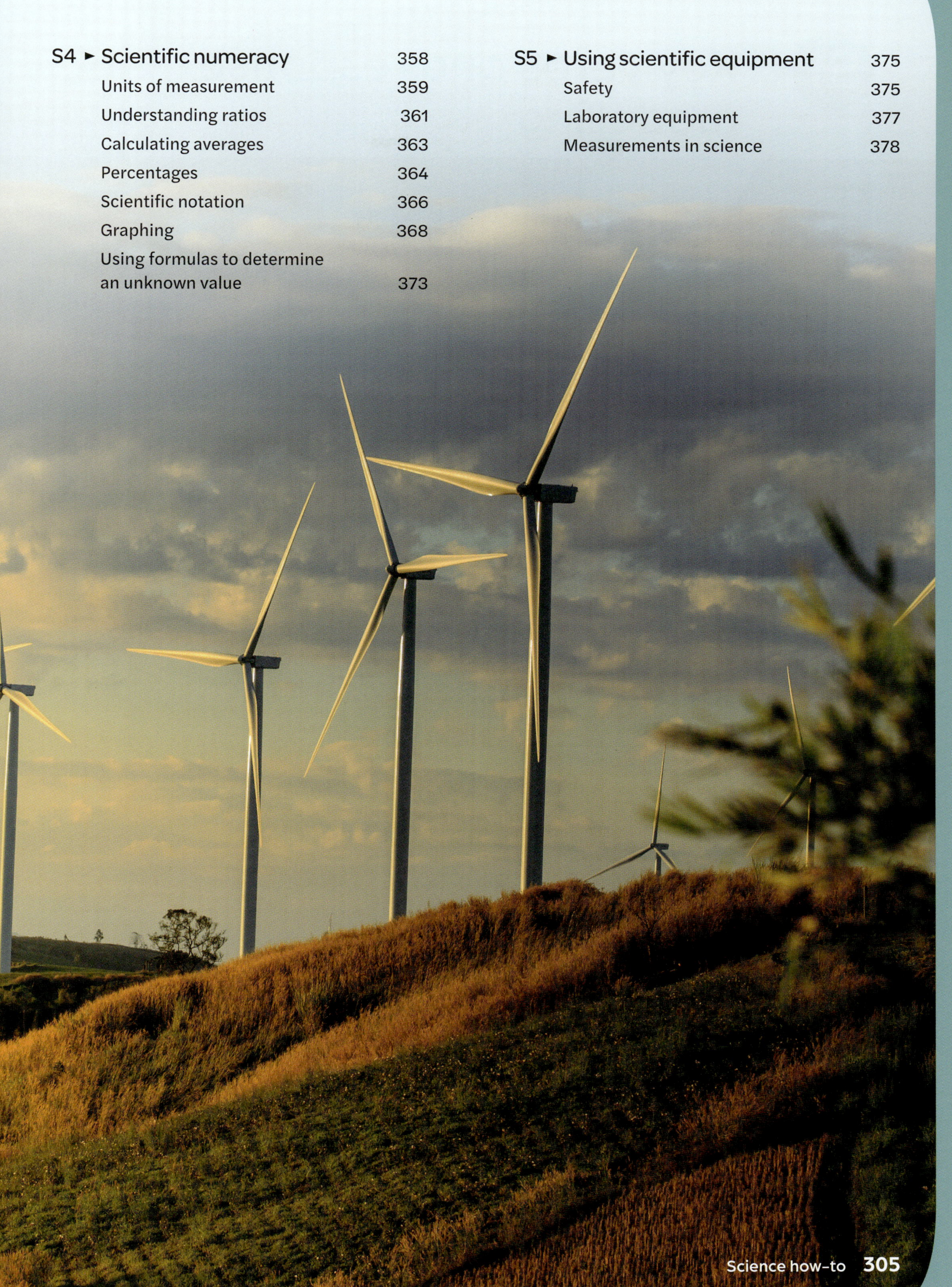

Key verbs and response modelling

Key verbs are the instruction words in a question. They tell you the approach to use when you answer.

The key verbs are listed alphabetically over the following pages. We have also included an example of how they would be used in a question and a sample response.

Account for	
Explanation	State reasons for; report on.
Sample question	Account for the movement of a ball down a slope.
Sample response	*A ball moves down a slope as a result of unbalanced forces. The force of gravity pushes down on the ball, causing it to accelerate down the slope. The force of friction between the ball and the slope is smaller than the force of gravity, so friction does not stop the ball from moving.*

Analyse	
Explanation	Identify components and the relationship between them. Draw out and relate implications.
Sample question	Analyse the link between Earth, the Sun and the seasons.
Sample response	*Earth's movement is responsible for the changing seasons. Earth is tilted on its axis. This means that during part of Earth's orbit around the Sun, the lower half of Earth is tilted closer to it. Being tilted towards the Sun causes the southern hemisphere to receive more direct sunlight and experience summer, while the northern hemisphere, receiving less sunlight, experiences winter.* *During the opposite part of Earth's orbit, the upper half of Earth is tilted towards the Sun. The tilt causes the northern hemisphere to experience summer, while the southern hemisphere experiences winter.* *Earth moves between these positions in its orbit, causing the tilt to be 'side on' to the Sun. The other seasons – spring and autumn – occur when Earth is moving from one hemisphere being tilted towards the Sun to the other being tilted towards the Sun.*

Apply

Explanation	Use, utilise, employ in a particular situation.
Sample question	Apply your understanding of Linnaean classification to the following organisms: • orca • ring-tailed possum.
Sample response	*The Linnaean classification system classifies organisms from broad to specific categories. The organisms above would be classified as follows:*

Category	Orca classification	Ring-tailed possum classification
Domain	Eukarya	Eukarya
Kingdom	Animalia	Animalia
Phylum	Chordata	Chordata
Class	Mammalia	Mammalia
Order	Cetacea	Diprotodontia
Family	Delphinidae	Pseudocheiridae
Genus	*Orcinus*	*Pseudocheirus*
Species	*orca*	*peregrinus*

Assess

Explanation	Make a judgement of value, quality, outcomes, results or size.
Sample question	Assess the effectiveness of using solar energy to reduce household power costs.
Sample response	*Solar energy is energy from sunlight that is converted by panels into usable electrical energy.* *For:* • *Using solar energy can reduce or remove a household's reliance on obtaining electrical energy from non-renewable power sources.* • *After the initial panel installation, solar energy does not cost anything to obtain, which means households will save money.* • *Households can get credit for any unused solar electricity they send back to the grid, which further reduces their bills.* *Against:* • *Installing solar panels is expensive. To completely remove dependence on the power grid, householders need also to install expensive solar batteries to store the energy.* • *Solar energy requires regular sunny weather to function effectively. Households may have minimal power after a series of cloudy or rainy days.* *Assessment: Even though it costs a lot to set up at the start, solar power is highly effective at reducing household power costs. This is because power bills are much lower when you use solar energy.*

Calculate

Explanation	Ascertain/determine from given facts, figures or information.
Sample question	Conduct a first-hand investigation to determine the distance travelled by a model car with wheels made from milk bottle caps. Calculate the average distance travelled over three trials.
Sample response	*The bottle cap wheel travelled 11 cm in trial 1, 9 cm in trial 2 and 13 cm in trial 3.* *This means the average distance travelled = (11 + 9 + 13) ÷ 3 = 11 cm.*

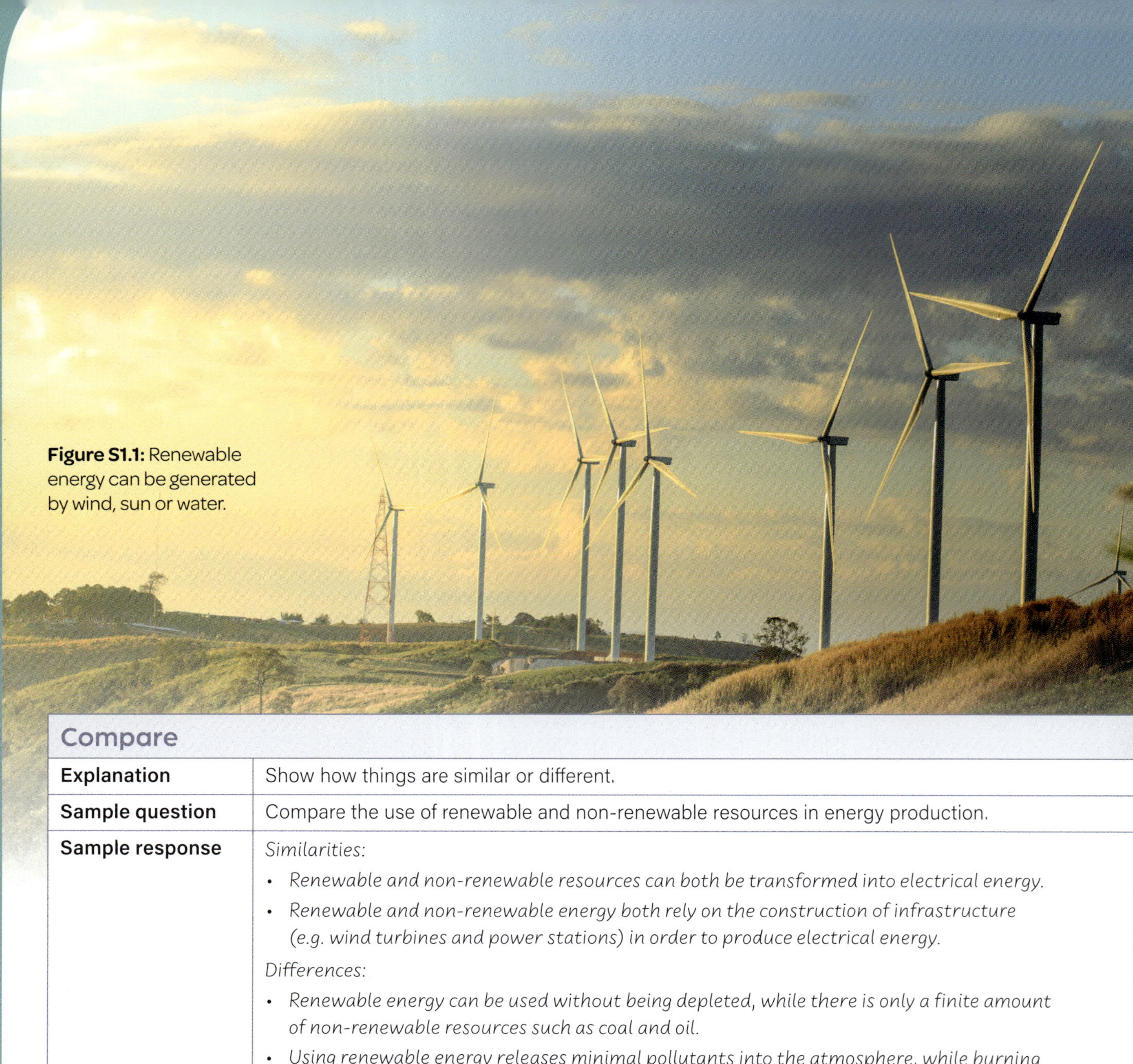

Figure S1.1: Renewable energy can be generated by wind, sun or water.

Compare	
Explanation	Show how things are similar or different.
Sample question	Compare the use of renewable and non-renewable resources in energy production.
Sample response	*Similarities:* • *Renewable and non-renewable resources can both be transformed into electrical energy.* • *Renewable and non-renewable energy both rely on the construction of infrastructure (e.g. wind turbines and power stations) in order to produce electrical energy.* *Differences:* • *Renewable energy can be used without being depleted, while there is only a finite amount of non-renewable resources such as coal and oil.* • *Using renewable energy releases minimal pollutants into the atmosphere, while burning non-renewable energy releases pollutants and greenhouse gases such as carbon dioxide.*

Construct	
Explanation	Make. Build. Put together items or arguments.
Sample question	Construct a poster, diagram or presentation to show your understanding of the solar system.
Sample response	Different versions of responses to this question are shown in 'Presenting scientific information' on pages 349–353.

Define	
Explanation	State the meaning and identify essential qualities.
Sample question	We learnt about forces in Chapter 2. Define the term 'balanced force'.
Sample response	*When forces of equal strength are pushing on an object in opposite directions.*

Describe	
Explanation	Provide characteristics and features.
Sample question	A solvent is a substance, usually a liquid, that can dissolve another substance to make a solution. Describe how water acts as a solvent.
Sample response	*Water is a liquid that is able to dissolve the particles of other substances (solutes); it can become a solution with those dissolved particles by breaking their chemical bonds.*

Discuss	
Explanation	Identify issues and provide points for and/or against.
Sample question	Discuss the use of non-renewable energy such as coal for large-scale electrical energy generation.
Sample response	*Coal has historically been the most widely used resource for power generation across Australia. It can provide reliable energy on a large scale, but using it has advantages and disadvantages.* *For:* • *Infrastructure in place at current power stations is set up to use and burn coal, meaning that it is an accessible and cheap method of providing power to many homes.* • *Most households are set up to receive electrical energy generated from already existing grids and power stations.* *Against:* • *Burning coal releases large amounts of carbon dioxide into the environment, contributing to global warming. It also produces particles and pollutants, such as sulfur dioxide, which cause acid rain.* • *Coal is a non-renewable fossil fuel, meaning that if we continue to use it to generate energy at a large scale, we will eventually run out of this resource.*

Distinguish	
Explanation	Recognise or note/indicate as being distinct or different from. To note differences between.
Sample question	Solids, liquids and gases are discussed in Chapter 4. Distinguish between liquids and gases.
Sample response	*Liquids contain vibrating particles that are next to one another and are able to 'slide' past one another. They will take the shape of any container they are put into. Gases have particles that vibrate and move around separately from one another, taking up the entire space of the container they are in, with large gaps between the particles.*

Evaluate	
Explanation	Make a judgement based on criteria. Determine the value of.
Sample question	Evaluate the use of wind energy as a power source taking into account levels of pollution and reliability of the energy source.
Sample response	*Wind energy is a renewable form of electricity that uses turbines to transform kinetic energy into electrical energy.* *For:* • *Wind energy is renewable, meaning it will not run out and it can be used continually.* • *Wind energy is 'clean' – it does not release unwanted pollutants into the atmosphere.* • *In favourable conditions, a single wind turbine is capable of supplying the power demands of over 300 homes.* *Against:* • *Current infrastructure does not support the wide-scale use of wind turbines, and constructing them is expensive.* • *Wind turbines rely on environmental conditions – they need to be constructed in windy areas that are obstruction free, so that they can spin and generate electrical energy.* *Judgement: Using wind energy as a power source is highly effective. While upfront costs to set up wind farms are high, the benefits outweigh these costs as they can power multiple homes and they reduce the amount of pollution in the atmosphere.*

Explain	
Explanation	Relate cause and effect. Make the relationships between things evident. Provide why and/or how.
Sample question	Explain how energy moves through an ecosystem. (Ecosystems are explained in Chapter 5.)
Sample response	*Energy is able to move through an ecosystem via living things. 'Producers' such as plants use energy from sunlight to produce glucose, which can be used as chemical energy. When a 'consumer', such as a grasshopper, eats a plant, it takes in and uses the plant's energy. Then, a consumer like a bird eats the grasshopper, and the energy is transferred again. When the bird dies and lands on the surface of the ground, decomposers such as bacteria break it down using the energy from the bird and form nutrients for plants to absorb while they are growing. This cycle continues, providing energy to all organisms in the ecosystem.*

Extrapolate	
Explanation	Infer from what is known.
Sample question	Looking at the graph, extrapolate what will happen when the temperature of the water is 70 °C.
Sample response	*The graph shows that the amount of salt that can be dissolved by the water increases as the temperature increases. At 70 °C, the water should dissolve 35 g of salt, based on the trendline of the graph.*

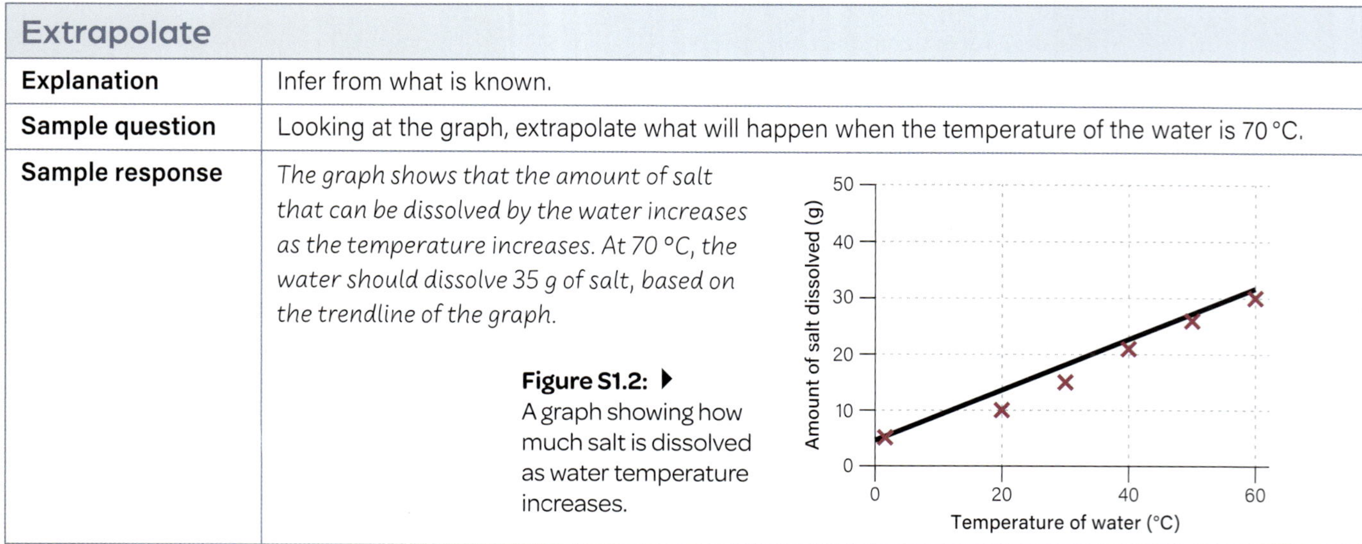

Figure S1.2: ▶ A graph showing how much salt is dissolved as water temperature increases.

Identify	
Explanation	Recognise and name.
Sample question	As discussed in Chapter 2, identify the term used to describe a push, pull or twist.
Sample response	*Force.*

Investigate	
Explanation	Plan, inquire into and draw conclusions about.
Sample question	n/a
Sample response	Investigating is covered in the 'Writing investigation reports' section on page 348.

Justify	
Explanation	Support an argument or conclusion.
Sample question	Justify the use of water filtration in waste management.
Sample response	*Water filtration should be used in waste management. Filtering water means we can easily remove large particles that are not dissolved. This means that waste products such as plastics can be removed from waterways before the water is reintroduced to the environment.* *Filtering water can also remove tiny particles of sewage from waste. Using pumps, sewage is sent through microfilters that are capable of removing tiny solids and even some harmful bacteria. This improves water quality and stops harmful agents from being released into the environment.*

Figure S1.3: Water filtration systems remove impurities and waste products.

Outline	
Explanation	Sketch in general terms; indicate the main features of.
Sample question	Outline what happens to the particles of a substance as it changes from a solid to a liquid.
Sample response	*The substance is melting – heat energy is being absorbed, causing the particles of the substance to vibrate faster and move further apart.*

Figure S1.4: We can test how sunlight affects plant growth.

Predict	
Explanation	Suggest what may happen based on available information.
Sample question	Predict the effect of sunlight on the growth of tomato plants.
Sample response	*If a tomato plant receives a lot of sunlight, then it will grow taller than tomato plants that receive little light.*

Propose	
Explanation	Put forward (for example, a point of view, idea, argument, suggestion) for consideration or action.
Sample question	Propose a reason for wearing safety glasses when using a Bunsen burner to heat water.
Sample response	*When water heats up, it starts to boil and changes state from a liquid to a gas. This means that it bubbles and may splash out of its beaker. Wearing safety glasses prevents the water from splashing into your eyes and causing damage.*

Recall	
Explanation	Present remembered ideas, facts or experiences.
Sample question	Recall the common characteristics of all living things.
Sample response	*All living things have the following characteristics:* • *They are made of cells.* • *They are made of molecules containing carbon.* • *They are able to grow, move and reproduce.* • *They are able to respond to stimuli.*

Summarise	
Explanation	Express, concisely, the relevant details.
Sample question	Summarise the two groups that animals can be classified into.
Sample response	*The kingdom of animals can be classified into two groups. One is vertebrates, which includes all animals that have a backbone, such as humans. The other is invertebrates, and this group consists of all animals that do not have a backbone, such as insects.*

Working scientifically

Key terms

accuracy: how close a measured value is to the true, exact value; how closely a recorded value matches the expected outcome of an investigation

correlation: a relationship between two factors or variables that shows them changing together, in either the same or opposite way

data: a collection of information gathered through observations, measurements, study or analysis

data point (datum): a single identified element in a dataset

dataset: a collection of data, often from numerous trials related to a single factor

dependent variable: the thing that is measured in a first-hand investigation

ethical: in science, minimising harm to those involved, and ensuring investigations are conducted honestly and data is collected and recorded accurately

evaluate: judge value based on scientific evidence

fair test: a test where all variables are kept the same, except for the independent variable and the dependent variable

hazard: something that can harm living things, objects or the environment

hypothesis: a suggested explanation or prediction of a scientific problem that can be tested with an investigation

independent variable: the thing that is deliberately changed in a first-hand investigation

pattern (data): when data repeats in a predictable way

plausible: could be reasonably accepted based on available evidence

precision: how close measured values are to each other within a dataset

relationship: a link between two factors

trend (data): when data moves in a general direction, usually up or down

valid: when an investigation meets its intended purpose

variable: a factor in an investigation or a model that can be changed, measured or controlled

Science has its own processes that help us question and test how the world around us works. Scientists use these skills to conduct investigations and test their ideas. Working scientifically allows you to collect information and test your ideas to discover how different parts of the world work.

Figure S2.1: Working scientifically

Understanding working scientifically

Working scientifically requires following eight fundamental processes:

1 observing
2 questioning and predicting
3 planning investigations
4 conducting investigations
5 processing data and information
6 analysing data and information
7 problem-solving
8 communicating.

This section will break down each process for you – where to start processes, how you can apply it as you learn more, and how mastering the skill will elevate your science learning.

The Learning Ladder and working scientifically

At the start of each chapter, each Learning Ladder lists two, three or four of the eight scientific processes and the five levels of progress for each of those processes.

Figure S2.2 shows the working scientifically Learning Ladder, with Step 1 being the first level and Step 5 the final, most challenging, level. You can only climb the ladder by mastering each step. This approach is called developmental learning and puts you in charge of your own learning progress.

Figure S2.2: Start at Step 1 and work your way up to mastery.

Learning Ladder

Steps in progression	Observing	Questioning and predicting	Planning investigations	Conducting investigations
5	I can use observations and measurements to answer questions.	I can formulate testable questions and predictions, considering variables and controls.	I can construct an appropriate plan before conducting a first-hand investigation.	I can conduct first-hand investigations and accurately record the collected data.
4	I can make inferences based on my observations.	I can make predictions to explain observations or phenomena.	I can develop appropriate scientific aims for a range of investigations.	I can collect accurate and reliable data from first-hand and second-hand investigations.
3	I can record observations and measurements accurately.	I can construct questions to investigate scientific concepts or problems.	I can distinguish between controlled, dependent and independent variables.	I can implement safe practices when using scientific equipment.
2	I can use scientific tools to enhance observations.	I can select questions to investigate scientific concepts or problems.	I can describe ways to reduce risks for a range of investigations.	I can use scientific equipment to conduct investigations and gather first-hand data and information.
1	I can make observations using my senses.	I can make predictions based on prior knowledge and observations.	I can identify appropriate materials and technologies to conduct investigations.	I can identify correct processes for conducting investigations.

Working scientifically processes

Each section of this book contains a Learning Ladder question block. The questions relate to both content and working scientifically processes. Question 1 corresponds to Step 1, Question 2 corresponds to Step 2, and so on.

The green headings mean these are working scientifically process questions. They will test you on a specific process, such as observing. There is a page reference to the Science how-to section, where you can get more information on what is required at each step of the process.

Again, the question numbers correspond to the steps on the Learning Ladder, so you can work to your level.

Learning Ladder

Cells and classification

1. Identify an organelle that is unique to plant cells.
2. Describe two organelles of plant cells.

Observing

see page 316

1. Can you use your senses to make observations about cell organelles? Explain your answer.
2. Identify which scientific equipment you could use to make observations of cells. Why would using this item improve your observations?
3. Outline how you could record your observations of cells.

In context

Draw a diagram of a bacterium cell, a plant cell and an animal cell. Label the diagram to compare the three cells by identifying their similarities and differences.

Success criteria

- I can describe the common organelles of plant and animal cells.

Next we will consider each working scientifically process and break down what is required at each step of the Learning Ladder to develop your skills and achieve process mastery.

Steps in progression	Processing data and information	Analysing data and information	Problem-solving	Communicating
5	I can process and interpret quantitative and qualitative data from a range of sources.	I can evaluate data and information for accuracy, reliability and validity.	I can evaluate problem-solving strategies used to solve an identified problem.	I can present scientific findings using appropriate conventions for specific audiences.
4	I can collect and present data in a range of appropriate formats.	I can draw conclusions based on patterns in data and information.	I can use given criteria to find solutions to scientific problems.	I can construct a range of appropriate scientific presentations based on first-hand and second-hand information and data.
3	I can construct appropriate graphs with headings and units for data.	I can explain relationships between datasets and information.	I can explain scientific problems and phenomena using cause-and-effect relationships.	I can use digital technologies to organise and present information and data.
2	I can construct appropriate tables with headings and units for data.	I can describe trends from collected data and information.	I can suggest solutions to familiar scientific problems.	I can select appropriate ways to communicate information.
1	I can identify data from graphs, tables and digital sources.	I can identify trends within given data and information.	I can identify scientific problems.	I can recognise scientific information.

Observing

Observations are the basis of scientific investigation. By making observations, we are able to conduct investigations, record **data** and make inferences from what we can observe. Scientists use their senses and specialised equipment to make observations.

Step 1 I can make observations using my senses

Our senses are the basis of all observations that we can make. Using the five senses, we are able to determine information about a first-hand investigation that can later be used to draw conclusions.

To use your senses, ask yourself:

- What am I trying to find out?
- How will I be able to observe this?

Table S2.1 shows this in practice.

Table S2.1: Using the five senses to make scientific observations

Question	Observation	Sense
Does the texture of a ramp affect the time taken by a car to travel down it?	There are rough and smooth surfaces	Touch
What gas is produced during a chemical reaction?	A strong rotten-egg smell comes out of the test tube when the chemicals are added	Smell
What substances can be classed as acids?	The vinegar turned red when the indicator was added	Sight
What happens as heat is added to water?	The water starts to bubble in the beaker	Sight and hearing
Does adding more sugar affect the batter of a cake?	The batter tastes sweeter when more sugar is added	Taste

Step 2 I can use scientific tools to enhance observations

Even though our senses can provide us with information, often they are not precise or accurate. To make precise and accurate observations, we need to use scientific equipment to record our measurements during investigations.

To decide what equipment to use when observing, ask yourself:

- What am I trying to observe?
- What equipment would allow me to measure accurately and precisely?

Figure S2.3: Adding indicator to colourless solutions can allow us to observe and identify their properties.

Considering three of the examples in Table S2.1, Table S2.2 shows some appropriate measurements and equipment.

Table S2.2: Determining the right scientific equipment

Question	Observation	Measurement	Scientific equipment
Does the texture of a ramp affect the time taken by a car to travel down it?	There are rough and smooth surfaces	The time taken by a car to travel on the different surfaces	Time can be measured precisely and accurately with a stopwatch
What substances can be classed as acids?	The vinegar turned red when the indicator was added	Solutions can be measured as acids or bases using a pH scale	Universal indicator and charts allow us to interpret colours as numbers
What happens as heat is added to water?	The water starts to bubble in the beaker	The temperature where water starts to change state	A thermometer can measure the temperature of the water when it is boiling

Figure S2.4: Stopwatches can be used to precisely and accurately record the time taken for a reaction to occur.

Step 3 I can record observations and measurements accurately

Once observations have been made using specific equipment, they need to be recorded accurately. To do this, the precise quantities need to be organised in an effective way so that they can be easily interpreted. One of the best ways to record observations is in a scientific table. To ensure your observations are accurate, include:

- headings showing what is being observed
- units of measurement that the observations are being recorded in.

For example, when recording data on temperature, the data should be neatly recorded in a column labelled 'Temperature (°C)'. More information on constructing scientific tables can be found in the 'Processing data and information' section on page 329.

Step 4 I can make inferences based on my observations

Collecting and recording observations then allows us to interpret the information we are collecting. This interpretation is an inference – a conclusion that is drawn based on the observations we have made.

To make an inference, you have to decide what your observations are telling you. To do this, ask:

- What am I observing?
- What measurements have I recorded?
- What do these measurements tell me about my investigation?

For example: In an investigation, a student was looking at how the texture of a ramp affects the time taken by a toy car to travel down its surface. Table S2.3 shows the observations the student recorded.

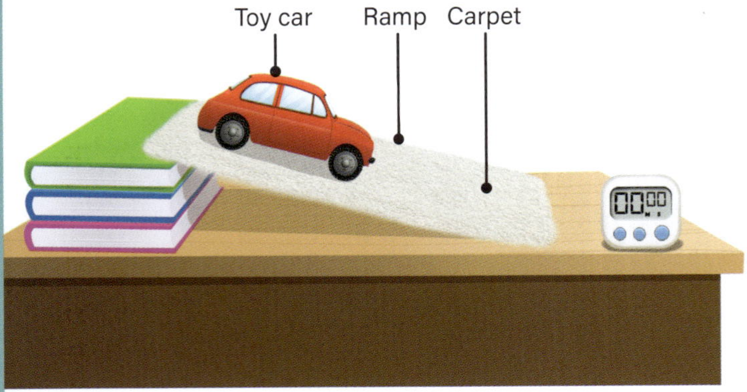

Toy car Ramp Carpet

Figure S2.5: Using scientific equipment to make observations can allow us to make inferences about the outcome of investigations.

Table S2.3: Investigation observations

Ramp surface	Time for toy car to travel down ramp (s)
Polished wood	5.3
Sandpaper	6.4
Carpet	7.8

Looking at the data, the student makes an inference that a rougher surface will cause the car to travel more slowly.

Step 5 I can use observations and measurements to answer questions

Observations can additionally be used to allow us to answer questions in investigations. To do this, you must look at the observations you have recorded and see which questions they enable you to respond to. When answering questions you need to consider:

- what the question is asking
- how your observations can be used to respond to the question.

For example, in an investigation, a student was making observations about water changing states to a gas. They recorded their observations in Table S2.4.

Table S2.4: Observations of water changing state

Water state	Observation	Temperature (°C)
Liquid	The water sits still inside the beaker over the Bunsen burner	25
Liquid	Some of the water begins to turn into a gas, forming small bubbles	70
Gas	The water rapidly bubbles, and all of the water is becoming a gas	100

They are then asked the question: At what temperature does water become a gas?

Using their observations, the student writes a response to the question: 'Water boils and becomes a gas at 100 degrees Celsius'.

Figure S2.6: Observations about temperature can allow us to answer questions about when different liquids change state.

Questioning and predicting

Questioning is the basis of all science. By asking questions, we can come up with ideas that we can test scientifically. Predicting scientific outcomes helps us to make better decisions. Predicting weather patterns can help people prepare for extreme temperatures, while predicting which illnesses might spread rapidly can help healthcare workers adapt their treatments. By making accurate predictions, scientists can solve real-world problems and improve the quality of human life.

Step 1 I can make predictions based on prior knowledge and observations

When you are writing questions and making predictions, you need to ensure you are proposing things that are **plausible**. To help you predict something plausible, you can think back to experiences you have had or match the investigation to a relevant scientific theory you already know.

For example, if you have a herb garden at home, you may have observed that the parsley grown in the shade does not grow as tall as the parsley in full sunlight. Add 'because' to your 'if, then' statement:

If a tomato plant gets more sun, then it will grow more because the parsley in full sun in my garden grows taller than parsley in the shade.

You can use this experience to predict how the amount of sunlight will affect the growth of tomatoes. Other examples of experiences that can help you make predictions are:

- slipping on a puddle of water on tiles but not on a similar-sized puddle on concrete
- noticing someone bigger than you is faster when you roll down a steep hill on your bikes
- seeing a hot-air balloon in the sky and being told the heat from the flames makes the balloon lift off the ground.

Figure S2.7: The observation of a hot-air balloon in the sky supports the prediction that heat can make objects rise.

Step 2 I can select questions to investigate scientific concepts or problems

Scientific questions contain one factor that we can change during an investigation, and one factor we can measure. Doing this means that we can test the question by setting up a first-hand investigation. We usually write scientific questions in the following format:

How does *changing one factor* affect the factor that *can be measured*?

When checking if a question can be investigated scientifically, you need to check that only *one* thing changes in the question, and there is *one* thing you can measure. Table S2.5 shows how.

Table S2.5: What to look for when setting up an investigation

Question	Thing changed	Thing you can measure	Can it be investigated?
How does the amount of sunlight a plant receives affect how tall it grows?	The amount of sunlight the plant gets	How tall the plant grows	Yes
How can I look after my plants at home?	Many things – anything you do to look after your plants	None	No

The first question in Table S2.5 *can* be investigated, as there is one thing we can change and one thing we can measure. The second question *cannot* be investigated, as there is nothing to measure and many things we can change.

Figure S2.8: The changed factor has an effect on the measured factor.

Step 3 I can construct questions to investigate scientific concepts or problems

We call the factors that we change and measure in the scientific question the **variables**. The factor that we change is called the **independent variable**. The factor that we measure is called the **dependent variable**. We want to see if there is a **relationship** between these two variables. In our scientific question structure, this looks like this:

How does the *independent* variable affect the *dependent* variable?

For example, we may have the following scientific problem: The tides at the beach change across the day, depending on where the Moon is in relation to Earth. So our question is:

How does the *location of the Moon* affect the *height of a tide* in the oceans?

The independent variable is the location of the Moon. The dependent variable is the height of the tides. The additional step is to identify the relationship. In our example, we are testing the relationship between location and height.

Figure S2.9: The Moon directly affects the tides of the oceans – but how?

Step 4 I can make predictions to explain observations or phenomena

The next level up is to use scientific knowledge to help you make predictions. To link your prediction to scientific knowledge, you must match the content of your knowledge to the investigation factors. Your 'If ... then' statement will then be followed by a 'because':

If a tomato plant gets more sunlight, *then* it will grow more, *because* sunlight helps plants to grow.

The scientific knowledge you use can be general – you do not need to include specific details. However, it must be relevant to the investigation and the predicted effect.

Step 5 I can formulate testable questions and predictions, considering variables and controls

To create testable questions and predictions, you need to be able to identify the independent and dependent variables, and some *controlled variables*, which are things that could be kept the same.

The following method will help you to formulate a scientific question:

a Read the question.
b Identify the independent variable.
c Identify the dependent variable.
d Identify some controlled variables.
e Make a judgement on whether or not this is a good scientific question.
f Suggest any improvements for the question, then rewrite it.

Table S2.6 shows how you can evaluate two different scientific questions.

Table S2.6: Evaluating scientific questions by identifying the variables

1 How does the *amount of sunlight* a plant receives affect how *tall it grows*?	
Independent variable	Amount of sunlight the plant receives
Dependent variable	Height of plant
Controlled variables	None
Judgement	This is a good scientific question that changes only one factor; however, it needs to include the controlled variables to make it better, such as the type of soil, the amount of water given and the temperature of the room, all of which must be kept the same.
Improvement	This question could be improved by controlling the variables – the conditions that plants are grown in. For example, how does the amount of sunlight a plant receives affect how tall it grows when it is in the same conditions, including soil type, amount of water, temperature and fertiliser provided?
2 How does the *location of the Moon* affect the *height of a tide* in the oceans?	
Independent variable	Moon location
Dependent variable	Tide height
Controlled variables	One – location of ocean in relation to the Moon
Judgement	This is an excellent scientific question that includes one changed and one measured variable, as well as a way to control the investigation. No improvements are required.

▲ **Figure S2.10:** There can be a variety of relationships between independent and dependent variables. Scientific knowledge and our own experiences can help us predict what that relationship will be.

Now that you can write and support a prediction generally, you can update your prediction with the specific variables of the investigation. This means we can also look at the relationship between the variables. Step 5 requires more detail on how you will change the independent variable to test your hypothesis, the relationship with the dependent variable and the predicted effect.

Once you have identified and understood your variables, include them in your prediction using the following template:

If [the independent variable and the specific change],
then [the dependent variable and the predicted effect].

So your prediction will become:

If the amount of sunlight a tomato plant receives is increased,
then its growth will increase by a similar amount.

Planning investigations

Planning will help you to come up with well-constructed investigations to answer scientific questions. A well-planned investigation should use appropriate scientific equipment, follow specific and ordered steps, and be considered a *fair test*.

Step 1 I can identify appropriate materials and technologies to conduct investigations

When planning an investigation, you need to select the pieces of equipment to suit your purpose. As well as listing the equipment you will need to conduct the investigation, you need to ask yourself what the investigation will measure.

For example, if you are trying to measure the distance a toy car will travel in an investigation, you will need a toy car, and a ruler or a tape measure. For time measurements, you may need a stopwatch, while temperature measurements would require a thermometer.

Go to the 'Laboratory equipment' section on page 377 to see basic equipment illustrated.

▲ **Figure S2.11:** When we perform investigations, we can measure temperature change using a thermometer. Here, two students monitor the temperature of an oil bath used to heat a reaction vessel.

Step 2 I can describe ways to reduce risks for a range of investigations

When conducting investigations, we must be able to identify **hazards**. To do this, you should look at the equipment you will use for your investigation and ask how this equipment could cause harm, and describe how this risk of harm could be reduced. For example, in an investigation, the equipment is:

- Bunsen burner
- box of matches
- water
- measuring cylinder
- beaker
- tripod
- gauze mat
- thermometer.

How could this equipment cause harm, and what can you do to reduce the risk?

- The Bunsen burner, box of matches and boiling water could cause burns if they touch skin. I can reduce the risk of harm by keeping my hands away from the flames of the Bunsen burner and match, and by keeping a safe distance from the boiling water and wearing a lab coat.
- The boiling water could hurt someone's eye if it splashes. I can reduce this risk by making sure we wear safety glasses.
- Someone could slip if water is spilled. I can reduce this risk by only adding water to the beaker after it is at the bench.
- The measuring cylinder, beaker and thermometer are made of glass. If they break, they could cut someone. I can reduce this risk by carrying them with two hands, and placing them towards the middle of the bench so they don't fall on the floor.
- The tripod and gauze mat could cause burns after they have been heated. I can reduce this risk by leaving them to cool down before packing away my equipment.

Step 3 I can distinguish between controlled, dependent and independent variables

When planning an investigation, you need to be able to determine the independent variable that will be changed, the dependent variable that will be measured and all the other variables that you will have to control (keep the same). You can determine your variables by asking the following questions:

- What am I trying to investigate?
- What one factor should I change?
- What one factor should I measure?
- What factors will I need to control?

Figure S2.12: A salt-water investigation – what will you change and what will you measure?

For example, imagine you are conducting a first-hand investigation to separate salt from a salt-water mixture and wonder if the amount of salt will affect the time it takes.

- *What am I trying to investigate?*
  How does the amount of salt in a salt-water mixture affect the time it takes to be separated?
- *What one factor should I change?*
  I can see from the question that the amount of salt will need to be changed – this will be my *independent variable.*
- *What one factor should I measure?*
  The question asks if the time will be different, so I should measure the time taken to separate the salt – this will be my *dependent variable.*
- *What factors will I need to control?*
  To separate the salt, I will crystallise it by boiling away the water. So, each time, I should use the same:
  - amount of salt water
  - basin to evaporate the salt
  - stopwatch to record the time
  - Bunsen burner with blue flame.

  These will be my *controlled* variables.

Figure S2.13: Our aim shows us the purpose of our investigation. Here, we are investigating the link between the slope of a surface and how far a toy car can travel.

Step 4 I can develop appropriate scientific aims for a range of investigations

A scientific aim links together the independent and dependent variables of a first-hand investigation. This statement is used to show the purpose of a first-hand investigation. We want to see if there is a relationship between these two variables. In our scientific aim structure, it looks like this:

To determine the effect of the independent variable on the dependent variable.

For example, we may have the following scientific problem: The distance a toy car travels changes, depending on the slope it is travelling down. So our aim is: To determine how the angle of a ramp impacts the distance a toy car will travel. The independent variable is the angle of the ramp. The dependent variable is the distance travelled by the toy car. In our example, we are testing the relationship between ramp angle and distance travelled.

Step 5 I can construct an appropriate plan before conducting a first-hand investigation

Before conducting an investigation, a plan should be constructed so that you know what you are going to do. Plans are written in the form of scientific methods. These should be written as a series of numbered steps. It should also be written in the *third person* (do not use *I*, *we* or *you*).

In addition, your investigation should include:

- the *independent and dependent variables* and how you will measure them
- the *controlled variables* and how you will keep them the same
- evidence of *repetition*
- a way to *record the results.*

As an example, imagine we are investigating the following question:

Does the force a toy car is pushed with affect how far it travels across a flat surface?

Our method or procedure would follow the process below:

1 Collect all equipment.
2 Find a flat surface and place a line of tape down to mark the start point of the car.
3 Place the car level with the starting line.
4 Gently push the car so that it starts to move.
5 Let the car travel along the surface until it comes to a stop.
6 Measure the distance the car travels from the start line in centimetres, using a tape measure.
7 Record distance in a results table.
8 Return car to the start line.
9 Repeat steps 3 to 8 two more times, applying the same force to the car.
10 Repeat steps 3 to 8 while applying medium and then firm force to the car to start its movement.

Figure S2.14: We are studying whether the force applied affects how far a toy car will travel along a flat surface.

Conducting investigations

When conducting investigations, you follow a series of steps while using the appropriate equipment and maintaining safety standards. This will result in reliable results to test your hypothesis.

Step 1 I can identify correct processes for conducting investigations

When conducting investigations, it is important to use correct processes. This ensures that your investigation will remain a **fair test**.

Processes in first-hand investigations include using appropriate scientific equipment, in the correct way. As an example, imagine we want to investigate the following question:

How can we change water from liquid to gas?

Once we know the question, we can select and set up equipment. Ask yourself:

- What do I need to measure?
- How do I need to measure it?
- What order do I need to do things in?

For example, to answer the question above, we would need to measure out a specific quantity of water. The correct process for measuring liquid is to use a measuring cylinder, and to make sure that the water level is accurately measured by getting the meniscus of the water to sit in line with the measurement we want, when the cylinder is sitting on a flat surface.

Figure S2.15: The correct process for measuring liquids is to add them to a measuring cylinder until the base of the meniscus (curved line) sits level with the amount we want to measure.

Step 2 I can use scientific equipment to conduct investigations and gather first-hand data and information

To conduct an investigation, you should:

- read through all the steps before starting
- collect all the pieces of equipment listed
- complete all parts of a step before moving to the next one.

As an example, imagine we want to investigate the following question:

How can we change water from liquid to gas?

The investigation lists the following procedure:

1 Measure 50 mL of water into 250 mL beaker using a 200 mL measuring cylinder.
2 Connect Bunsen burner to gas tap.
3 Place tripod and gauze mat over Bunsen burner, and set beaker on top.
4 Light match and hold it over Bunsen burner, then turn gas tap on.
5 Place thermometer into beaker to measure temperature.
6 Record temperature where water started to change state from liquid to gas.

Figure S2.16: Scientists follow specific steps to set up Bunsen burners for first-hand investigations.

We should read through every step. Then we need to collect all the equipment:

- Bunsen burner
- water
- beaker
- gauze mat
- box of matches
- measuring cylinder
- tripod
- thermometer.

Finally, we need to complete all parts of each step before moving onto the next one. So, for the first step of our investigation, we would measure the water and pour it into the beaker.

Step 3 I can implement safe practices when using scientific equipment

Risks you describe when planning an investigation can be structured in a table like Table S2.7. This shows us what we need to do in the investigation to reduce the risk of injury. By following the steps in the table, we are implementing the safe practices we have identified and keeping everyone safe.

Table S2.7: Reducing risks when using a Bunsen burner

Hazard	Risk	Strategy
Bunsen burner	High likelihood of burns from flame if it contacts the skin or hair	• Place Bunsen burner in the middle of the lab bench and away from hands. • Always have Bunsen burner on yellow safety flame when not in use. • Wear lab coat and tie back long hair.
Boiling water	High likelihood of eye damage if it splashes into the eye	• Stand away from the beaker of water while it is boiling. • Wear safety glasses to create a barrier between the water and the eyes.
Glass beaker and measuring cylinder	Very high likelihood of cuts if glassware breaks	• Never run while carrying glassware. • Carry glass with two hands. • Place glassware away from the edge of the lab bench to prevent it from being knocked onto the floor.

Step 4 I can collect accurate and reliable data from first-hand and second-hand investigations

As well as being safe, your investigations must also be **ethical** – that is, you must conduct your investigation with integrity, and ensure that data is collected accurately and recorded honestly. To check that an investigation is safe and ethical, you must ensure:

- you or your teacher has conducted an appropriate risk assessment for the investigation
- the equipment used is appropriate for what you are testing
- the investigation is conducted honestly; no one tries to skew the test or measurement to obtain a certain result
- the data is recorded exactly as it is measured.

Figure S2.17: Using safety gear minimises the likelihood that someone will get hurt during an investigation.

You can use a checklist to help you confirm that your investigation is safe and ethical.

SAFE AND ETHICAL INVESTIGATION CHECKLIST

- ☐ I made a risk assessment that identifies at least three hazards and how to prevent them from happening.
- ☐ I chose equipment that allows me to conduct the investigation and measure results accurately.
- ☐ I recorded the data I collected exactly as measured in a results table.

When you repeat a test or investigation and the data you collect is the same or similar every time, your data is reliable. To collect reliable data, you should:

- conduct the investigation and record the data
- repeat the investigation at least twice more and check that the data is the same each time, within a reasonable margin
- use your repeated trials to generate average data values.

When your investigation results match with the results conducted in other investigations on the same topic, your data is accurate. To collect accurate data, you should:

- check that the equipment is appropriate for the investigation
- use the equipment to precisely measure quantities of materials used in the investigation
- compare your results to other, secondary, sources and check that the data is the same, within a reasonable margin.

Step 5 I can conduct first-hand investigations and accurately record the collected data

As well as being safe, your investigations must also ensure that data is collected accurately and recorded honestly. To check that data is recorded accurately, you must ensure:

- a results table is constructed to record data from the independent and dependent variables of the investigation
- the equipment used is appropriate for what you are testing
- the investigation is conducted honestly; no one tries to skew the test or measurement to obtain a certain result
- the data is recorded exactly as it is measured.

You can use a checklist to help you confirm your data is recorded accurately.

ACCURATE DATA RECORDING CHECKLIST

- ☐ I constructed a results table that allows me to place data about the independent and dependent variables.
- ☐ I chose equipment that allows me to conduct the investigation and measure quantities precisely and accurately.
- ☐ I recorded the data I collected exactly as measured in a results table.

Processing data and information

Information generated during a scientific investigation should be recorded in an organised way, so you can process it. Processing your data means presenting information in a way that clearly displays what you have found. This will help you to identify and understand the data you have collected.

Step 1 I can identify data from graphs, tables and digital sources

The first step in processing data is simply recognising data that has already been organised into graphs or tables. When information is on display, you need to read what is presented carefully. Start with the title. What does it tell you? Then move onto headings and units. Ask yourself, what type of information am I looking at? And, what does the data tell me?

Table S2.8 is organised in a way that clearly displays information. From the title, we can recognise that the data will show us how plant growth is affected by sunlight. The headings indicate that measurements were taken in millimetres, and the values in the table show the growth of the plants in various amounts of sunlight.

Table S2.8: Effect of sunlight on plant growth

Plant and environment	Initial plant height (mm)	Growth after 1 week (mm)	Growth after 2 weeks (mm)	Growth after 3 weeks (mm)
Plant 1: no sunlight	181	0	–1	–3
Plant 2: indirect sunlight	175	1	2	4
Plant 3: direct sunlight	178	2	3	5

Step 2 I can construct appropriate tables with headings and units for data

Once you are comfortable recognising information from a table or graph that already contains data, you can prepare your own table to record data for a particular investigation using the following instructions:

a Identify the independent and dependent variables. Use these to write a descriptive title for the table. The descriptions representing the independent variable are listed in the left column and the dependent variable in the other columns. For example, in Table S2.8 we list the plants and how much sunlight they receive down the left column (independent variable). Then we have three headings for measurements at different times at the top of the other columns (dependent variable).

b Determine how many rows and columns are required to construct the table. The number of **datasets** indicates the number of rows required (plus one row for headings). In Table S2.8, we have three datasets: one for each of plants 1, 2 and 3. The number of columns depends on how many times you choose to take measurements. This could be based on time intervals, different items to measure or the number of trials you conduct. The example above uses weekly time intervals across three weeks, including the initial height. Units should be included in the headings when the data is a form of measurement; for example, (cm) or (mL).

c Construct a table with the appropriate number of rows and columns. Ensure they are evenly spaced and make sure there is enough room to write the information clearly.

d Write headings to represent the variables. The left column includes the plant number, as well as the specific plant environment. Ensure your headings give the reader a clear picture of the data collected, including the units of measurement used.

Figure S2.18: Because plants grow slowly, it makes sense for you to measure in millimetres (mm) over three weeks.

The following example shows how to construct a table based on the investigation question:

How does the type of wheel – bottle cap, compact disc (CD) or washer – affect the distance travelled by a model car?

a Identify the independent and dependent variables. Use these to write a descriptive title for the table.

*The independent variable is the type of wheel, because this is the thing that changes. We will measure the distance travelled, so that is the dependent variable. A good title is '**Effect of wheel type on distance travelled**'.*

b Determine how many rows and columns you need.

There are three types of wheels, so there will need to be three rows for recording data plus heading rows. We can measure distance in centimetres (cm). We won't measure the time it takes. We think it is best to run three trials and then determine an average distance. One column is required for each trial, as well as one for the average, plus one for the wheel types.

c Construct a table with the appropriate number of rows and columns.

d Write headings to represent the variables. Remember: the independent variable belongs in the left column and the dependent variable goes along the top, with its unit of measurement.

Now you're ready to conduct the investigation and record the data, as shown in Table S2.9.

Table S2.9: Effect of wheel type on distance travelled

Wheel type	Distance travelled (cm)			
	Trial 1	Trial 2	Trial 3	Average
Bottle cap	11	9	13	$(11 + 9 + 13) \div 3 = 11$
CD	33	31	35	33
Flat washer	26	20	32	26

Note that in the cells where we show the averages, we only show the process for calculating the average once. For more information, go to the 'Calculating averages' section on page 363.

Step 3 I can construct appropriate graphs with headings and units for data

The next step in recording and processing is to take your data and show it in a graph. Graphs are useful for displaying data, as they can be easier to understand and interpret. The three graphs we will use most often in science are the column graph, the line graph and the scatter plot, shown in Figure S2.19.

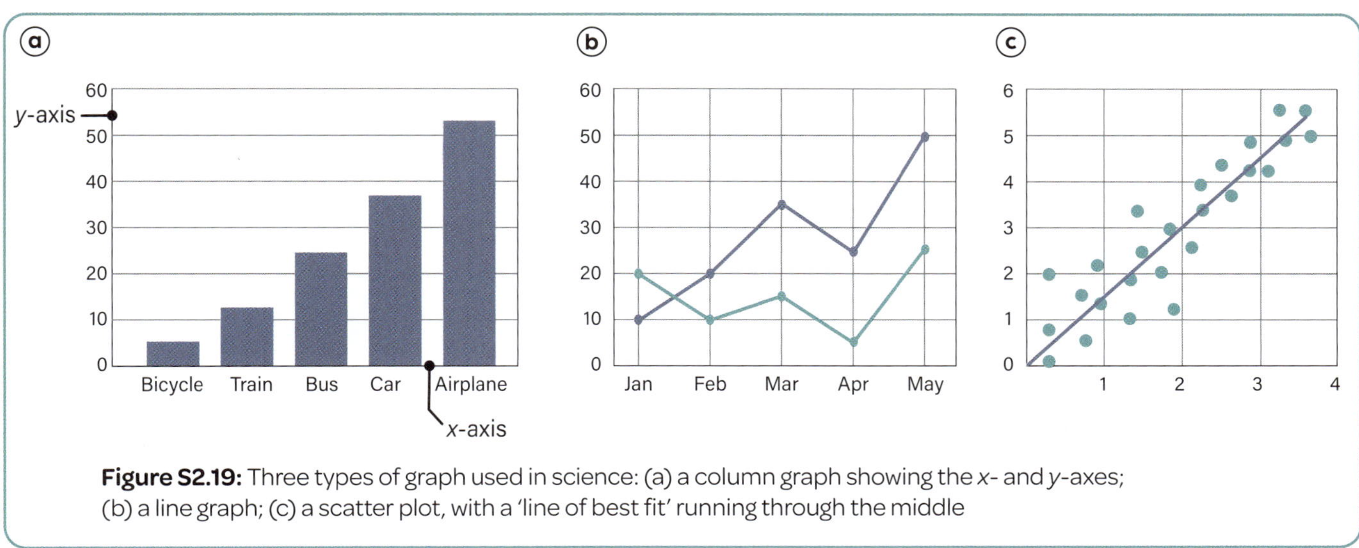

Figure S2.19: Three types of graph used in science: (a) a column graph showing the *x*- and *y*-axes; (b) a line graph; (c) a scatter plot, with a 'line of best fit' running through the middle

Figure S2.20 is a column graph of the results of the model car investigation. The annotation describes how to construct the graph.

(a) Determine the range of the axes based on the data you have collected.

Write the dependent variable on the *y*-axis (vertical axis) and the independent variable on the *x*-axis (the horizontal axis).

Plot the measurements along the *y*-axis, and the wheel type along the *x*-axis. The lowest data value from our averages is 11. The range of the *y*-axis must begin below this value. We could choose 10, 5 or 0, but it is good to start at 0 for simplicity. The largest data value is 33, so the highest value on the *y*-axis must be higher than that, such as 35.

(b) Use a ruler to draw the axes and select the space between values. Since the *y*-axis will be spread from 0 to 35, it makes sense to count by fives up the axis. Avoid writing every number between 1 and 35 as this will make the graph too cluttered.

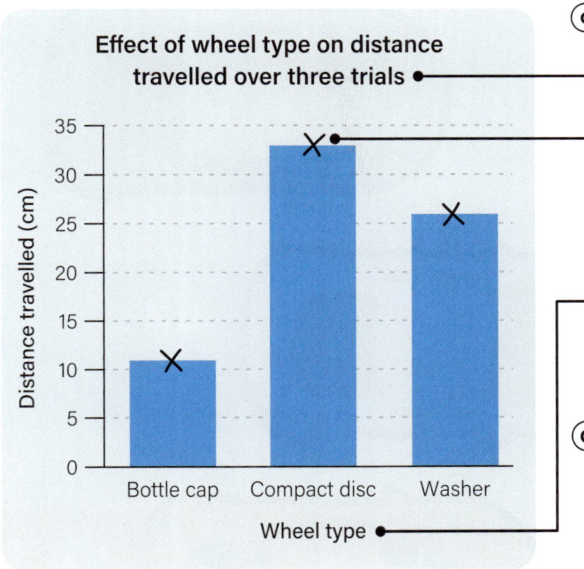

Figure S2.20: A column graph of the results of a model car investigation

(c) Write the title, label the axes and plot your points on the graph – the graph title can be the same as the title of your table. Be sure to include the independent and dependent variables, not only in the title, but also on the axes labels. Include the unit of measurement, if needed, on each axis label.

(d) Draw your graph – a column graph requires you to draw and shade the columns. (A line graph means you connect the points in a line, while a scatter plot graph requires you to draw a 'line of best fit' that represents an average of your points.)

Categorical factors like the wheel type are best represented in a column graph, as shown in Figure S2.21. If measuring continuous time is involved, a line graph with connected points is best. If both the independent and dependent variable values are numbers, a scatter plot is best. A scatter plot includes a 'line of best fit', to show the **trend**.

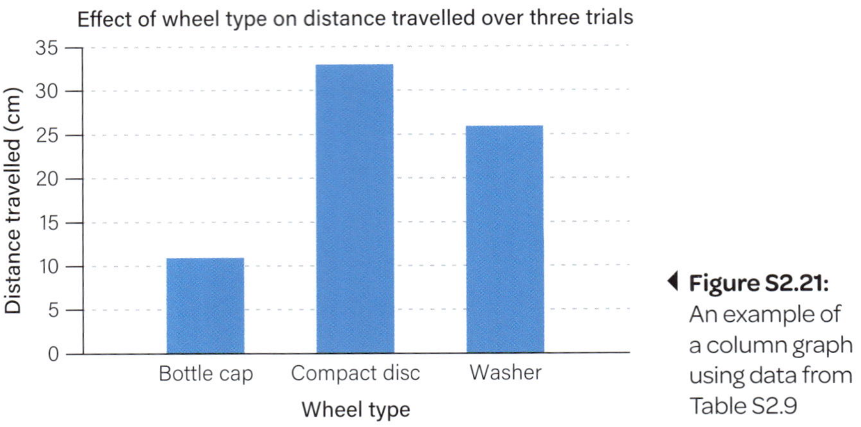

Figure S2.21: An example of a column graph using data from Table S2.9

Step 4 I can collect and present data in a range of appropriate formats

When choosing how to display your findings, you need to select the right visual for your information, just like picking the right tool for a job. Figures S2.22 to S2.26 show some ways you can visually communicate information.

Figure S2.22: An infographic showing the model car investigation results

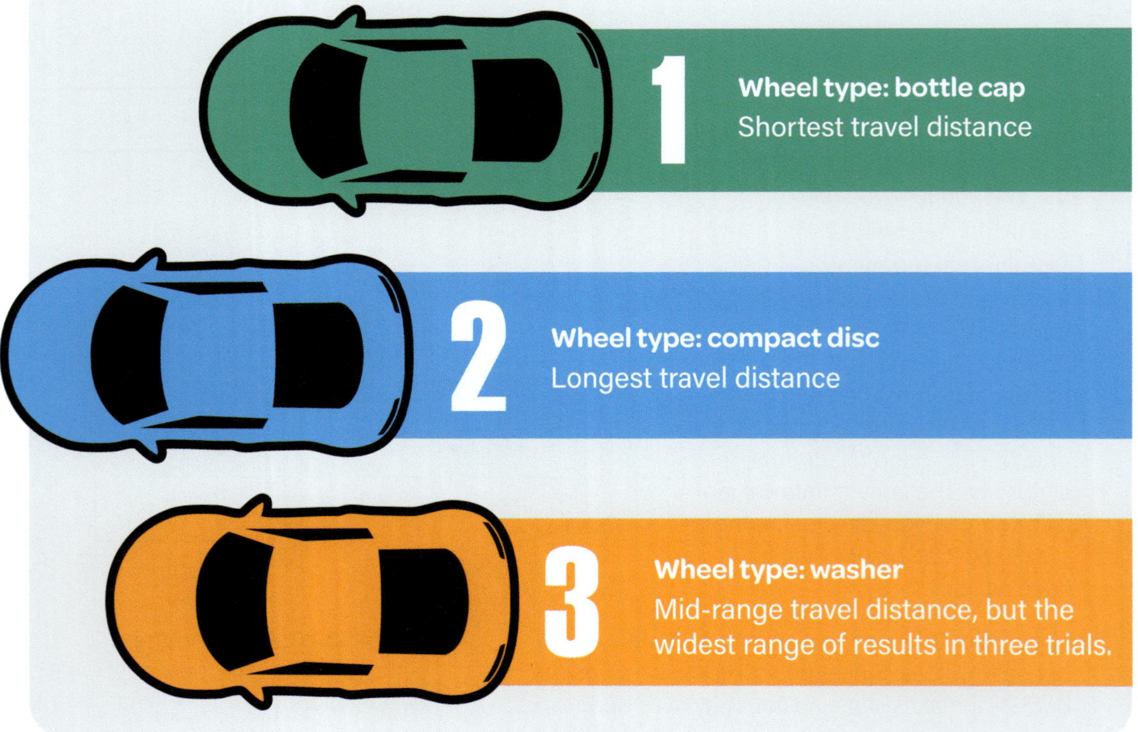

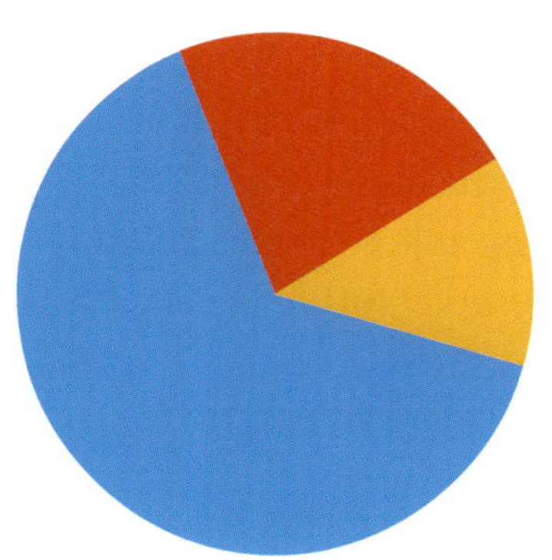

◀ **Figure S2.23:** A pie chart is best for looking at the parts that make up a whole.

◀ **Figure S2.24:** A Venn diagram is used to compare similarities and differences.

▲ **Figure S2.25:** A word cloud is a creative way to show different proportions.

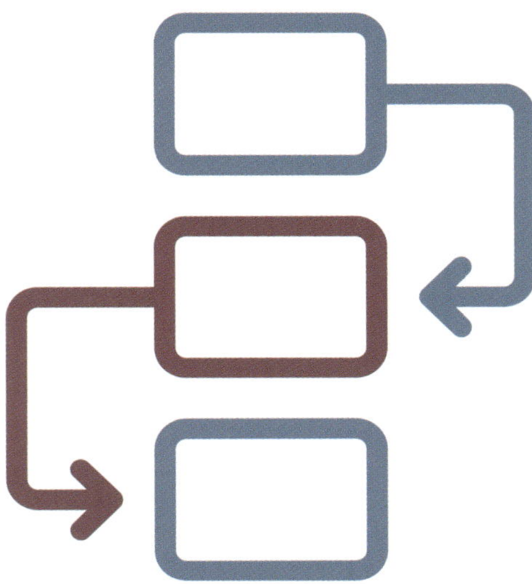

▲ **Figure S2.26:** A flow diagram can be used to show how one thing moves or transforms to another.

Step 5 I can process and interpret quantitative and qualitative data from a range of sources

The most challenging aspect of processing data is being able to understand what it means, no matter how it is presented. In steps 2, 3 and 4, the focus was on presenting data using tables, graphs and other visual formats. Now that you have worked your way through the different modes of representation, you should be able to interpret what is being presented, no matter the type of data or the format in which it is presented.

Not only are you able to process raw data into tables and graphs, but you can also understand what the data is communicating. To display your skill at this level, you should be able to explain how the raw data should be or has been adapted to communicate a variety of possible investigation results in a clear and focused way.

Consider Table S2.8 on page 329. Try to summarise the important aspects of the data, without being prompted by questions like those from Step 1. The summary could be in the form of a written paragraph or simply a discussion with your peers about the story that this data shares with the audience.

Analysing data and information

Analysing data is like being a detective for numbers! We need to make sense of information to reveal the key findings of our investigations. Being able to analyse data helps people to make informed decisions, solve problems and predict future trends.

Step 1 I can identify trends within given data and information

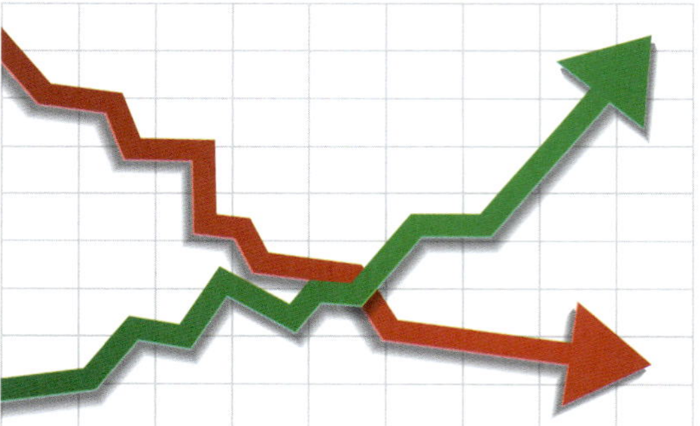

Figure S2.27: The green line is trending up while the red line is trending down.

A **trend** is shown by the line or bars in your graph moving in a general direction, usually up or down, as shown in Figure S2.27. A **pattern** is when the data repeats in predictable ways. Sometimes your data has no clear trend, so your visualisation will show random values all over, or there may be a sideways trend (a horizontal line).

Figure S2.28 is a graph of snake sightings. It does not show a trend across the whole year. It shows a trend downwards from January to June and a trend upwards from July to December.

To recognise patterns, look for recurring values or repeating shapes. For example, Figure S2.28 does not show a year-wide trend, so the next thing you look for is a pattern. It is clear that there are more snake sightings in the warmer months, and not many at all in winter. This is a pattern that we would expect to see year after year.

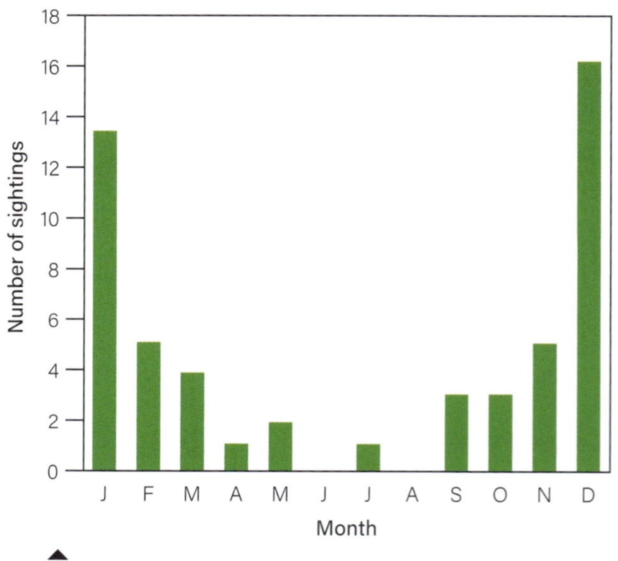

Figure S2.28: Snake sightings, average per month

Figure S2.29: The number of snake sightings increases with average temperature across the year.

We already identified the trends within the pattern of S2.28: that snake sightings trend downwards in the first half of the year and upwards in the second half. The trends alternate in a repeating pattern each year.

Step 2 I can describe trends from collected data and information

Our next step is to link any trends to the variables of the investigation. The relationship between the variables in an investigation is called the **correlation**. Identifying the correlation enables you to then describe the relationship between the dependent and independent variables:

- *Positive correlation*: the relationship between variables is direct. This means that if the value of one variable increases, the value of the other variable also increases, and vice versa.

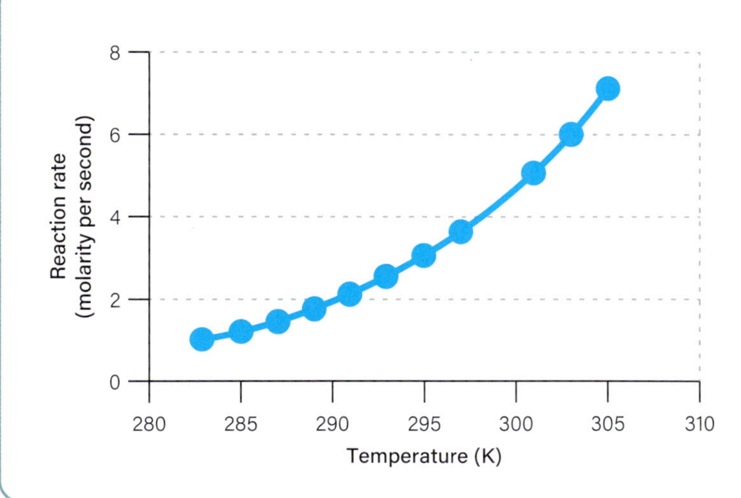

This graph shows a positive correlation. As temperature increases, so does the rate of the chemical reaction. The relationship between temperature and rate of reaction is direct.

Figure S2.30: A line graph of reaction rates in an investigation, showing a positive correlation.

- *Negative correlation*: the relationship between variables is indirect. This means that if the value of one variable changes in one direction, the value of the other variable does the opposite.

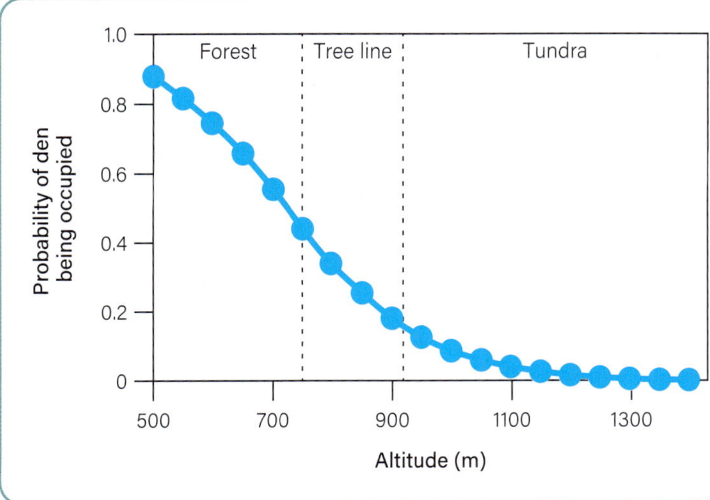

This graph shows a negative correlation. The likelihood of the fox den being occupied decreases as the altitude of the habitat increases. If one goes up, the other goes down. This displays an indirect relationship between whether a den is occupied and its altitude.

Figure S2.31: A line graph of fox den occupation rates, showing a negative correlation.

- *No correlation*: there is no apparent relationship between the variables.

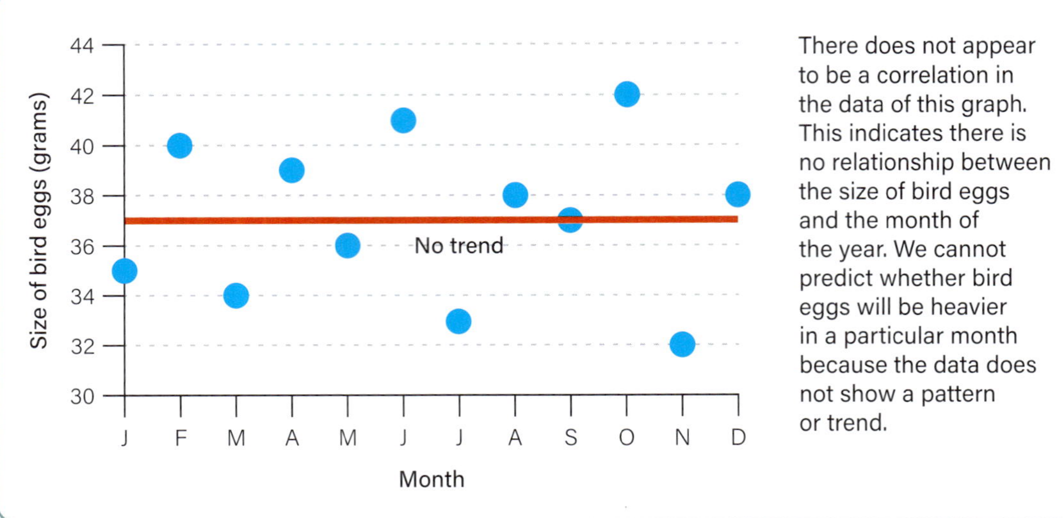

Figure S2.32: A graph of the mass of bird eggs and month laid

Step 3 I can explain relationships between datasets and information

To explain relationships between variables, you must draw on your scientific knowledge and theoretical understandings. After identifying a relationship, the next step is to ask: Why is one variable affected by another variable in this way?

For example, in Figure S2.31 the negative correlation between the fox den being occupied and altitude is linked to habitat theory and the effect of altitude on living conditions for the fox. As altitude increases, temperature and access to food decreases. These are reasons for why the den is less likely to be occupied at high altitudes, especially in winter.

Often, the information used to explain relationships between variables is aligned with the information used to inform predictions made at the start of an investigation.

Step 4 I can draw conclusions based on patterns in data and information

As scientists, we can draw conclusions based on the data that emerges from an investigation or study by:

a stating the relationship between the specific variables, based on the data

b replacing the terms specific to the investigation with general terminology, so it can be applied in a variety of contexts.

For example, if you investigate the effects of friction on the time it takes an object to reach the bottom of a ramp, you could use a toy car as the object, and ramp surfaces with varying amounts of friction, such as glass and wood. The results would show that the car reaches the bottom more quickly when rolled down the glass ramp than the wooden ramp. You can then use the points above to draw a conclusion.

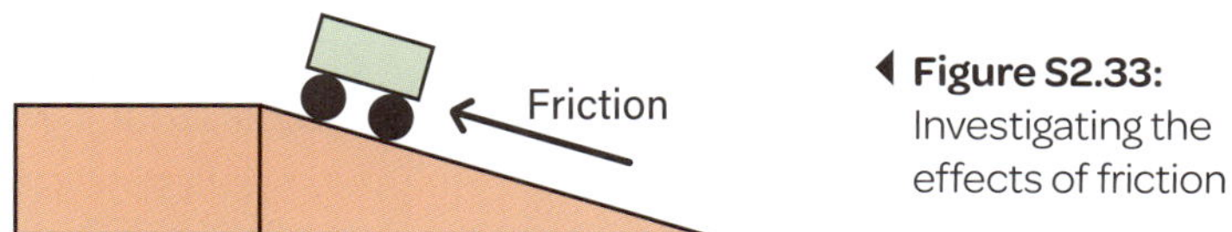

Figure S2.33: Investigating the effects of friction

a State the relationship between the specific variables, based on the data.

When the car rolls down the wooden ramp with more friction, it stops sooner than when it rolls down the glass ramp with less friction.

b Replace the terms specific to the investigation with general terminology so the statement can be applied in a variety of contexts.

When an object slides across a surface with more friction, it will come to a stop sooner than when it slides over a surface with less friction.

We have drawn a conclusive statement from the results. The term *car* was replaced with *object*, and instead of stating the *specific surface* used in the experiment, we generalised to the *amount of friction* – more or less. Table S2.10 provides some more examples of drawing conclusions.

Table S2.10: Converting investigation results into broader conclusions

Relationship between variables	Conclusion
When carbon dioxide gas from a chemical reaction is captured in a balloon, then heated in a hot water bath, the size of the balloon increases.	When the temperature of gas particles increases, the volume of the substance expands.
As the number of trees planted in eucalypt forests increases, the population of ring-tailed possums also increases.	Revegetation efforts increase native animal populations by restoring natural habitats.
The more a water barrel contains, the more people are required to push it 2 metres.	The greater the mass of an object, the more force is required to move it from a stationary position.

This does not mean that all our conclusions are correct. It is simply an opportunity to say what we know at that time and to investigate further.

Step 5 I can evaluate data and information for accuracy, reliability and validity

The most challenging aspect of analysing information is determining the quality of the data itself – is it *high quality* or *low quality*? To do this, you must understand the terms **accuracy**, **reliability** and **validity**, and be able to connect them to the data you generate in an investigation.

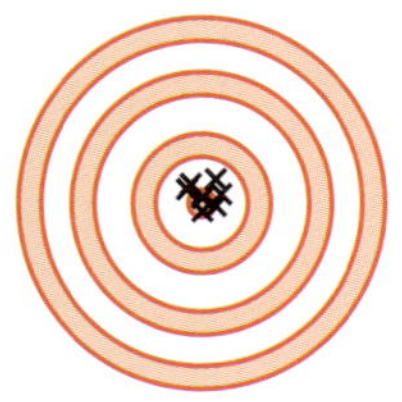

High accuracy
High precision

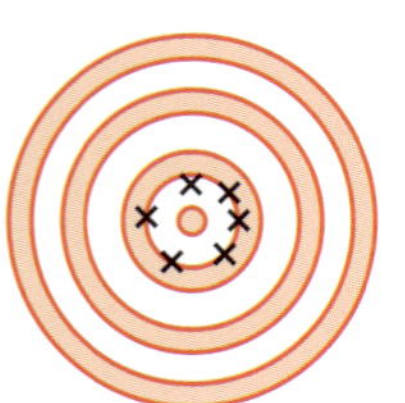

High accuracy
Low precision

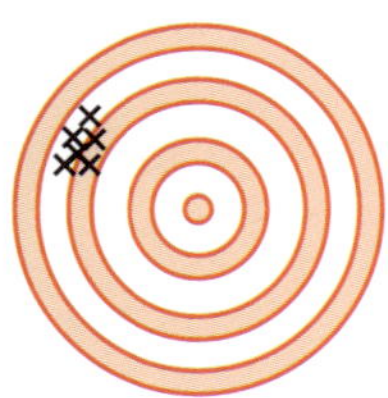

Low accuracy
High precision

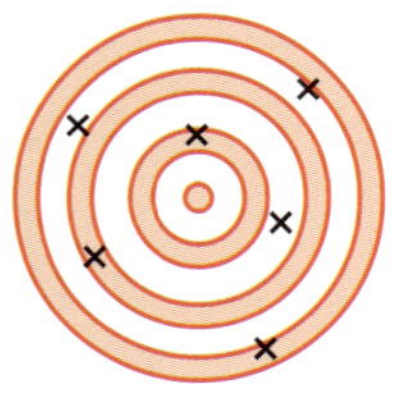

Low accuracy
Low precision

▲ **Figure S2.34:** Accuracy and precision are *not* the same thing.

Accuracy

Accuracy refers to how close your results are to the true value. When analysing information, the only way to know if it is accurate is to compare it to known theoretical values shown in your textbook or in scientific publications. You may have to conduct a little research to determine if the values are accurate or not.

Reliability

Data and information that is reliable means that we can trust that if we repeated the investigation, we could produce similar results. Reliable results indicate that the method used produces consistent and dependable results. Reliability considers both accuracy and **precision**.

Precision is a measure of how similar values are to each other within a dataset. If you do multiple trials of the same test, you would expect the results to be mostly the same. If the values are very different from one trial to the next, your results are not precise.

However, if we conduct enough trials, the imprecise values will average out to a value that is near the true or accurate value. Therefore, the more precise your values are, and the more trials you do, the more reliable your results will be.

Validity

If an investigation is valid, it means the method measures what it intends to measure. This allows us to make generalisations from our conclusion. It also allows us to apply our learnings to different situations. Validity considers both accuracy of measurements and the control of variables.

In order for results to be valid, we must design a fair test that tests one variable at a time, while controlling all other variables that could affect the results. This is how we can determine a cause-and-effect relationship. If more than one variable changes, we can't be sure what caused the effect, which means the test would not be considered fair and the results would not be valid.

By considering the accuracy of measurements, the precision of values within each dataset, and the method and materials used to isolate and control variables, you are able to assess data and information for reliability and validity.

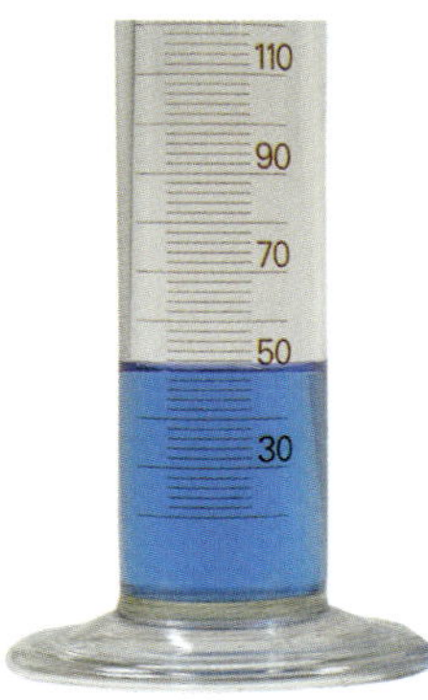

◀ **Figure S2.35:** Which of these containers has a volume closest to 50 mL? Measurements are always subject to errors.

Problem-solving

Problems are questions about phenomena or issues that need to be answered. After asking scientific questions, scientists use problem-solving skills to help find answers that explain identified problems.

Step 1 I can identify scientific problems

A scientific problem is an identified issue or phenomenon that can be explained through investigation. This means that for a problem to be scientific there has to be a specific change that has a measured effect.

To check if a problem is scientific, you can ask:

- Does this problem address an issue or phenomenon?
- Is there a specific change that can be made to allow a result to be measured?

Table S2.11 shows an example.

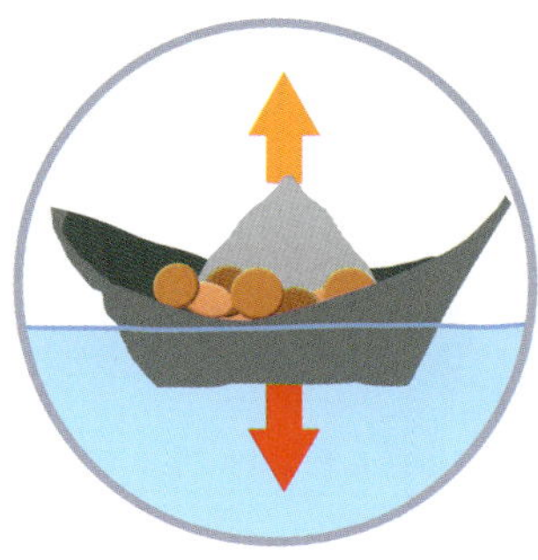

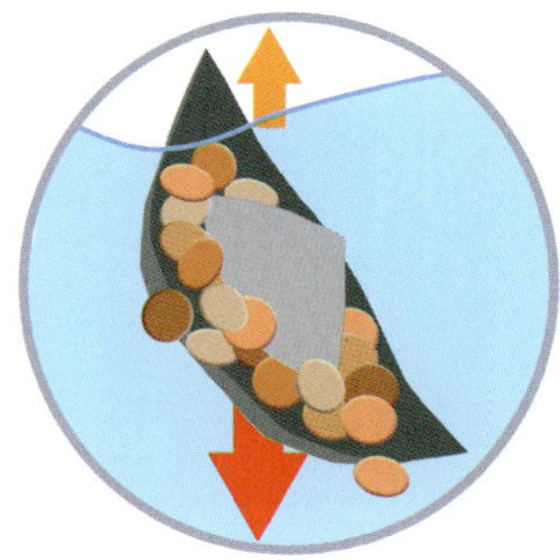

Figure S2.36: The problem of how much paper boats are capable of carrying can be investigated scientifically.

Table S2.11: Assessing whether a problem is scientific

Problem	Issue or phenomenon	One changed and one measured factor?	Scientific problem?
A heavy box cannot be lifted by hand	How forces work	No. There are multiple factors to consider.	No
A boat will sink if it has to carry too much mass	Buoyancy forces	Yes • Changed – mass the boat carries • Measured – whether the boat floats	Yes

Step 2 I can suggest solutions to familiar scientific problems

Familiar problems are problems that we have experienced before and have knowledge about. We can use our knowledge to suggest ways to solve these problems. To do this, previous ways we have seen these problems solved need to be considered. Ask yourself:

- When have I seen a problem like this before?
- How was this problem solved?

For example, imagine you notice that some plants in your house are not growing, even though they are in nice pots and they are watered regularly.

- Where have I seen this before?
  In school, we tested growing plants in different places, and the plants in the cupboard didn't grow because they needed sunlight.

Figure S2.37: Using knowledge from previous situations can help us to solve problems – like how we can get our house plants to grow.

- How was this problem solved?

  Placing the plants in the sunlight allowed them to start growing.

This leads you to suggest moving the pots into a sunny area so that the plants get sunlight and can grow.

Step 3 I can explain scientific problems and phenomena using cause-and-effect relationships

There is a link between scientific phenomena and their explanations. Once we have solved a problem that demonstrates this link, we can represent it by explaining the link between a cause and its effect.

When writing about these links, we structure our writing in a specific way. This structure allows us to show the link between the cause of a problem and its effect. This can be done in three sentences.

1 State what the problem was.
2 State what the cause of the problem was.
3 Link the two ideas together by stating how the problem was solved.

For example, after you have moved the plants that were not growing into the sunlight, they improved, and they started to grow after a week. Now they are bushy and healthy.

You write an explanation for this problem:

The house plants were not growing because they were in a dark part of the house. Plants need sunlight to grow. So, when the plants were moved into the sunlight, they were able to photosynthesise and produce glucose, which allowed them to start growing again.

Step 4 I can use given criteria to find solutions to scientific problems

When solving problems, we can apply criteria to help us find a solution. In a question, we may be given criteria to use. In other cases, we may need to identify which criteria we think will best help us solve the problem. Either way, we must use these criteria when working on a solution. With complicated problems, we may need to use research from secondary sources to help us understand and apply the criteria.

When working on a problem, ask yourself:

- What is the problem I need to solve?
- Have I been given criteria to use, and what are they?

  If no criteria are given, then ask:
- Can I identify any criteria that can help me to solve the problem?

  Once you have clearly identified any criteria, then ask:
- How will the criteria help me to solve the problem?

Now you can solve the problem using the criteria to inform your solution.

For example, a student was given the following problem to solve: Investigate buoyancy by designing a boat that can carry maximum mass, taking into account the shape of the boat's hull and the type of material used to construct the boat.

What is the problem I need to solve?

I need to design a boat that can carry the most mass possible.

Have I been given criteria to use, and what are they?

I can see there are two criteria to use. The first criterion is the shape of the boat, and the second criterion is the material used to make it.

How will the criteria help me to solve it?

- *I can design a boat and investigate using different shapes for the boat's hull, such as a long thin shape or a wide short shape, or shapes with a deeper hull below the water, then see if it changes the mass the boat can carry.*
- *I can change the material used to make the boat and see if it changes the mass that can be carried.*

Will secondary sources help me to identify anything I can test?

Research from a secondary source tells me that a boat with a wider deck can hold more mass. If I widen my boat's design, I might be able to solve the problem of carrying more mass.

The student created some models to test different boat designs and used the criteria to make the following conclusion:

After testing some model boats, it is clear that a wider hull on a boat can carry more mass. The material used to construct the boat was able to hold more mass when it was lightweight but strong, and did not change shape when mass was put on it.

Step 5 I can evaluate problem-solving strategies used to solve an identified problem

We need to select the most effective strategies when trying to solve a scientific problem. To do this, we should assess the available problem-solving options.

To assess the effectiveness of a strategy, we consider the benefits and limitations of each strategy, then judge which one is the most effective. For example, we could have the following problem:

I have a sample of sand, salt and iron that is a mixture. I want to use the salt and iron separately for different tasks, but they are mixed together.

We can use two possible methods to separate the mixture:

- decanting, centrifugation and chromatography
- magnetic separation, filtration and evaporation.

Table S2.12 shows how we can assess the benefits and limitations of each method.

Table S2.12: Assessing the best strategy to separate a mixture

Strategy	Benefits	Limitations
Decanting, centrifugation and chromatography	• Decanting would allow me to filter out the sand and iron if I dissolved the salt in water.	• Centrifugation and chromatography would not effectively separate the materials in the mixture. • The salt would be dissolved in water, so I would not be able to use it.
Magnetic separation, filtration and evaporation	• Iron is magnetic and could therefore be separated from the dry mixture. • Filtering the sand and salt would leave me with just salt water. • Evaporating the water would mean that the salt would crystallise out.	• Unless I add heat, evaporating all of the water off the salt could take a long time.

Judgement:

Based on the benefits and limitations, the second strategy of separation techniques would be the most effective way to solve this problem. This is because the separation techniques in this strategy allow the three components of the mixture to be individually separated, ready for use.

Figure S2.38: Weighing up the benefits and limitations of various problem-solving strategies can help you to decide on the best strategy, such as which techniques to use to separate a mixture.

Communicating

Communicating is essential to science – scientists need to share the ideas they have and the findings they have discovered. When we communicate, we need to consider what we are trying to say, and who we are trying to say it to. This allows us to present our ideas and findings using specific methods and language, depending on our audience.

Step 1 I can recognise scientific information

In science, specific scientific language is used to communicate findings or ideas to other scientists. Scientific language can help to describe specific features or qualities, and allow us to communicate information from a first-hand investigation.

To recognise scientific information, you need to consider the following:

- Do the words used provide information about the science behind this idea or concept?
- Are the words specific and used correctly in context?

For example, if we want to identify scientific information on the quokka, we can study the following passage:

The quokka (Setonix brachyurus) is a mammal that is part of the same family as the kangaroo and the wallaby, and part of kingdom Animalia. Quokkas are herbivores, living off native shrubs and grasses on Rottnest Island in Western Australia. Quokkas are currently listed as a 'vulnerable' species, because they are eaten by predators, usually introduced species such as cats.

The passage describes part of the classification of the quokka, as well as its diet, habitat and predators. Specific classification and biology terms are used, and they correctly describe the quokka from a scientific perspective. This information can be recognised as scientific.

Figure S2.39: Describing the features of a quokka helped scientists to classify it.

Step 2 I can select appropriate ways to communicate information

You can communicate your scientific findings in many different ways. To decide on a suitable method of communication, ask the following questions:

- What information do I have?
- What do I want to communicate to the audience?
- What format will display this information most clearly?

Table S2.13 shows some examples of how to apply these questions to make your decision.

Table S2.13: Choosing formats for scientific communication

Information	Communication goal	Format
Data from a first-hand investigation	What I found out while investigating a scientific question.	Scientific table or graph
A complete first-hand investigation	How I did the investigation and what the results of the investigation were	Scientific report
What causes an eclipse	How the movement of the Sun, Earth and Moon cause an eclipse	Scientific poster with text and diagrams
Water movement	How water moves through the different stages of the water cycle	Scientific diagram with labels
Bacteria structure	What a bacteria cell looks like and contains	Scientific model

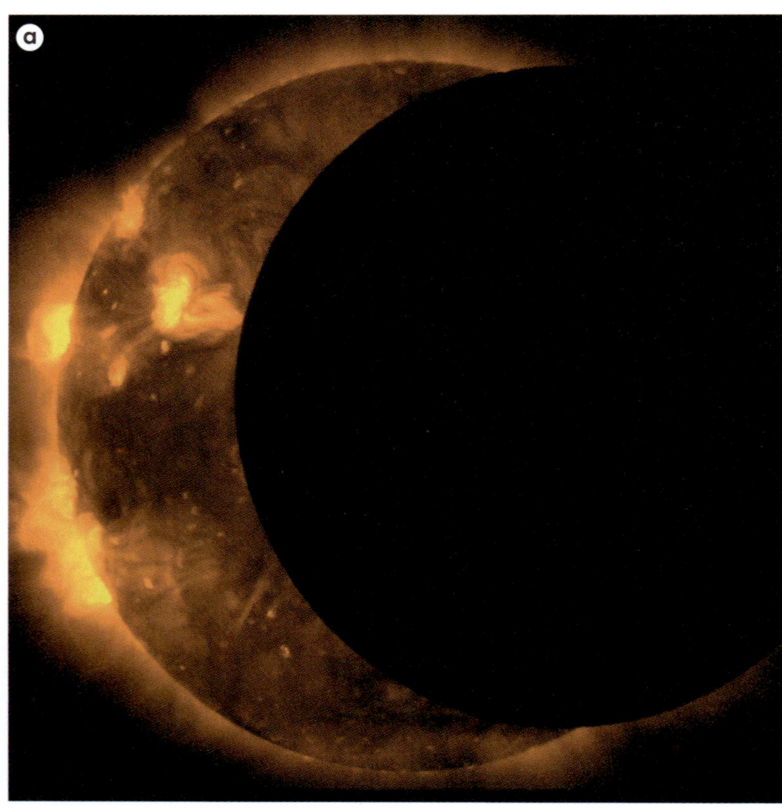

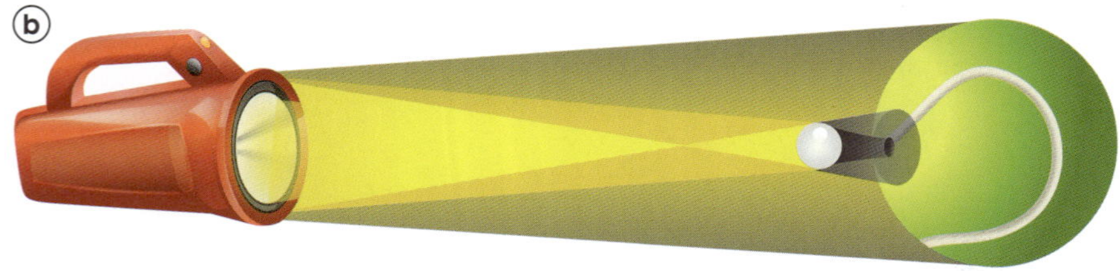

Figure S2.40: (a) A solar eclipse; and (b) how it could be modelled to scientifically communicate what causes this phenomenon

Step 3 I can use digital technologies to organise and present information and data

Once you have selected an appropriate way to communicate your information, it is important to be able to represent any data you have digitally. Using digital technologies means that information can be presented neatly and succinctly, so that it is easy for audiences to understand. Table S2.14 provides a range of examples.

Table S2.14: Digital presentation formats

Information	Presentation format	Example communication presentation method
Data from a first-hand investigation	Scientific table or graph	Google Sheets or Microsoft Excel
A complete first-hand investigation	Scientific report	Google document or Microsoft Word document, with typed information
What causes an eclipse	Scientific poster with text and diagrams	Canva or slides to visually present information
Water movement	Scientific diagrams with labels	Canva, slides or Lucidchart software to create a diagram with labels
Bacteria structure	Scientific model	Canva or Lucidchart software to create two-dimensional models

Figure S2.41: Digital technologies help us to neatly lay out information ready for presenting.

Step 4 I can construct a range of appropriate scientific presentations based on first-hand and second-hand information and data

After selecting a method of communication, and a digital technology to present it, you can then look at how to construct a presentation. When considering how to present scientific information in a chosen format, you can ask the following questions:

- What is the main point I want to get across?
- What information do I need to include in this format?
- What is the best way to present this information in my chosen format?

Once you have answered these questions, you can start to put your information into your selected format. Make sure you are using correct scientific terminology and information relating to your topic. More detailed examples of how to construct each type of presentation can be found in the 'Scientific writing' section on page 347.

Step 5 I can present scientific findings using appropriate conventions for specific audiences

Now that you can determine different formats, you should select information and text that matches the audience you are trying to communicate with. To help you figure out *how* to tailor your communication, consider the:

- age of your audience
- knowledge level of your audience
- idea or scientific principle you want to communicate
- best way to tell your audience about this principle.

Table S2.15 shows how the same information can be presented to two different audiences.

Table S2.15: Tailoring your message for different audiences

Age	Knowledge level	Idea	Communication
Eight-year-old children	Very limited	How to conserve water sustainably	Summarise the water cycle and encourage household water-saving measures on a poster.
Adults	High – environmental scientists	How to conserve water sustainably	Present research on ideas to prevent surface evaporation from lakes and dams as an article or oral presentation.

We can then write a response for each audience. For the eight-year-olds, we could include the following text on a simple poster.

In the water cycle, rain falls and then runs off into storage areas like dams and lakes. We use some of this water in our homes. To help us use less water, we can take shorter showers or shallow baths, and turn off taps while cleaning our teeth. These actions mean every household uses less water overall, and this is also known as using water sustainably.

The adult scientists would expect more sophisticated language, and the information could be displayed using specific diagrams and figures.

When water that has precipitated runs off into storage areas such as dams and lakes, it is subject to surface evaporation, reducing the amount available for household consumption. Placing products such as evaporation-reducing shade balls on the surfaces of these areas will assist in minimising surface evaporation, reducing water loss and increasing availability for household usage, even in times of drought.

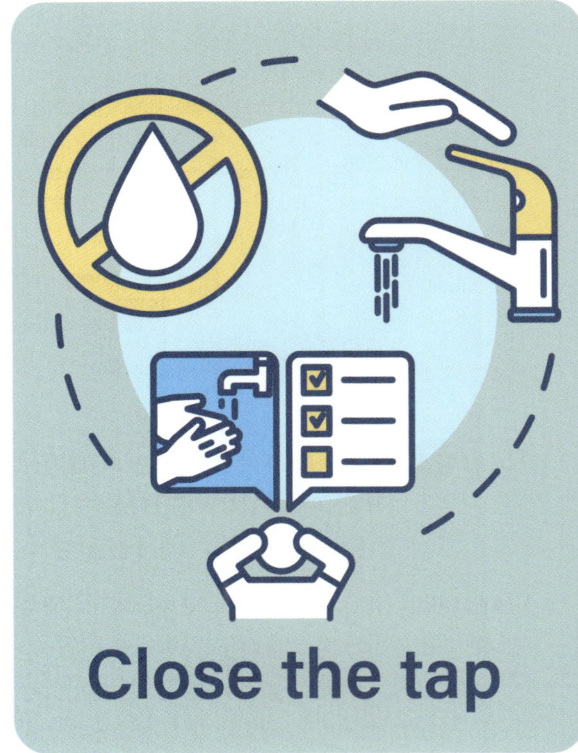

Figure S2.42: A poster showing sustainable water usage in the home – this would be a suitable communication method to share ideas with younger audiences who have less scientific knowledge.

Scientific writing

'Scientific writing' means applying your writing skills to your science studies to help you develop your understanding and communicate what you have learnt.

Key terms

citation: a way of giving credit to a source, usually in the same text where the information appears

plagiarise: to copy someone else's work and present it as your own

Figure S3.1: Scientific writing is fundamental to science communication.

Table S3.1: Investigation report sections matched to their working scientifically processes

Section	Skill
Title	Questioning and predicting
Introduction • Background • Aim • Hypothesis	Questioning and predicting Planning investigations
Risk assessment	Planning investigations
Materials	Planning investigations
Method	Conducting investigations
Results	Processing data and information
Discussion and conclusion	Analysing data and information
References	Analysing data and information

Writing investigation reports

Writing a clear and concise report after your investigation is complete will help other people to understand your work. Table S3.1 provides an overview of the sections that are often included in a standard investigation report, as well as the science inquiry skill used in each section.

Your investigation reports should typically have a similar structure to the one shown below.

The title is clear and uses plain language. Many scientists write their title as a research question.

How does the amount of direct sunlight affect plant growth?

Set the scene. If there is scientific information or context for the investigation, summarise it here.

Introduction

Plants require sunlight to produce their food using photosynthesis. Some plants get direct sunlight, while others get indirect sunlight. It would be interesting to know how different plants are affected by different amounts of sunlight, to help us make decisions about where to put garden beds for certain types of plants.

The background often leads into the objective of the investigation. Use your research question to write the aim starting with a verb, such as: 'To investigate ...'

Aim

To investigate whether the amount of direct sunlight affects growth of a certain species of plant.

If your investigation is an experiment, you should predict what you think will happen based on scientific theory. A good hypothesis will include the independent variable and the predicted effect on the dependent variable.

Hypothesis

If a plant is placed in direct sunlight, then it will grow more than a plant in indirect or no light, because plants require sunlight to produce food required for growth.

List all materials and equipment with amounts and sizes as simple bullet points.

Materials

- *3 plants of the same species and of similar size*
- *250 mL beaker*

The method shows how the investigation was conducted. Number the steps, and write in the past tense and in the third person.

Methods should be written like a recipe – simple, clear and detailed.

Method

1. *Each plant was labelled 1, 2 or 3 and measured to obtain a starting height for each one. The heights were recorded in a simple table.*
2. *Plant 1 was placed in a dark cupboard.*
3. *Plant 2 was placed near a window where it could receive indirect sunlight.*
4. *Plant 3 was placed outside in direct sunlight but sheltered from rain.*
5. *The heights of the plants were measured every week for three weeks and the growth was recorded in the results table.*
6. *The plants were watered the same amount every three days.*

Results

Include recorded data such as the results table. Include any visual elements, such as photos or graphs, here.

Table 1: *Effect of sunlight on plant growth*

Plant environment	Initial height (mm)	Growth after 1 week (mm)	Growth after 2 weeks (mm)	Growth after 3 weeks (mm)
Plant 1: no sunlight	181	0	−1	−3
Plant 2: indirect sunlight	175	1	2	4
Plant 3: direct sunlight	178	2	3	5

Discussion

Use the discussion to analyse the results and identify a key finding. Evaluate the method to identify the quality and limitations of the data.

Start by summarising the data, using amounts in your table. Identify any trends or patterns that could lead to your key finding. Link your finding to your scientific understandings.

Identify any potential errors here and suggest improvements to try to control them.

As the results in Table 1 show, the plant that had the most growth was the plant in direct sunlight (Plant 3). The plant in direct sunlight had the highest growth, with 2 mm after the first week, 3 mm after the second week and 5 mm after the third week, compared to the plant in no sunlight (Plant 1), which had no growth, then shrank and lost growth in weeks 2 and 3. The results make sense because sunlight is crucial in photosynthesis – the process by which plants transform sunlight into usable energy.

One source of error is that two of the plants were inside, and one was outside, which may have had an impact on growth. The method could be improved by requiring all plants to be outside, each with a different amount of shade coverage. There may have also been some errors to do with accurately measuring the plants. The method could be improved by including photos of the plants against the same ruler backdrop, to confirm the accuracy of measurements.

Conclusion

Conclude the report by responding to the aim. Mention whether the results supported or refuted your hypothesis. Do not present any new information.

The conclusion can also be the closing paragraph of the discussion.

The results of this investigation show that the amount of direct sunlight does impact on the growth of this species of plant. The investigation supported the hypothesis that if this species of plant is put in direct sunlight, it will grow more than a plant of that species in indirect or no light. This is due to more sunlight being available for photosynthesis, which is how plants grow.

References

References show the source of any information you used that was not your own, including for the background.

BBC, 2019, 'Photosynthesis', BBC Bitesize Articles, accessed 10 January 2024, mea.digital/gsnsw4_9_1.

Presenting scientific information

The ability to choose and create an appropriate presentation format allows you to transform complex ideas into engaging concepts and to share your discoveries in ways that will best reach your audience. Presenting scientific information effectively is a powerful tool that can help shape the future of science.

Posters

The purpose of a scientific poster is to communicate information on a particular topic or to display the findings of an investigation in a quick and engaging way. When information is shared on a poster, it should be easy for people to see the main idea as they are walking past. A display of scientific posters might include dozens of posters lined up, one after another. This is why it is important for your main finding to be the central focus.

A poster includes the same sections as an investigation report. However, each section includes only the most important information. Anyone who wants to learn more about the investigation details can read the full report. Figure S3.2 shows the basic framework for a scientific poster.

Figure S3.2: A simple framework for the structure of a scientific poster.

Investigation title and student name		
Introduction **Materials & method**	Statement of the **main finding of the investigation**, summarised into one sentence. **Results** Including data, graphs, photos or diagrams.	**Discussion** **Conclusion**
References		

Figure S3.3 gives an example of how we might present the findings from our investigation report on plants and sunlight. We could add visual interest by adding photos of the plants.

Figure S3.3: A simple scientific poster

How does the amount of direct sunlight affect plant growth?

Introduction

Plants require sunlight to produce their food using photosynthesis. This investigation aims to explore how the amount of sunlight affects plant growth. It is hypothesised that 'If a plant is placed in direct sunlight, then it will grow more than a plant in indirect or no light.'

Materials & method

Three plants were placed in different locations: in a dark cupboard, near a window with indirect sunlight, and outside in direct sunlight but sheltered from rain. All the plants were watered the same. Plant height was measured every week for three weeks.

Plants with ***direct sunlight grow more*** *than plants of the same species with indirect or no light.*

Results data

Table 1: Effect of sunlight on plant growth

Plant environment	Initial height (mm)	Growth after 1 week (mm)	Growth after 2 weeks (mm)	Growth after 3 weeks (mm)
Plant 1: no sunlight	181	0	−1	−3
Plant 2: indirect sunlight	175	1	2	4
Plant 3: direct sunlight	178	2	3	5

Discussion

The plant that had the most growth (5 mm) was the plant in direct sunlight. The plant in no sunlight had no growth in week 1, then shrank in weeks 2 and 3. The results make sense because sunlight is crucial in the process by which plants transform sunlight into usable energy – photosynthesis.

One source of error is that two of the plants were inside, and one was outside, which may have had an impact on growth. The method could be improved by requiring all plants to be outside.

Conclusion

The investigation supported the hypothesis that if this species of plant is put in direct sunlight, it will grow more than a plant of that species in indirect or no light.

Reference: *BBC, 2019, 'Photosynthesis', BBC Bitesize Articles, accessed 10 January 2024, mea.digital/gsnsw4_9_1.*

Articles

The purpose of a scientific article is to communicate the findings of an investigation or to summarise research into secondary sources. A scientific article is divided into specific sections, each with a purpose to communicate specific information relating to the scientific concept being studied.

1 Heading – a relevant title that is a statement of what the article is about
2 Author – the name of the person writing the article
3 Abstract – a summary of what was done or what was found in the study
4 Introduction – some research on and background to the topic, linked to a hypothesis
5 Method – the methods scientists used to investigate the topic
6 Results – the results found about the topic
7 Discussion – a detailed analysis of the results, linked to the background
8 References – a list of your sources. See the 'Referencing' section on page 356 for more information.

Figure S3.5 provides an example framework for a science article.

▲ **Figure S3.4:** High-quality scientific articles can be reviewed by other scientists and published in scientific journals.

▼ **Figure S3.5:** The structure of a science article on plant growth

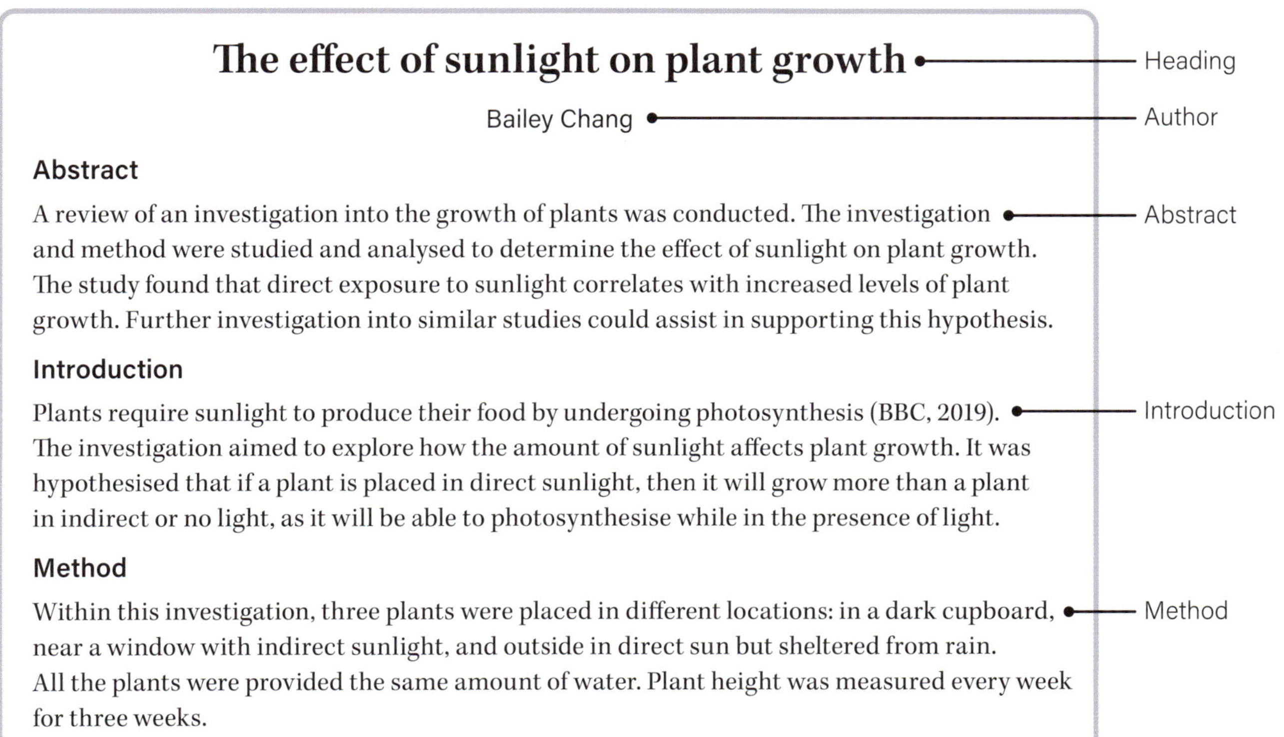

The effect of sunlight on plant growth

Bailey Chang

Abstract

A review of an investigation into the growth of plants was conducted. The investigation and method were studied and analysed to determine the effect of sunlight on plant growth. The study found that direct exposure to sunlight correlates with increased levels of plant growth. Further investigation into similar studies could assist in supporting this hypothesis.

Introduction

Plants require sunlight to produce their food by undergoing photosynthesis (BBC, 2019). The investigation aimed to explore how the amount of sunlight affects plant growth. It was hypothesised that if a plant is placed in direct sunlight, then it will grow more than a plant in indirect or no light, as it will be able to photosynthesise while in the presence of light.

Method

Within this investigation, three plants were placed in different locations: in a dark cupboard, near a window with indirect sunlight, and outside in direct sun but sheltered from rain. All the plants were provided the same amount of water. Plant height was measured every week for three weeks.

Results

Results

It was found that sunlight has a direct impact on plant growth. Table 1 shows the amount of plant growth in different levels of sunlight.

Table 1: Effect of sunlight on plant growth

Plant environment	Initial height (mm)	Growth after 1 week (mm)	Growth after 2 weeks (mm)	Growth after 3 weeks (mm)
Plant 1: no sunlight	181	0	−1	−3
Plant 2: indirect sunlight	175	1	2	4
Plant 3: direct sunlight	178	2	3	5

Discussion

Discussion

The results show that sunlight exposure had a direct impact on plant growth levels. It was found that the plant in direct sunlight had the most growth (5 mm). The plant in no sunlight had no growth in week 1, then shrank in weeks 2 and 3. The results make sense because sunlight is crucial in the process by which plants transform sunlight into usable energy – photosynthesis.

One source of error is that two of the plants were inside, and one was outside, which may have had an impact on growth. The method could be improved by requiring all plants to be outside. This would make the investigation valid.

References

References

BBC, 2019, 'Photosynthesis', BBC Bitesize Articles, accessed 10 January 2024, mea.digital/gsnsw4_9_1.

Presentations

The purpose of a scientific presentation is to communicate the findings of an investigation or article quickly and concisely. You need to summarise the key methods and findings clearly, as your audience may view several presentations in one sitting.

Include the same sections that you would in a scientific article to provide your audience with an overview of the concepts. Those who are interested can then read the full report.

To create a clear oral presentation, use the following tips:

- Pick a theme and stick with it.
- Use carefully selected images that help you to make your point.
- Ensure the summary on each slide is large enough to be read from the back of the room.
- If you use slide transitions, only use one type for the whole presentation as too many can be distracting.

Introduction

Plants require sunlight to produce their food using photosynthesis. This investigation aims to explore how the amount of sunlight affects plant growth. It is hypothesised that if a plant is placed in direct sunlight, then it will grow more than a plant in indirect or no light.

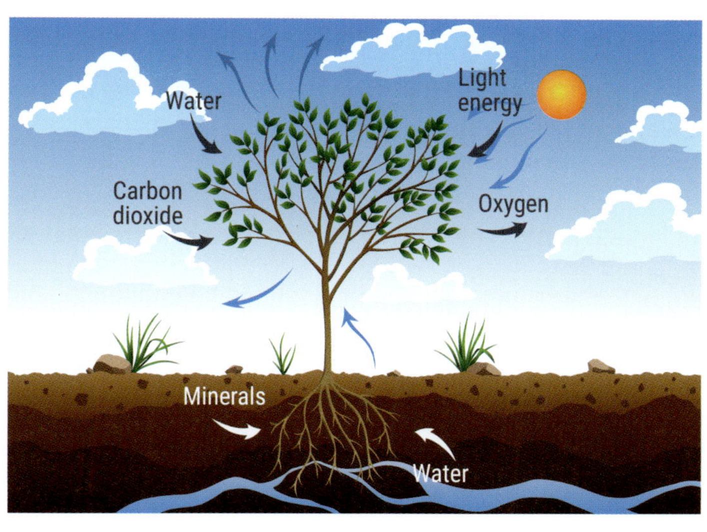

Figure S3.6: An example of a slide layout for a scientific presentation.

Models

The purpose of a scientific model is to provide a representation of a scientific concept that could be difficult to see or understand. Scientific models can show microscopic or large-scale ideas that are impossible to view with the naked eye, or they can help us to visualise the parts and functions of something; for example, a model of how the human eye works.

To construct a good scientific model, use the following criteria:

- Does this model contain all of the relevant components of the concept it is showing?
- Is each section of the model clearly labelled?
- Can the model be easily interacted with (visually or physically) by the audience?

Figure S3.7a aims to inform the audience about the structure of a specific atom. However, it is missing specific labels. To improve the information it communicates and make it a better scientific model, labels should be added. Figure S3.7b is appropriate as it also includes labels of the atom and its structures and components.

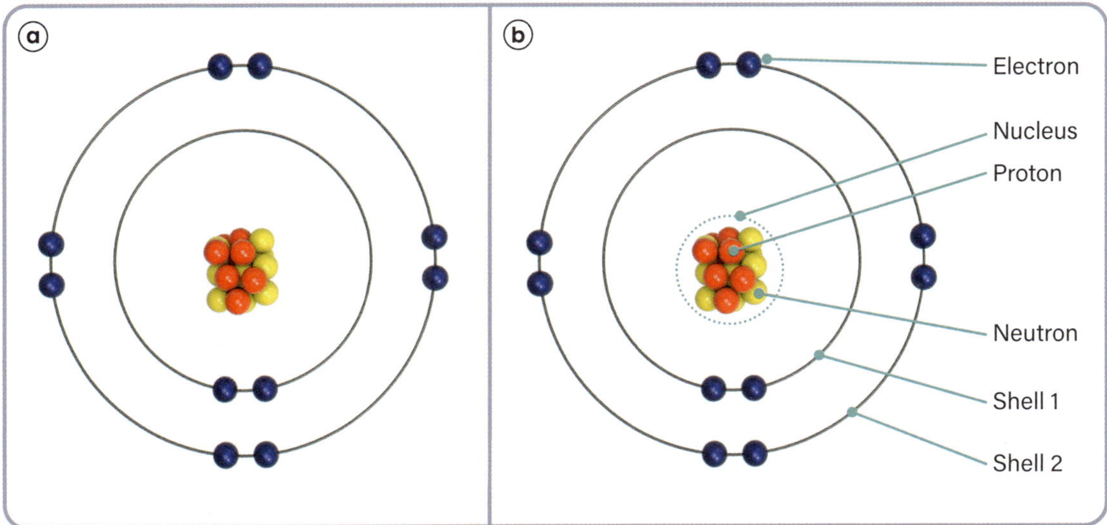

Figure S3.7: (a) A model of an atom of neon that is missing labels; (b) an improved model of an atom of neon including labels

Conducting scientific research

CRAAP

When conducting scientific research, we need to confirm that the information is valid. Even if a source looks credible on the surface, it may not have valid information. One way that we can check this is by using the CRAAP test.

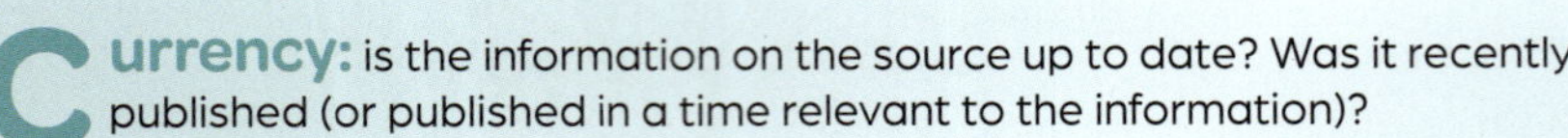

Currency: is the information on the source up to date? Was it recently published (or published in a time relevant to the information)?

Relevance: does the information answer the question I am asking? Is it written in a way that I can understand?

Authority: does the author of the text have experience in this branch of study? Do they have appropriate qualifications to be giving out information on the topic?

Accuracy: where is the information from? Can I find similar information across multiple sources? Is there evidence supporting the information?

Purpose: why is this particular author or company publishing this information? Are they unbiased or are they trying to sell a particular viewpoint? Is the information fact or opinion?

Source: Meriam Library, California State University, Chico, mea.digital/gsnsw4_9_2

A secondary source of information must meet all of these criteria to be considered a valid source. If the source meets the criteria, use it to support your scientific writing. For example, when looking at the question 'How is water used in Australia?', we can analyse two sources of information about water set out in Table S3.2. The table shows us that the first source meets all of the CRAAP criteria, making it valid, while the second source only meets two criteria and should be disregarded.

Table S3.2: Evaluating research sources for the topic of water

Source	Source 1: 'Water Use in Australia, 2022–23'	Pass?	Source 2: 'My Opinion on Water Use in New Zealand'	Pass?
Currency	Updated recently, with the date of last update clearly stated e.g. *last updated: 7 June 2023* – current	☑	Updated recently, with the date of last update clearly stated e.g. *last updated: 13 June 2024* – current	☑
Relevance	Provides information on the distribution and use of water in Australia – highly relevant	☑	Provides information on water use, but not in Australia – related, but not relevant for our research question	☑
Authority	Reputable scientific organisations or research bodies that state clearly their qualifications and methods e.g. *Geoscience Australia – a government department focused on water research* – good authority	☑	No listed qualifications, or qualifications are from organisations that don't exist. The author might be an expert in a different field but isn't qualified to be considered an expert on water	☒
Accuracy	Similar information can be found across multiple other reputable sources and is supported by evidence	☑	Very few other sources include similar information; claims are not supported by evidence or are rejected by other experts in the field	☒
Purpose	To educate Australians on where water is and how it is used, using current peer-reviewed scientific research	☑	To share the opinion on water use of one person who is not an expert in the field	☒

Types of information

When conducting research for investigations and articles, it is important to remember to include a variety of primary and secondary sources.

Primary sources are original, or first-hand, material and can include:

- letters or diaries
- photos
- research data
- laboratory notes
- data collected during first-hand investigations.

Secondary sources are materials that evaluate primary sources. These include:

- scientific reports
- journal articles.

There are also *tertiary sources* of information, which include both primary and secondary sources. Tertiary sources include:

- textbooks and encyclopaedias
- general-knowledge websites (make sure the authors are trusted and the information is current).

Some sources of information are considered *non-scientific*. You should not include these in your reports or articles. These include:

- your personal stories – back up your information with evidence!
- blogs or articles that are not based in fact
- personal opinions.

A well-constructed scientific report or article will include a combination of sources. This shows that you have researched widely to ensure that you include valid and trusted information to inform your audience.

Researching information

When researching a scientific topic, sources include:

- textbooks, books, magazines and journals available from a library or online collection
- search engines such as Google; however, do not just use the information that pops up initially – visit the sites listed and ensure they fulfil the CRAAP test
- specific search areas such as Google Scholar, which will show scientific journal articles relating to your topic.

When searching for information, remember:

1. go to sources that contain valid information – check each one against the CRAAP test
2. choose three or four key words about your topic and use them when entering search terms into Google or Google Scholar; this should increase the number of relevant sites and articles that appear
3. save a copy of the URL (web address) for each website you use, and note down the details or take photos of every textbook or physical source you use. You will need this information when referencing your work at the end of your writing.

Figure S3.8: There are many places to access valid sources of information, such as Google Scholar. All sources of information should be referenced and cited in your written work.

Referencing

When researching and writing scientific material, you must acknowledge the sources of information you use. This is known as referencing. Referencing shows the audience that your information is supported by research, and that you have been careful not to **plagiarise**, or copy, the work of others. At the end of your report or article, you collect the details of all your sources in a reference list.

A reference list should be in alphabetical order and follow a specific format. The most commonly used format in Australia is 'author–date'. Write each entry as follows:

1 surname of author(s) and first initial (or an organisation's name)
2 year the book or article was published
3 book title in *italics* or <u>underlined</u>; or article title in quote marks, with details of the journal
4 the title of the web page or the name of the publisher of the book
5 for digital content, the date you accessed it and the URL.

For example, below are references for a web article and a journal article.

BBC, 2019, 'Photosynthesis,' BBC Bitesize Articles, accessed 10 January 2024, mea.digital/gsnsw4_9_5.

Hilty J, Muller B, Pantin F and Leuzinger S, 2021, 'Plant Growth: The what, the how and the why,' *New Phytologist*, Volume 232, Issue 1, accessed 10 January 2024, mea.digital/gsnsw4_9_6.

Writing evidence-based essays

As part of your science studies, you will be required to write essays. Your essays must be supported by evidence. This type of scientific writing is usually informative or argumentative. In either case, you will need to structure your essay. One way to do this is to write an essay with five or more paragraphs or sections, consisting of:

- an introduction paragraph
- three or more body paragraphs
- a concluding paragraph.

Each body paragraph should follow the TEEL approach. The TEEL acronym reminds you of what to include in each paragraph of your essay:

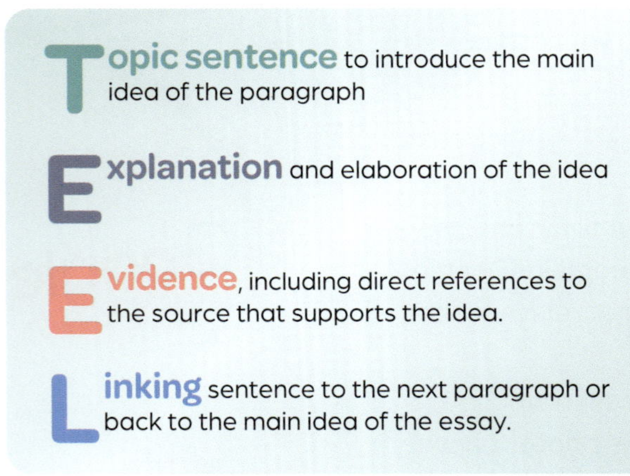

Topic sentence

Explanation/elaboration

Evidence/examples

Linking sentence

Figure S3.9: Each TEEL paragraph should build on the topic and link to the next idea or the essay as a whole.

Writing an informative essay

Informative scientific essays communicate scientific facts and ideas with a neutral tone, and cover all aspects of a topic. These essays are sometimes called research papers because they summarise multiple sources of research available on a particular topic, without including any opinions or personal narrative. There are many ways to structure an informative essay, such as the TEEL approach.

Writing a scientific argument

A scientific argument presents a position on a particular topic. Instead of all aspects of the research being communicated evenly, the research is evaluated and used by the writer to form a position.

Before you begin writing your argument, you will need to develop an evidence base by conducting research on the topic and evaluating each source. This will help you to choose your position.

For example, if the topic is nuclear power, you should find multiple sources that discuss positive and negative aspects of nuclear energy. Read each research piece carefully and critically. Review and evaluate each source by asking the following questions:

- Is this piece of evidence relevant?
- Is this piece of evidence written by a reputable person or group? (If not, it still may be useful as you could use it to criticise the opposing position.)
- What position does it take?
- Is it valid in taking this position?
- If it is not valid, what are the problems with it?

Use the answers to these questions to decide which position you intend to argue scientifically.

Finally, once you have established your position, you need to write your argument. You can follow the standard essay structure, using five or more paragraphs and the TEEL approach for each paragraph. Usually, each of your three body paragraphs should provide a new reason why your position is valid.

Scientific numeracy

Key terms

chemical formula: an expression of the elements that make up a chemical compound, usually presented as a ratio using letters and numbers; for example, H_2O

conversion factor: a number used to change one unit of measurement to another

density: how heavy something is for its size; mass divided by volume

equation: a mathematical statement that shows that two things are equal; for example, $2x + 6 = 14$ is an equation that needs to be solved so that $2x + 6$ does actually equal 14

exponent: the superscript value to the right of a number that says how many times to use the number in a multiplication; for example, when we write 10^3, '3' is the exponent. It means we need to multiply 10 by itself 3 times

mathematical formula: a rule or principle that helps you to find the answer to a question or understand the relationship between variables

mean: a measure of centre (an average) calculated by adding all the numbers together and dividing by how many numbers there are

median: the middle number in a set of numbers when they are arranged in order

mode: the number that appears most frequently in a set of numbers

power: the end product obtained by multiplying a quantity by itself one or more times; for example, 2 to the *power* of 3 is $2 \times 2 \times 2$, which is 8, so the *power* of 2^3 is 8

range: in a set of numbers, a measure of spread between the highest number and the lowest number

ratio: a way of comparing like quantities without units

scale factor: the ratio between corresponding measurements of an object and a copy of that object

scientific notation: a way to write very large or very small numbers in a simple form

subscript: a letter or number written slightly below and to one side of another; for example, '2' in H_2O

superscript: a letter or number written slightly above and to one side of another; for example, '2' in 8^2

'Scientific numeracy' means having the knowledge and skills to apply mathematics to your science studies and understanding the science you see in everyday life. It involves using mathematical knowledge to understand, interpret and assess scientific knowledge and information. We encourage you to look at the *Science 7–10 Data Book* to help you to apply this knowledge to scientific concepts you are working on.

Figure S4.1 Scientific numeracy is the application of maths to science studies.

Units of measurement

When we come across numbers in science, we need to identify:

- what the number is measuring
- the symbol used to represent the measurement
- what unit the number is measured in.

For example, you read that a ball takes 10 seconds to roll down a ramp, and you identify that:

- *time* is the measurement
- the symbol used is *t*
- the unit of measurement used is seconds.

In science, we use an international standard system of units for all types of measurements. These are called 'SI units'. We use some of these units every day, such as using metres to measure distance or minutes to measure time, but some others are only used in advanced science and maths. For example, in this book we use the common unit degrees Celsius (°C) for temperature rather than the SI unit of kelvin (K).

Some numbers do not have units. If you swim two laps of a pool, the length can be measured in metres, but the number of laps will just be '2' with no units because it is a general quantity.

Table S4.1 shows measurements you will come across in this text, their symbol and unit.

Table S4.1: Some common measurements, symbols and units

Measurement	Symbol	Commonly used unit at this level
Length or distance	*d*	metre (m)*
Time	*t*	second (s)*
Speed or velocity	*v*	metres per second (m/s)*
Mass	*m*	gram (g), kilogram (kg)*
Volume	*V*	cubic metre (m^3)*
Capacity		millilitre (mL)
Density	*D*	grams per cubic centimetre (g/cm^3)
Temperature	*T*	degrees Celsius (°C)
Force	*F*	newton (N)*
Energy	*E*	joule (J)*
Number of ...	*N*	no units

*SI unit

Tip 1: **Capitalisation matters**. Notice that **d** is the symbol for distance, while *D* is the symbol for density. If you mix up the lower and upper cases, it will change the meaning.

Tip 2: **Symbols are not the same as units**. Notice that *m* in *italics* is the symbol for 'mass', while 'm' in regular text is the unit for metres. Symbols are used to represent variables in an equation, while the unit tells us *how* the variable is measured. You should be able to distinguish between symbols and units.

Converting between units

Sometimes you will need to change measurements from one kind of unit to another. You can do this if you know the **conversion factor**. For example, if you are converting minutes into seconds, the conversion factor is 60 seconds (60 s), which is the number of seconds in a minute. At other times, you may need to look up the conversion factor. For example, on Earth a kilogram is about the same as 9.8 newtons, so the conversion factor from kg to N is 9.8. Once you know the conversion factor, you then need to multiply or divide.

If the new unit you are converting to is larger, you must divide by the conversion factor. If the new unit is smaller, then you multiply by the conversion factor.

For example, converting minutes to seconds is from a larger unit to a smaller unit, so you multiply by 60. Converting from seconds to minutes is from a smaller to a larger unit, so divide by 60. Worked example 1 shows more examples of conversions.

Worked example 1: Converting between units

1 Convert 2.5 kilometres to metres.
Answer: 2.5 × 1000 = 2500 **m**

- There are 1000 metres in a kilometre, so the conversion factor is 1000.
- This conversion is from a big unit (**km**) to a small unit (**m**), so you need to multiply (**×**).
- Remember to include the unit you are converting to; in this case, metres (**m**).
- If converting in the opposite direction, from m to km, you divide by 1000.

2 Convert 150 grams to kilograms.
Answer: 150 ÷ 1000 = 0.15 **kg**

- There are 1000 grams in a kilogram, so the conversion factor is 1000.
- This conversion is from a small unit (**g**) to a big unit (**kg**), so you need to divide (÷).
- Remember to include the unit you are converting to; in this case, kilograms (**kg**).
- If converting in the opposite direction, from kg to g, you multiply by 1000.

3 Convert 4 hours to seconds.
1 hour = 60 minutes
1 minute = 60 seconds
1 (hour) × 60 (minutes) × 60 (seconds) = 3600 **s**
Answer: 4 × 3600 = 14 400 **s**

- You need to do two conversions here. Hours to minutes and then minutes to seconds. Start with calculating 1 hour in seconds.
- These conversions are both from big units to small units, so you need to *multiply* (**×**). The conversion factors are both 60.
- Remember that there are 4 hours, so we need to multiply by 4 for the total number of seconds. Remember to include the unit you are converting to; in this case, seconds (**s**).

Figure S4.2: Conversion charts for length, time, volume and mass ▼

However, if you want to keep it simple, you can use conversion charts as shown in Figure S4.2, which provides the conversion factor and shows whether to multiply or divide, depending on which direction you are converting to. Worked example 2 has some examples.

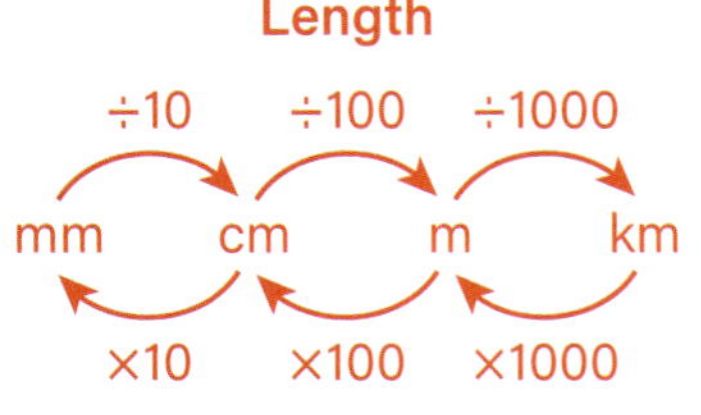

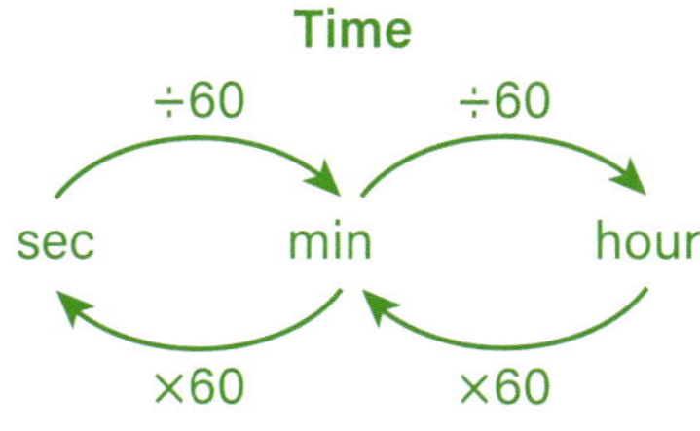

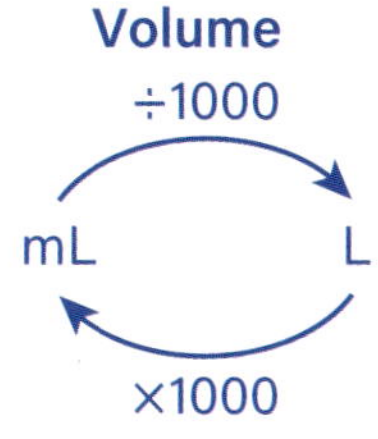

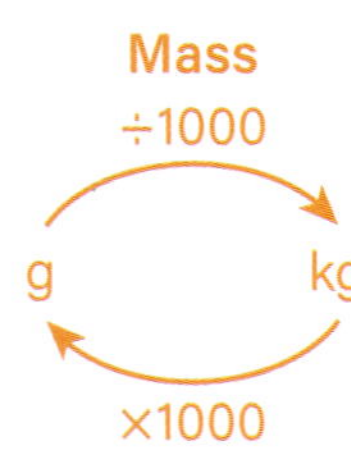

Worked example 2: Converting between units using conversion charts

1 Convert 2.5 km to metres.
Hint: you are converting to a *smaller unit*; therefore, you need to *multiply*.
The conversion from the chart is to multiply by 1000:
2.5 × 1000 = 2500 m

2 Convert 150 g to kilograms.
Hint: you are converting to a *larger unit*; therefore, you need to *divide*.
The conversion from the chart is to divide by 1000:
150 ÷ 1000 = 0.15 kg

3 Convert: 4 hours to seconds.
The conversion for 1 hour to seconds is to multiply by 60 to get minutes, then multiply by 60 again to get seconds. We also need to multiply by 4 to get 4 hours.
4 × 60 × 60 = 14 400 s

Understanding ratios

Ratios are a way of comparing like quantities without units. For example, if a pancake recipe asks for one part milk and two parts flour, that is a ratio of 1 to 2. We can write this as 1:2. Ratios do not require units – no matter how much milk you use, you must always use twice as much flour (× 2).

To identify a ratio, first count the total of each category and write the two totals (for example, '10 to 20'), then replace the word 'to' with a colon, so '10:20'. Next, simplify the ratio by dividing both sides by a common factor. So, divide by 10 to get a ratio of 1:2.

For example, a black Labrador has eight puppies. Six are black and two are chocolate brown. Including the mother, there are nine dogs in total. Table S4.2 helps us to identify the different ratios for this example.

Figure S4.3: The colours of Labrador puppies in a litter can be expressed as ratios.

Table S4.2: Identifying ratios

Ratio	Initial count	Simplify	Answer
Black to chocolate puppies	Black = 6 Chocolate = 2 The ratio is 6:2.	Divide both numbers by the common factor of 2. 6 ÷ 2 = 3 and 2 ÷ 2 = 1 The simplified ratio is 3:1.	The ratio of black to chocolate puppies is 3:1.
Chocolate to black puppies	Switch the numbers around, as the question now asks about chocolate puppies first. The ratio is 2:6.	Divide by 2. 2:6 = 1:3	The ratio of chocolate to black puppies is 1:3.
Black to total puppies	There are 6 black puppies and 8 puppies in total. The ratio of black puppies to total puppies is 6:8.	Divide by 2. 6:8 = 3:4	The ratio of black to total puppies is 3:4.
Total to chocolate puppies	There are 8 puppies in total, and 2 chocolate puppies. The ratio is 8:2.	Divide by 2. 8:2 = 4:1	The ratio of total to chocolate puppies is 4:1.
Chocolate puppies to black dogs	Black dogs = 7 (including the mother). The ratio is 2:7.	This cannot be simplified any further.	The ratio of chocolate puppies to black dogs is 2:7.

Ratios in science

We see ratios all the time in science. A chemical compound, for example, can be represented using a **chemical formula** based on the ratio of elements. We can use the chemical formula and the ratio to solve mathematical problems. The **subscript** to the right of each element is part of the ratio of atoms in the substance.

Water is a substance with the chemical formula H_2O. 'H' is the symbol for the element hydrogen and 'O' is the symbol for the element oxygen. A molecule of water contains 2 hydrogen atoms and 1 oxygen atom. This can be written as the ratio 2:1.

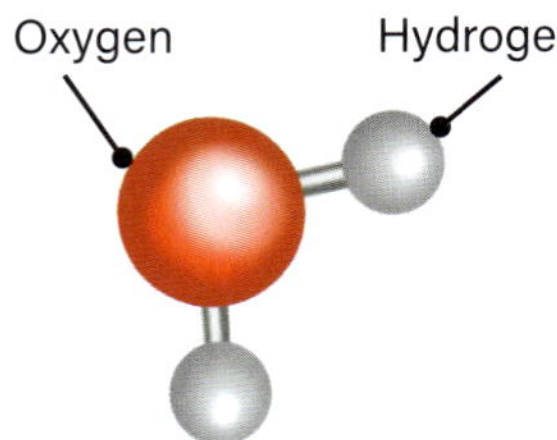

Figure S4.4: This model of water clearly shows the 2:1 ratio of hydrogen to oxygen.

Worked example 3: Ratios in a chemical mixture

How many carbon atoms will there be if there are 40 hydrogen atoms?

1 C_3H_8.
The ratio of carbon atoms to hydrogen atoms is 3:8.
- This means there are 3 carbon atoms and 8 hydrogen atoms in a molecule of propane.
- The coloured subscript numbers indicate how many atoms there are in the molecule.

2 Turn 3:8 into a fraction.
$$\frac{\text{carbon atoms}}{\text{hydrogen atoms}} = \frac{3}{8}$$
- Ratios can be written as fractions.

3 Work out how many carbon atoms we will have if we have 40 hydrogen atoms given the ratio is 3:8.
$40 \div 8 = 5$
- Divide 40 by 8 to find the **scale factor**.
- The scale factor is 5.

4 Multiply the *numerator* x 5 to get the number of carbon atoms.
$3 \times 5 = 15$
- The numerator is 3.
- Multiply 3 by 5 (because 5 is the scale factor).

5 Answer: There will be 15 carbon atoms if there are 40 hydrogen atoms.

Ratios can also be used to carry out calculations based on relative comparisons like percentages, as shown in Worked example 4.

Worked example 4: Calculating energy transfer in a food chain

A diagram of a food chain is shown in Figure S4.5. Each level up in a food chain only receives about 10 per cent of the energy from the level below. For example, grass would have to store 10 joules (J) of energy for a grasshopper to receive 1 J of energy when it eats the grass. The ratio is therefore 1:10, and the scale factor from a lower level to the next level is 10.

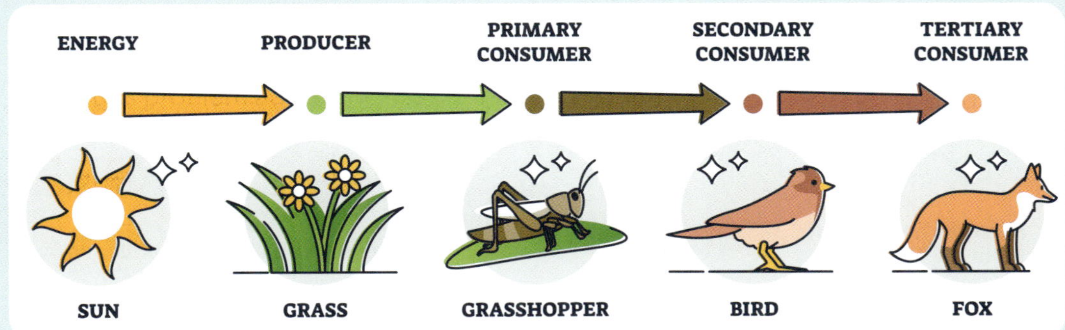

Figure S4.5: Energy moves up a food chain.

Question: How much energy would a grasshopper need to store to provide a bird with 630 joules of energy?

630 x 10 = 6300 J

Answer: The grasshopper needs to store 6300 J of energy to provide the bird with 630 J of energy.

- A bird needs 630 J of energy from a grasshopper, but only 10% of the grasshopper's energy store passes up the food chain to the bird.
- This 10% is the *scale factor.*
- Therefore, you multiply 630 by 10.

Calculating averages

Mean, *median*, *mode* and *range* are terms you will often hear when working with data. Mean, median and mode are different ways to measure the centre of the data. Each are used in different situations for different purposes, depending on what you are trying to understand. Range is used to understand the spread of numbers in a dataset. We will go through each of them in turn.

Mean

The **mean** (often called the average) can be used when many measurements are taken of the same thing and you want a value near the centre of all the measurements. This value is often required in scientific investigations when you have conducted multiple trials. The mean is often used in visual representations of the data.

The mean equals the sum of all values divided by the total number of values.

Here is a set of values: 5, 8, 9, 11, 12.

To arrive at the mean or average of the set of values, add them together: 5 + 8 + 9 + 11 +12 = 45. Then divide 45 by the number of values (in this case, 5 values) to calculate the mean.

The mean is 45 ÷ 5 = 9.

See Worked example 5 on the next page.

Worked example 5: Calculating the mean

Newcastle recorded the following temperatures over four consecutive days in January: 29 °C, 39 °C, 35 °C and 33 °C. Calculate the mean of this data.

1 Mean = sum of all the values (add the values up) divided by the total number of data values

- Identify the formula for calculating the mean.

2 $\frac{29+39+35+33}{4}$

- Substitute the known values into the formula.
- There are four temperature readings, so the total number of data values is 4.

3 $\frac{136}{4} = 34$

- Solve for the mean.

4 Answer: the mean is 34 °C.

- Remember to include any units of measurement (in this case, °C) in your answer.

Median

The **median** is the middle value in a group of values. In the set of values 5, 8, 9, 11, 12, the middle value is 9. Therefore, the median is 9.

Mode

The **mode** is the most frequent value in a dataset. If you are asked to identify the mode, look for the value that occurs the most often. For example, in the group of values 2, 4, 5, 2, 6, 2, the mode is 2 because it occurs 3 times.

Range

The **range** is the difference between the smallest and largest values in a dataset. In the group of values 5, 8, 9, 11, 12, the smallest value is 5 and the largest value is 12. The range is the difference between them. 12 – 5 = 7. Therefore, 7 is the range of this dataset.

Percentages

Percentages tell you what part of the whole you have. One hundred per cent (100%) is the whole thing, so percentages should always add up to 100. Percentages are another way to represent a fraction or decimal, as shown in Table S4.3. A percentage is converted to a decimal by dividing by 100.

Table S4.3: Percentage can be represented in fraction or decimal form

Percentages	0%	10%	20%	30%	40%	50%	60%	70%	80%	90%	100%
Decimals	0	0.1	0.2	0.3	0.4	0.5	0.6	0.7	0.8	0.9	1
Fractions	$\frac{0}{10}$	$\frac{1}{10}$	$\frac{2}{10}$	$\frac{3}{10}$	$\frac{4}{10}$	$\frac{5}{10}$	$\frac{6}{10}$	$\frac{7}{10}$	$\frac{8}{10}$	$\frac{9}{10}$	$\frac{10}{10}$

Use the following basic formula to calculate percentage:

$$\text{Percentage} = \frac{\text{the part of something you want the percentage of}}{\text{the whole thing}} \times 100$$

$$\% = \frac{\text{part}}{\text{whole}} \times 100$$

If someone wants to eat three pieces of this pizza, what percentage of the pizza will they eat?

1 3 out of 10 pieces were eaten.

2 Percentage = $\frac{\text{part}}{\text{whole}} \times 100$

3 Percentage = $\frac{3 \text{ [pieces]}}{10 \text{ [total pieces]}} \times 100$

4 Percentage = 0.3 × 100 = 30%

Table S4.3 also shows you that

$30\% = \frac{3}{10} = 0.3$.

The person will eat 30% of the pizza if they eat 3 slices out of 10.

Figure S4.6: If a pizza is cut into 10 equal pieces, each piece represents 10 per cent.

Worked example 6: Calculating percentages

There is 41 500 000 km^3 of fresh water on Earth, but only 6 225 000 km^3 is available to use. What percentage of fresh water on Earth is available?

Step	Explanation
1 Percentage = $\frac{\text{part}}{\text{whole}} \times 100$	Identify the formula for calculating percentage.
2 The part = available fresh water = 6 225 000 km^3 The whole = total fresh water = 41 500 000 km^3	Identify and label the values given in the question.
3 Percentage = $\frac{6\,225\,000}{41\,500\,000} \times 100$ Percentage = 0.15 x 100 = 15%	Substitute the given values into the formula. You might need to use your calculator.
4 Answer: Available freshwater is 15%	The percentage symbol (%) must always accompany a percentage value.

This formula can be adapted to suit a variety of situations, as shown in Worked example 7.

Worked example 7: Energy efficiency and percentages

Your hair dryer uses 28 000 joules of energy but only puts out 15 960 joules of blowing energy. What is the percentage efficiency of your hair dryer?

1 Percentage = $\frac{\text{part}}{\text{whole}} \times 100$

Identify the formula for calculating percentage.

2 Efficiency = $\frac{\text{part}}{\text{whole}} \times 100$

Efficiency = $\frac{\text{blowing energy}}{\text{total}} \times 100\%$

Adapt the formula to suit the problem. We are trying to work out the *efficiency* of the hair dryer.

3 The part = blowing energy = 15 960 J

The whole = total energy = 28 000 J

Identify and label the values given in the question.

4 Efficiency = $\frac{15\,960}{28\,000} \times 100\%$

Efficiency = 0.57 x 100 = 57%

Substitute the given values into the formula. You might need to use your calculator to work this out.

5 Answer: The efficiency of the hairdryer is 57%

The percentage symbol (%) must always accompany a percentage value.

There is a lot more to learn about percentages and how they are used in science, including how to calculate percentage error. You will learn more in Stage 5.

Scientific notation

Scientific notation, sometimes called standard form, helps us to write very large or very small numbers in a simpler way. Every number can be written in scientific notation as the product (product = multiply, or times) of two numbers that are:

- a decimal greater than or equal to 1 and less than 10
- a power of 10 written as an **exponent**. (An example of an exponent is 10^3, which is the same as 10 × 10 × 10. The '3' means that the 10 is multiplied by itself 3 times.)

2.56×10^3 is an example of scientific notation. It is a different way of writing the number 2 560. It is the same as writing 2.56 × 10 × 10 × 10.

Scientific notation becomes really useful when you work with very big and very small numbers. For example, 2.56×10^7 is the same as writing 2.56 × 10 × 10 × 10 × 10 × 10 × 10 × 10, which equals 25 600 000. It is faster and easier to write 2.56×10^7. The bigger or smaller the number, the more useful it becomes.

In Worked examples 8 and 9, we will explore writing large and small numbers in scientific notation.

Worked example 8: Scientific notation for large numbers

1 Write 65 000 000 in scientific notation.

- 65 million is a very large number.

2 Turn 65 000 000 into a decimal > 1 and < 10.

- To write a number in scientific notation, you need two things: a decimal greater than or equal to 1 and less than 10, and a **power** of 10. To create your decimal, you need to move the decimal point at the end of 65 000 000 left 7 times (think of 65 000 000 as 65 000 000.0)

6.5 0 0 0 0 0 0.

It becomes 6.5 000 000.

3 Answer: 65 000 000 in scientific notation is 6.5×10^7.

- The number of times you move the decimal point to the left becomes the power you need to write your number as in scientific notation. You moved the decimal point 7 times, so the power is 10^7. It is a positive power.
- Remember: When you are expressing a very large number in scientific notation, move the decimal point *left* to create your decimal. Your power will be *positive*.

Worked example 9: Scientific notation for small numbers

1 Write 0.000 009 8 in scientific notation.

- 0.000 009 8 is a very small number less than 1.

2 Turn 0.000 009 8 into a decimal ≥ 1 and < 10. It becomes:

- To write this in scientific notation, you must have the same two things: a decimal greater than or equal to 1 and less than 10, and a power of 10.

0.0 0 0 0 0 9.8

It becomes 9.8.

3 Answer: 0.000 009 8 in scientific notation is 9.8×10^{-6}.

- The number of times you move the decimal point to the right becomes the power you need to write your number as in scientific notation. But because your number is less than 1, your power is ***negative***. You moved the decimal point right 6 times, so the power is 10^{-6}.
- Remember: When you are expressing a very small number in scientific notation, move the decimal point *right* to create your decimal. Your power will be *negative*.

If you want to convert your numbers back from scientific notation into decimal notation, you reverse the method. If the power is positive, move the decimal point to the right to make the number larger; and if the power is negative, move the decimal point to the left to make the number smaller:

3.2×10^4 = move the decimal point 4 places to the right = 32 000

3.2×10^{-4} = move the decimal point 4 places to the left = 0.00032

Table S4.4 compares several examples of decimal notation to scientific notation.

Table S4.4: Comparing decimal and scientific notation

Decimal notation	Scientific notation
6	6×10^0 (10 to the power of zero is 1.)
500	5×10^2
4323.7	4.3237×10^3
0.004 323 7	4.3237×10^{-3}
−23 000	-2.3×10^4
5 830 000 000	5.83×10^9
0.9	9×10^{-1}
0.000 000 0351	3.51×10^{-8}

Graphing

Graphing data is an excellent way to visually represent the quantitative information you have gained from a first-hand investigation. The type of data will determine the type of graph to make. In first-hand investigation reports, this is usually a scatter plot or a column graph.

Figure S4.7: Different graph formats help us to make sense of information in different ways.

Scatter plots

Scatter plots are used when two pieces of data are directly related and show trends, such as a change over time. When looking at trends, a line graph is useful as it can show increases and decreases very clearly. For example, in a particular city, the percentage of new cars sold that had airbags was recorded each year. The results were 18% in 2018, 24% in 2019, 32% in 2020, 45% in 2021 and 60% in 2022. This data could be displayed in a line graph.

Remember that the independent variable (changed factor) should go on the horizontal axis, the *x*-axis. In this case, that would be the year the car was sold. The dependent variable (measured factor) should go on the vertical axis, the *y*-axis. This would be the percentage of cars with airbags. Each axis should have a label and an even scale.

Plotting points

Once you have drawn an even scale, you can plot the points on the graph:

1 Locate the relevant data on the *x*-axis (e.g. year 2018).
2 Locate the relevant data on the *y*-axis (e.g. percentage 18%).
3 Trace on the graph to where the two points meet.
4 Where the points intersect, plot the point by drawing an 'x' on the spot.
5 Continue plotting points for each piece of data.

Line of best fit

Once all of the **data points** are on your graph, you can add a trend line, known as the 'line of best fit'. This line will go through the middle of most of the points, but will not necessarily connect the dots between each. A line of best fit can be straight (ruled) or curved, depending on the data. Use the following instructions to draw a line of best fit:

1 Look at all the points on the graph. Do they make a straight or curved line?
2 For a straight line, place a ruler on the graph, through the points. Move it until most of the points are close to or touching the line. Carefully rule the line on the graph, continuing past the final point to the end of the *x*-axis.
3 For a curved line, carefully sketch a single line curve through the data points, and continue it to the very top of the graph, or edge of the *x*-axis.

Your finished graph should look similar to Figure S4.8.

Figure S4.8 shows a scatter plot for the data given in the example. From this graph, we can see a *trend*. A trend describes what the data is showing. In this case, cars sold with airbags are increasing as the years increase. This is a positive trend.

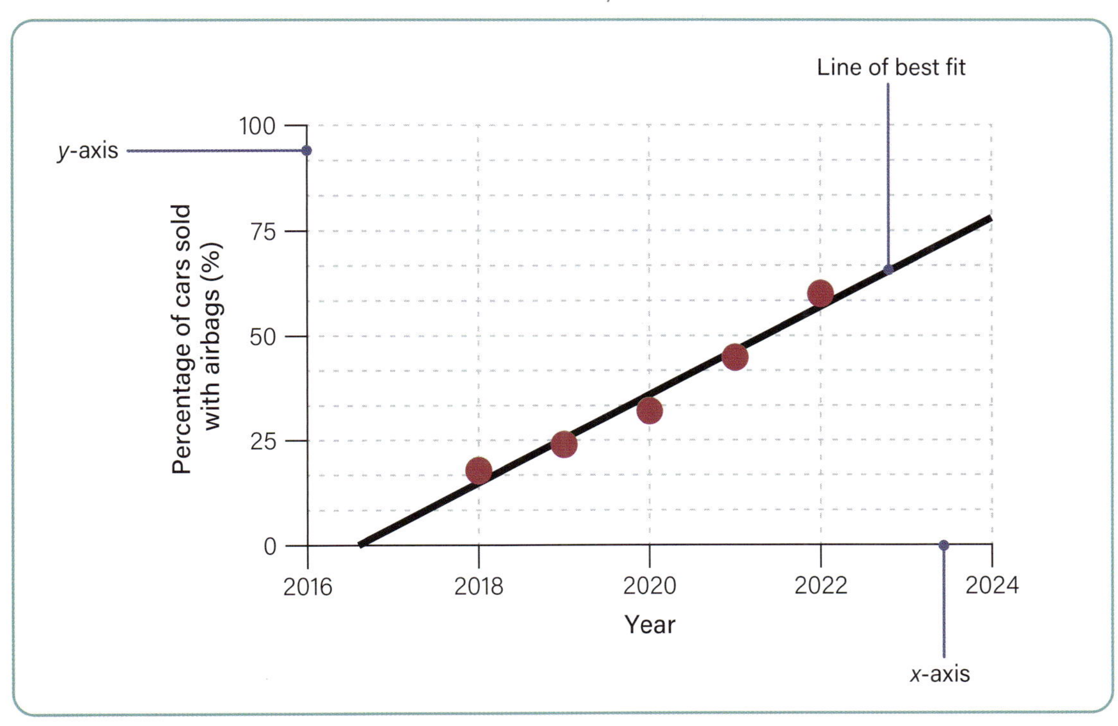

Figure S4.8: Percentage of cars sold with airbags (%) per year, 2016–2024

Column graphs

Column graphs are used to compare separate categories of the same type of data. This means that there is a distinction between the independent variables – each is its own separate category. However, they are still measured against the same criterion.

For example, in City X the average daily maximum temperature for the month was recorded for every month of a year (2024). The results were 33 °C in January, 30 °C in February, 25 °C in March, 19 °C in April, 17 °C in May, 12 °C in June, 8 °C in July, 10 °C in August, 13 °C in September, 20 °C in October, 27 °C in November and 31 °C in December. This data could be displayed in a column graph.

Remember that the independent variable (changed factor) should go on the horizontal *x*-axis. In this case, it is the month of the year. The dependent variable (measured factor) should go on the vertical *y*-axis. This is the temperature in degrees Celsius (°C). Each axis should have a label and an even scale.

Adding columns

Once you have drawn an even scale, add the columns to the graph:

1 Locate the relevant data on the *x*-axis, such as the month of January.
2 Locate the relevant data on the *y*-axis, such as the temperature of 33 °C.
3 Trace on the graph to where the two points meet.
4 Where the points intersect, put a small pencil mark.
5 Using a ruler, extend the point out on each side of the mark, and then down to the *x*-axis to make a column.
6 Continue for each piece of data on the *x*- and *y*-axes.

A completed column graph from the data above should look similar to Figure S4.9.

Note: If you prefer, the data represented as a column graph here could also be represented as a line graph.

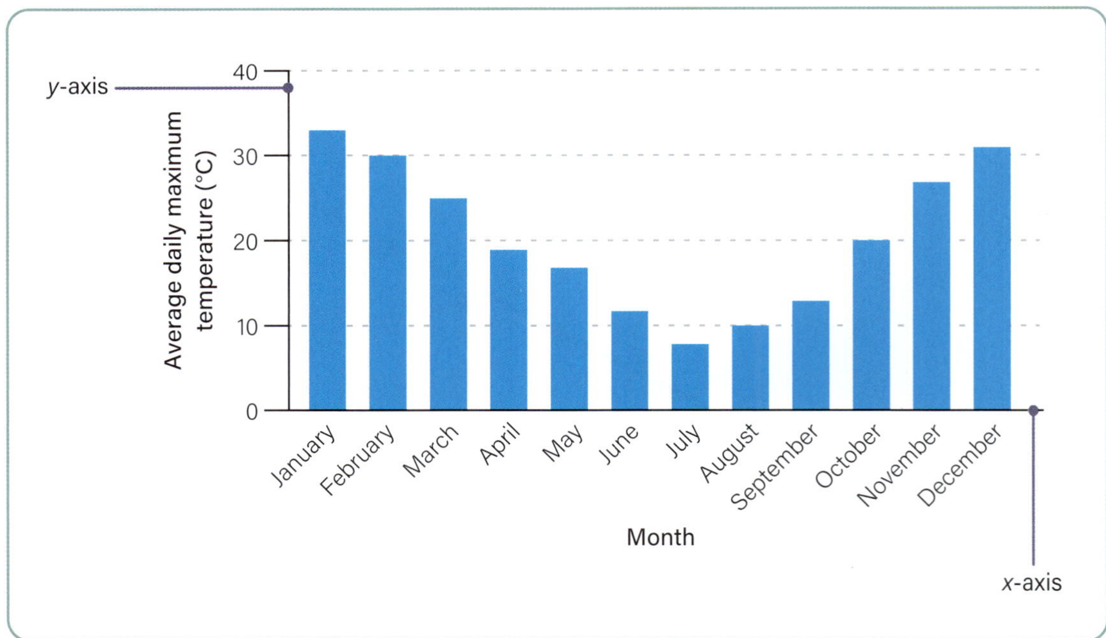

Figure S4.9: Average daily maximum temperature across month of the year in City X, 2024

Making pie charts

A pie chart is a visual representation of parts of a whole. The 'pie' is a circle where each 'piece' represents a fraction or percentage. Therefore, the whole circle represents the whole thing: 100 per cent.

To make a pie chart, the size of each 'piece' should reflect the percentage it represents. So, if one of the parts is 25 per cent, that piece should take up a quarter of the pie, which is 25 per cent.

When there are lots of parts, It can be difficult to estimate the size of each piece by hand. Ideally, you should use a computer program like Microsoft Excel, or an app for making pie charts, to ensure your pieces are the right size for each part they represent, but if you are creating your pie chart by hand, you can estimate. A helpful way to approach this is to divide your circle into eight equal pieces. Each piece is 12.5 per cent of the circle, which will help you make your estimates.

Draw a circle.

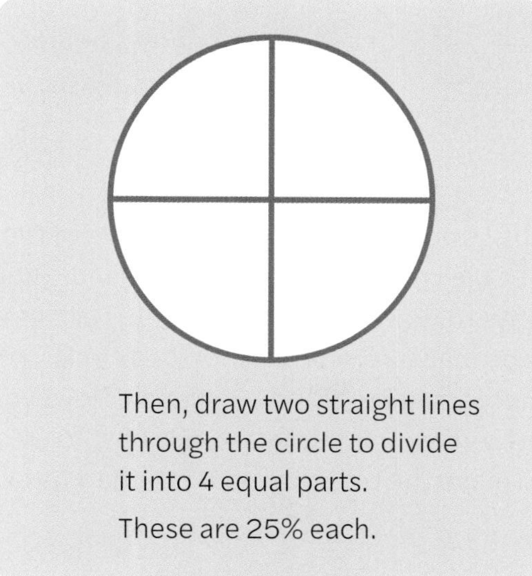

Then, draw two straight lines through the circle to divide it into 4 equal parts.

These are 25% each.

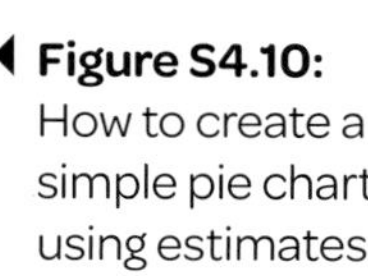

Figure S4.10: How to create a simple pie chart using estimates

Next, draw two more diagonal lines that divide each of the 4 pieces into 2 equal parts, to make 8 equal pieces in total.

These are 12.5% each.

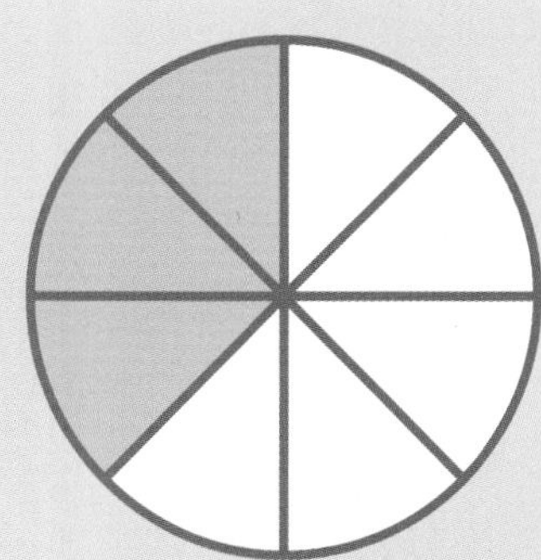

Next, shade the pieces as needed to estimate the size each piece should be for your pie chart.

For example, a pie chart with two parts, 37.5% and 62.5%, would look like the circle shown.

Let's say the percentages you want to represent in a pie chart are 25%, 12% and 63%.

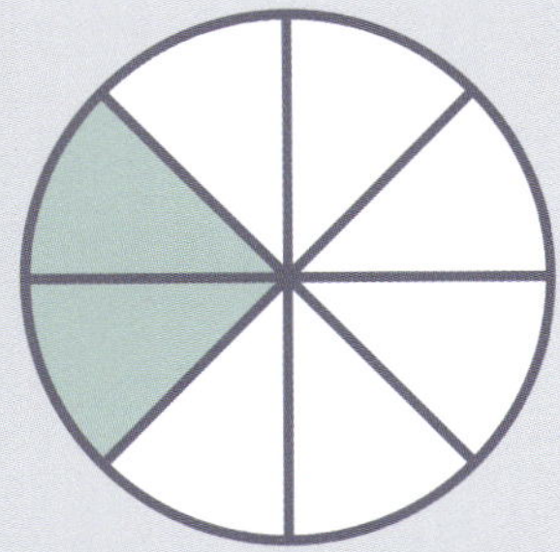

- Start with the pie divided into 8 pieces.
- If any of the parts are exactly 12.5% or multiples of 12.5%, shade those first.
- 25% is 12.5 × 2, so this is a good part to shade first in your pie chart, using 2 pieces to represent 25%.

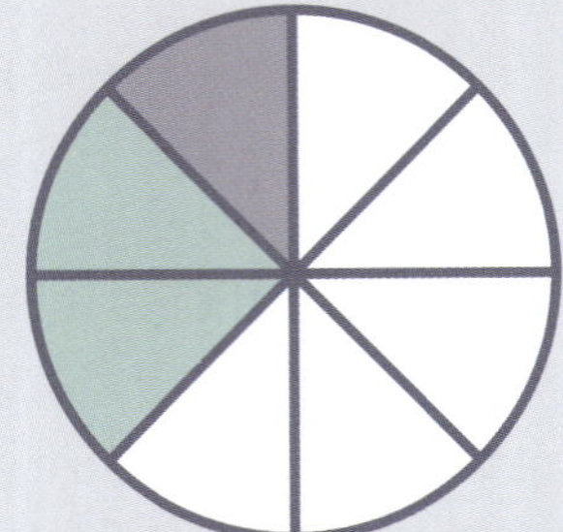

- The next value to represent in your pie chart is 12%.
- Since each piece is 12.5%, you will estimate the part to be slightly smaller than one piece of the pie.
- Shade this in a different colour to the 25% part.

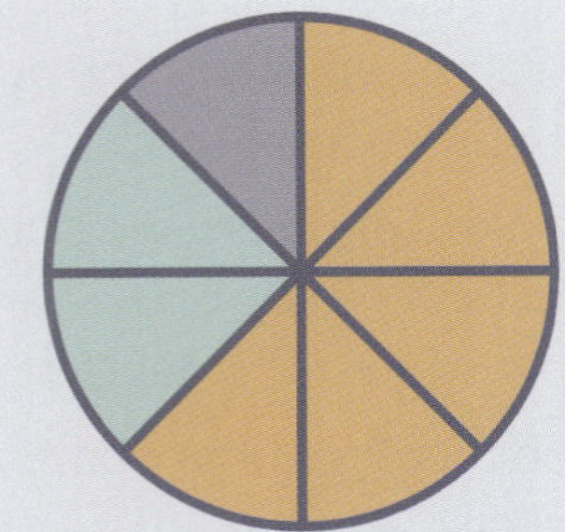

- The remaining part should be 63% of the circle. This is about 5 pieces (5 × 12.5% = 62.5). So, shade 5 more pieces, plus the tiny bit or the piece left over from the 12% part. Use a third colour for the third part of the pie.
- You have created this by hand, so it may not be exact, but using the 12.5% pieces to estimate will help your pie chart be as accurate as possible when digital tools are not available.

▲ **Figure S4.11:** Representing three amounts in a pie chart using estimates

This is one way to estimate the size of the pieces of a pie chart when drawing by hand. Explore other ways to draw pie charts by hand, such as by using a tape measure, a protractor, or a circle with 10 equal pieces that are 10 per cent each. However, when you need to formally present your information, it is best to create the pie chart digitally, to make the pieces of your pie chart as exact as possible.

Worked example 10: Drawing pie charts

To create each sector, first add up the total distance. In our investigation results, Table S2.9 on page 330, the total distance all three wheels travelled is 70 cm. Using a calculator, divide each distance by the total distance, then multiply it by 360 degrees to find the angle for each sector, as shown in Table S4.5.

Table S4.5: Average distance and sector angles

Wheel type	Average distance (cm)	Sector angle
Bottle cap	11	11 ÷ 70 × 360 = 56 degrees
CD	33	33 ÷ 70 × 360 = 170 degrees
Flat washer	26	26 ÷ 70 × 360 = 134 degrees

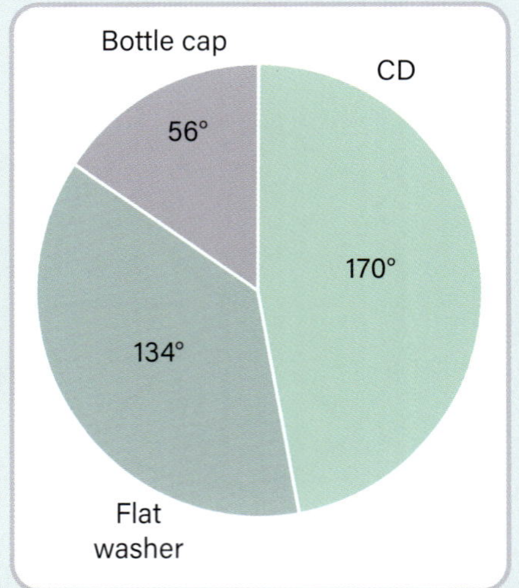

▲ **Figure S4.12:** The finished chart clearly shows each part of the whole distance.

Draw a circle, locate the centre and draw a radius from the top to the centre. Measure the largest angle from that radius line to make the first sector (in this case, 170 degrees). Starting from the new line, measure the next angle and repeat until finished. If your data is given as percentages, then find that percentage of 360 degrees to determine the angles.

Using formulas to determine an unknown value

Formulas are used to show the mathematical relationship between variables. Often, you will be given all the values of variables in a particular **equation** except for one (the unknown). You can use the formula to solve the equation to find the unknown. However, before you apply a formula, you need to determine which formula will be best to use.

Follow these steps to identify and use scientific formulas to determine an unknown value:

1. Read the question carefully. Identify and label the values that are given in the question, and the variable you are trying to determine.
2. Consider the variables you have listed. Identify the formula that includes those variables.
3. Rearrange the formula, as needed, based on your unknown variable, and substitute in known values.
4. Solve for the unknown, keeping the same units as indicated by the given values, or convert units if required by the question.

To practise these steps, we will look more closely at a specific formula. Density is a measure of how heavy something is compared to how much space it takes up. This relationship can be shown mathematically:

$$\text{Density} = \frac{\text{mass}}{\text{volume}}$$

When we replace the words with their symbols (in *italics*), it is now a formula equation:

$$\rho = \frac{m}{V}$$

Worked example 11 will help you to understand how to use this formula to determine an unknown value

Worked example 11: Choosing a formula to determine an unknown value

You pour a cup of tea and measure a volume of 30 mL of honey into it. Your tea now weighs (has a mass of) 42 g more than before the honey was added. What is the density of the honey?

1 mass, $m = 42$ g
volume, $V = 30$ g/cm^3
density, $\rho = ?$

- Collect the information you have been given:
You know the volume of honey and the mass of the tea. You don't know the density of the honey. That is what you are trying to work out.
Note that 1 cm^3 = 1 mL

2 $\rho = \frac{m}{V}$

- Choose the formula that allows you to use the information you have collected. The formula is for density.

3 $\rho = \frac{42}{30}$

- Calculate the density of the honey by substituting the known values into the formula. (You can use your calculator.)

4 Answer: $\rho = 1.4$ **g/cm³**

- $42 \div 30 = 1.4$
- Remember to use the same units as the given values in your answer.

The density formula can be rearranged to determine mass or volume, if you know the other two variables. For example, if you are solving for mass, the formula is rearranged to $m = \rho \times V$. If you know the mass and density, the formula to solve for volume is $V = m/\rho$.

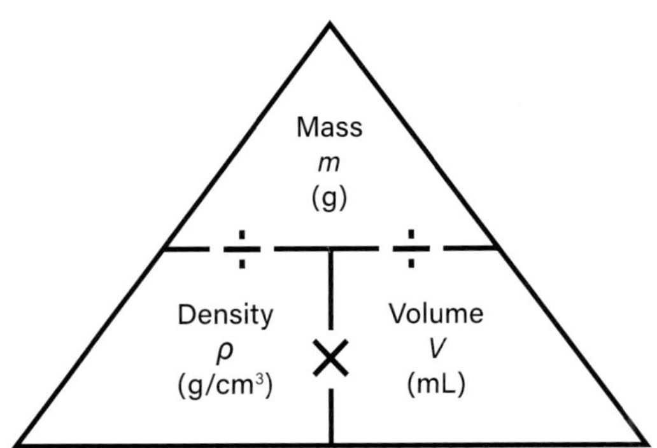

Using scientific equipment

Good science involves curiosity, a desire to understand the world, and a passion for improvement and innovation. Good science also means conducting investigations safely and accurately, which includes knowing your equipment and how to use it. In this section, we will familiarise ourselves with common equipment.

Key terms

concave: hollowed or rounded inwards

meniscus: the curve seen at the top of a liquid in its container

parallax error: the apparent shift in something's position when it is viewed from different angles

surface tension: where the molecules at the surface of a liquid are more attracted to each other than to the air above the liquid

Safety

Conducting investigations in a safe manner is key to good science. Follow laboratory safety rules to ensure you minimise risk. Use the acronym PIECE to remember these safety rules.

Figure S5.1: Working with chemicals requires following safety information and wearing personal protective equipment (PPE).

Protective equipment: Wear safety glasses, gloves and a lab coat to protect you from any chemical splashes or spills.

Instructions: Listen to and read all instructions carefully before you start your investigation.

Equipment: Inspect all equipment to make sure you have the correct items for your specific investigation. Ensure equipment is intact and working properly.

Consumption: Food and drink can become contaminated. Never bring or consume food or drink in the laboratory.

Energy: Manage your energy levels so that you are walking sensibly and holding equipment securely while moving around the lab.

For more information on safety in science, see Section 1.6, pages 14–15.

Pictograms

Pictograms are a part of science safety. They are used to label chemicals with their known hazards. Figure S5.2 shows some pictograms you are likely to see in the laboratory.

Pictograms are used across many industries, so you may recognise them from cleaning products used in your home. Pay close attention to the pictograms on the chemicals you use in the laboratory and follow any warnings. Some chemicals may not be hazardous, so you will not see any pictograms. Others may have several hazards and contain several pictograms for a single substance.

There are many more pictograms than the ones shown here, so be sure to ask your teacher if you come across an unknown pictogram.

Figure S5.2: Common warning pictograms

Flammables

Gases under pressure

Acutely toxic

Burns skin, damages eyes, corrosive to metals

Explosives

Oxidisers

Toxic to aquatic environment

Health hazard

(Chronic health hazards, denoted by the health hazard pictogram, include carcinogens, reproductive toxins, mutagens, specific target organ toxicants, and aspiration toxicants.)

Acutely toxic (harmful)

(Other health hazards, denoted by the exclamation mark pictogram, include skin, eye and respiratory irritation, allergic skin reactions, drowsiness and dizziness.)

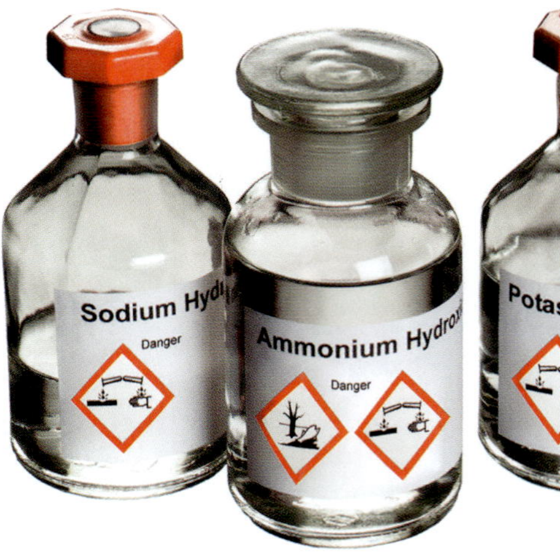

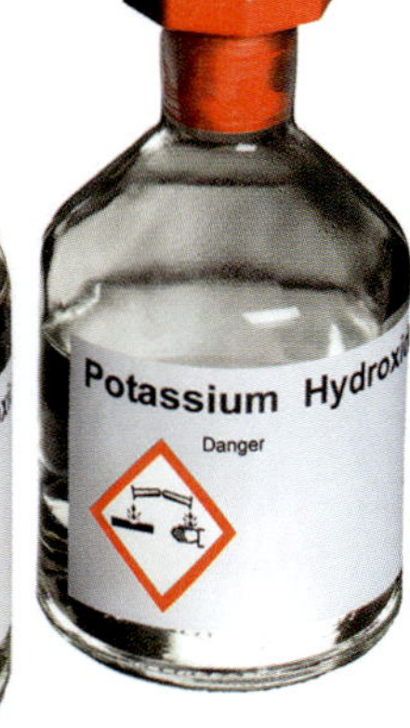

Figure S5.3: Pictograms show the chemicals in your laboratory that are hazardous.

Laboratory equipment

Get to know common laboratory equipment

By learning the names and uses of laboratory equipment, you can select and use the correct equipment for any investigation. Some of the most common equipment is shown here.

Note: You should already be familiar with Bunsen burner use and safety. For a refresher, see page 15.

Figure S5.4: Laboratory equipment

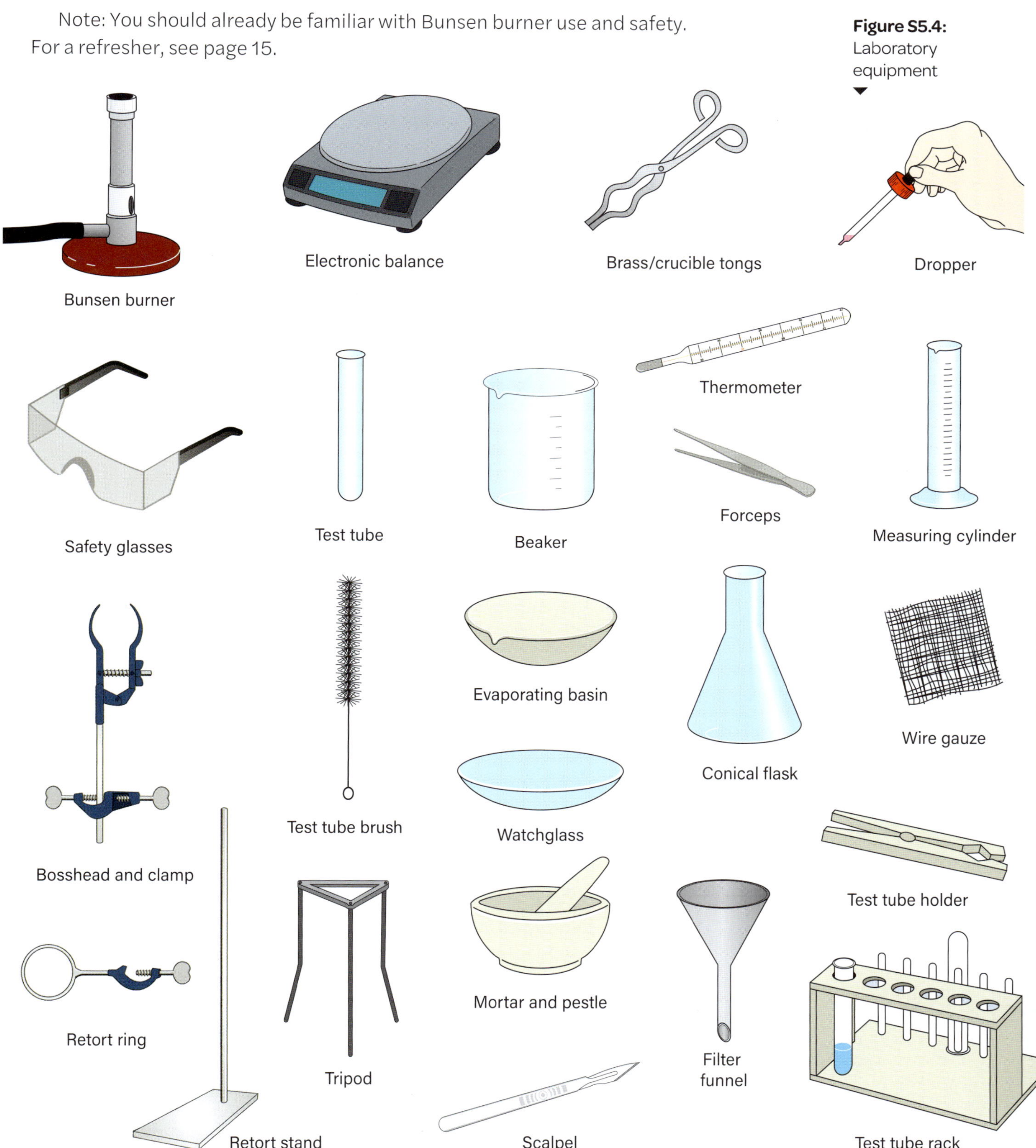

Measurements in science

As discussed earlier, scientists use specialised equipment to measure temperature and volume. In the school laboratory, we measure temperature in degrees Celsius (°C) using thermometers. We also measure volume in litres (L) and millilitres (mL) using measuring cups and beakers. By measuring things carefully, scientists can report their findings reliably.

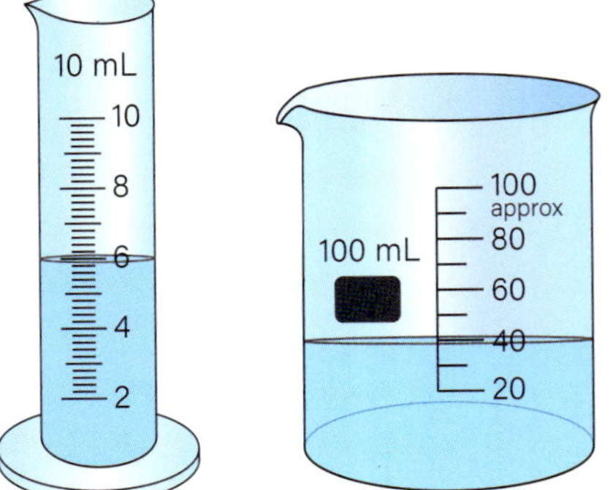

▲ **Figure S5.5:** Beakers and measuring cylinders

Volume measurements

Reading volume measurements is not as easy as it might seem. It requires an understanding of how liquids can curve and the ability to take this into account. You need to use specific equipment, which will help you measure accurately.

Equipment for measuring volume

Beakers and measuring cylinders are the most common pieces of equipment you will use to measure volume. As they are filled with liquid, the measurement reading increases.

When precision matters always use the smallest measuring cylinder available that can measure the required amount. For example, if you need to measure 8.0 mL of something, it is best to use a 10 mL measuring cylinder to get the most accurate result.

Observing the meniscus

Because of **surface tension**, liquids do not form a flat surface in a relatively small container. Instead, a curve forms, which is called the **meniscus**. You should take volume readings at eye level, at the very bottom of a **concave** meniscus, as shown in Figure S5.6. It is obvious that the bottom of the meniscus is between the 36 mL and 37 mL increment lines, which is recorded with certainty. The place value after the increment marks – in this case, the tenth place (.0) – should be estimated. Therefore, the measurement for this volume is 36.5 mL, where the .5 is uncertain.

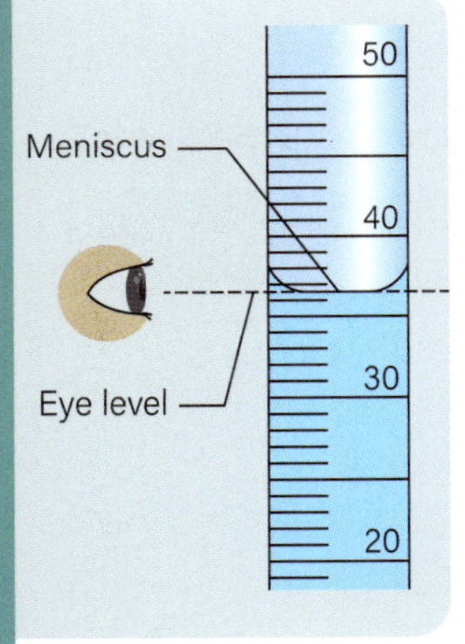

▲ **Figure S5.6:** Take volume readings at eye level.

If you do not take the reading at eye level, **parallax error** will occur. If this happens repeatedly, you will have a systematic error in your measurement, because the error will be the same for all readings. Figure S5.7 shows how reading the meniscus above eye level results in a low reading as compared to the true value, and reading the meniscus from below gives a higher value.

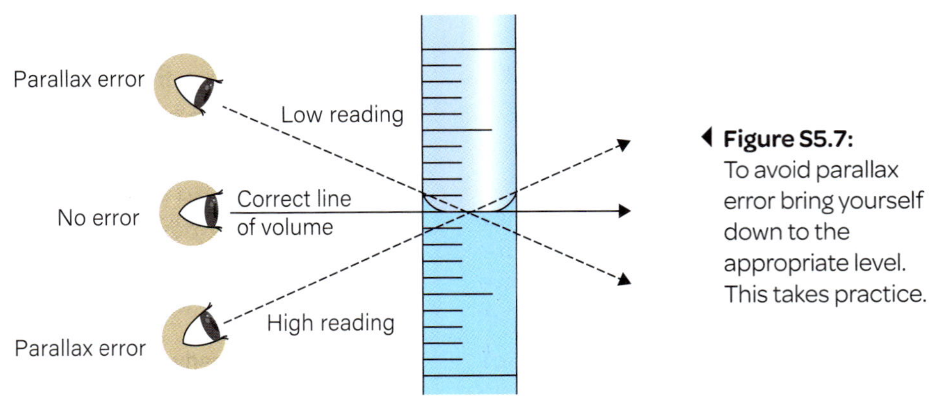

◀ **Figure S5.7:** To avoid parallax error bring yourself down to the appropriate level. This takes practice.

Light microscope

The most common type of microscope in the school science laboratory is a light microscope. It allows you to study samples by shining a bright light through an extremely thin slice of material. The image is magnified by the microscope's lenses, which you look through.

The eyepiece of a light microscope already magnifies samples by 10 times. The objective lens, which is lower down and usually rotates, then magnifies samples by a further amount. To identify the total magnification, multiply the eyepiece magnification (10) by the objective lens magnification.

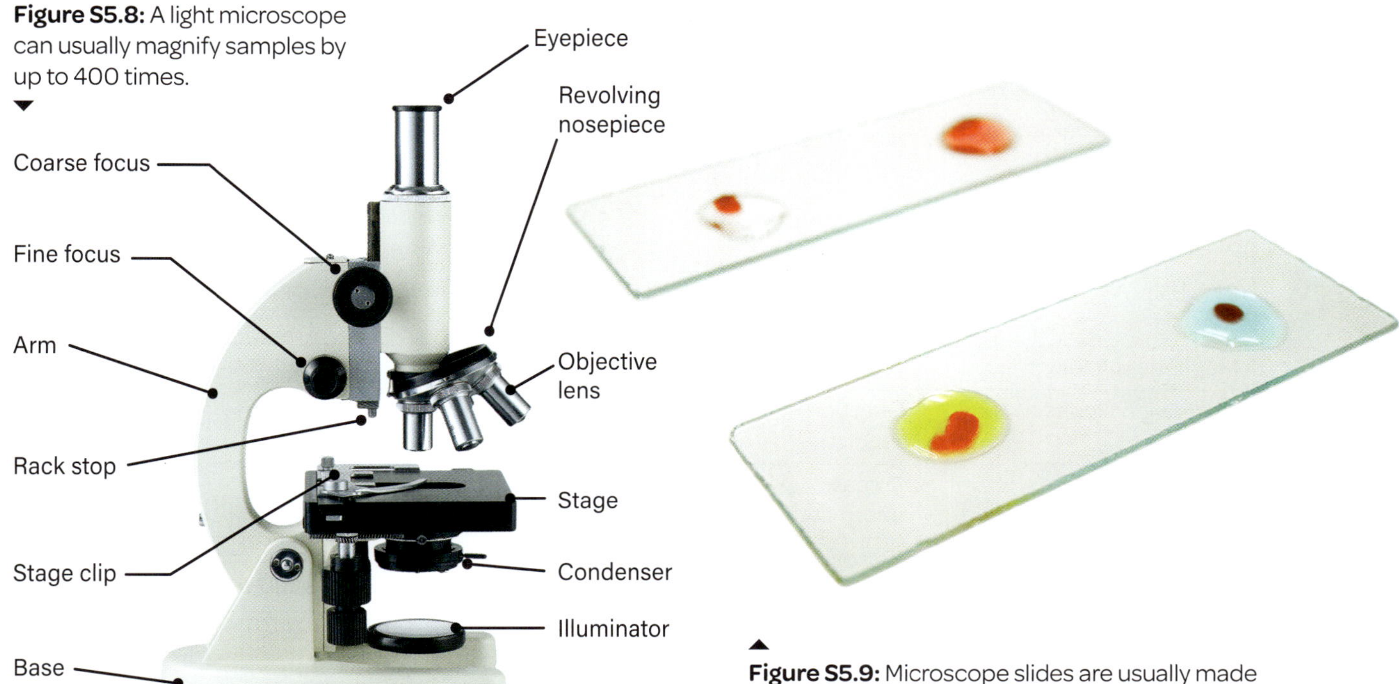

Figure S5.8: A light microscope can usually magnify samples by up to 400 times.

Figure S5.9: Microscope slides are usually made of glass and are very fragile – treat them carefully.

Setting up and using a light microscope

1 Place your microscope on the bench or table, making sure that it is not too close to the edge and that the arm is facing you.

2 Plug your microscope in and turn it on. The illuminator (light) will come on.

3 Lower the stage as far as it can go, using the coarse focus knob.

4 Consider each of the rotating objective lenses. Often they are different colours and have the magnification written on them.

5 Start with the lowest magnification – this is usually 4× (four times). Your microscope eyepiece already has a 10× magnification on its own, so when coupled with the 4× eyepiece, what you are looking at will be 40 times the actual size.

6 Carefully insert the microscope slide onto the top of the stage and hold it firmly under the stage clip. Position it so the object you need to see is in the middle of the stage.

7 Look through the eyepiece and slowly bring the stage upwards towards you, using the coarse focus knob. This can take some time and everything will look bright and fuzzy until you get a glimpse of the slide as it comes into focus.

8 When the slide is roughly in focus, use the fine focus to turn it into a clear image.

9 Increase the magnification by changing the objective lens to 10×, 20× or 40×.

10 40× is usually the highest available magnification and can be tricky to find and focus on. It can also bring the lens extremely close to the slide, enough to crack and break it – monitor this carefully.

Investigations

Investigation number	Investigation title	Working scientifically process	Syllabus outcome	Working scientifically content focus	Teacher demon-stration
Focus area 1: Observing the universe					
1.1	Comparing observations	Observing	SC4-WS-01	Making observations using your senses and scientific tools	
1.8	Using scientific equipment to make observations	Observing	SC4-WS-01	Using scientific tools to enhance observations	
1.10	Investigating how colour affects temperature	Conducting investigations	SC4-WS-04	Using scientific equipment	
1.13	Modelling eclipses	Conducting investigations	SC4-WS-04	Implementing safe practices when using scientific equipment	
Focus area 2: Forces					
2.1	Balloon rockets	Processing data and information	SC4-WS-05	Constructing a table to gather first-hand information	
2.2	Blowball	Analysing data and information	SC4-WS-06	Describing trends in collected data	
2.3	Magnetic shielding	Problem-solving	SC4-WS-07	Suggesting solutions to problems	
2.4	Charging balloons	Questioning and predicting	SC4-WS-02	Proposing hypotheses	
2.6	Measuring gravity	Processing data and information	SC4-WS-05	Constructing graphs	
2.8	Magnetic accelerator	Processing data and information	SC4-WS-05	Constructing a table to gather first-hand information	
2.9	Building an electric motor	Questioning and predicting	SC4-WS-02	Constructing a scientific research question	
2.10A	Making a catapult	Problem-solving	SC4-WS-07	Using criteria to solve problems	
2.10B	Modelling a seesaw	Problem-solving	SC4-WS-07	Identifying cause-and-effect relationships	
2.10C	Pulley investigation	Problem-solving	SC4-WS-07	Solving familiar problems	
2.10D	Using ramps	Problem-solving	SC4-WS-07	Identifying scientific problems	
Focus area 3: Cells and classification					
3.2	Classifying items	Conducting investigations	SC4-WS-04	Collecting and recording data	
3.4	Classifying supermarket items	Communicating	SC4-WS-08	Communicating information	
3.5	Investigating features of marine animals	Observing	SC4-WS-01	Using scientific tools to enhance observations	
3.7	Investigating structural adaptations	Observing	SC4-WS-01	Recording observations	
3.9	Examining cells under a microscope	Conducting investigations	SC4-WS-04	Implementing safe practices when using scientific equipment	
3.10	Examining cell structures	Observing	SC4-WS-01	Recording observations	
3.11A	Investigating respiration	Conducting investigations	SC4-WS-04	Using scientific equipment	

Investigation number	Investigation title	Working scientifically process	Syllabus outcome	Working scientifically content focus	Teacher demon-stration
3.11B	Energy from food	Conducting investigations	SC4-WS-04	Implementing safe practices when using scientific equipment	
Focus area 4: Solutions and mixtures					
4.1	Compressing liquids and gases	Planning investigations	SC4-WS-03	Identifying scientific variables	
4.2	Changing states of water	Problem-solving	SC4-WS-07	Identifying cause-and-effect relationships	
4.3	Exploring density	Planning investigations	SC4-WS-03	Distinguishing between variables	
4.4	Calculating density	Planning investigations	SC4-WS-03	Distinguishing between variables	
4.5	Solutes and solvents	Planning investigations	SC4-WS-03	Identifying variables	
4.6A	Water as a solvent	Conducting investigations	SC4-WS-04	Using scientific equipment	
4.6B	Solubility and temperature	Conducting investigations	SC4-WS-04	Implementing safe practices	
4.7A	Calculating the concentrations of solutions	Conducting investigations	SC4-WS-04	Collecting accurate data	
4.7B	Preparing dilutions	Conducting investigations	SC4-WS-04	Using scientific equipment	
4.8A	Classifying substances	Problem-solving	SC4-WS-07	Using criteria to find solutions to problems	
4.8B	The Tyndall effect	Problem-solving	SC4-WS-07	Suggesting solutions to familiar problems	
4.9	Separating colours in dyes by paper chromatology	Problem-solving	SC4-WS-07	Explaining phenomena using cause-and-effect relationships	
4.10	Purifying muddy water	Conducting investigations	SC4-WS-04	Using scientific equipment	
4.11A	Evaporating a solution	Conducting investigations	SC4-WS-04	Implementing safe practices	
4.11B	Growing crystals	Conducting investigations	SC4-WS-04	Using scientific equipment	
4.11C	Demonstrating distillation	Conducting investigations	SC4-WS-04	Identifying correct processes	✓
4.14	Purifying oily water	Planning investigations	SC4-WS-03	Identifying appropriate materials	✓
Focus area 5: Living systems					
5.3	Breathing rate and exercise	Processing data and information	SC4-WS-05	Collecting and presenting data	
5.4	Dissecting a heart	Questioning and predicting	SC4-WS-02	Constructing a scientific research question	
5.7	Water transport in plants	Questioning and predicting	SC4-WS-02	Formulating testable questions (hypotheses) considering variables and controls	

Investigation number	Investigation title	Working scientifically process	Syllabus outcome	Working scientifically content focus	Teacher demon-stration
5.8	Dissecting a flower	Communicating	SC4-WS-08	Constructing a scientific presentation	
5.9	Identifying community members in a terrestrial ecosystem	Communicating	SC4-WS-08	Presenting scientific findings	
5.11	Measuring abiotic factors	Questioning and predicting	SC4-WS-02	Constructing a scientific research question	
5.13	Algal balls	Processing data and information	SC4-WS-05	Collecting and presenting data	
Focus area 6: Periodic table and atomic structure					
6.1	Comparing metals and non-metals	Processing data and information	SC4-WS-05	Constructing a table to gather information	
6.2	Comparing the properties of elements, compounds and alloys	Analysing data and information	SC4-WS-06	Identifying and describing trends in collected data	
6.4	Flame test	Analysing data and information	SC4-WS-06	Drawing conclusions that are based on patterns in data and information	
6.6A	Modelling atomic structure alongside the periodic table	Processing data and information	SC4-WS-05	Identifying data from tables	
6.6B	Investigating the reactivity of metals	Processing data and information	SC4-WS-05	Constructing a table to organise data collected in a first-hand investigation	
6.7	Investigating the different properties of a compound and one of its elements	Analysing data and information	SC4-WS-06	Explaining relationships between observations and information	
6.9	Modifying metals	Analysing data and information	SC4-WS-06	Evaluating experimental methods	
Focus area 7: Change					
7.1	Rolling balls	Observing	SC4-WS-01	Using observations and measurements to answer questions	
7.2A	Conduction: Heat energy transfer in a solid	Planning investigations	SC4-WS-03	Constructing investigation plans	
7.2B	Convection: Heat energy transfer in a liquid	Planning investigations	SC4-WS-03	Identifying variables	
7.3	Energy transformations	Conducting investigations	SC4-WS-04	Identifying and managing risks	
7.6	Law of conservation of mass	Conducting investigations	SC4-WS-04	Using scientific equipment to conduct investigations	
7.7A	Modelling respiration and photosynthesis	Planning investigations	SC4-WS-03	Identifying scientific materials	
7.7B	Corrosion of iron	Planning investigations	SC4-WS-03	Identifying and managing risks	
7.7C	Preventing corrosion	Planning investigations	SC4-WS-03	Identifying investigation variables	
7.8	Exothermic and endothermic reactions	Conducting investigations	SC4-WS-04	Collecting accurate first-hand and second-hand data	
7.9	Modelling Earth's structure	Observing	SC4-WS-01	Using scientific tools to enhance observations	

Investigation number	Investigation title	Working scientifically process	Syllabus outcome	Working scientifically content focus	Teacher demon-stration
7.10	Modelling seafloor spreading	Observing	SC4-WS-01	Making inferences based on your observations	
7.11A	Modelling slab pull	Conducting investigations	SC4-WS-04	Implementing safe practices	
7.11B	Observing convection currents	Conducting investigations	SC4-WS-04	Using scientific equipment	
7.12	Slinky waves	Planning investigations	SC4-WS-03	Constructing scientific aims	
7.13A	Viscosity of lava	Conducting investigations	SC4-WS-04	Collecting accurate data	
7.13B	Wax volcano	Conducting investigations	SC4-WS-04	Implementing safe practices	
7.15A	Observing minerals	Observing	SC4-WS-01	Using senses to make observations	
7.15B	Extracting copper	Observing	SC4-WS-01	Making inferences based on observations	
7.16	Modelling the rock cycle	Conducting investigations	SC4-WS-04	Implementing safe practices when using scientific equipment	
7.17A	Cooling rate and crystal size	Conducting investigations	SC4-WS-04	Using scientific equipment to conduct investigations	
7.17B	Observing igneous rocks	Conducting investigations	SC4-WS-04	Collecting accurate data	
7.18	Modelling contact metamorphism	Planning investigations	SC4-WS-03	Constructing scientific methods	
7.19A	Modelling the formation of sandstone	Observing	SC4-WS-01	Recording observations accurately	
7.19B	Observing sedimentary rocks	Observing	SC4-WS-01	Using senses to make observations	
7.20	Making 'fossils'	Planning investigations	SC4-WS-03	Identifying variables	
Focus area 8: Data science 1					
8.1	Pin drop	Analysing data and information	SC4-WS-06	Identifying trends in data	
8.2	Have you been influenced?	Problem-solving	SC4-WS-07	Solving scientific problems using cause-and-effect relationships	
8.3	Can you believe your eyes?	Problem-solving	SC4-WS-07	Suggesting solutions to scientific problems	
8.4	Modelling bridges	Problem-solving	SC4-WS-07	Using criteria to solve problems	
8.5	Trends in the weather	Analysing data and information	SC4-WS-06	Describing trends from data and information	
8.6	Modelling the big bang	Problem-solving	SC4-WS-07	Identifying scientific problems	
8.7A	Reaching great heights	Problem-solving	SC4-WS-07	Providing solutions to problems	
8.7B	Height and arm span	Problem-solving	SC4-WS-07	Explaining scientific problems using cause-and-effect relationships	
8.8	Analysing rainfall	Analysing data and information	SC4-WS-06	Explaining relationships between datasets and information	
8.9	Modelling the yowie	Problem-solving	SC4-WS-07	Explaining scientific phenomena	

Investigation 1.1

Comparing observations

Process: Observing

Focus: Making observations using your senses and scientific tools

Our observations can be enhanced by using scientific equipment. Which scientific tools can help you make detailed observations of items from your school's playground?

Hint 1: *What features can you describe using your senses?*

Hint 2: *How does magnifying something help you make detailed observations?*

Aim

To make and compare observations of leaves using your senses and scientific equipment

Materials

- plant leaves (collected from the playground)
- magnifying lens
- microscope

Method

1 Copy the results table into your notebook, adding a title.

2 Collect two different types of leaves from the school playground.

3 a Inspect the leaves using your senses. What observations can you make about them?

 b Record your observations in the results table.

4 a Inspect the leaves using the magnifying lens. Have your observations changed?

 b Record any new observations in the results table.

5 a Place the leaves under a microscope. Observe the leaves at high magnification. Can you make any new observations?

 b Record any new observations in the results table.

Questions

1 Which senses did you use to make observations of your leaves?

2 How did using the magnifying lens change your observations?

3 How did you make sure that your observations were recorded accurately in the results table?

4 What inferences can you make about the leaves based on your observations?

5 Suggest how scientific equipment can sometimes limit our observations. (Hint: Think about the observations you made with your senses.)

Conclusion

Copy and complete:

'The results show that: *(respond to the aim).*'

Figure I1.1: Try to find leaves from native plants, like these leaves from a eucalyptus tree.

Table I1.1: Results

Plant leaf	Observations using your senses	Observations using a magnifying lens	Observations using a microscope

Investigation 1.8

Using scientific equipment to make observations

Process: Observing

Focus: Using scientific tools to enhance observations

A key component of first-hand investigations is making precise observations. Selecting and using appropriate equipment allows us to measure observations and record quantitative (numerical) information. From this data, we can draw conclusions.

Hint 1: What observations will be useful to make during this investigation?

Hint 2: What equipment enables you to precisely measure your observations?

Aim

To investigate the temperature at which water boils

Materials

- heatproof mat
- Bunsen burner
- tripod
- gauze mat
- retort stand with bosshead clamp
- safety glasses
- 100 mL measuring cylinder
- tap water
- 250 mL beaker
- thermometer
- matches
- stopwatch
- heatproof gloves

AN OPEN FLAME AND BOILING WATER ARE HAZARDS. BE CAREFUL. IF YOU BURN YOURSELF, TELL YOUR TEACHER IMMEDIATELY AND PLACE THE BURNT AREA UNDER COLD RUNNING WATER FOR 20 MINUTES.

Method

1. Copy the results table into your notebook, adding a title and as many rows as needed.
2. Set up the heatproof mat, Bunsen burner, tripod, gauze mat and retort stand with a bosshead clamp (see Figure I1.8 on the next page). Put on the safety glasses.
3. Using the measuring cylinder, measure 50 mL of cold tap water. Pour the water into the beaker. Place the beaker on the gauze mat above the Bunsen burner.
4. Record the initial temperature of the water using the thermometer. Record this temperature in the results table.
5. Use the matches to light the Bunsen burner. Turn to the blue flame. Start the stopwatch.
6. Observe the water. When the water starts to boil (bubble rapidly), turn off the Bunsen burner. Stop the stopwatch.
7. a. Record the temperature of the boiling water in the results table.
   b. Record the time it took for the water to boil in the results table.

Table I1.8: Results

Trial number	Cold water temperature (°C)	Boiling water temperature (°C)	Time taken to reach final temperature (minutes)

continues ▶

8 Repeat the experiment twice to see if the results are consistent.

9 Wearing heatproof gloves, hold the beaker with both hands and pour the water into the sink.

Questions

1 How did you use your senses to make observations about the changes occurring to the water?

2 What piece of equipment allowed you to quantitatively measure your observations of the water?

3 Explain how you accurately recorded this quantitative data.

4 What conclusion did you draw from the data you collected?

5 What is happening to the water when it starts to boil, from a scientific perspective?

Conclusion

Copy and complete:
'The results show that: *(respond to the aim)*.'

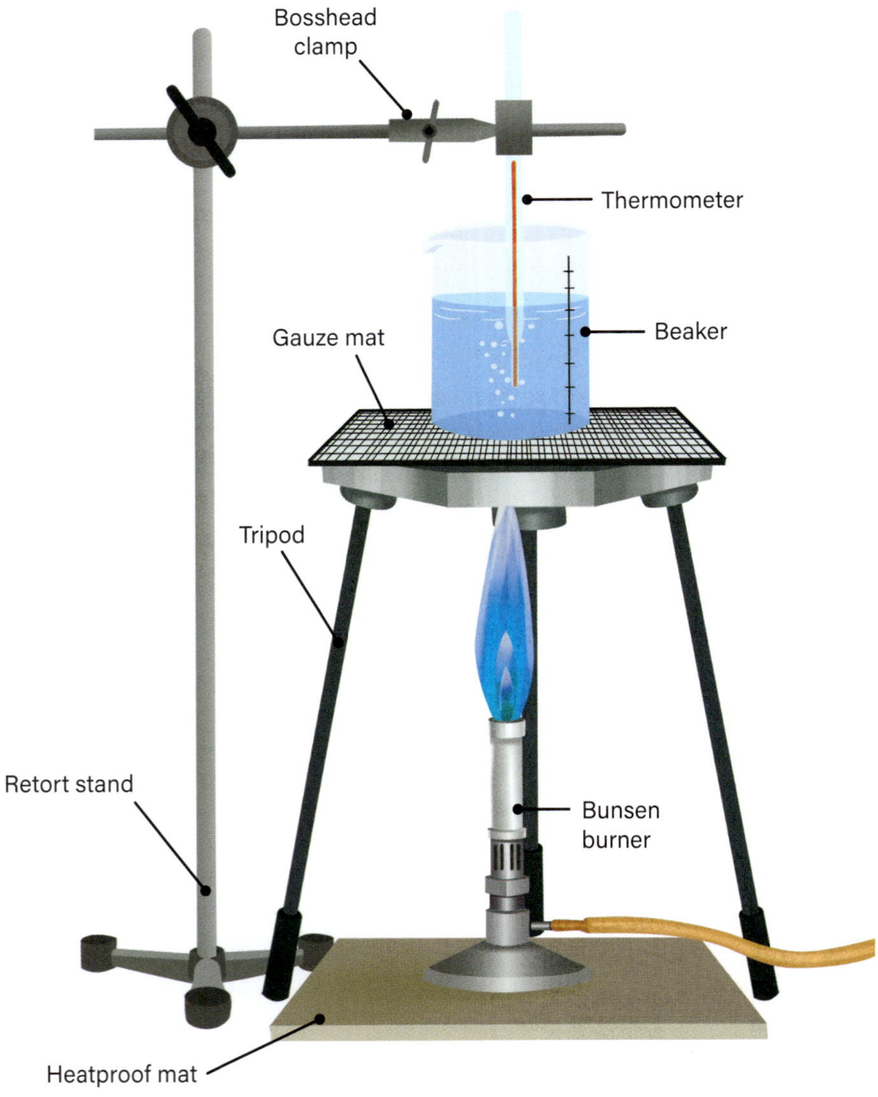

Figure I1.8: For this investigation, set up the materials as shown in this diagram.

Investigation 1.10

Investigating how colour affects temperature

Process: Conducting investigations

Focus: Using scientific equipment

Measurements taken during a scientific investigation should be accurate. This accuracy ensures the data collected is of a high quality. While conducting this investigation, make sure you are using appropriate pieces of scientific equipment.

Hint 1: What would be the best piece of equipment to use, based on the data you are collecting?

Hint 2: How precisely can you record your data in the results table?

Aim

To investigate how colour affects the transfer of heat energy

Materials

Per group (5 groups):

- 1 piece of coloured paper (there are 5 pieces of coloured paper – black, white, red, blue and green – for the 5 groups; each group receives a different colour)
- 2 test tubes
- sticky tape

BE CAREFUL USING THE HEAT LAMP. IF YOU BURN YOURSELF, TELL YOUR TEACHER IMMEDIATELY AND PLACE THE BURNT AREA UNDER COLD RUNNING WATER FOR 20 MINUTES.

- tap water
- test tube rack
- 2 thermometers
- heat source (sunlight or a heat lamp)

Method

1. Copy the results table (on the next page) into your notebook, adding a title.
2. Tightly wrap the piece of coloured paper around one of the test tubes and secure with sticky tape. Leave the other test tube uncovered as a control.
3. Fill both test tubes with the same amount of water. Place the test tubes in the test tube rack.
4. Place a thermometer in each test tube. Allow enough time for the temperature to stabilise.
5. Record the temperature of the water in each test tube. Write the results in the first row of the results table.

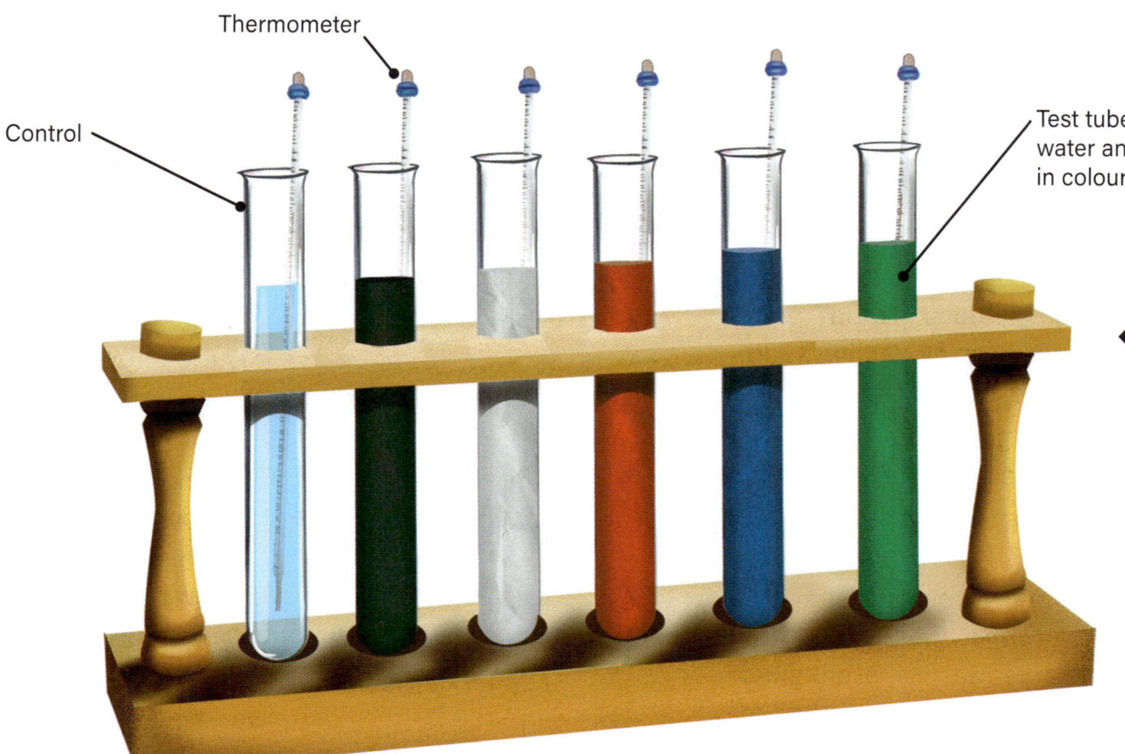

Figure I1.10: For this investigation, different groups will investigate the effects of different colours of paper.

continues ▶

6 Leave the test tubes in the sunlight, or near a heat lamp. Record the temperature of the water in each test tube every 5 minutes for half an hour.

7 Share your results with the other groups. Write the results of the other groups in the results table.

Questions

1 Identify a process that made sure this investigation was conducted correctly.

2 Describe how one piece of scientific equipment allowed you to collect precise results from this investigation.

3 What can you do to stay safe while conducting this investigation?

4 Construct a graph that represents the data you collected in the results table.

Conclusion

Copy and complete:
'The results show that: *(respond to the aim).*'

Table I1.10: Results

Time (minutes)	Temperature (°C)					
	Control test tube	**Black paper**	**White paper**	**Red paper**	**Blue paper**	**Green paper**
0						
5						
10						
15						
20						
25						
30						

Investigation 1.13

30 min

Modelling eclipses

Process: Conducting investigations

Focus: Implementing safe practices when using scientific equipment

When conducting any first-hand investigation, it is important to identify potential hazards and come up with strategies to minimise the chances of these dangers causing harm. **Answer Question 1 before you start this investigation.**

Hint: *What equipment could cause you harm if used improperly? How could this be prevented?*

Aim

To investigate, using a model, how solar and lunar eclipses are explained by the relative positions of the Sun, Moon and Earth

Materials

- sticky tape
- 2 cardboard tubes or rolls of paper
- scissors
- aluminium foil
- coathanger wire (length: 20 cm)
- polystyrene ball (diameter: 10–12 cm)
- ping-pong ball (or small polystyrene ball)
- square of heavy card
- retort stand with bosshead and clamp
- torch

Method

1. Make a series of small (2 cm), even, vertical cuts around each end of one of the cardboard tubes.
2. Bend the cut pieces of both ends of the cardboard tube outwards and stand the tube upright. Tape one end of the tube to one of the pieces of heavy card to create a base for the model.
3. Tape the base to the polystyrene ball. This is the model of Earth.
4. Insert one end of the wire vertically into the Earth model.
5. Make two small holes in the ping-pong ball for the wire to pass through.
6. Cover the ping-pong ball with aluminium foil. This is the model of the Moon.
7. Bend the wire to secure the Moon. The Moon's equator should be level with Earth's equator.
8. Adjust the torch using the retort stand so that it is in line with Earth's equator. The other cardboard tube can be used to direct the light.
9. Dim the lights in the classroom.
10. Follow the instructions below and on the next page to model a solar eclipse and a lunar eclipse.

To model a solar eclipse (see Figure I1.13A):

1. Stand facing the torch and swing the wire around until the Moon casts a shadow on Earth. The Moon is now positioned between Earth and the Sun. The Moon is blocking the Sun's light and preventing the light reaching Earth.
2. Show that the shadow moves by slowly rotating the wire.
3. Bend the wire up so that the Moon only partially blocks the Sun.

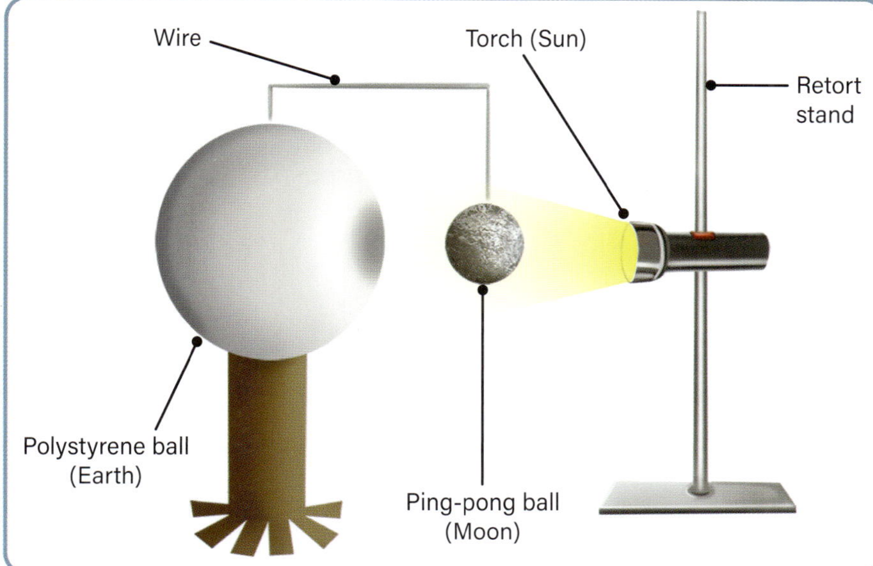

Figure I1.13A: To model a solar eclipse, set up the materials as shown in this diagram.

continues ▶

To model a lunar eclipse (see Figure I1.13B):

1 Stand facing the torch and swing the wire around until the Moon is completely behind Earth. Earth is now positioned between the Moon and the Sun. Earth is blocking the Sun's light and preventing the light reaching the Moon.
2 Bend the wire so that only part of Earth's shadow falls on the Moon, to model a partial lunar eclipse.

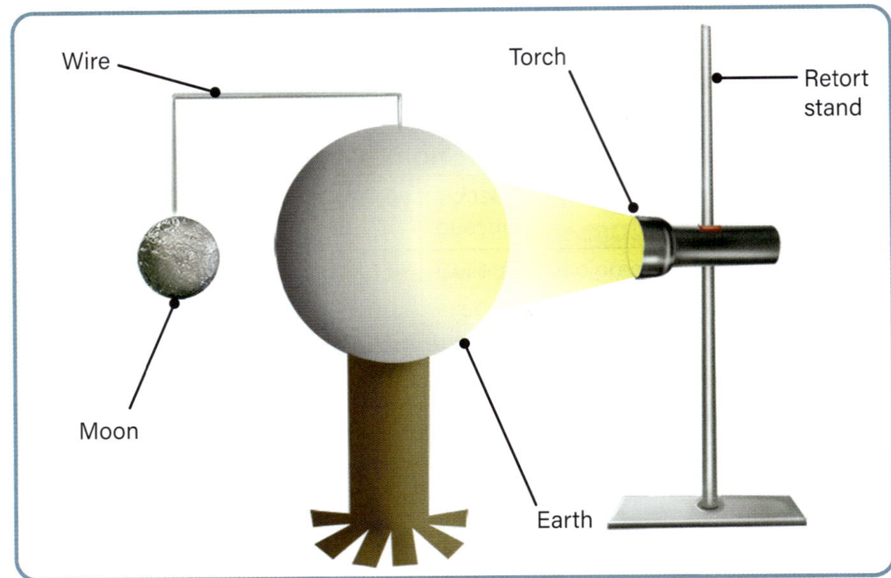

▲ **Figure I1.13B:** To model a lunar eclipse, set up the materials as shown in this diagram.

Questions

1 Conduct a risk assessment for this first-hand investigation.
   a Copy the risk assessment table below into your notebook. Write your answers to Questions 1b to 1d in the table.
   b **Hazards**: Identify two hazards (things that could harm you) in this investigation.
   c **Risk**: Decide what the chances are of each hazard causing you harm.
   d **Management strategy**: What can you do to protect yourself from being harmed by each hazard?
2 a Which object is in shadow in a solar eclipse?
   b Which object is in shadow in a lunar eclipse?
3 a In a solar eclipse, which object casts the shadow?
   b In a lunar eclipse, which object casts the shadow?
4 a Does everyone in the world see solar eclipses? Who cannot see them?
   b Does everyone in the world see lunar eclipses? Who cannot see them?
5 Do you see solar eclipses at night or during the day?
6 During a solar eclipse, what would you see if you stood on the Moon and looked at Earth?
7 a What is the phase of the Moon during a solar eclipse?
   b What is the phase of the Moon during a lunar eclipse?
8 Imagine you lived on Mars. Undertake research to find out if you would be able to see a solar eclipse and a lunar eclipse.

Conclusion

Copy and complete:
'The results show that: *(respond to the aim)*.'

Table I1.13: Risk assessment

Hazard	Risk (low, medium or high risk)	Management strategy

Investigation 2.1

Balloon rockets

Process: Processing data and information

Focus: Constructing a table to gather first-hand information

What do you think makes a balloon travel? To investigate this, you need to identify all the variables involved in making a balloon travel, and establish which of these variables can be changed (independent variables) and which need to be kept the same (control variables). **Answer Question 1 before you start this investigation.**

Aim

To investigate which variables make a balloon rocket travel furthest

Materials

- masking tape
- paper straw
- balloon
- scissors
- string

Method

1. Using a piece of masking tape, stick a straw to the balloon.
2. Use the scissors to cut a piece of string and then thread it through the straw.

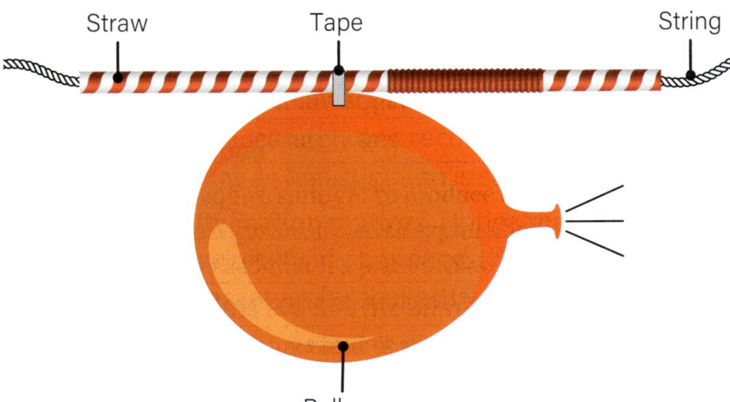

Figure I2.1: For this investigation, set up the materials as shown in this diagram.

3. Inflate the balloon and hold it closed.
4. Release the balloon so that it travels along the string. Record the distance the balloon travels in your results table.
5. Write a list of all the variables involved in making the balloon travel along the string.
6. a Choose one variable to change (an independent variable). Keep the other variables the same (the controlled variables).
   b Predict if the balloon will travel further than in the original test.
7. Repeat Steps 1–4.
8. Record the results in your results table. Repeat the test for as many variables as you can think of.

Questions

1. Construct a results table in your notebook to record the data you will collect in this investigation.
   Hints:
   - Read the materials and method to determine what data you will be collecting.
   - List the independent variables in the left column.
2. Identify the different forces involved with your balloon rocket:
   a Which are direct forces?
   b Which are indirect forces?
3. Explain the effect each of the forces has on the movement of the balloon.
4. Present your data in a graph.
5. Describe how your results compare with your predictions. Explain why they were the same or different.

Conclusion

Copy and complete:
'The results show that: (*respond to the aim*).'

DO NOT SHARE THE BALLOON BETWEEN REPETITIONS. EACH STUDENT MUST HAVE THEIR OWN BALLOON.

Investigation 2.2

Blowball

Process: Analysing data and information

Focus: Describing trends in collected data

Write your observations (qualitative data) for each aspect of the investigation. Using the information gathered, you can describe and explain the relationships found.

Aim

To investigate the effects of unbalanced forces on a ping-pong ball

Materials

- 4 paper straws
- table
- ping-pong ball

Method

1 Gather into groups of four. Each member of the group takes one straw and stands at a different side of the table.

2 Place the ping-pong ball in the centre of the table.

3 Take turns to blow through your straws to move the ball around the table.

4 Have one group member blow the ball across the table while the others blow on it in the opposite direction.

5 Have another group member blow the ball across the table while the others blow on it from the sides.

6 Divide into two pairs, with each pair on a different side of the table. As a pair, try to blow the ball across the table and off the side while the other pair does the same.

7 Record your observations (qualitative data) in your notebook. In particular, describe how different sized forces affected the motion of the ping-pong ball.

DO NOT SHARE STRAWS. EACH STUDENT MUST HAVE THEIR OWN STRAW.

Questions

1 When were the forces on the ball balanced?

2 Describe what happened to the speed and direction of the ball when you blew on it as it moved towards you.

3 What happened when you blew on the ball as it moved at a right angle to you?

4 Draw a force diagram showing the forces that were acting on the ball as you blew it away from you.

5 Review the observations you wrote in your notebook. Describe and explain any trends or relationships you can see in the data.

Conclusion

Copy and complete:

'The results show that: (*respond to the aim*).'

Figure 12.2: In this investigation, you move objects by blowing air through a straw.

Investigation 2.3

Magnetic shielding

Process: Problem-solving

Focus: Suggesting solutions to problems

Magnetic fields can damage sensitive electronic equipment, such as laptops and mobile phones. How can you block magnetic fields to protect your equipment?

Aim

To test which materials can block a magnetic field

Materials

- 50 g mass
- paperclip
- cotton thread
- Blu Tack, sticky tape or plasticine
- bar magnet
- retort stand, bosshead and clamp
- sheets of different material such as cardboard, aluminium, plastic, steel, tin, wood, a ceramic tile, copper

Method

1 Copy the results table into your noteboook, adding a title and as many rows as needed.

2 Set up the equipment as shown in Figure 12.3.

3 Find the maximum distance that can be left between the paperclip and the magnet before the paperclip drops.

4 Insert each different material between the magnet and the paperclip. Record what happens in the results table.

Questions

1 In terms of the strength of the magnetic field, explain why the paperclip fell when the distance between the paperclip and the bar magnet increased.

2 Which materials blocked the magnetic field and which materials did not?

3 Which other materials do you think block magnetic fields?

4 In scrapyards, electromagnets are used to sort the scrap material. Suggest which materials are removed from the scrap pile by the magnet.

Table 12.3: Results

Material	Paperclip stayed or dropped

Conclusion

Copy and complete:

'The results show that: *(respond to the aim)*.'

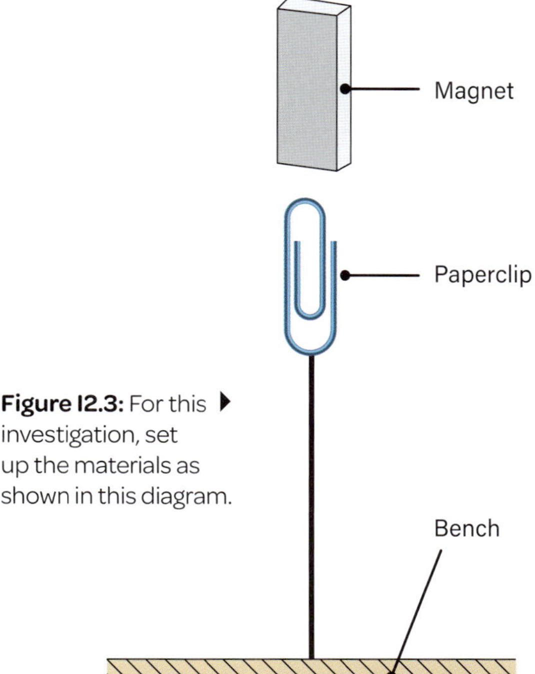

Figure 12.3: For this investigation, set up the materials as shown in this diagram.

Investigation 2.4

Charging balloons

Process: Questioning and predicting

Focus: Proposing hypotheses

Part of a scientific investigation includes proposing a hypothesis , which is a suggested explanation or prediction of a scientific problem that can be tested with an investigation. **Answer Question 1 before you start this investigation.**

Hint 1: What do you know about how charged particles behave?

Hint 2: What changes might you observe as charge is added to objects?

Aim

To investigate the behaviour of charged objects

Materials

- tissues
- plastic rod
- piece of furry fabric
- 2 balloons
- 2 × 30 cm lengths of string

Method

1. Tear the tissues into small squares of 1 cm × 1 cm.
2. Rub the plastic rod with the furry fabric vigorously for 1 minute.
3. Place the tip of the plastic rod close to, but not touching, the tissue squares. Record your observations.
4. Inflate the balloons and tie a 30 cm length of string to each one.
5. Rub the balloons against your hair for 1 minute. Record your observations.
6. Repeat Step 5 with a partner, so that each of you now holds a charged balloon. Hold each balloon by the string and move them close together (but not touching). Record your observations.

Questions

1. Read the method. What do you think will happen to the tissue squares and the balloons? This is your hypothesis. Write your hypothesis in your notebook using the format below.
   If ... then ... because ...
2. What did you observe happening to the tissue squares?
3. What did you observe happening to the balloons?
4. Explain the behaviour of the tissues and balloons in terms of electrostatic charge. Did this match your hypothesis?

Conclusion

Copy and complete:
'The results show that: (*respond to the aim*).'

Figure I2.4: Step 5 of this investigation: rub the balloons against your hair.

Investigation 2.6

Measuring gravity

Process: Processing data and information

Focus: Constructing graphs

Graphing data shows the relationship between factors.

Hint 1: *What type of graph should be used for the data gathered in this investigation?*

Hint 2: *How can you make sure your data goes onto the correct graph axis?*

Aim

To investigate the strength of gravitational force

Materials

- 5 × 50 g masses
- 5 N spring scale
- mass hanger or carrier

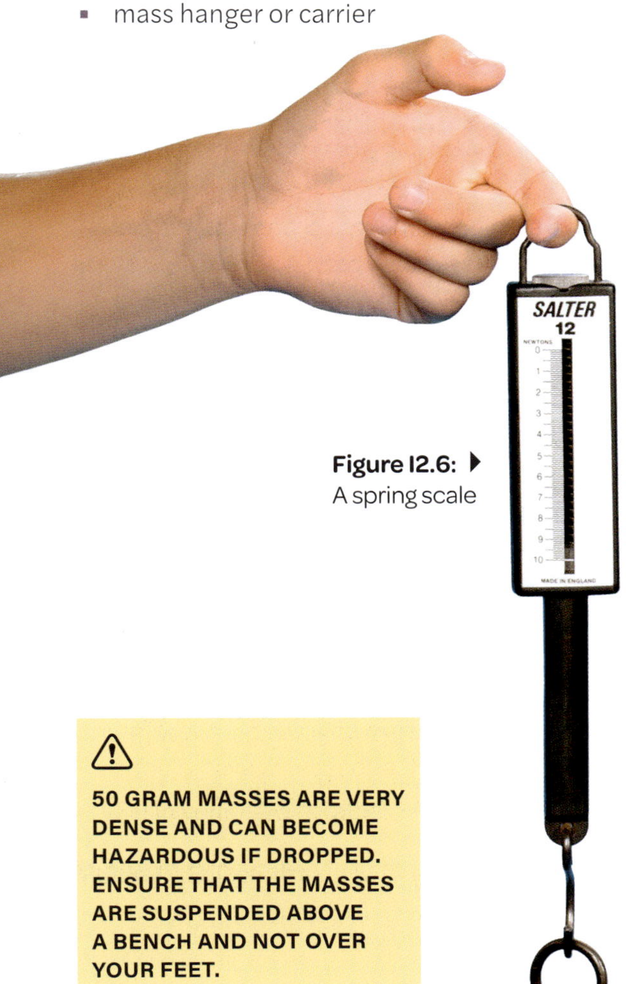

Figure 12.6: A spring scale

50 GRAM MASSES ARE VERY DENSE AND CAN BECOME HAZARDOUS IF DROPPED. ENSURE THAT THE MASSES ARE SUSPENDED ABOVE A BENCH AND NOT OVER YOUR FEET.

Method

1. Copy the results table into your notebook, adding a title and as many rows as needed.
2. a. Attach a 50 g mass to the bottom of the spring scale.
   b. Convert the mass into kilograms. Record this information in the results table.
   c. The reading on the spring scale shows the weight of the mass (measured in newtons (N)). Record the reading on the spring scale in the results table.
3. Repeat Step 2, adding another 50 g mass each time, until there is a total mass of 250 g on the spring scale.
4. For each reading, calculate the strength of gravity by dividing the weight (N) by the mass (kg). Record the gravitational strength in the results table.

Questions

1. Is there a pattern in your gravitational strength calculations?
2. What can you deduce about the strength of gravity on Earth?
3. How would this investigation be different if you conducted it on the Moon?
4. Construct a graph that shows the relationship between mass and weight, from the data collected in your results table.

Conclusion

Copy and complete:

'The results show that: (*respond to the aim*).'

Table 12.6: Results

Mass (kg)	Weight (N)	Gravitational strength

Investigation 2.8

Magnetic accelerator

Process: Processing data and information

Focus: Constructing a table to gather first-hand information

Data collected during first-hand investigations needs to be organised in a way that is easy to understand. A simple way to effectively organise collected data is in a results table.

In this investigation, you will test which orientation of the large magnets is the best to accelerate the small magnets the furthest along the accelerator chute. **Answer Question 1 before you start this investigation.**

Aim

To determine the arrangement of magnets in a magnetic accelerator that makes a pair of magnets move the furthest

Materials

- 2 pieces of square-end wooden dowel (30 cm long)
- clear plastic ruler
- Blu Tack
- compass
- 2 large neodymium magnets
- 2 small neodymium magnets

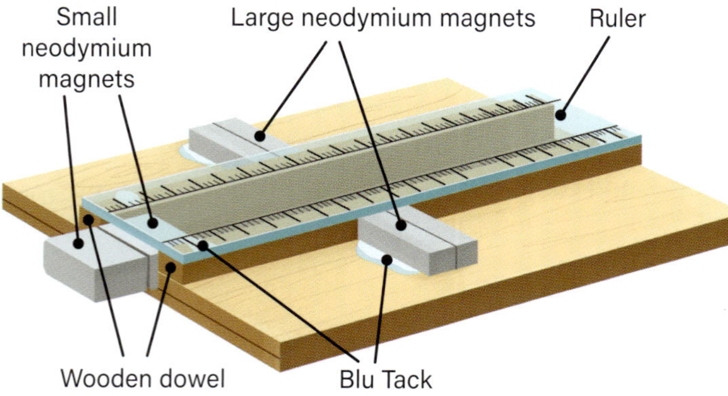

▲ **Figure 12.8:** For this investigation, set up the materials as shown in this diagram.

Method

1. Stick the two pieces of wooden dowel to the ruler with Blu Tack to make an 'accelerator chute'. Make sure the space between the pieces of dowel is large enough for a pair of small magnets to pass through but not so wide that the magnet can turn around while in the chute.
2. Add two pieces of Blu Tack to the outside of the accelerator chute.
3. Use a compass to determine the poles of the large neodymium magnets.
4. Place the large magnets on either side of the accelerator chute, as shown in Figure 12.8. Make sure the poles of the large magnets are facing the chute.
5. Connect the two small magnets. Release the small magnets along the chute and record how far they travel.
6. Repeat Step 5 after changing the orientation of the large magnets.

Questions

1. Construct a results table in your notebook to record the data you will collect in this investigation.
   Hints:
   - Read the list of materials and the method to work out what data you will be collecting.
   - Show the independent variable (the design of the accelerator chute) in the rows of the table.
   - Show the dependent variable (the distance travelled by the small magnet) in the columns of the table.
2. Using the data you have collected, explain which orientation of the large magnets was better. Include a description of how the magnets were arranged.
3. Where could this principle of acceleration be useful?

Conclusion

Copy and complete:
'The results show that: *(respond to the aim)*.'

Investigation 2.9

Building an electric motor

Process: Questioning and predicting

Focus: Constructing a scientific research question

To construct a research question (also called a hypothesis), turn the aim of this investigation into a question that asks about the scientific problem or concept you are investigating.

Aim

To build and investigate a simple electric motor

Materials

- 50 cm insulated copper wire (approx. 20 mm diameter)
- 15 mm PVC pipe
- wooden board
- sandpaper
- 2 metal paperclips
- D-cell battery
- electrical tape
- Blu Tack
- bar or round magnet

Method

1. Wrap the insulated copper wire around the PVC pipe several times to make a coil, leaving the ends sticking out. Make sure the length of the coil and ends is more than 8 cm (2 cm longer than the length of a D-cell battery). Also ensure that the coil ends are very straight so that the coil can spin without resistance.
2. Carefully remove the wire from the PVC pipe and place it on a wooden board. Use the sandpaper to scrape away the plastic coating from the wire.
3. Bend one end of each paperclip so they form a closed loop. Tape the paperclips to the battery as shown in Figure I2.9.
4. To keep the battery in place, you can secure it to the bench using Blu Tack.
5. Slide the coil of wire through the paperclips so it is supported within the loops. Make sure the coil can spin freely.
6. Hold the magnet over the top of the coil so that the coil spins.

Questions

1. What happened when you spun the coil?
2. Explain what happened to the coil in terms of magnetic fields.
3. a. With a partner, brainstorm ways that you can alter your electric motor to make it spin faster.
   b. Select one variable to investigate further. Construct a question to investigate this variable (that is, propose a hypothesis).
4. Explain how a simple electric motor works.

Conclusion

Copy and complete:

'The results show that: *(respond to the aim).*'

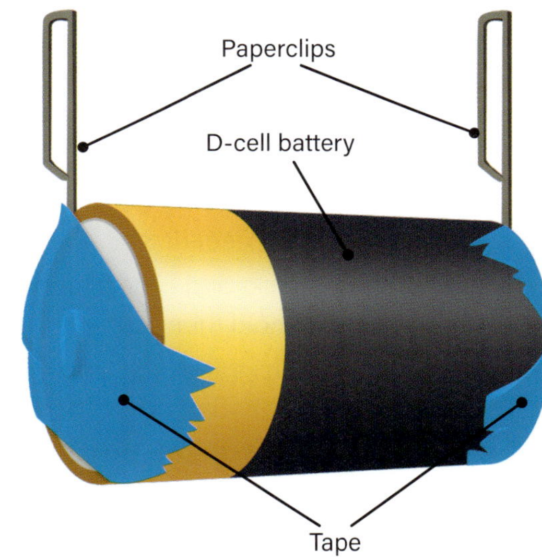

Figure I2.9: For this investigation, set up the equipment as shown in this diagram.

Investigation 2.10A

Making a catapult

Process: Problem-solving

Focus: Using criteria to solve problems

Within scientific investigations, we construct a question to test. After concluding our tests, problems can arise that require us to alter our initial designs. Here, the distance the polystyrene ball needs to travel is increased, which creates a new problem to solve.

Hint: *How will the change in parameters affect the design of the catapult?*

Aim

To investigate the types of levers involved in loading and firing a catapult

Materials

- 7 popsticks
- 4 rubber bands
- small polystyrene ball
- metre ruler (or a tape measure)

Method

1. Copy the results table into your notebook, adding a title and as many rows as needed.
2. Form a stack of five popsticks and wrap a rubber band around each end to hold them together.
3. Stack the other two popsticks and bind them together at one end using another rubber band.
4. Separate the two popsticks at the end without the rubber band and slide the stack of five in between the two sticks. The upper stick should make an angle of about 25 degrees with the horizontal.
5. Wrap the final rubber band around the centre of the structure to hold it all together.
6. Place a polystyrene ball on the raised end of the sloping stick. Launch the ball by pressing down and releasing the stick.
7. Measure the distance travelled by the polystyrene ball and record this information in the results table. Repeat this several times.

Questions

1. Identify the forces acting in this catapult.
2. Describe the type of lever that is being used.
3. Explain what changes you would make to the catapult if you wanted to make the polystyrene ball travel further.
4. In what context would this type of lever be useful and help solve a problem?
5. Sometimes, we may need a catapult to launch an object higher or faster.
   a. Consider ways you would alter the design of the catapult to allow for this.
   b. Propose situations where these new designs would be useful.

Table I2.10A: Results

Attempt number	Distance travelled

Conclusion

Copy and complete:
'The results show that: *(respond to the aim)*.'

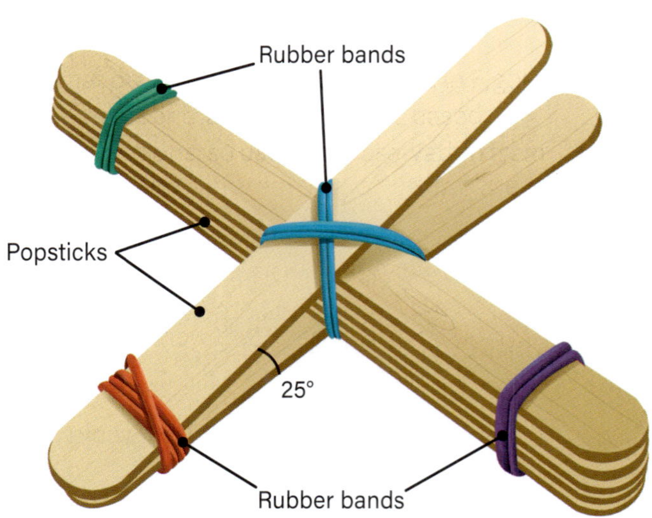

Figure I2.10A: For this investigation, set up your catapult as shown in this diagram.

Investigation **2.10B**

Modelling a seesaw

Process: Problem-solving

Focus: Identifying cause-and-effect relationships

When we make a change in an investigation, it can affect the results. In this investigation, we are *causing* the movement of the mass and measuring the *effect* it has on our seesaw. The placement of the mass directly affects the movement.

Hint 1: What is causing the seesaw to move up and down? Consider the forces and where along the seesaw they are acting on it.

Hint 2: How could the seesaw be made to balance and not move?

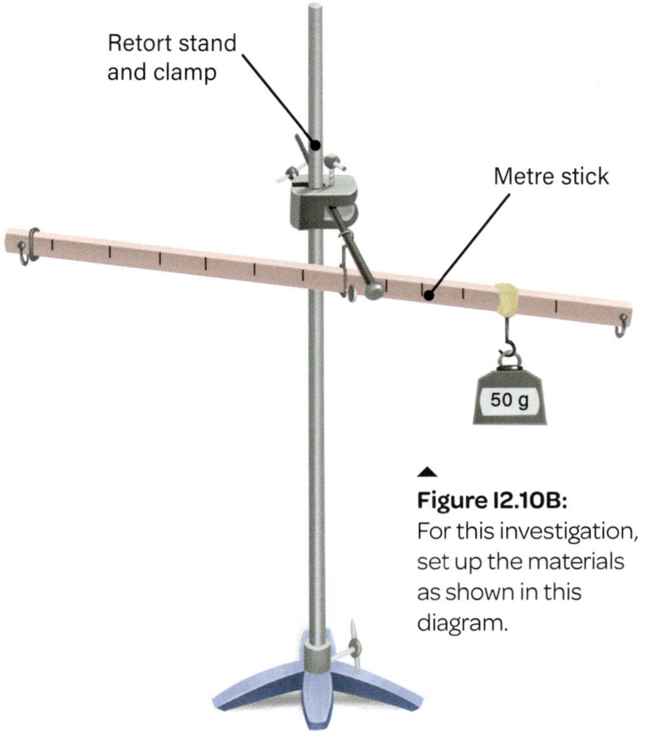

Figure 12.10B: For this investigation, set up the materials as shown in this diagram.

Aim

To investigate factors that balance a seesaw

Materials

- hanger clamps
- metre stick
- retort stand and clamp
- 4 × 25 g masses
- masking tape

Method

1. Copy the results table into your notebook, adding a title and as many rows as needed.
2. Using a hanger clamp, attach a metre stick so it hangs, balanced, from a clamp on a retort stand (see Figure 12.10B). This is the fulcrum of the lever.
3. Tape two 25 g masses 40 cm from the balance point of the stick. This is the load force.
4. Hang two 25 g masses as a counterweight from the other end of the stick. Move the counterweight until the stick is balanced. This is the effort force.
5. In the results table, record the distance from the centre of the metre stick to the position where the mass balances the metre stick.
6. Remove the hanging counterweight. Repeat Steps 4 and 5, replacing the counterweight with a 100 g mass, a 200 g mass and a 500 g mass. Record your results in the table.

Questions

1. Draw a force diagram and label the forces that are acting in this model of a seesaw.
2. Describe the effect that increasing the mass of the counterweight (effort force) had on how close it needed to be to the fulcrum.
3. Describe when you may have experienced having to adjust people on a seesaw to make it balance.

Table 12.10B: Results

Mass (g)	Distance from fulcrum (cm)
50	
100	
200	
500	

Conclusion

Copy and complete:

'The results show that: *(respond to the aim).*'

Investigation 2.10C

Pulley investigation

Process: Problem-solving
Focus: Solving familiar problems
The results of first-hand investigations can be used to help us solve everyday problems. Pulleys make it easier to lift heavy objects, and different pulleys are appropriate in different settings.

Aim

To determine how to use a pulley to make it easier to lift a load

Materials

- metre ruler or board, with two hooks, one on either side of the centre (or 2 retort stands and clamps)
- 3 pulleys with hooks
- block of wood (10 cm long by 10 cm wide) with a hook screwed into one end
- 1 N spring scale
- 2 pieces of heavy string or butcher cord (1.5 m and 3 m)

Method

1 Copy the results table into your notebook, adding a title and as many rows as needed.

Part 1

2 Hang the block of wood from the spring scale. In the results table, record the force required to suspend the wood.

3 Support the board between two desks, as shown in Figure I2.10C1. Hang a pulley from one of the hooks on the board. Put the block of wood on the floor under the pulley. String the cord through the pulley and tie it to the block of wood.

4 Mark the cord at the point where it exits the pulley.

5 Tie a loop in the loose end of the cord. Hook the scale to the loop. Gently pull down on the scale until the block of wood rises slowly.

6 Record the reading on the scale.

7 Keep pulling until the block is 10 cm off the floor. Mark the string at the point where it now exits the pulley. Record how far the string was pulled.

Figure I2.10C1: Pulley set-up for Part 1

Part 2

8 Attach a pulley to the hook just to the right of the centre of the board.

9 Attach the block of wood onto a second pulley using the hook on the end of the block.

10 Tie the cord to the hook just to the left of the centre of the board. String the cord down through the pulley attached to the block of wood, then up and through the fixed pulley on the board, and then back towards the floor (see Figure I2.10C2).

11 Mark the cord at the point where it exits the pulley system.

12 Hook the scale to the loop on the free end of the string. Gently pull down on the scale until the block of wood rises.

13 As the block rises, record the reading on the scale.

14 Measure and record how far the cord is pulled down to raise the block 10 cm.

Figure 12.10C2: Pulley set-up for Part 2

Part 3

15 Attach a pulley to each of the hooks on the board.

16 Attach a pulley to the top of the block of wood.

17 Run the string from the top of the pulley attached to the block of wood, up through the pulley on the left, back down and through the lower pulley, and then up and through the second pulley on the right. Run the loose end down towards the floor. See Figure 12.10C3.

18 Repeat Steps 11–14.

Figure 12.10C3: Pulley set-up for Part 3

Questions

1 Which pulley set-up required the most effort to lift the block of wood? Explain why.

2 Which pulley set-up required the least effort to lift the block of wood? Explain why.

3 What difference was there in the distance the cord was pulled in Parts 1, 2 and 3?

4 Which pulley system would you recommend to lift a load at a shipping dock? Justify your choice.

Table 12.10C: Results

Pulley system	Force applied to lift wood (N)	Distance cord was pulled (cm)
1		
2		
3		

Conclusion

Copy and complete:
'The results show that: *(respond to the aim).*'

Investigation 2.10D

Using ramps

Process: Problem-solving

Focus: Identifying scientific problems

An inclined plane or ramp is the simplest kind of machine. In this activity, you will investigate whether lifting or using an inclined plane makes work easier.

Hint: *What problem is the ramp solving?*

Aim

To measure the force required to lift a block vertically and to lift it using ramps of different slopes

Materials

- string
- spring scale
- smooth, flat board (10 cm wide, 50–100 cm long)
- smooth block of wood measuring 10 cm long by 10 cm wide (with a hook screwed into one end)
- 6 textbooks (or a retort stand and clamp)
- metre ruler
- protractor

Method

1. Copy the results table into your notebook, adding a title and as many rows as needed.
2. Use a loop of string to tie the spring scale to the block of wood.
3. Let the block hang freely. Lift the block from the desktop to the height of the stack of three books. As you lift, note the force recorded on the spring scale. Record that force.
4. Record the height, in metres, from the desk to the top of the stack of books.
5. Incline the board from the desk to the top of a stack of three books.
6. Place the block of wood at the bottom of the ramp.
7. Pull the block of wood up the ramp slowly. As you do this, note the force reading on the spring scale. Record the effort used.
8. Add another three books to the stack. This will put the ramp at a different angle. Repeat Steps 3–7. Record your results in the table.

Figure I2.10D: For this investigation, set up the materials as shown in this diagram.

Questions

1. In each trial, how does the force used to slide the block up the inclined plane compare with the force needed to lift the block to the same height?
2. What effect does increasing the angle of the ramp have on the force needed to slide the block up the ramp? Justify your answer.
3. What changes could you make to improve this machine?
4. List three situations in which using a ramp is useful.
5. How do wheelchair ramps, switchbacks, propellers and screws use an inclined plane?

Table I2.10D: Results

Angle of ramp (°)	Distance along ramp (cm)	Effort to pull trolley up ramp (N)
0		

Conclusion

Copy and complete:

'The results show that: *(respond to the aim)*.'

Investigation 3.2

Classifying items

Process: Conducting investigations

Focus: Collecting and recording data

The data you collect from a first-hand investigation can be quantitative (numerical information) or qualitative (written descriptions and observations).

Hint 1: How can you record quantitative data?

Hint 2: What observations caused you to record the data in this way?

Aim

To investigate different ways of classifying living things

Materials

- notebook
- 20 pieces of paper
- camera

Method

1. Copy the results table into your notebook, adding a title and as many rows as needed.
2. Visit an area near your school or home where there are animals and plants. This could be a natural space, such as a beach or an area of bushland, or a built space, such as a park.
3. Record at least 10 different plants that you find in the area. Photograph each one, if possible.
4. Record at least 10 different animals that you find in the area. Photograph each one, if possible. If the animals belong to other people, such as pets being walked, ask permission before taking a photo.
5. In class, print each photo and label it with the name of the organism, or write the name of each animal or plant you recorded on an individual piece of paper.
6. Sort the organisms you found into three to five different groups. The organisms in each group should have at least one thing in common.
7. Fill in your results table.

Questions

1. Identify a process you used to conduct this investigation. For example, how did you record the plants you found?
2. What scientific equipment did you use in this investigation? How did it help you to gather data?
3. Explain one risk associated with this investigation. How did you minimise the chances of this risk causing harm?
4. How did you make sure that the data you recorded was accurate?
5. Compare your results table with the tables of three other students. Did they use the same groups as you? Suggest reasons for any differences.

Conclusion

Copy and complete:

'The results show that: *(respond to the aim).*'

Figure 13.2: Plants can be classified by their leaves.

SOME PLANTS AND ANIMALS CAN BE HAZARDOUS. THEREFORE, DO NOT TOUCH ANY PLANTS OR ANIMALS.

Table 13.2: Results

Group number	Organisms in the group	Features shared by the organisms in the group

Investigation 3.4

Classifying supermarket items

Process: Communicating

Focus: Communicating information

An important part of scientific investigations is effectively communicating information to specific audiences. Scientists use classification keys to communicate information about objects and living things.

Hint: *If you cannot visit a supermarket for the purposes of this activity, look online or use your imagination!*

Aim

To investigate the design and use of classification keys

Materials

- notebook
- ruler
- pen
- supermarket items (first-hand observations or photos)

Method

1 Working in pairs, select one of the following groups of supermarket items:
   - fruits and vegetables
   - bakery items
   - dairy items
   - snack foods.

2 Design a classification system for the items in the group. Your key should include four levels of classification.

3 Visit a supermarket and apply your key to 20 different items. Record how each item is classified according to your key.

Questions

1 Is your classification key a branching key or dichotomous key? Explain your answer.

2 How would you explain your classification key to a classmate? Which words, images, tables or diagrams would you use to help them understand how you have grouped the items?

3 Use a digital technology to present your classification of the 20 items.

4 Compare your classification key to those of other students. How can you improve your key?

Conclusion

Copy and complete:
'The results show that: *(respond to the aim).*'

Figure 13.4: How could fruits and vegetables be classified based on their physical characteristics?

Investigation 3.5

Investigating features of marine animals

Process: Observing

Focus: Using scientific tools to enhance observations

Making observations of dead animals can assist scientists to correctly classify them. Scientists use specialised equipment to prepare specimens and to improve their observations of specimens.

Hint: Which scientific tools could you use to view the internal features of the organisms?

Aim

To investigate the physical and skeletal features of different marine animals

Materials

- disposable apron
- disposable gloves
- safety glasses
- 1 whole fish
- 1 large whole prawn
- 1 squid
- 3 dissecting boards
- magnifying lens
- dissecting kit (dissection scissors, blunt forceps and 2 blunt probes)

Method

1. Working in groups of three or four, put on an apron, gloves and safety glasses, then place the fish, prawn and squid on separate dissection boards.
2. Observe the external features of the fish and record your observations.
3. Open the fish's mouth and look inside, using the magnifying lens to magnify the details. While looking in the mouth, use a dissection probe to open the gill covers on the head. Record your observations.
4. Carefully cut open the fish lengthwise so you can see its skeleton. Record your observations.
5. Use a probe to open the prawn's mouth. Look inside through the magnifying lens. Try to identify whether the prawn has gills and, if so, where they are located. Record your observations.
6. Feel the outside of the prawn, then peel it and cut it in half. Record your observations.
7. Use a probe to open the squid's mouth. Look inside through the magnifying lens. Try to identify whether the squid has gills and, if so, where they are located. Record your observations.
8. Feel the outside of the squid, then peel the skin off and cut it in half. Record your observations.

Questions

1. a. Which of the three marine animals are vertebrates (with skeletons)?
   b. Which of the three marine animals are invertebrates (without skeletons)?
2. How did using the magnifying lens improve your observations of the three marine animals?
3. Draw three diagrams showing the three marine animals. Show and label the external and internal features of each specimen.
4. How are the gill structures of the three animals different?
5. Identify one physical feature that all three marine animals have.
6. Construct a classification key to sort these three marine animals based on the skeletons and gills.

Conclusion

Copy and complete:
'The results show that: *(respond to the aim)*.'

Figure 13.5: Using scientific tools can improve our observations of organisms like this fish.

WEAR SAFETY GLASSES AND GLOVES. BE CAREFUL WHEN USING CUTTING IMPLEMENTS. DISPOSE OF ALL MATERIALS AS DIRECTED BY YOUR TEACHER. WASH YOUR HANDS AFTER THE INVESTIGATION.

Investigation 3.7

Investigating structural adaptations

Figure 13.7: The common snake-necked turtle has many structural adaptations.

Process: Observing

Focus: Recording observations

It is important to record observations clearly and accurately. Recording qualitative data (written descriptions and observations) in a table is a good way to make sure that it is easy for others to understand the data.

Hint 1: *How much information should go into each section of the table?*

Hint 2: *In which column of the table do you record your observations? In which column of the table do you record your inferences?*

Aim

To investigate the structural adaptations of different organisms

Materials

- a range of preserved specimens
- magnifying lens

Method

1. Copy the results table into your notebook, adding a title and as many rows as required.
2. Select five specimens to inspect.
3. Write the name of each specimen in the results table.
4. Use the magnifying lens to observe the features of each specimen.
5. For each specimen, identify one structural adaptation. Record this information in the results table.
6. Why does the organism have this structural adaptation? Record a reason in the results table.

Questions

1. Look at Figure 13.7. Identify one structural adaptation.
2. Describe how using the magnifying lens improved your observations of your chosen specimens.
3. How did you make sure that you recorded your data accurately in the results table?
4. What do you infer about the turtle in Figure 13.7? Remember that an inference is based on observations, so use your observation from Question 1.

Conclusion

Copy and complete:

'The results show that: *(respond to the aim).*'

Table 13.7: Results

Organism's name	Organism's structural adaptation	How the structural adaptation helps the organism to survive

Investigation 3.9

Examining cells under a microscope

Process: Conducting investigations

Focus: Implementing safe practices when using scientific equipment

When working with glassware and biological materials, it is important to use safe practices to minimise the risk of harm occurring.

Hint 1: What materials could cause harm if used incorrectly?

Hint 2: What procedures minimise the chance of harm occurring?

Hint 3: More information about assessing risks in first-hand investigations is available in the Science how-to section on page 323.

Aim

To investigate the cells in onion tissue under a microscope

Materials

- slice of onion
- light microscope
- microscope slides and coverslips
- knife and chopping board
- tweezers (or forceps)
- dropper bottle of weak iodine solution (1.3 per cent concentration)
- blunt dissecting needle (or a probe or a sharp pencil)
- paper towel

Figure 13.9: Steps 2, 6 and 7 of this investigation

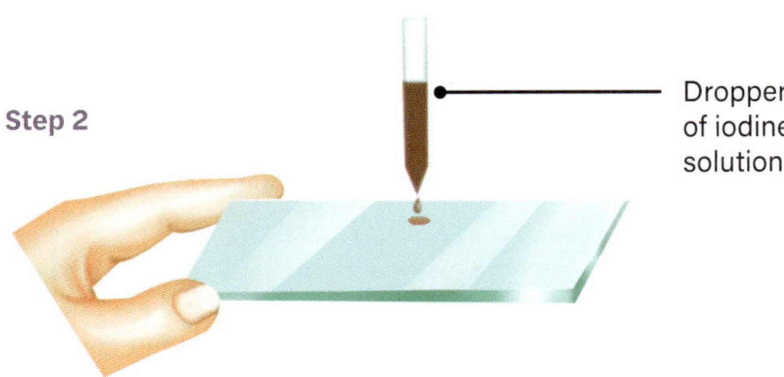

BE CAREFUL USING A KNIFE. IF YOU CUT YOURSELF, TELL YOUR TEACHER IMMEDIATELY AND SEEK FIRST AID.

Method

1. Watch your teacher demonstrate how to prepare a microscope slide.
2. Place one drop of iodine solution in the centre of the microscope slide.
3. Using the knife and chopping board, cut a small piece from an onion ring.
4. Using your fingernails or tweezers, carefully peel away the membrane from the inner curve of the onion piece.
5. Place this membrane on the drop of iodine on the slide. Try to prevent the membrane curling. Place a second drop of iodine solution on top of the onion membrane.

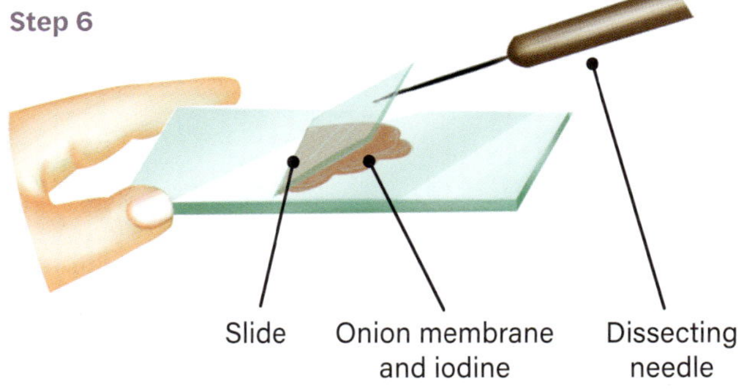

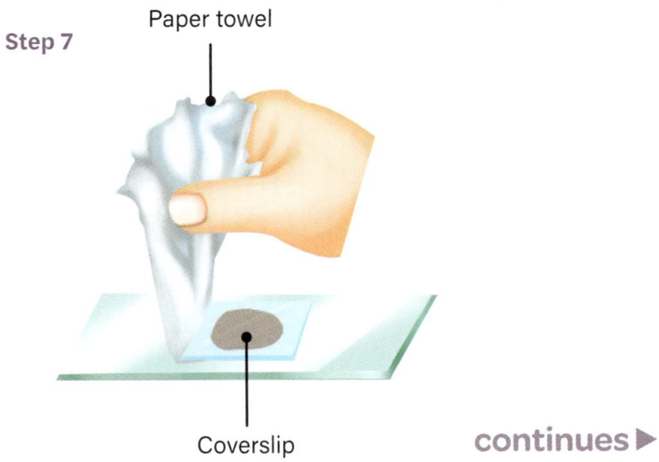

continues ▶

6 Use the blunt dissecting needle to lower a coverslip carefully and slowly over the onion membrane.

7 Use a corner of a paper towel to gently soak up any extra liquid that squeezes out from under the coverslip.

8 Examine the slide under the microscope. Record your observations.

Questions

1 Why was iodine used in this investigation?

2 Describe how you brought your onion slide into focus using the light microscope.

3 Copy the risk assessment table into your notebook. Write your answers to the following questions in the table.

 a **Hazards**: Identify three things that could harm the person conducting this investigation.

 b **Risks**: Decide whether you think there is a low, medium or high chance that each hazard will harm the person conducting this investigation.

 c **Management strategies**: What steps can you take to make sure each hazard does not harm the person conducting this investigation?

4 a Draw a scientific diagram of the onion cell as you see it under the microscope.

 b Record the magnification of the image.

 c Give your diagram a heading.

Conclusion

Copy and complete:
'The results show that: *(respond to the aim)*.'

Table 13.9: Risk assessment for Investigation 3.9

Hazard	Risk (low, medium or high)	Management strategy

Investigation 3.10

Examining cell structures

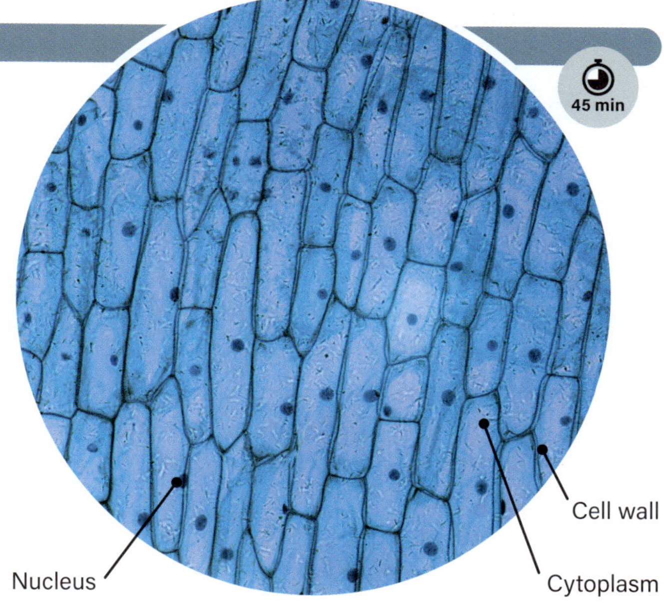

▲ **Figure 13.10:** Plant cells at 400x magnification

Process: Observing

Focus: Recording observations

Using microscopes allows us to see different types of cells in detail. To record our observations, we can draw cell diagrams.

Hint: How can you find the magnification level of a microscope?

Aim

To investigate the structures of different cells

Materials

- light microscope
- prepared microscope slides of different cells (cells from animals, plants, fungi and bacteria)

Method

1. Copy the results table into your notebook, adding a title.
2. Select one prepared slide from each category (cells from animals, plants, fungi and bacteria).
3. a Place the slide with animal cells under the microscope and observe the cells.
   b Record the cell type, magnification and your observations in the results table.
4. Draw a cell diagram of the animal cell. Hints:
   - Include a heading naming the cell type.
   - Include labels that point out the features of the cell.
5. Repeat this process with the prepared slides of plant, fungus and bacteria cells.

Table 13.10: Results

Cell type	Magnification	Observations

Questions

1. What observations could you make about the cells on the slides before they were under the microscope?
2. Describe how using the microscope improved your observations of the cells.
3. Explain how you recorded your observations of the cells.
4. Think about your observations of the plant and animal cells. What can you infer about plant and animal cells from your observations?
5. What is the difference in the size of the cells you observed? Rank the four types of cells from the smallest to the largest.

Conclusion

Copy and complete:
'The results show that: *(respond to the aim)*.'

Investigation 3.11A

Investigating respiration

Process: Conducting investigations

Focus: Using scientific equipment

Using scientific equipment to measure exact quantities and temperatures of variables in first-hand investigations improves the accuracy and reliability of the results.

Hint 1: *Remember that respiration is a chemical reaction that converts glucose (a type of sugar) to energy.*

Hint 2: *Which piece of equipment should you use to measure the different quantities of water in this investigation?*

Aim

To identify which temperature is the most suitable for yeast respiration

Materials

- 3 × 500 mL empty, plastic, single-use soft-drink or water bottles with lids (or borosilicate reagent bottles)
- 3 sticky labels (or masking tape)
- 10 g yeast
- 25 g sugar
- metric teaspoon
- filter funnel
- 150 mL measuring cylinder
- 3 beakers, large enough to hold more than 100 mL of water each
- 300 mL tap water
- ice
- kettle
- thermometer
- marker pen
- 3 balloons
- small tape measure (2 m or 5 m)
- matches

Figure I3.11A1: When using a kettle to heat water, do not touch the steam or the boiling water as these substances will burn your skin.

Method

1 Copy the results table into your notebook, adding a title.

2 Using the sticky labels, label the three bottles: Bottle 1, Bottle 2, Bottle 3.

3 Add 2 teaspoons of yeast and 5 teaspoons of sugar to each empty bottle using a filter funnel.

4 Prepare three beakers of water:
- Beaker 1: 100 mL of water at 10 °C
- Beaker 2: 100 mL of water at 35 °C
- Beaker 3: 100 mL of water at 60 °C

Use the ice and the kettle to adjust the temperature of the water in the beakers. Measure the water using the measuring cylinder.

5 a Pour the contents of Beaker 1 into Bottle 1.

b Pour the contents of Beaker 2 into Bottle 2.

c Pour the contents of Beaker 3 into Bottle 3.

6 Replace the lids on the three bottles. Shake each bottle to mix the contents. Use the pen to mark the level of the mixture in each bottle.

7 Remove the lids of the three bottles. While you hold each bottle, have a classmate stretch a balloon over the mouth of each.

8 Use the tape measure to measure the distance the yeast mixture rises every 5 minutes for 30 minutes. Record your observations in the results table.

9 Remove the balloon that has expanded to the largest size. Hold the end of the balloon closed so the gas does not escape.

10 Have a classmate light a match. Holding the balloon closed, move the end of the balloon close to the flame. Release the gas in the direction of the flame. Record what happens.

Table I3.11A: Results

Bottle	Water temperature (°C)	Growth of yeast mixture (cm)						Observations
		5 min	10 min	15 min	20 min	25 min	30 min	
1								
2								
3								

Questions

1 Which of the following is the correct way to measure 1 teaspoon of yeast?

A Fill the teaspoon most of the way to the top.

B Fill the teaspoon exactly to the top without going over the top.

C Heap the teaspoon with yeast.

2 Why should you use a tape measure rather than a ruler in this investigation?

3 When you are pouring boiling water from the kettle, what should you do to make sure you do not burn yourself?

4 How did you make sure that you accurately measured and recorded the height of the mixtures in the bottles?

5 Describe three ways you measured in the different parts of the investigation to make sure the measurements stayed the same.

Conclusion

Copy and complete:

'The results show that: *(respond to the aim).*'

THE OPEN FLAME OF A MATCH AND BOILING WATER ARE HAZARDS. NEVER LIGHT A MATCH UNLESS INSTRUCTED TO DO SO BY YOUR TEACHER. BE CAREFUL USING MATCHES AND POURING BOILING WATER. IF YOU BURN YOURSELF, TELL YOUR TEACHER IMMEDIATELY AND PLACE THE BURNT AREA UNDER COLD RUNNING WATER FOR 20 MINUTES.

Figure I3.11A2: Yeast comes in different forms.

Investigation 3.11B

Energy from food

Process: Conducting investigations

Focus: Implementing safe practices when using scientific equipment

All investigations have some degree of risk. To reduce the chances of harm occurring, strategies need to be put in place to manage the hazards in an investigation. This process – of identifying hazards, determining the chance of harm occurring, and coming up with strategies to reduce the likelihood of harm occurring – is called a risk assessment.

Hint 1: Which aspects of this investigation could harm the scientist conducting it?

Hint 2: More information about assessing risks in first-hand investigations is available in the Science how-to section on page 323.

Aim

To investigate which foods release the most energy when burnt

Materials

- piece of bread
- selection of other foods (for example, pasta, cheese, potato, apple, broccoli, carrot, meat, fish)
- knife
- test tubes (1 per food item)
- 10 mL measuring cylinder
- water
- retort stand with bosshead and clamp
- thermometer
- Bunsen burner
- 2 heatproof mats
- evaporating basin
- tripod
- gauze mat
- matches
- probe (wooden handle with sharp metallic end) or metal tongs

Method

1. Copy the results table into your notebook, adding a title and as many rows as needed.
2. Cut all the food into pieces of the same size that will fit into a test tube.
3. Add 5 mL of water to a test tube. Secure the test tube upright using the retort stand, bosshead and clamp (see Figure I3.11B).
4. Place a thermometer inside the test tube and measure the initial temperature. Record this temperature in the results table.
5. On your bench, set up the Bunsen burner on a heatproof mat, with the tripod and the gauze mat. Light the Bunsen burner and turn it to the blue flame.
6. Hold the piece of bread with the probe or tongs. Heat the bread over the flame of the Bunsen burner until it ignites (catches on fire).
7. Place the piece of burning bread into the evaporating basin. Note the highest temperature the water reaches. Record this temperature in the results table.
8. Calculate the change in the temperature of the water (deduct the initial temperature from the highest temperature).
9. Repeat Steps 3–8 with other foods. Ensure you turn off the Bunsen burner at the end of this investigation.

Table I3.11B1: Results

Food item	Initial temperature (°C)	Highest temperature (°C)	Change in temperature (°C)
Bread			

Questions

1 Describe the correct process for setting up a Bunsen burner.

2 How did using the probe or tongs allow you to safely conduct the investigation while still collecting data?

3 Copy the risk assessment table into your notebook. Write your answers to the following questions in the table.

   a **Hazards**: Identify three things that could harm the person conducting this investigation.

   b **Risks**: Decide whether you think there is a low, medium or high chance that each hazard will harm the person conducting this investigation.

   c **Strategies**: What steps can you take to make sure each hazard does not harm the person conducting this investigation?

4 a Identify three things that could go wrong during this investigation. These errors could mean the results are not accurate.

   b Suggest how these three errors could be avoided if this investigation is repeated.

5 a What was the highest temperature the water reached?

   b Which food was under the water when you recorded the highest temperature?

   c Why do you think the water reached the highest temperature for that particular food?

Conclusion

Copy and complete:

'The results show that: *(respond to the aim).*'

Table I3.11B2: Risk assessment for Investigation 3.11B

Hazard	Risk (low, medium or high)	Strategy

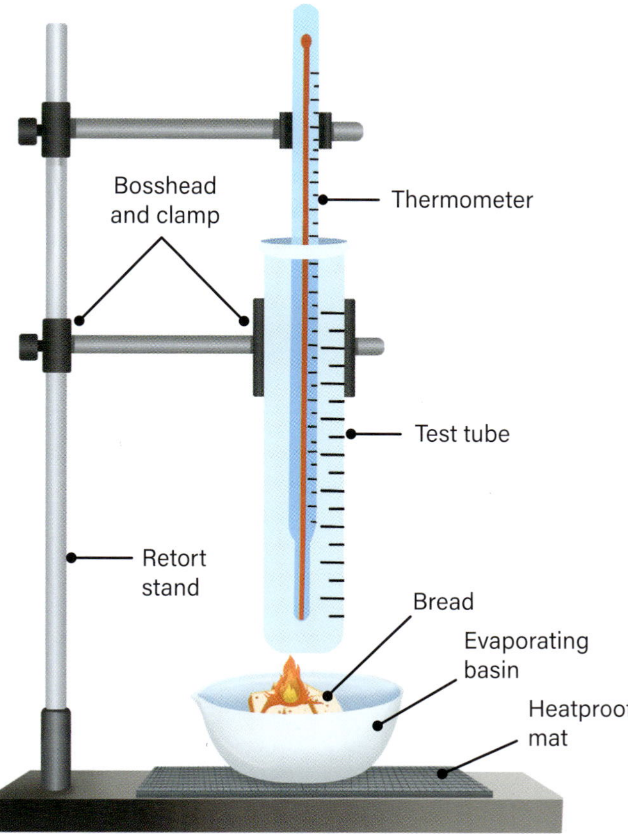

AN OPEN FLAME IS A HAZARD. BE CAREFUL WHEN USING A BUNSEN BURNER. IF YOU BURN YOURSELF, TELL YOUR TEACHER IMMEDIATELY AND PLACE THE BURNT AREA UNDER COLD RUNNING WATER FOR 20 MINUTES.

Figure I3.11B: For this investigation, set up the materials as shown in this diagram.

Investigation 4.1

Compressing liquids and gases

Process: Planning investigations

Focus: Identifying scientific variables

In an investigation, there should be one independent and one dependent variable. This is to make sure that the changed variable is responsible for any changes in what we measure. What are we changing and measuring in this investigation?

Aim

To investigate how much liquid and gas can be compressed

Materials

- 100 mL beaker
- plastic syringe
- tap water

Method

1 Copy the results table into your notebook, adding a title and as many rows as needed.

2 Fill the beaker with water.

3 Draw some water from the beaker into the syringe so that it is about half full.

4 Hold your finger over the nozzle of the syringe so that water cannot come out, then try to push in the plunger. Are you able to compress the water? Record your observations in the results table.

5 Empty the water from the syringe and pull back the plunger so that the syringe is half full of air.

6 Again, hold your finger over the nozzle of the syringe and try to push in the plunger. Are you able to compress the air? Record your observations in the results table.

Questions

1 For this investigation, identify:

 a the independent variable.

 b the dependent variable.

 c two controlled variables.

2 What do the results tell you about how much liquid can be compressed (compressibility)?

3 What do the results tell you about the compressibility of gas?

4 Explain the results you recorded, with reference to the particles of liquids and gases.

5 How could you investigate the compressibility of solids using an ice cube?

Conclusion

Copy and complete:

'The results show that: *(respond to the aim)*.'

Table I4.1B: Results

Material	Initial reading (mL)	Final reading (mL)	Difference in size (mL)

Figure I4.1: In this investigation, you will use a syringe to try to compress liquid and gas.

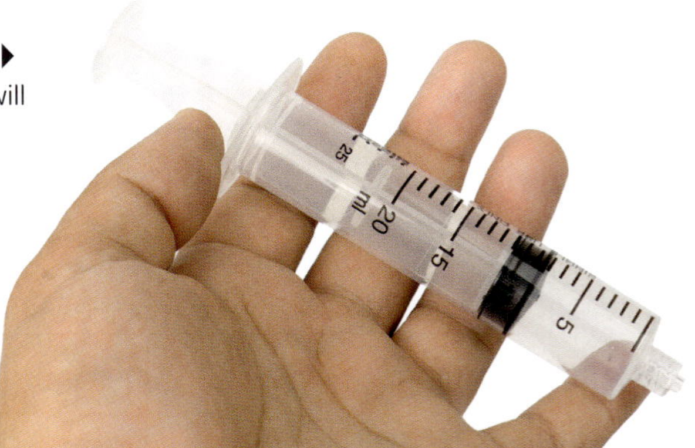

Investigation 4.2

Changing states of water

Process: Problem-solving

Focus: Identifying cause-and-effect relationships

During investigations, the cause of changes can sometimes be determined by making observations. How does the Bunsen burner cause the change of state in the water?

Hint 1: Complete the risk assessment to plan how you are going to manage the risks of this investigation. Answer Question 1 before you start this investigation.

Hint 2: What is the Bunsen burner adding to the water, and how does this cause the change of state?

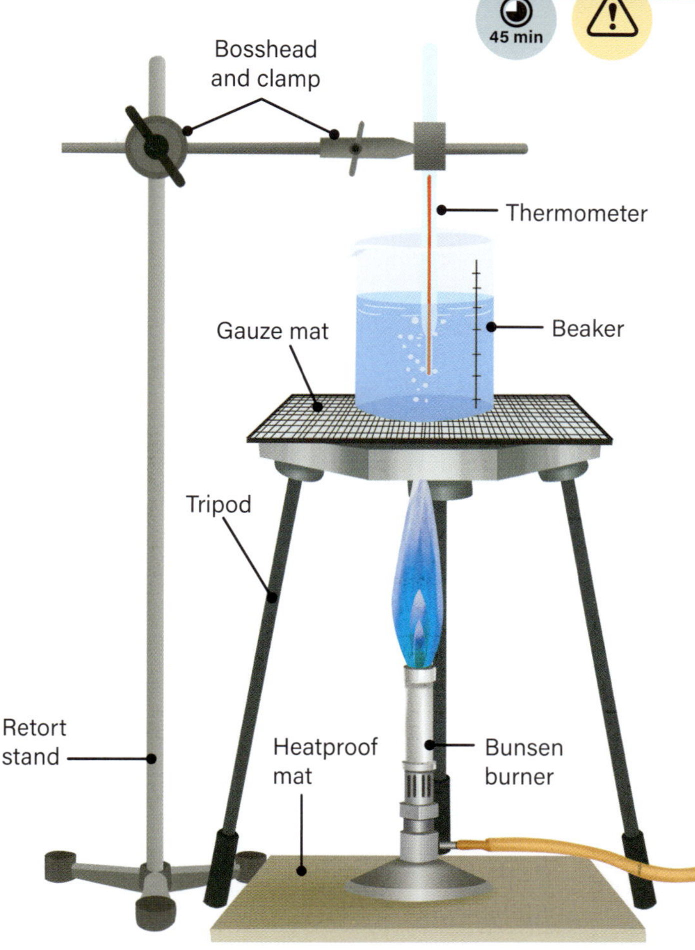

Figure I4.2: Make sure the thermometer does not touch the beaker!

Aim

To determine the temperature of water at each change of state

Materials

- 1 cup full of ice cubes
- 250 mL beaker
- tap water
- thermometer (or data logger with temperature probe)
- Bunsen burner
- heatproof mat
- gauze mat
- retort stand and clamp
- matches
- stopwatch
- heatproof gloves

Method

1. Copy the results table (on the next page) into your notebook, adding a title and as many rows as needed.
2. Add ice cubes up to the 100 mL mark on the beaker.
3. Add water to the beaker so that it surrounds the ice cubes up to the 100 mL mark.
4. Use the thermometer to measure the temperature of the water. Record this information in the results table.
5. Set up the thermometer, Bunsen burner, heatproof mat, gauze mat, retort stand and clamp (see Figure I4.2).
6. Light the Bunsen burner and turn to the blue flame. Heat the water. Ensure the collar of the Bunsen burner is turned so the airhole is open. Start timing immediately with the stopwatch.
7. Use the thermometer to measure the temperature of the water every 30 seconds. Record this temperature in the results table.

AN OPEN FLAME AND BOILING WATER ARE HAZARDS. BE CAREFUL. IF YOU BURN YOURSELF, TELL YOUR TEACHER IMMEDIATELY AND PLACE THE BURNT AREA UNDER COLD RUNNING WATER FOR 20 MINUTES.

continues ▶

8 Continue measuring and recording the temperature of the water until the liquid has boiled for 3 minutes. Ensure you turn off the Bunsen burner.

9 The beaker, gauze mat and other heated equipment will be very hot. Leave the equipment to cool and only handle while using heatproof gloves.

Table 14.2A: Results

Time (seconds)	Temperature (°C)
0	
30	
60	
90	
120	
150	
180	

Questions

1 Copy the risk assessment table into your notebook.

a **Hazards**: Identify three things that could harm the person conducting this investigation.

b **Risks**: Decide whether you think there is a low, medium or high chance that each hazard will harm the person conducting this investigation.

c **Management strategies**: What steps can you take to make sure each hazard does not harm the person conducting this investigation?

2 Which three pieces of equipment did you need in order to successfully conduct this investigation? Why did you need each item?

3 For this investigation, identify:

a the independent variable.

b the dependent variable.

c two controlled variables.

4 a What temperature was the water when the ice completely melted?

b Do you think this is the melting point of water? Explain why or why not.

Conclusion

Copy and complete:
'The results show that: *(respond to the aim).*'

Table 14.2B: Risk assessment

Hazard	Risk (low, medium, high)	Management strategy

Investigation 4.3

Exploring density

Process: Planning investigations

Focus: Distinguishing between variables

Scientific investigations aim to find the link between a specific change you make (independent variable) and the results you collect (dependent variable). To achieve this, you must control all the other factors in an investigation to make sure they are not influencing your results.

Hint 1: *What are the independent and dependent variables?*

Hint 2: *How can you make sure you are keeping the other factors the same?*

Aim

To investigate the relative densities of some household liquids and objects

Materials

- digital scales
- 100 mL measuring cylinder
- water (with food dye, to make it easier to see)
- honey
- corn syrup
- dishwashing liquid
- vegetable oil
- salt water
- isopropyl alcohol
- safety glasses
- gloves

Method

1. Copy the results table (on the next page) into your notebook, adding a title.
2. Use the digital scales to weigh the empty 100 mL measuring cylinder. Record its mass in grams in your notebook.
3. Pour 100 mL of water into the measuring cylinder. In the results table, record the volume of the water in millilitres.
4. Weigh the measuring cylinder with the water. Be careful not to spill the water. In your notebook, record its mass in grams.
5. Work out the mass of the water by subtracting the weight of the empty measuring cylinder. In the results table, record the mass of the water.
6. Thoroughly clean and dry the measuring cylinder, then repeat Steps 2–5 with the other substances.

WEAR SAFETY GLASSES AND GLOVES WHEN HANDLING ISOPROPYL ALCOHOL.

Figure I4.3: ▶ When substances of different densities are placed in the same container, the less dense substances stay at the top and the most dense substances settle at the bottom.

continues ▶

Questions

1 Which pieces of equipment did you need in order to successfully conduct this investigation? Why did you need each item?

2 For this investigation, identify:
  a the independent variable.
  b the dependent variable.
  c two controlled variables.

3 Rank the liquids from most dense liquid to least dense.

4 If you placed all the liquids in one container (as in the image in Figure 14.3 on the previous page), what order would they settle in, based on your results? Why is this?

5 a Calculate the density of each substance. Record this information in the results table.
  b Do the densities you calculated in Question 5a match the SI values in the last column of the table? Give a reason you might have calculated different values. (Hint: Think about your controlled variables.)

Conclusion

Copy and complete:
'The results show that: *(respond to the aim)*.'

Table 14.3: Results

Substance	Volume (mL)	Mass (g)	Density from experiment $\left(\frac{\text{mass}}{\text{volume}}\right)$	Density SI value (g/cm^3)
Water				1.000
Honey				1.440
Corn syrup				1.370
Dishwashing liquid				1.200
Isopropyl alcohol				0.789
Vegetable oil				0.910
Salt water				1.150

Note: SI stands for the International System of Units, which are the standard measurements used by scientists around the world.

Investigation 4.4

Calculating density

Process: Planning investigations

Focus: Distinguishing between variables

Scientific investigations aim to find the link between a specific change you make (independent variable) and the results you collect (dependent variable). To achieve this, you must effectively control all other factors so that they do not influence your results.

Hint 1: *What are the independent and dependent variables?*

Hint 2: *How can you make sure you are keeping the other factors the same?*

Hint 3: *Note that* $1\ cm^3 = 1\ mL$

Aims

- To determine the volume and mass of regular-shaped and irregular-shaped objects
- To calculate the densities of these objects using the formula $\rho = m/V$

Materials

- 1 cm cube
- digital scales
- water
- 100 mL plastic measuring cylinder
- marble: 1 cm diameter
- cylinder: 1 cm diameter
- rock: 1 cm diameter
- cork
- eraser: 1 cm diameter
- piece of wood: 1 cm diameter
- paperclip
- nail
- zinc metal (irregular in shape): 1 cm diameter

Method

1 Copy the results table (on the next page) into your notebook, adding a title.

2 Weigh the cube on the digital scales. In the results table, record the mass of the cube in grams.

3 Pour water into the measuring cylinder, up to the 50 mL mark.

4 Place the cube into the measuring cylinder. Be careful not to spill the water.

5 a Does the cube float or sink? Record this information in the results table.

b In the results table, in the 'Final volume' column, record the volume of the water with the cube in it.

6 Work out the volume of the cube using the following formula:

final volume (water + cube) − initial volume (water)

Record the volume of the cube in the results table.

7 Repeat Steps 2–6 with the other objects.

8 Using the density formula ($\rho = m/V$), calculate the density of each object.

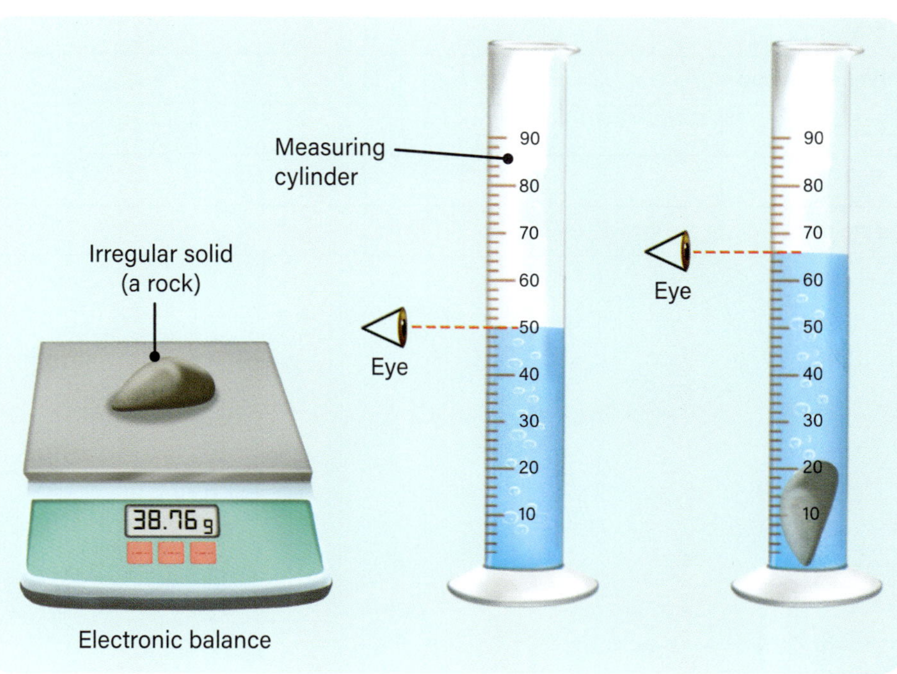

Figure 14.4: For this investigation, set up the equipment to measure the volume of the objects as shown in this diagram.

continues ▶

Questions

1 Which pieces of equipment did you need in order to successfully conduct this investigation? Why did you need each item?

2 For this investigation, identify:

   a the independent variable.

   b the dependent variable.

   c two controlled variables.

3 a Rank the objects from the most dense object to the least dense object.

   b Was this ranking what you expected? Why or why not?

Conclusion

Copy and complete:
'The results show that: *(respond to the aim)*.'

Table 14.4: Results

Object	Mass of object (g)	Float or sink?	Initial volume (mL)	Final volume (mL)	Density from experiment ($D = m/V$) (g/cm^3)
Cube					
Marble					
Cylinder					
Rock					
Cork					
Eraser					
Piece of wood					
Paperclip					
Nail					
Zinc metal					

Solutes and solvents

Process: Planning investigations

Focus: Identifying variables

When conducting an investigation, it is important to change only one factor so that its effect can be measured. Identify the independent (changed) and dependent (measured) variables of this investigation.

Hint 1: What factor are you changing?

Hint 2: What factor are you measuring?

Aim

To investigate which liquids will dissolve salt

Materials

- safety glasses
- nitrile gloves
- 5 test tubes
- marker pen
- test tube rack
- spatula
- digital scales
- 25 g sodium chloride (table salt) (NaCl)
- 50 mL measuring cylinder
- 20 mL water
- funnel
- glass stirring rod
- 20 mL 0.1 mol/L hydrochloric acid (HCl) in a small sealed dropper or reagent bottle
- 20 mL oil
- 20 mL milk
- 20 mL 0.1 mol/L sodium hydroxide (NaOH) in a small sealed dropper or reagent bottle

Method

1. Copy the results table into your notebook, adding a title.
2. Put on the safety glasses and gloves.
3. Collect 5 test tubes and label them 1–5 with a marker pen. Place them in a test tube rack.
4. Using the spatula and digital scales, measure 5 g of salt into each test tube.
5. Use the measuring cylinder to measure 20 mL of water. Using the funnel, pour the water into test tube 1 and use the stirring rod to stir the mixture.
6. Observe whether the salt dissolves. Record your observations in the results table. If the salt dissolves, place a tick in the 'Salt is soluble' column. If the salt does not dissolve, place a tick in the 'Salt is insoluble' column.
7. Determine whether or not water is a solvent for salt. (If the salt dissolved, water is a solvent for it.) If water is a solvent for salt, place a tick in the 'Acts as a solvent?' column.
8. Repeat Steps 4–6 with the other liquids.

Table I4.5: Results

Test tube	Liquid	Salt is soluble	Salt is insoluble	Acts as a solvent?
1	Water			
2	Hydrochloric acid			
3	Oil			
4	Milk			
5	Sodium hydroxide			

Questions

1. For the test tubes where you observed a solution, identify the solute and solvent.
2. For this investigation, identify:
   a. the independent variable.
   b. the dependent variable.
   c. two controlled variables.

Conclusion

Copy and complete:

'The results show that: *(respond to the aim)*.'

HYDROCHLORIC ACID IS A HAZARD. BE CAREFUL. WEAR SAFETY GLASSES AND GLOVES DURING THIS INVESTIGATION TO PREVENT ACID FROM BURNING YOUR SKIN OR SPLASHING IN YOUR EYES. DISPOSE OF ALL LIQUIDS AS DIRECTED BY YOUR TEACHER.

Investigation 4.6A

45 min

Water as a solvent

Process: Conducting investigations

Focus: Using scientific equipment

Before conducting a first-hand investigation, you need to ensure that you have an appropriate plan in place. How are you going to ensure that you will use the scientific equipment correctly to achieve accuracy?

Hint 1: Have you made sure the initial reading of your scale is zero?

Hint 2: What equipment will enable you to collect the right type of data?

Figure I4.6A: Make sure your digital scales read 0 (zero) before measuring the 5 g of each substance.

Aim

To investigate which substances dissolve in water

Materials

- 5 x 100 mL beakers
- marker pen
- 100 mL measuring cylinder
- water
- spatula
- digital scales
- 5 g table salt (sodium chloride (NaCl))
- glass stirring rod
- 5 g sugar (sucrose)
- 5 g baking soda (sodium hydrogen carbonate ($NaHCO_3$))
- 5 g chalk (calcium carbonate ($CaCO_3$))
- 5 g sand

Method

1. Copy the results table into your notebook, adding a title.
2. Collect 5 beakers and label them 1–5 with a marker pen.
3. Use the measuring cylinder to measure 50 mL of water into each beaker.
4. Use the spatula and the digital scales to measure 5 g of salt. Add the salt to beaker 1. Stir the mixture with the stirring rod.
5. Observe whether the salt dissolves. Record this information in the results table by placing a tick in the 'Soluble in water' or 'Insoluble in water' column.
6. Repeat Steps 4 and 5 for the other substances.

Table I4.6A: Results

Beaker number	Substance	Soluble in water	Insoluble in water
1	Salt		
2	Sugar		
3	Baking soda		
4	Chalk		
5	Sand		

Questions

1. For the beakers in which you observed a solution, identify the solute and solvent.
2. Which pieces of equipment did you need in order to successfully conduct this investigation? Why did you need each item?
3. For this investigation, identify:
   a. the independent variable.
   b. the dependent variable.
   c. two controlled variables.
4. Describe how you could tell if a solution had been formed or not.

Conclusion

Copy and complete:

'The results show that: *(respond to the aim).*'

Investigation **4.6B**

Solubility and temperature

Process: Conducting investigations

Focus: Implementing safe practices

Before conducting a first-hand investigation, you need to ensure that you have an appropriate plan in place. How are you going to ensure your safety during this practical? **Answer Question 1 before you start this investigation.**

Aim

To investigate which substances dissolve in water at different temperatures

Materials

- marker pen
- 4 beakers
- measuring spoons
- table salt (sodium chloride (NaCl))
- sugar (sucrose)
- sodium hydrogen carbonate ($NaHCO_3$)
- sand
- 250 mL of 20 °C water in a large beaker
- 250 mL of 40 °C water in a large beaker
- thermometer
- stirring utensil (spoon or stirring rod)
- stopwatch

Method

1. Copy the results table (on the next page) into your notebook, adding a title.
2. Use a marker pen to label four beakers as 'salt', 'sugar', 'baking soda' and 'sand'.
3. Add 1 teaspoon of salt to the beaker labelled 'salt'. Repeat for the other three substances.
4. In your notebook, record the appearance of each substance. Note the colour, texture and any other observable characteristics.
5. Adjust the temperature of the water in one beaker to 20 °C by adding hot or cold water. Confirm the temperature using the thermometer.
6. Use a measuring cylinder to add 50 mL of water to each beaker. Start the stopwatch.
7. Stir gently with a spoon to see if the substances dissolve.
8. In the results table, record which substances dissolve and which do not.
9. For substances that dissolve, record the time it takes for them to dissolve completely.
10. Repeat Steps 2–8 using water at 40 °C.

Questions

1. Copy the risk assessment table (on the next page) into your notebook.
   a. **Hazards**: Identify three things that could harm the person conducting this investigation.
   b. **Risks**: Decide whether you think there is a low, medium or high chance that each hazard will harm the person conducting this investigation.
   c. **Management strategies**: What steps can you take to make sure each hazard does not harm the person conducting this investigation?
2. Which pieces of equipment did you need in order to successfully conduct this investigation? Why did you need each item?
3. How does temperature affect the solubility of substances?
4. For each substance, compare its initial state with its dissolved state. What physical changes did you observe?
5. For the substances that dissolved, use the data to construct an appropriate graph of solubility at different temperatures.

Conclusion

Copy and complete:
'The results show that: *(respond to the aim).*'

continues ▶

Table 14.6B1: Risk assessment

Hazard	Risk (low, medium, high)	Management strategy

Table 14.6B2: Results

Substance	Water at 20 °C		Water at 40 °C	
	Dissolve or not dissolve	Time taken to dissolve	Dissolve or not dissolve	Time taken to dissolve
Salt				
Sugar				
Baking soda				
Sand				

Investigation 4.7A

Calculating the concentrations of solutions

Process: Conducting investigations

Focus: Collecting accurate data

Data from an investigation needs to be collected accurately; this means the measurements must be precise. How can you ensure precision when conducting this investigation?

Hint: *Think about the equipment you are using when obtaining quantities of solute and solvent.*

Aim

To determine the composition of different substances based on their properties

Materials

- digital scales
- 50 g sodium chloride
- 4 × 50 mL beakers
- 50 mL measuring cylinder
- 100 mL tap water
- glass stirring rod
- Bunsen burner
- heatproof mat
- tripod
- gauze mat
- matches

Figure 14.7A: Equipment like digital scales can make your investigations more precise and accurate.

Method

1 Copy the results table (on the next page) into your notebook, adding a title.

2 a Use the digital scales to measure 5 g of salt.

b Place the salt in a beaker.

3 a Use the measuring cylinder to measure 20 mL of water.

b Place the water in the beaker with the salt.

continues ▶

AN OPEN FLAME IS A HAZARD. BE CAREFUL. IF YOU BURN YOURSELF, TELL YOUR TEACHER IMMEDIATELY AND PLACE THE BURNT AREA UNDER COLD RUNNING WATER FOR 20 MINUTES.

4 Stir the salt and water with a stirring rod until no salt remains visible.

5 If salt is still visible after stirring, place the beaker over a Bunsen burner on a tripod and gauze mat.

6 a Use the matches to light the Bunsen burner and turn to the blue flame.

b Heat the salt and water while stirring with the stirring rod until the salt is dissolved.

7 Allow the solution to cool before removing the beaker from the gauze mat. Observe: Is all the salt still dissolved?

8 Repeat Steps 2–7, increasing the mass of salt by 5 g each time, until 20 g of salt has been dissolved to form a saltwater solution.

Questions

1 How did you make sure you put the right amount of salt into the beaker?

2 Why was it important to use a measuring cylinder to measure the water?

3 Provide three reasons why the data you collected in this investigation was accurate.

4 Which of your solutions were supersaturated solutions? Justify your choice.

5 Using the information in the results table, calculate the concentration of each of your saltwater solutions. Record these values in the results table.

Conclusion

Copy and complete:
'The results show that: *(respond to the aim)*.'

Table 14.7A: Results

Mass of sodium (g)	Volume of water (L)	Saltwater solution concentration (g/L) (mass/volume)
5	0.02	
10	0.02	
15	0.02	
20	0.02	

Investigation 4.7B

Preparing dilutions

Process: Conducting investigations

Focus: Using scientific equipment

Data from an investigation needs to be collected accurately; this means the measurements must be precise. How can you ensure precision when conducting this investigation?

Hint: *Think about the equipment you are using when obtaining quantities of solute and solvent.*

Aim

To prepare a range of dilutions of cordial and to compare colour intensities

Materials

- marker pen
- 6 large test tubes (minimum: 20 mm x 150 mm)
- test tube rack
- 10 mL plastic measuring cylinder
- 30 mL cordial concentrate
- water

Method

1. Copy the results table into your notebook, adding a title.
2. Use the marker pen to label the 6 test tubes 1–6. Place the test tubes in a test tube rack.
3. Use the measuring cylinder to accurately measure 10 mL of cordial concentrate and pour it into test tube 1.
4. Use the measuring cylinder to accurately measure 10 mL of water and pour it into test tube 1. Remember to rinse the measuring cylinder to remove traces of cordial between trials.
5. For the other test tubes, repeat Steps 3 and 4, but change the volume of cordial to:
   - test tube 2: 8 mL cordial; 10 mL water
   - test tube 3: 6 mL cordial; 10 mL water
   - test tube 4: 4 mL cordial; 10 mL water
   - test tube 5: 2 mL cordial; 10 mL water
   - test tube 6: 0 mL cordial; 10 mL water.
6. In the results table, record the colour intensities of the liquids in the test tubes, using the words 'darkest', 'dark', 'light' and 'lightest'.

Questions

1. Identify the steps to correctly use the measuring cylinder to measure the correct amount of cordial and water. (Hint: You need to be aware of the meniscus.)
2. Why was it important to rinse the measuring cylinder between trials?
3. Describe how this investigation models changes in concentrations of solutions.
4. Which of your solutions is the most concentrated and which is the least concentrated? Justify your choice using the %v/v calculation.

Conclusion

Copy and complete:

'The results show that: *(respond to the aim)*.'

Table I4.7B: Results

Test tube	Volume of cordial (mL)	Volume of water (mL)	Colour intensity (darkest, dark, light and lightest)	%v/v $\left(\frac{\text{volume of solute (mL)}}{\text{volume of solution (mL)}} \times 100\right)$
1	10	10		
2	8	10		
3	6	10		
4	4	10		
5	2	10		
6	0	10		

Investigation 4.8A

Classifying substances

Process: Problem-solving
Focus: Using criteria to find solutions to problems

In this investigation, you are required to classify substances based on their properties. To determine the classification, you need to follow the criteria that determine what an element, compound or mixture is. What are these criteria?

Hint: *How many particles are in each type of substance? Are they all the same?*

Aim

To determine the composition of different substances based on their properties

Materials

All materials displayed in clear, glass, sealed containers:

- 15 g table salt
- 100 mL water
- glass microscope slide
- water, sand and pebbles
- magnesium metal strip
- quartz crystal
- zinc metal strip
- graphite powder
- copper sulfate solution

Method

1. Copy the results table into your notebook, adding a title and as many rows as needed.
2. Locate the substances from the materials list in your classroom. (The substances are labelled.) Do not open the containers or touch the substances.
3. As a group, observe the first substance you find.
4. Based on your observations, decide if the substance is an element, a compound or a mixture.
5. Record your decision in the results table and give a reason for your decision.

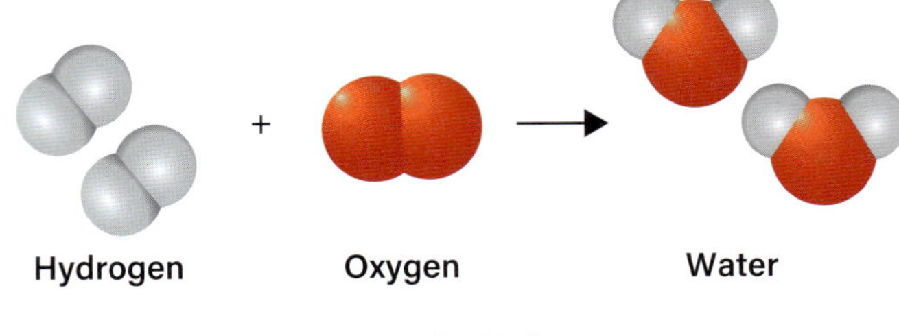

Figure I4.8A: Elements are made of only one type of atom, whereas compounds and mixtures are made up of more than one type of atom. How can you use observation to tell the difference?

Table I4.8A: Results

Substance	Element, compound or mixture?	Reason for classification

Questions

1. Which of the following problems did this investigation address?
   - **A** What is the difference between different atoms?
   - **B** How can I tell the difference between elements, compounds and mixtures?
   - **C** Which substances are the easiest to separate?
2. How many substances did you classify as elements?
3. **a** How many substances did you classify as mixtures?
   b Were they all easy to identify as mixtures? Why or why not?
4. Think back to the problem you identified in Question 1. What criteria did you use to solve this problem when you were classifying the different substances?

Conclusion

Copy and complete:
'The results show that: *(respond to the aim).*'

Investigation **4.8B**

45 min

The Tyndall effect

Process: Problem-solving

Focus: Suggesting solutions to familiar problems

The Tyndall effect allows us to identify a mixture as a colloid or suspension. This can help us to solve problems when considering uses for materials.

Hint: When would we want to remove particles from mixtures before using them?

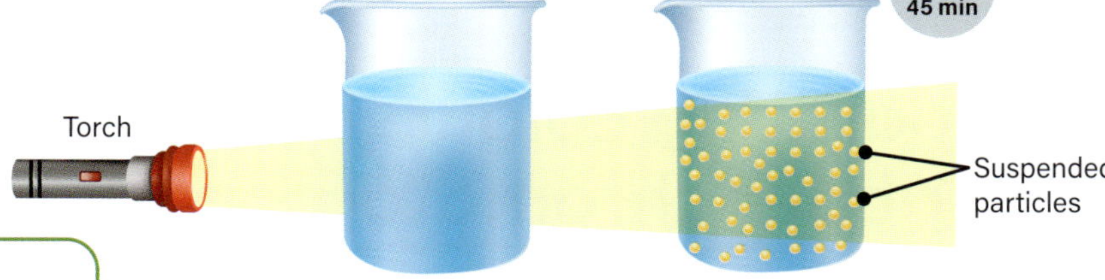

Figure I4.8B: Mixtures containing suspended particles display the Tyndall effect (a visible, scattered light beam).

Aim

To identify three types of mixtures – solutions, suspensions and colloids – by using the Tyndall effect

Materials

- test tube rack
- 6 labelled test tubes containing prepared mixtures of pure water and the following substances:
  - milk
  - soluble starch
  - food colouring
  - sugar
  - flour
  - salt (sodium chloride (NaCl))
- glass stirring rod
- torch

Method

1 Copy the results table into your notebook, adding a title.

2 Observe the mixtures. Record a description of each mixture in the results table.

3 a Use the stirring rod to stir each mixture.

b In the results table, record which mixtures separate on standing.

4 Make the room as dark as possible or use an under-bench cupboard. For each mixture that does not separate on standing, shine a torch on it. In your results table, describe whether the mixture exhibits the Tyndall effect (it scatters light) (see Figure I4.8B).

5 Classify each mixture as a solution, suspension or colloid. Record these classifications in the results table.

Questions

1 The effect of light scattering by particles is known as the __________ __________.

2 a If a mixture is left to stand and then separates, it is a __________.

b If a mixture does not separate on standing and the Tyndall effect is not seen, the mixture is a __________.

c If a mixture does not separate on standing and exhibits the Tyndall effect, the mixture is a __________.

3 a Identify the piece of equipment used to measure precise quantities in this investigation.

b Describe how this precision improves the investigation.

Conclusion

Copy and complete:

'The results show that: *(respond to the aim)*.'

Table I4.8B: Results

Test tube	Mixture	Brief description	Separates on standing?	Exhibits Tyndall effect?	Classification (solution, suspension or colloid)
1	Milk				
2	Soluble starch				
3	Food colouring				
4	Sugar				
5	Flour				
6	Salt				

Investigation 4.9

Separating colours in dyes by paper chromatography

Process: Problem-solving

Focus: Explaining phenomena using cause-and-effect relationships

Conducting investigations can help scientists to solve problems by showing the way that certain processes work. This can often be represented in writing, where a specific process is linked to the effect that occurs. What is the effect of this separation technique?

Hint 1: What scientific concept is linked to chromatography?

Hint 2: How could this link help scientists to solve real-world problems?

Aim

To investigate how chromatography can be used to separate colours

Materials

- different coloured confectionery; for example, jelly beans, Skittles, Smarties
- watch glass
- eyedropper (or a couple of plastic pipettes)
- tap water
- filter paper
- 150 mL beaker

Method

1. a Place a sample of the confectionery on a watch glass.
   b Use the eyedropper to put 2 drops of water on the sample.
2. When the colour in the confectionery starts to dissolve, use the eyedropper to collect some of the coloured liquid.
3. Place 1 drop of the coloured liquid in the middle of the filter paper. Then place the filter paper on an empty beaker.
4. Add 1 drop of water directly over the top of the coloured liquid and observe what happens. Record your observations.
5. Repeat Steps 1–4 using a different coloured confectionery.
6. Describe what happens to the sample as the solvent (water) is added.

Questions

1. Describe how you used the filter paper to determine the different colours in each of the dyes.
2. Which confectionery colours are single substances and which ones are mixtures?
3. How many different dyes are needed to make all the confectionery colours?
4. Explain the link between particle size and the colour results you can see.
5. Discuss when this separation technique could be used in the real world to help scientists solve problems.

Conclusion

Copy and complete:
'The results show that: *(respond to the aim)*.'

NEVER EAT ANYTHING IN A SCIENCE LABORATORY.

Investigation **4.10**

Purifying muddy water

Process: Conducting investigations

Focus: Using scientific equipment

Scientific equipment is used for specific purposes during first-hand investigations. In this investigation, how should the materials be used to ensure that the muddy water mixture is properly separated?

Hint 1: Are there holes in the filter paper that the residue could pass through?

Hint 2: Did you let the mixture settle before decanting?

Aim

To purify muddy water by decanting and filtration

Materials

- 250 mL beaker
- 125 mL tap water
- 5 g soil
- glass stirring rod
- 5 g sand
- filter paper
- filter funnel to suit conical flask
- 250 mL conical flask

Method

1. Copy the results flowchart (on the next page) into your notebook.
2. Half fill the beaker with water.
3. Place the soil into the beaker. Stir with a stirring rod until a suspension forms.
4. Place the sand into the beaker and allow the sand to settle to the bottom.
5. Fold the filter paper as shown in Figure I4.10A.
6. Set up the filter paper, filter funnel and conical flask as shown in Figure I4.10B:
   a. Place the folded filter paper into the filter funnel.
   b. Place the filter funnel into the conical flask.
7. Carefully pour the mixture from the beaker into the conical flask, without disturbing the sand on the bottom of the beaker.
8. Record your observations in the results flowchart.

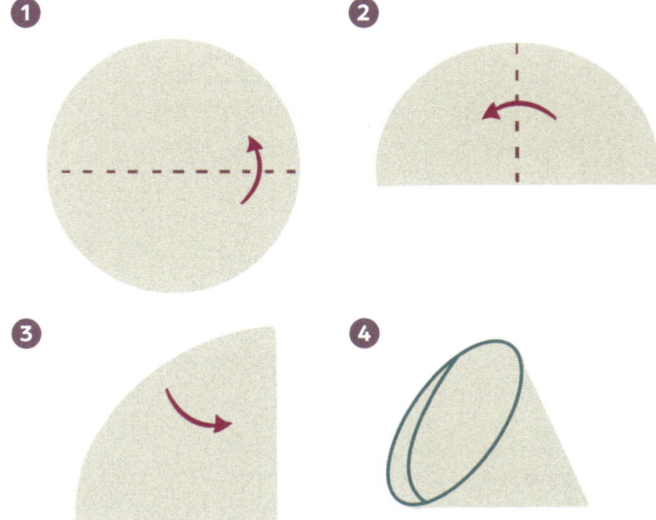

Figure I4.10A: For this investigation, follow these steps to fold the filter paper: (1) Fold the filter paper in half. (2) Fold it in half again. (3) Open it so it forms a cone shape with three layers on one side and one on the other.

Figure I4.10B: For this investigation, set up the materials as shown in this diagram.

continues ▶

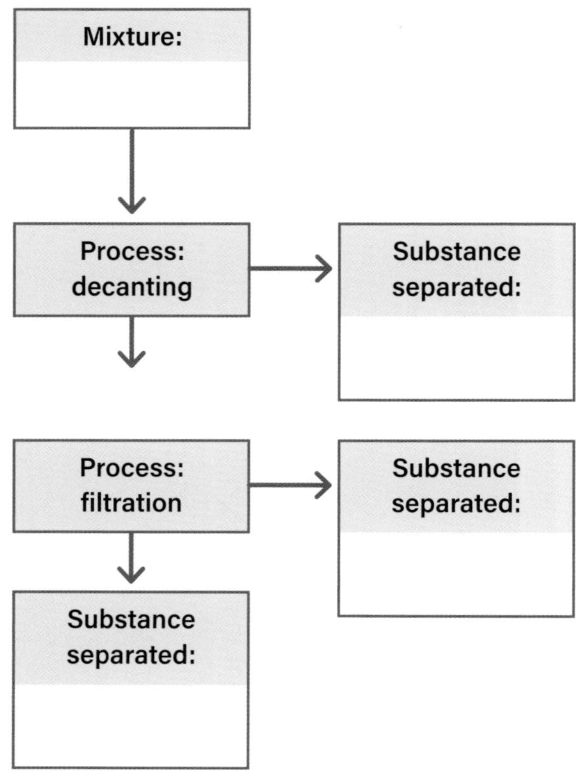

▲
Figure 14.10C: Results flowchart

Questions

1 Identify why the filter paper enabled you to separate the soil and water.

2 Which pieces of equipment did you need in order to successfully conduct this investigation? Why did you need each item?

3 Identify which component formed the:

   a suspension.

   b sediment.

   c filtrate.

   d residue.

4 Describe the physical properties that allowed you to separate the soil and the sand.

Conclusion

Copy and complete:
'The results show that: *(respond to the aim)*.'

Investigation 4.11A

Evaporating a solution

Process: Conducting investigations

Focus: Implementing safe practices

Before conducting a first-hand investigation, you need to ensure that you have an appropriate plan in place. How are you going to ensure your safety during this practical? **Answer Question 1 before starting this investigation.**

Aim

To investigate what happens when a solution evaporates

Materials

- safety glasses
- 5 mL copper sulfate solution
- evaporating basin
- Bunsen burner
- heatproof mat
- tripod
- gauze mat
- matches

Method

1. Place the copper sulfate solution into an evaporating basin.
2. Set up the materials as shown in Figure I4.11A.
3. Put on safety glasses, and nitrile gloves if needed. Light the Bunsen burner and turn to the blue flame.
4. In your notebook, record your observations as the water evaporates.
5. Turn off the Bunsen burner before the last few drops evaporate.

Table I4.11A: Risk assessment

Hazard	Risk (low, medium, high)	Management strategy

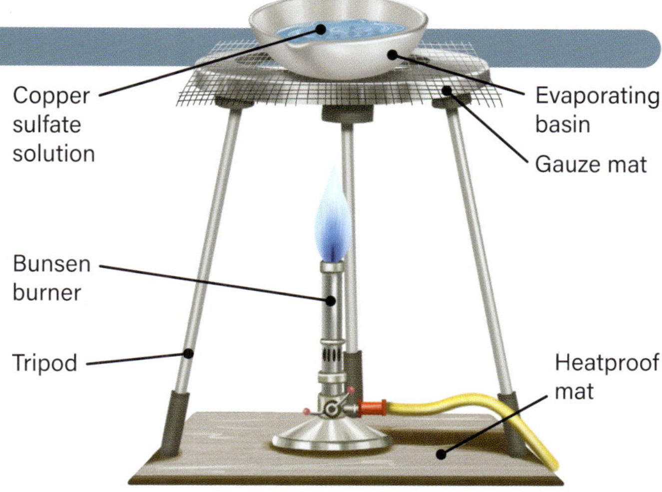

▲ **Figure I4.11A:** For this investigation, set up the materials as shown in this diagram.

Questions

1. Copy the risk assessment table into your notebook.
   a. **Hazards**: Identify three things that could harm the person conducting this investigation.
   b. **Risks**: Decide whether you think there is a low, medium or high chance that each hazard will harm the person conducting this investigation.
   c. **Management strategies**: What steps can you take to make sure each hazard does not harm the person conducting this investigation?
2. a. Describe why the Bunsen burner was important to the evaporation process.
   b. Could evaporation have been achieved using a different method?
3. Describe the substance before and after evaporation occurred.
4. Compare the initial state and the final state of the mixture. What physical changes did you observe? What happened to the water?

Conclusion

Copy and complete:

'The results show that: *(respond to the aim)*.'

AN OPEN FLAME IS A HAZARD. BE CAREFUL. IF YOU BURN YOURSELF, TELL YOUR TEACHER IMMEDIATELY AND PLACE THE BURNT AREA UNDER COLD RUNNING WATER FOR 20 MINUTES.

COPPER SULFATE IS A HAZARD. IF YOU COME INTO CONTACT WITH THE SOLUTION, TELL YOUR TEACHER IMMEDIATELY AND WASH IT OFF YOUR SKIN. ALWAYS WEAR SAFETY GLASSES.

Investigation 4.11B

Growing crystals

Process: Conducting investigations

Focus: Using scientific equipment

Scientific equipment helps to ensure that precise quantities of materials can be collected so that the investigation can be effectively completed. How will you measure quantities in this investigation?

Hint: Look at what you need to measure. How can you ensure you obtain a precise quantity?

Aim

To investigate growing crystals from solutions at different temperatures

Materials

- safety glasses
- nitrile gloves
- 100 mL measuring cylinder
- cold water from tap
- hot water from tap
- 2 × 100 mL beakers
- 30 g copper sulfate (solid)
- 2 glass stirring rods
- cotton thread
- 2 paperclips
- digital scales

Method

1. Use the measuring cylinder to measure 40 mL of hot water. Pour the water into a beaker.
2. Add 30 g of copper sulfate to the beaker containing the hot water. Stir with a stirring rod until the solid dissolves.
3. Pour half of the volume into another beaker. Cool this beaker by running cold water on the outside until it reaches room temperature.
4. Tie a piece of cotton thread to each stirring rod and attach a paperclip at the other end. Hang the strings with paperclips inside each beaker.
5. Leave the beakers for several days to allow crystals to grow.
6. In your notebook, describe what you can see inside each beaker.

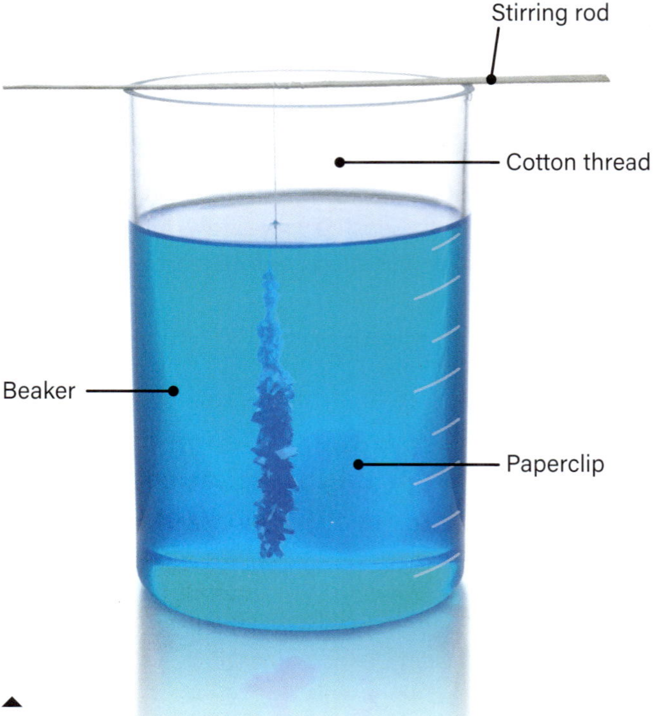

Figure I4.11B: In this investigation, you will grow crystals inside two beakers.

Questions

1. Which pieces of equipment did you need in order to successfully conduct this investigation? Why did you need each item?
2. Describe two ways you used scientific equipment in this investigation to make sure you measured precise quantities of the materials.
3. Describe the substance before and after crystallisation.
4. Compare the sizes of the crystals in the two beakers. How could you make the crystals bigger?

Conclusion

Copy and complete:

'The results show that: *(respond to the aim)*.'

Investigation 4.11C

Demonstrating distillation

TEACHER DEMONSTRATION

Process: Conducting investigations

Focus: Identifying correct processes

Before conducting a first-hand investigation, it is important to understand the processes involved. What processes are important in successfully separating mixtures using distillation?

Hint 1: *How is the Bunsen burner used?*

Hint 2: *What process makes sure the heat is evenly distributed?*

Aim

To investigate separating a salt solution using distillation

Materials

- Bunsen burner
- gauze mat
- matches
- Liebig condenser
- thermometer
- distillation flask
- retort stand with 2 bossheads and 2 clamps
- receiving flask
- 150 mL beaker containing salt solution
- marble chips

Method

1. Your teacher will set up the materials, as shown in Figure I4.11C.
2. Your teacher will pour the salt solution into the distillation flask, and then add a few marble chips so that it heats evenly.
3. Your teacher will turn on the water through the condenser, boil the mixture in the flask and collect the water that comes out of the condenser in a receiving flask.

Questions

1. Identify the components separated as residue and distillate.
2. Why is it important to use the blue flame when heating materials with a Bunsen burner?
3. Marble chips were added to the solution prior to boiling. What was the effect of these chips?
4. Describe the changes of state that happened during the distillation.
5. Explain why the salt and water could be separated.

Conclusion

Copy and complete:
'The results show that: *(respond to the aim).*'

AN OPEN FLAME IS A HAZARD. BE CAREFUL. IF YOU BURN YOURSELF, TELL YOUR TEACHER IMMEDIATELY AND PLACE THE BURNT AREA UNDER COLD RUNNING WATER FOR 20 MINUTES.

continues ▶

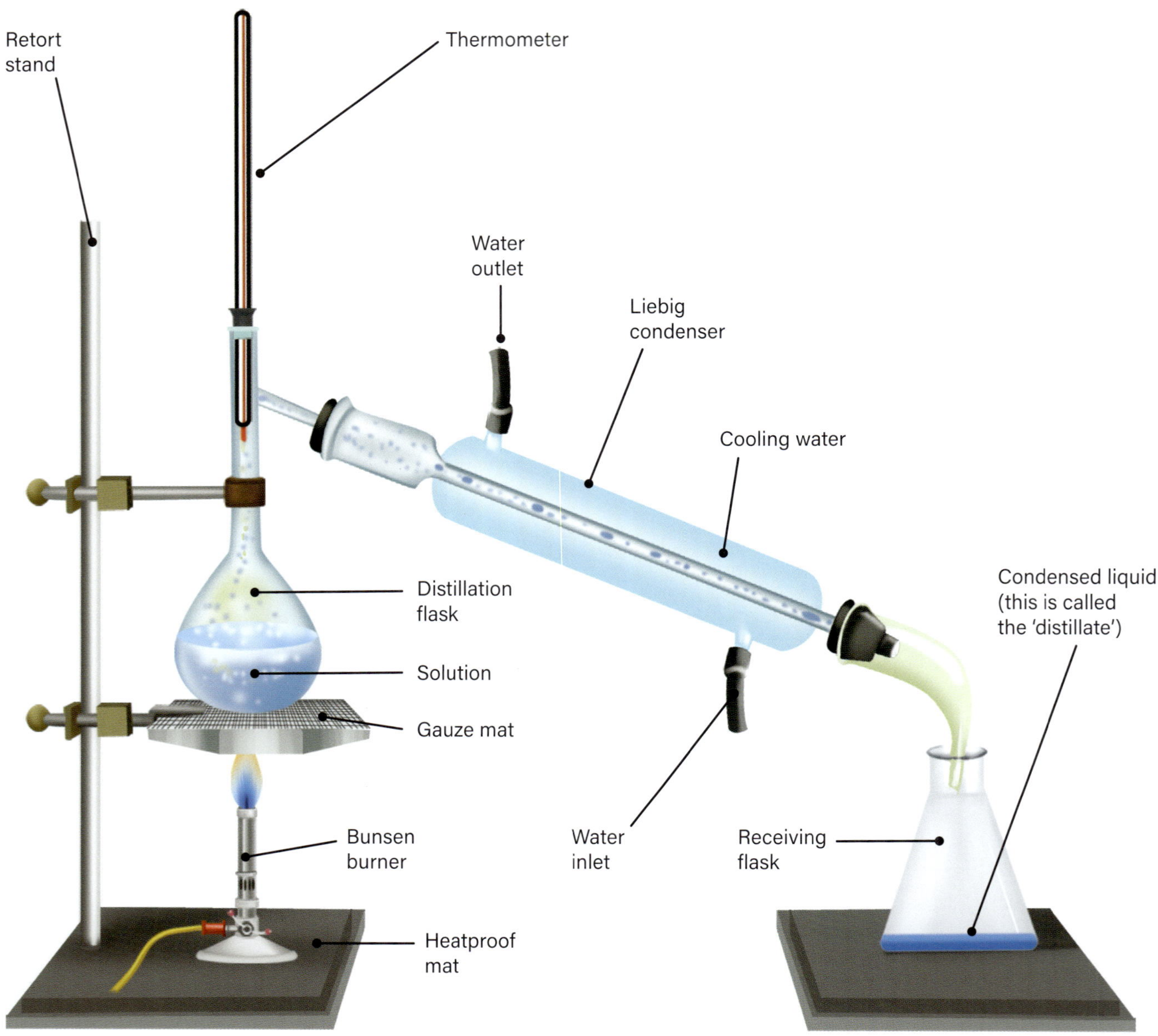

Figure 14.11C: For this investigation, the equipment will be set up as shown in this diagram.

Investigation 4.14

Purifying oily water

TEACHER DEMONSTRATION

Process: Planning investigations
Focus: Identifying appropriate materials
Before starting a first-hand investigation, it is important to identify the equipment that will allow the investigation to be effectively conducted. What equipment is vital to the separation process?
Hint: *How do the properties of oil and water enable their separation?*

Aim

To use a separating funnel to separate oil and water

Materials

- 2 × 25 mL measuring cylinders
- 20 mL water
- 20 mL oil
- 50 mL separating funnel
- retort stand with bosshead and clamp
- retort ring
- 2 × 100 mL conical flasks
- tap

Method

1. Your teacher will set up the materials, as shown in Figure 14.14.
2. Your teacher will measure 20 mL of water and transfer it to the separating funnel.
3. Your teacher will measure 20 mL of oil and transfer it to the separating funnel.
4. Your teacher will stopper the funnel and shake it. Next, they will place the funnel back into the retort ring. Observe the mixture separating and record your observations in your notebook.
5. After two obvious layers have formed, your teacher will slowly open the tap and collect the bottom layer in a flask. Observe the transfer of the contents to a measuring cylinder and measure the volume of the liquid collected.
6. Once your teacher pours out the remaining liquid, measure the volume collected. Record your observations in your notebook.

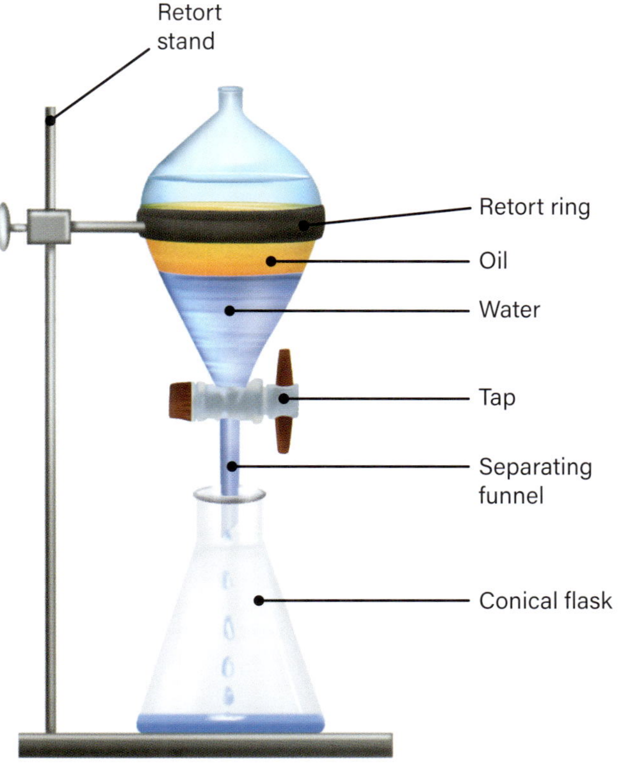

Figure 14.14: For this investigation, the materials will be set up as shown in this diagram.

Questions

1. Describe how the process of the separation was controlled using the tap.
2. Suggest why a separating funnel was used to perform this investigation.
3. What happened to the mixture when the separating funnel was shaken?
4. Explain why the mixture settled after some time.
5. Consider the volumes you started with for each liquid and the volumes you recovered at the end. Based on these quantities, how efficient is separating a mixture using a separating funnel?

Conclusion

Copy and complete:
'The results show that: *(respond to the aim).*'

Investigation 5.3

Breathing rate and exercise

Process: Processing data and information
Focus: Collecting and presenting data
In investigation reports, there is a results section that includes the data collected; this is often presented as a table, graph or image. Choosing how to represent your data so it can be clearly communicated to someone reading your investigation report is an important skill.
Hint: *There are many ways to visually present the data you have collected, including bar charts, line graphs and pie charts.*

Aim

To investigate the effect of exercise on breathing rate, heart rate, perspiration and body temperature

Materials

- body temperature thermometer
- heart rate monitor (if available)
- skipping rope
- watch (or a stopwatch)

Method

1. Copy the results table into your workbook, adding a title.
2. Find a partner to work with and an area that is suitable to exercise in.
3. Observe your partner's skin tone (redness) (look at their face and arms). Rate their skin tone using a scale of 1–4, where 1 is normal, 2 is slightly red, 3 is red, and 4 is very red. Record your rating in the results table in the '0 (rest)' row.
4. Observe your partner's level of perspiration. Rate their perspiration level using a scale of 1–4, where 1 is no sweating, 2 is slightly sweating, 3 is sweat building up under the armpits, and 4 is extremely sweaty. Record your rating in the results table in the '0 (rest)' row.
5. Use the body temperature thermometer to measure your partner's body temperature. Record this information in the results table in the '0 (rest)' row.
6. a Use the heart rate monitor to count how many times your partner's heart beats in 15 seconds. Record this information in the results table in the '0 (rest)' row.
   b Use the heart rate monitor to measure your partner's heart rate, which is the number of beats per minute. Record this information in the results table in the '0 (rest)' row.

   If you do not have a heart rate monitor, you can measure your partner's pulse by placing your index and third fingers on their neck to the side of their windpipe. Count how many times you feel a pulse in 15 seconds, then multiply by four to get the number of heart beats per minute.
7. a Count how many breaths your partner takes in 15 seconds. Record this information in the results table in the '0 (rest)' row.
   b Calculate your partner's breathing rate, which is the number of breaths they take in 1 minute. To do this, multiply the number of breaths they take in 15 seconds by 4. Record your partner's breathing rate in the results table in the '0 (rest)' row .
8. Your partner exercises vigorously (for example, skipping rope or doing jumping jacks) for 5 minutes.
9. While your partner is exercising, measure and record their skin tone, level of perspiration, heart rate and breathing rate every minute. Measure and record their temperature if you have a thermometer that allows you to do this (for example, an infrared thermometer).
10. Take these measurements again every minute for 5 minutes after your partner stops exercising.

Table 15.3: Results

Time (minutes)	Skin tone (redness) (1–4)	Perspiration level (1–4)	Body temperature (°C)	Heartbeats in 15 seconds	Heart rate (beats per minute)	Breaths in 15 seconds	Breathing rate (breaths per minute)
0 (rest)							
1							
2							
3							
4							
5 Stop exercise							
6							
7							
8							
9							
10							

Questions

1 Describe how your partner's breathing rate changed throughout the investigation.

2 What changes did you observe in your partner's:

- a skin tone (redness)?
- b perspiration level?
- c body temperature?
- d heart rate?

3 Explain how changes in skin tone, perspiration level and body temperature during and after exercise are linked to homeostasis.

4 If you repeat an exercise routine regularly, you become fitter and your heart rate and breathing rate do not rise as high as they did before you exercised regularly. Suggest why this occurs.

5 a Construct line graphs of your data, showing how the variables changed over time.

- b Use your graphs to identify the relationship between breathing rate and heart rate.
- c Suggest a reason for the relationship you identified in Question 5b, considering that the heart is not part of the respiratory system.

Conclusion

Copy and complete:
'The results show that: *(respond to the aim).*'

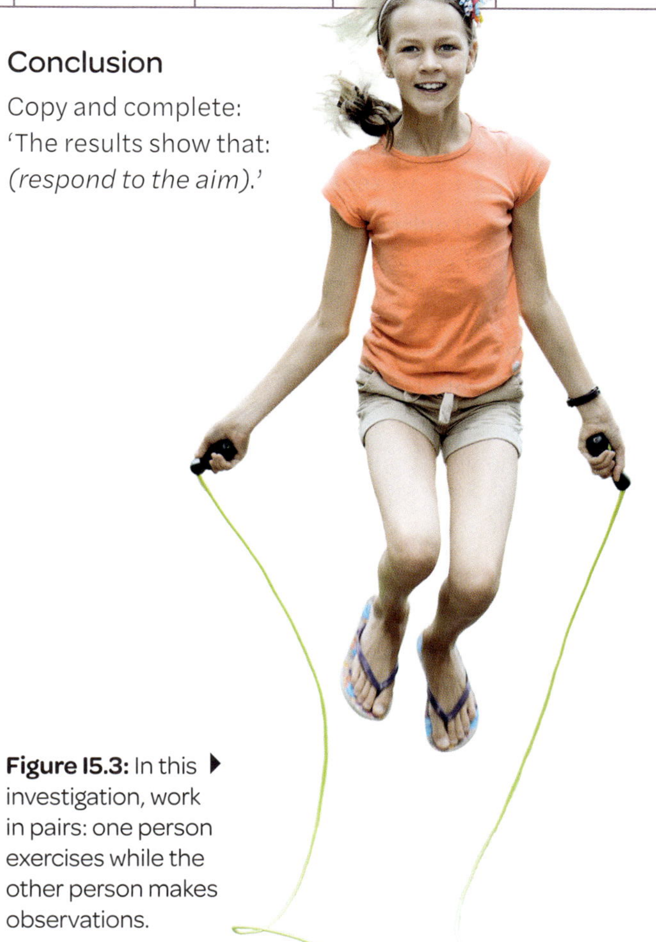

Figure 15.3: In this investigation, work in pairs: one person exercises while the other person makes observations.

Investigation 5.4

Dissecting a heart

Process: Questioning and predicting

Focus: Constructing a scientific research question

Turn the aim of this investigation into a question that asks about the scientific problem or concept you are investigating. This is called a research question.

Hint 1: Your research question can also be used as a title for an investigation report.

Hint 2: Refer to 'Questioning and predicting' in the Science how-to section on page 319.

Aim

To investigate the structure of a mammalian heart

Materials

- safety glasses
- disposable plastic aprons
- gloves
- newspaper (or butchers paper)
- dissecting board
- sheep or cow heart
- dissection kit (dissection scissors, blunt forceps and blunt probes)
- pipe cleaners
- pins
- labels
- disinfectant

Method

1. Place some newspaper on the bench and the dissecting board on top of it.
2. Wearing safety glasses, an apron and gloves, carefully examine the heart. The blood vessels around the outside are the coronary arteries and veins; these take blood to and from the heart muscle.
3. Feel either side of the heart. One side should feel thicker than the other. The thicker side is the left, and the smaller and thinner side is the right.
4. Look for the blood vessels on the top of the heart. Stick in pipe cleaners or your fingers to see where they lead to.
5. Place the heart on the dissecting board with the apex (pointed end) towards you, and the right (thinner) side of the heart to your left.

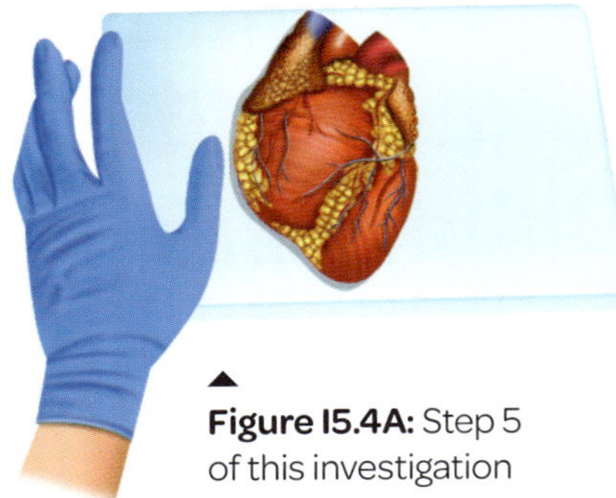

Figure 15.4A: Step 5 of this investigation

6. Use the dissection scissors to carefully make a cut around the outside of the heart.
7. You should now be able to open the heart and see inside both the right and left sides.

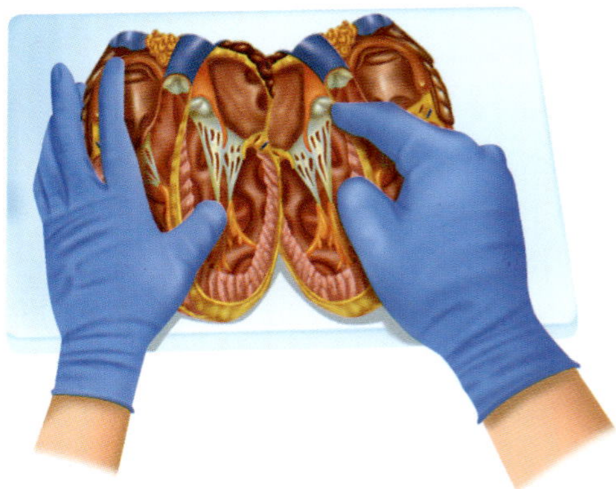

Figure 15.4B: Step 7 of this investigation

8 Carefully examine the inside of the heart. Identify the valves between the atria and ventricles of each side. These are controlled by tendons called *chordae tendineae*, which are commonly known as the 'heartstrings'. Try pulling these.

9 Identify the septum; this is the structure that separates the left and right sides of the heart.

10 Use the pins and labels to:

 a label the structures of the heart: left and right side, atria, ventricles, aorta, vena cava and valves.

 b indicate the direction of oxygenated and deoxygenated blood flow.

11 Take a photo of the labelled, dissected heart.

12 When you have completed your dissection, carefully clean up and disinfect your work area.

Questions

1 Compare the size of the left and the right sides of the heart. Why is there such a size difference?

2 Compare the size of the atria and the ventricles. Why is there such a size difference?

3 a What is the function of the valves in the heart?

 b What might occur if the valves did not function properly?

4 What is the function of the septum in the heart?

Conclusion

Copy and complete:
'The results show that: *(respond to the aim)*.'

WEAR SAFETY GLASSES AND GLOVES. BE CAREFUL WHEN USING CUTTING IMPLEMENTS. DISPOSE OF ALL MATERIALS AS DIRECTED BY YOUR TEACHER. WASH YOUR HANDS AFTER COMPLETING THE INVESTIGATION.

Investigation 5.7

Water transport in plants

Process: Questioning and predicting

Focus: Formulating testable questions (hypotheses) considering variables and controls

Before you formulate a hypothesis, identify the independent, dependent and controlled variables in the investigation. The independent variable is the one thing that you purposefully change. The dependent variable is what you measure. The controlled variables are all the things you need to keep the same. **Answer Question 1 before you start this investigation.**

Hint: Refer to 'Questioning and predicting' in the Science how-to section on page 319.

Aim

To investigate how water is transported in plants

Materials

- 3 × 200 mL beakers
- 200 mL water
- food colouring
- knife
- cutting board
- 3 celery sticks, with leaves still attached, cut to 20 cm lengths
- gloves
- magnifying lens

Method

1. Label the three beakers A, B and C.
2. a Add 100 mL of water to beaker A.
   b Add 100 mL of water to beaker B.
   c Leave beaker C empty.
3. Add enough food colouring to beaker A to make the water a dark colour.
4. Using a knife and a cutting board, cut off 2 cm from the bottom of each stick of celery.
5. Place one celery stick in each beaker. Leave them to stand overnight.
6. The next day, observe the leaves of the three celery sticks. Record your observations in your notebook.
7. Wearing gloves, remove the celery that was in the coloured water (beaker A) and use the knife to slice it crosswise and lengthwise. Use the magnifying lens to observe the pathway of the food colouring.
8. In your notebook, draw the celery from beaker A.

Questions

1. Write a hypothesis for this investigation using the following sentence:
   'It can be hypothesised that if *(something to do with your independent variable)*, then *(something to do with your dependent variable).*'
2. What was the purpose of beakers B and C?
3. Explain what happened to the celery in beaker A in terms of the plant's water transport system.
4. a What result did you obtain from beaker C?
   b Why do you think this happened?
5. Describe how water is transported in plants, using key terms, written observations or diagrams from this investigation to support your answer.

Conclusion

Copy and complete:
'The results show that: *(respond to the aim).*'

BE CAREFUL USING A KNIFE. IF YOU CUT YOURSELF, TELL YOUR TEACHER IMMEDIATELY AND SEEK FIRST AID.

Figure I5.7: Make sure the liquid in beaker A is a dark colour.

Investigation 5.8

Dissecting a flower

Process: Communicating

Focus: Constructing a scientific presentation

Scientists can use modelling and simulations to simplify and explain ideas. Models make things easier to understand and visualise, which helps us to make predictions. However, there can be limitations when using models. How might this investigation be limited or improved if you used an artificial model or digital simulation, rather than a real flower?

Hint: *Refer to 'Communicating' in the Science how-to section on page 343.*

Aim

To investigate the physical features of flowering plants

Materials

- gloves
- safety glasses
- magnifying lens
- 1 large, whole flower (such as a lily or tulip)
- dissecting board
- dissecting kit (dissection scissors and blunt forceps)

WEAR SAFETY GLASSES AND GLOVES. BE CAREFUL WHEN USING CUTTING IMPLEMENTS. DISPOSE OF ALL MATERIAL AS DIRECTED BY YOUR TEACHER. WASH YOUR HANDS AFTER THE INVESTIGATION.

Method

1. In groups of 3 or 4, use a magnifying lens to observe the external features of a flower. Record your observations in your notebook.
2. Locate both parts of the flower's stamen (the anther and filament). Record your observations in your notebook.
3. Record your observations from Steps 3a and 3b in your notebook.
   a. Locate all three parts of the flower's pistil (stigma, style and ovary).
   b. Wearing gloves and safety glasses, place the flower on the dissecting board. Using the dissection scissors, cut open the ovary and count the number of eggs inside.

Questions

1. Why is it important for the anther to be towards the top of the flower?
2. How many eggs were inside the plant's ovary? How could you tell they were eggs?
3. Explain how pollen moves from the male to the female reproductive parts of the flower.

Conclusion

Copy and complete:

'The results show that: *(respond to the aim)*.'

Figure 15.8: Use the magnifying lens to closely observe the flower. Remember to wear safety glasses and gloves!

Investigation 5.9

Identifying community members in a terrestrial ecosystem

Process: Communicating

Focus: Presenting scientific findings

Scientists can use modelling and simulations to simplify and explain ideas. Models make things easier to understand and visualise, which helps us to make predictions. However, there can be limitations when using models. How might this investigation be limited or improved if you used an artificial or simulated ecosystem, rather than the real thing?

Hint: *Refer to 'Communicating' in the Science how-to section on page 343.*

Aim

To investigate different species in a terrestrial ecosystem

Materials

- 4 witches hats (optional)
- species identification resources for your area
- thick gardening gloves
- magnifying lens
- coloured pencils

Method

1. Copy the results table into your notebook, adding a title and as many rows as needed.
2. Select a small ecosystem at your school, no larger than 20 metres by 20 metres. You can mark out an area using witches hats or other physical markers.
3. In the results table, list all the different plants in the ecosystem.
4. Put on the gloves.
5. Use the magnifying lens to look for small organisms living on the plants. Look on the underside of leaves, and in the branches and grass. Add the names of the organisms you find to the results table.
6. Look for more small organisms by carefully turning over fallen leaves and looking under rocks and in the soil. Add the names of the organisms you find to the results table.
7. Sit quietly and identify the different organisms you can see and hear in the ecosystem. Add the names of these organisms to the results table.
8. a Draw each organism you identified.
   b Make note of any interesting behaviour or relationships between the organisms that you observed.

Questions

1. Identify each organism in the results table as a producer, consumer or decomposer.
2. Create a food web of the organisms you observed in the ecosystem. You may need to conduct some extra research to do this.
3. What types of relationships between organisms did you observe?

Conclusion

Copy and complete:

'The results show that: *(respond to the aim).*'

Table I5.9: Results

Location: ____________________________

Date: ________________ Time: ____________

Organism name	Drawing of the organism	Observations of behaviour, including any relationships observed

Investigation 5.11

Measuring abiotic factors

Process: Questioning and predicting

Focus: Constructing a scientific research question

Turn the aim of this investigation into a question that asks about a scientific problem or concept you are investigating. This is called a research question.

Hint 1: *Your research question can also be used as a title for an investigation report.*

Hint 2: *Refer to 'Questioning and predicting' in the Science how-to section on page 319.*

Aim

To investigate the abiotic factors in a terrestrial (land-based) ecosystem

Materials

- thermometer
- trowel
- soil multiprobe (pH, light, moisture)

Method

1 Copy the results table (on the next page) into your notebook, adding a title and as many rows as needed.

2 Select a small ecosystem in your school.

3 Measure the air temperature 1 metre above the ground in the sunlight by holding the thermometer in the air for several minutes until the reading stabilises. Record the air temperature in the sunlight in the results table.

4 Measure the air temperature 1 metre above the ground in the shade. Record the air temperature in the shade in the results table.

5 Measure the temperature of the surface of the soil by carefully placing the end of the thermometer in the leaf litter or on top of the soil. Leave the thermometer for a few minutes to stabilise. Record the surface soil temperature in the results table.

6 Measure the temperature of the soil 5 cm below the surface. Use the trowel to carefully bury the thermometer bulb about 5 cm deep. Leave the thermometer for a few minutes to stabilise. Record the soil temperature in the results table.

7 Measure the soil moisture 5 cm below the surface. Switch the soil multiprobe to the 'moisture' setting. Carefully push the end of the probe about 5 cm into the soil. Leave the probe for a few minutes to stabilise. Record the soil moisture value in the results table.

8 Measure the pH of the soil 5 cm below the surface. Leave the multiprobe in the soil and carefully switch the setting to 'pH'. Leave the probe for a minute to stabilise. Record the soil pH value in the results table.

Figure I5.11: In this investigation, you will use a probe like this one.

continues ▶

9 Measure the light intensity in the soil 5 cm below the surface. Leave the multiprobe in the soil and carefully switch the setting to 'light/LUX'. Leave the probe for a minute to stabilise. Record the soil light intensity value in the results table.

10 a Use the trowel to collect a soil sample from about 5 cm deep.

b In the results table, describe the soil colour and texture. Is it coarse, gritty or fine? Does it contain organic matter? You may wish to view the sample under a microscope.

Table 15.11: Results

Air temperature in the sunlight (°C)	
Air temperature in the shade (°C)	
Soil temperature, surface (°C)	
Soil temperature, 5 cm deep (°C)	
Soil moisture, 5 cm deep	
Soil pH, 5 cm deep	
Light intensity, 5 cm deep (lux)	
Soil colour	
Soil texture	

Questions

1 Compare your results with those of your classmates. Did you and your classmates conduct your investigations in the same or different parts of your school? How might this have affected the results each student obtained?

2 Propose how your results might be different if you repeated this investigation at night, 6 hours earlier or later, or in 6 months.

3 a Research the temperature, light, water and soil requirements of at least two plant species in the same ecosystem.

b How do these values correlate with your results?

Conclusion

Copy and complete:
'The results show that: *(respond to the aim).*'

Investigation 5.13

Algal balls

Process: Processing data and information

Focus: Collecting and presenting data

It is important that scientists clearly present the data they collect in first-hand investigations so other scientists can understand the results. An excellent way to clearly present data is in a graph or chart.

Hint 1: There are many different charts and graphs that you can use to visually represent data, including bar charts, line graphs and pie charts.

Hint 2: *Refer to 'Processing data and information' in the Science how-to section on page 329.*

Aim

To investigate how different light levels affect the rate of photosynthesis, using algal balls

Materials

- algal balls
- 4 small sample jars with lids
- 10 mL measuring cylinder
- hydrogen carbonate indicator solution (0.0025% bromothymol blue works well)

Method

1 Copy the results table into your notebook, adding a title.

2 Carefully place an equal number of algal balls in each sample jar.

3 Use the measuring cylinder to measure an equal amount of hydrogen carbonate indicator solution into each sample jar, ensuring the algal balls are covered.

4 Place the lids on the jars.

5 Place the jars in different places around your classroom that get different amounts of sunlight. They could be placed on a windowsill, on a shelf away from the window or in a cupboard.

6 Leave the jars until the next day.

7 Compare the colour of each solution to the colour indicator chart (Figure 15.13).

8 Record the results of this investigation in the results table.

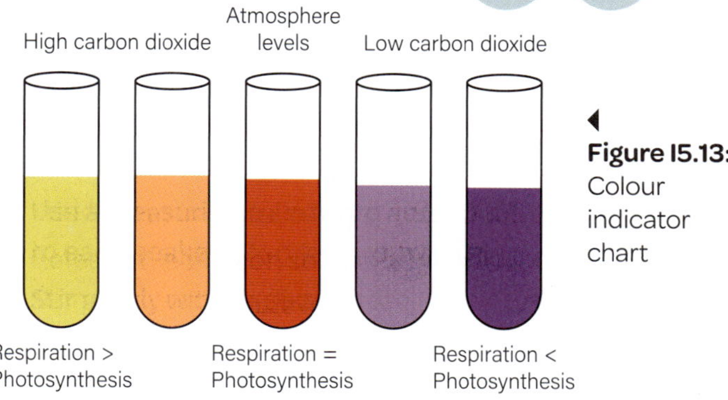

Figure 15.13: Colour indicator chart

9 Rank the sample jars from 1 to 4: 1 is the jar in which the *most* photosynthesis took place, and 4 is the jar in which the *least* photosynthesis took place. Write your rankings in the results table.

Table 15.13: Results

Sample jar location	Colour	pH	Rank (1–4)

Questions

1 Consider the photosynthesis reaction.
  a What are the reactants?
  b What are the products?
  c What else is required?

2 a In which sample jar did the *most* photosynthesis take place? How do you know this?
  b In which sample jar did the *least* photosynthesis take place? How do you know this?

3 Write a method for an investigation that uses algal balls to test the effect of one of these variables on photosynthesis:
  a wavelength of light (colour of light)
  b temperature
  c size of algal balls.

4 a Turn the data in your results table into a chart or a graph.
  b Identify any patterns or trends in your data.

Conclusion

Copy and complete:
'The results show that: *(respond to the aim).*'

Investigation 6.1

Comparing metals and non-metals

Process: Processing data and information

Focus: Constructing a table to gather information

It is important that the information you collect is organised. One of the best ways to organise scientific data is in a table. Based on the materials and method required for this investigation, you will construct a table to collect data before you begin this investigation.

Hint: *If you get stuck, refer to 'Processing data and information' in the Science how-to section on page 329.*

Aim

To investigate the properties of metals and non-metals

Materials

- selection of samples of metals, non-metals and metalloids (for example, carbon (graphite), silicon, aluminium, zinc, copper, glass, steel or iron, rubber)
- steel sewing needle
- 12 V power supply
- 12 V light bulb
- 3 connecting wires
- alligator clips

Method

1 Construct a results table in your notebook to record the data you will collect in this investigation. Hints:
   - List the names of the different samples (the independent variable) in the left column.
   - Add the properties you are testing (see the questions below) to the table. Write each property as a heading for a column across the top of the table.
   - Include a column where you write your predictions about whether each sample is a metal or a non-metal.

2 Choose one of the samples. In the results table, record your observations of its appearance (for example, its colour, how shiny it is).

3 Try to bend the sample to test its malleability. Record your observations in the results table.

4 Scratch the sample using the steel sewing needle. Is the sample shiny underneath? Record your observations in the results table.

5 Connect the electric circuit as shown in Figure I6.1.

6 Switch on the power supply to 12 V. Observe whether the light bulb turns on. If it does, the sample conducts electricity. Record your observations in the results table.

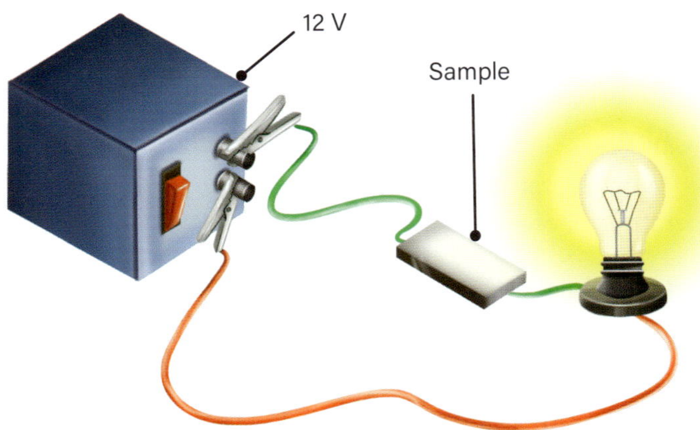

Figure I6.1: For this investigation, set up the electric circuit as shown in this diagram.

Questions

1 Construct a two-column table. Categorise each of the materials tested in this investigation as a metal or a non-metal.

2 What properties did all the metals have in common?

3 Were there any properties common to both metals and non-metals?

4 Predict the properties of tungsten (metal) and iodine (non-metal).

Conclusion

Copy and complete:

'The results show that: *(respond to the aim).*'

ELECTRICITY IS A HAZARD. BE VERY CAREFUL WHEN USING ELECTRICITY.

Comparing the properties of elements, compounds and alloys

Process: Analysing data and information

Focus: Identifying and describing trends in collected data

When you collect first-hand data, it is important to be able to recognise any trends or patterns in the information. In this investigation, you will collect qualitative data (written descriptions and observations). The information collected should be organised in a way that allows trends and patterns to be identified and described in relation to the aim of the investigation.

Hint: More information about identifying and describing trends in data is available in the Science how-to section on page 334.

Aim

To investigate and compare the properties of elements in their pure, compound and alloy forms

Materials

- samples: several trios of substances in pure, compound and alloy form (10 mL of liquid sample, 5 g of granular sample, 1 piece of solid sample); for example:
  - copper metal, copper sulfate, brass
  - graphite, ethanol, carbon steel
  - iron metal, iron oxide (pigment), cast iron
- 12 V power supply
- 12 V light bulb
- 3 connecting wires
- alligator clips
- 2 paperclips (electrodes)
- watch glass (for granular samples)
- spatula
- 10 mL measuring cylinder
- 100 mL beakers (1 for water and 1 for each liquid sample)
- 50 mL tap water

DO NOT POUR COPPER SULFATE DOWN THE DRAIN AS IT IS A TOXIN. YOUR TEACHER WILL TELL YOU HOW TO DISPOSE OF THIS SUBSTANCE.

Method

1 Copy the results table (on the next page) into your notebook, adding a title and as many rows as needed. You can modify the results table to suit the samples you will be testing. This may make it easier to identify trends in the data you collect. An example has been provided in the table to help you.

2 Choose one of the trios of substances.

3 Conduct the following investigations to determine the properties of each of the three substances. Record your observations in the results table.

   a **Appearance**: What does the substance look like? Observe its colour, lustre (shine) and physical state (solid, liquid or gas).

   b **Malleability**: Can the substance be bent and shaped? For bulk solid samples, try to bend the sample. Is the substance malleable (can you bend and shape it)? Is it brittle (does it break when you try to bend it)?

   c **Hardness**: How hard is the substance? Scratch the surface of the sample with the other solid samples. Record which samples leave a scratch on which substances.

   d **Conductivity**: Does electricity flow through the substance? To test this, connect the electric circuit as shown in Figure I6.2 (on the next page). Turn on the power supply to 12 V. Test each substance as instructed below. Observe whether the light bulb turns on. If it does, the substance conducts electricity.

      i **Bulk solid sample**: Connect the alligator clips to either side of the sample to complete the circuit.

continues ▶

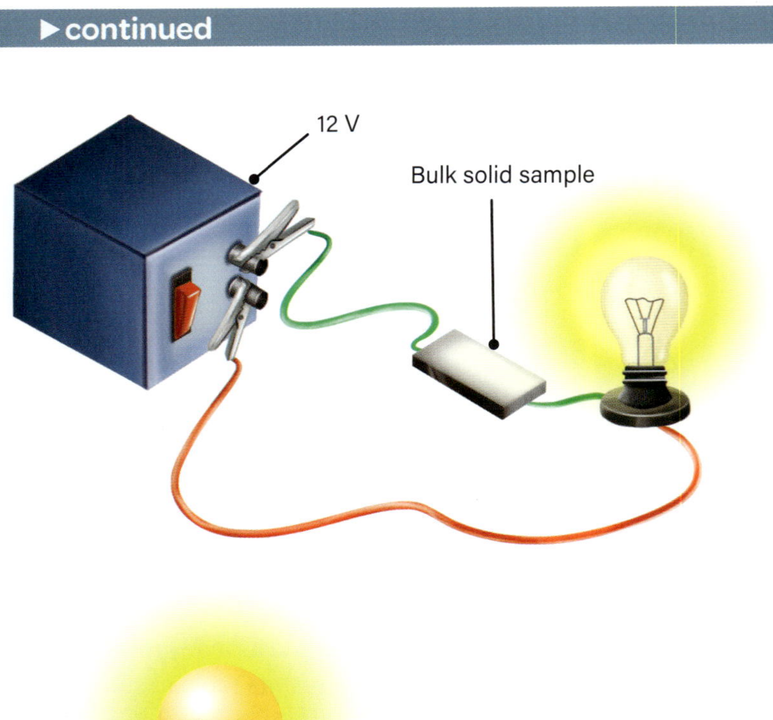

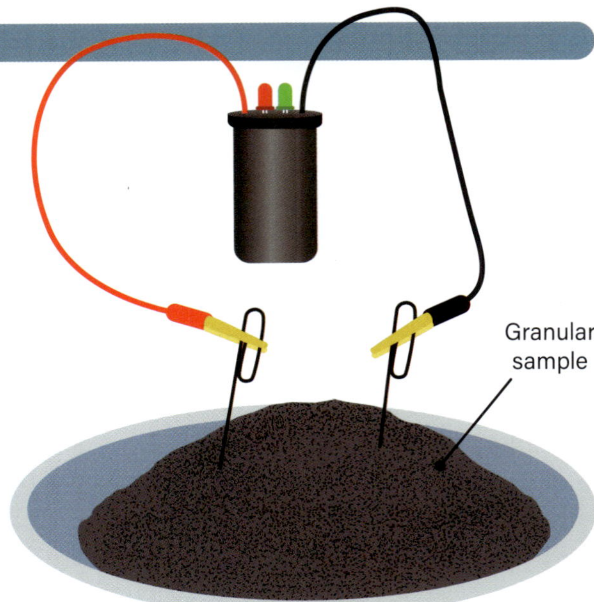

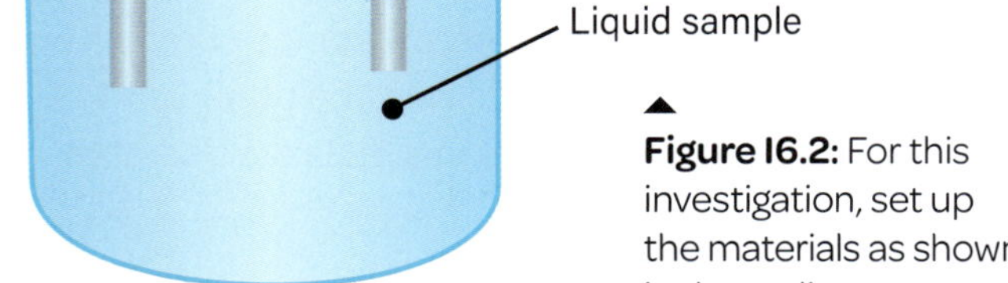

▲
Figure 16.2: For this investigation, set up the materials as shown in these diagrams.

ii **Granular solid sample**: Connect each alligator clip to an unfolded paperclip. Insert each paperclip into the powder sample, making sure the paperclips do not touch.

iii **Liquid sample**: Connect each alligator clip to an unfolded paperclip. Insert each paperclip into the liquid sample. Make sure the paperclips do not touch each other or the sides of the beaker.

e **Solubility**: Can the substance be dissolved in liquid? Place the solid sample, a spatula full of granular sample, or 10 mL of liquid sample into the beaker of water. Observe whether the sample dissolves or not.

Questions

1 Identify any trends in the data you have collected.

2 Describe any unexpected results. Why were these results not what you expected?

3 What other properties could you test?

4 Imagine you test the substances for the properties you identified in Question 3. Do you expect to see any trends in the data you would collect?

Conclusion

Copy and complete:
'The results show that: *(respond to the aim)*.'

Table 16.2: Results

Element	Form	Appearance	Malleability	Hardness	Conductivity	Solubility
Copper	Pure element (e.g. copper wire)					
	Compound (e.g. copper sulfate)					
	Alloy (e.g. brass)					

Investigation 6.4

Flame test

Process: Analysing data and information

Focus: Drawing conclusions that are based on patterns in data and information

When you write a formal investigation report, there is always a conclusion section that summarises the investigation by responding to (or answering) the aim. To do this, you need to draw a conclusion that is consistent with the data you have collected.

Aim

To use the distinctive colours produced by metallic ions in a flame test to identify two unknown metallic ions

Materials

- set of 0.1 mol/L metal chloride solutions, each solution in a 250 mL beaker (lithium chloride, strontium chloride, calcium chloride, barium chloride, copper chloride, cobalt chloride, sodium chloride, potassium chloride, phosgene)
- 2 unknown solutions
- Bunsen burner
- heatproof mat
- matches
- 12 popsticks
- tongs
- cobalt glass plate

Method

1 Copy the results table into your notebook, adding a title.

2 Light the Bunsen burner and turn it to the blue flame.

AN OPEN FLAME IS A HAZARD. BE CAREFUL. IF YOU BURN YOURSELF, TELL YOUR TEACHER IMMEDIATELY AND PLACE THE BURNT AREA UNDER COLD RUNNING WATER FOR 20 MINUTES. DISPOSE OF WASTES AS INSTRUCTED BY YOUR TEACHER.

3 a Dip the end of a clean popstick into one of the solutions until it is saturated.

b Hold the soaked end of the popstick in the hottest part of the flame, ensuring you do not set it on fire. You can hold your popstick with tongs.

4 Observe the colour of the flame. Make sure the popstick does not catch fire. Otherwise, you will not be able to reuse it to double-check the colour later.

5 Carefully record your observations in the results table. Be as accurate as possible. Your description of the colour must be specific enough to distinguish this metal ion from the other ions tested.

6 Repeat Steps 3–5 for each solution. Use a new, clean popstick to test each solution.

7 When you examine the sodium and potassium ions, first look at their colour unaided, then look through a cobalt glass plate. Record both observations separately in your results table.

Table 16.4: Results

Metal ion	Colour of the flame
Lithium	
Strontium	
Calcium	
Barium	
Copper	
Cobalt	
Sodium	
Sodium (with cobalt glass)	
Potassium	
Potassium (with cobalt glass)	
Unknown 1	
Unknown 2	

continues ▶

Figure 16.4: Different metallic ions produce different coloured flames.

8 When you have tested all the known solutions and can accurately describe the colour of each metal ion, obtain two different, unknown solutions from your teacher. Determine which metal ions are present by performing a flame test and comparing this data to your previous data.

Based on your observations, identify the two unknown ions.

Unknown ion 1 is: ______________________

Unknown ion 2 is: ______________________

Questions

1 Which ions produced similar colours in the flame tests?

2 What was the purpose of using the cobalt glass?

3 Explain how the colours observed in the flame tests are produced.

4 a Did you correctly identify the two unknown solutions?

  b What factors contributed to your correct or incorrect identification of the unknown solutions?

5 State three problems that may arise when using flame tests for identification purposes in a laboratory.

Conclusion

Copy and complete:

'The results of this investigation show that: *(respond to the aim).*'

Investigation 6.6A

Modelling atomic structure alongside the periodic table

Process: Processing data and information

Focus: Identifying data from tables

Chemists always have access to a periodic table, and it is important that they can identify the data in that table. In this investigation, you will use the periodic table to compile information about a variety of elements and their atoms. You will also use the periodic table to create models.

Hint: *Refer to 'Processing data and information' in the Science how-to section on page 329.*

Aim

To use the periodic table to produce atomic models of the atoms of the first 20 elements of the periodic table and to explore how the periodic table is organised

Materials

- a copy of the periodic table
- 1 large paper plate
- 40 small lollies (20 of one colour and 20 of a different colour) (or pieces of playdough or plasticine)
- camera

Method

1 Copy the results table (on the next page) into your notebook, adding a title and 20 blank rows. Number the rows from 1 to 20 in the 'Atomic number' column.

2 By consulting the periodic table, in the second column of the results table, write the names of the first 20 elements of the periodic table.

3 By consulting the periodic table, answer the following questions for each element in the results table. Write your answers to the questions in the table in your notebook.

   a Identify the element's symbol.

   b Identify the number of protons.

   c Identify the number of electrons.

   d In the periodic table, which period is the element in?

   e In the periodic table, which group is the element in?

Note: The number of neutrons in an atom's nucleus cannot be determined from the periodic table, so we will ignore neutrons for the purpose of this activity.

4 On the paper plate, draw five circles (see Figure I6.6A on the next page). The paper plate represents an atom. The innermost circle represents the nucleus. The second, third, fourth and fifth circles represent electron shells 1, 2, 3 and 4.

5 The lollies represent protons (one colour) and electrons (the other colour). Start with hydrogen. Using your paper plate 'atom' and your lolly 'subatomic particles', create an atomic model of a hydrogen atom. Take a photo. Repeat this process for elements 2 to 20 of the periodic table.

Note: When you are placing electrons in the electron shells, fill the shell closest to the nucleus first (shell 1), then shells 2, 3 and 4. Remember that electrons are distributed across electron shells in a 2, 8, 8, 2 pattern (see page 201).

6 After completing each model, in the results table, record the number of electron shells that have electrons, and the number of electrons in the valence shell.

For example, a neutral magnesium atom has 12 protons and 12 electrons. Place 12 lollies in the nucleus to represent the protons. Place two electrons in electron shell 1; place eight electrons in electron shell 2; then place the remaining two electrons in electron shell 3. This means that a magnesium atom has three electron shells and two electrons in its valence shell (the shell furthest from the nucleus).

NEVER EAT ANYTHING IN A SCIENCE LABORATORY.

continues ▶

Table 16.6A1: Results

Atomic number	Element name	Symbol	Number of protons in an atom	Number of electrons in an atom	Period of the periodic table	Group of the periodic table	Number of electron shells that have electrons	Number of valence electrons
1								
2								
3								

Questions

1 a What trends or patterns have you noticed in your results table?

b Link your observations to the organisation of the periodic table.

2 Which part of atomic structure is responsible for the properties of an element? Provide a reason for your response.

3 Electron configuration becomes more complicated after calcium.

a Copy Table 16.6A2 into your notebook. Without drawing the atoms, use the periodic table to complete the table.

b Provide a general reason for your responses.

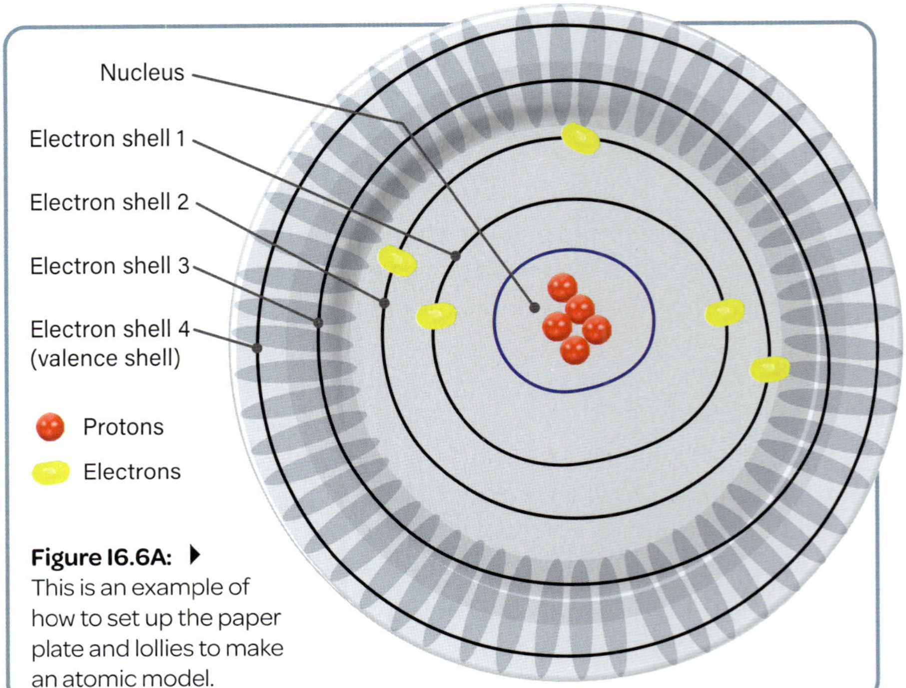

Figure 16.6A: ▶ This is an example of how to set up the paper plate and lollies to make an atomic model.

Table 16.6A2: Question 3a answers

	Predict the number of valence electrons	Predict the number of electron shells
iodine		
krypton		
tin		
barium		
potassium		

4 a Arrange the atomic model photos you took of the first 20 elements of the periodic table into a digital table structured like the periodic table.

b Label the groups and periods.

c Highlight the valence electrons and shell patterns that you observe.

Conclusion

Copy and complete:
'The results show that: *(respond to the aim)*.'

Investigation 6.6B

Investigating the reactivity of metals

Process: Processing data and information

Focus: Constructing a table to organise data collected in a first-hand investigation

It is good practice to construct a results table before you start an investigation. You can use the materials and the method to help you figure out what to label the rows and columns of your table.

Hint: If you get stuck, refer to 'Processing data and information' in the Science how-to section on page 329.

Aim

To investigate the reactivity of different metals – aluminium (Al), magnesium (Mg), zinc (Zn) and iron (Fe) – and to consider the positions of these elements in the periodic table

---

In this investigation, we will be testing the reactivity of different metals in hydrochloric acid. If a reaction occurs, hydrogen gas will be released from the hydrochloric acid, and the chlorine from the acid will combine with the metal to form a metal chloride salt. This is the word equation for a reaction that may or may not occur between aluminium and hydrochloric acid:

aluminium + hydrochloric acid →
hydrogen gas + aluminium chloride

Materials

- 4 test tubes
- test tube rack
- safety glasses
- gloves
- 10 mL measuring cylinder
- 40 mL of 1 mol/L hydrochloric acid
- funnel
- sandpaper
- a variety of samples of metals (aluminium, magnesium, zinc, iron)

Method

1. Construct a results table in your notebook to collect the data from this investigation. Use the checklist (Table 16.6B) to help you. Make sure you include a column to identify the reactivity of each metal.
2. Place the test tubes in the test tube rack.
3. Wearing safety glasses and gloves, measure 10 mL of 1 mol/L hydrochloric acid. Using the funnel, place 10 mL of acid into each test tube.
4. Use the sandpaper to clean the surface of each metal sample.
5. Place the aluminium sample into the first test tube. Observe how the metal reacts with the hydrochloric acid. Record your observations in your results table.
6. Repeat Step 5 with each of the other metal samples.
7. Consider the position of each metal in the periodic table. Think about this position and how each metal reacted with the acid.

Table 16.6B: Checklist for constructing a scientific table

☑ Scientific table checklist

- ☐ The table has a title
- ☐ Each column in the table has a heading
- ☐ Units of measurement are included
- ☐ The independent variable (the selection of samples of metals to test) is placed in the left column

Questions

1. Using the data you have collected and organised in your table, list the metals in order from the most reactive to the least reactive.
2. Do your results match the general trend in the periodic table in relation to the reactivity of metals? Explain your answer.
3. Write a word equation for the chemical reaction that may occur when magnesium is combined with hydrochloric acid.

Conclusion

Copy and complete:
'The results show that: *(respond to the aim)*.'

HYDROCHLORIC ACID IS A HAZARD. WEAR SAFETY GLASSES AND GLOVES DURING THIS INVESTIGATION TO PREVENT ACID FROM BURNING YOUR SKIN OR SPLASHING IN YOUR EYES. DISPOSE OF ALL LIQUIDS AS DIRECTED BY YOUR TEACHER.

Investigation 6.7

Investigating different properties of a compound and one of its elements

Process: Analysing data and information

Focus: Explaining relationships between observations and information

When you write an investigation report, there is always a discussion section that includes your analysis and explanation of the data you collected. This is where you explain your results by linking them to what you already knew about the science of what you are studying.

Hint 1: Before the reaction, the magnesium is a pure metal; after the reaction, the substance is a non-metal.

Hint 2: Refer to 'Analysing data and information' in the Science how-to section on page 334.

Aim

To investigate how the properties of a compound are different from the properties of the elements from which it is made

Materials

- Bunsen burner
- heatproof mat
- tripod
- pipeclay triangle
- crucible and lid
- 5 cm magnesium ribbon
- matches
- brass (crucible) tongs

Method

1 Set up the Bunsen burner, mat, tripod, pipeclay triangle and crucible as shown in Figure 16.7.

2 Look closely at the magnesium ribbon. Record your observations (for example, its colour, shininess and feel) in your notebook.

3 Place the magnesium ribbon inside the crucible and put on the lid.

4 Light the Bunsen burner. Heat the crucible containing the magnesium. Using tongs, lift the lid every 5–10 seconds to ensure enough air is in the crucible. **Do not look directly at the magnesium ribbon.**

5 The heated magnesium will combine with the oxygen in the air to create magnesium oxide, a compound. After the magnesium has finished reacting, observe the magnesium oxide. Record your observations about its physical properties in your notebook.

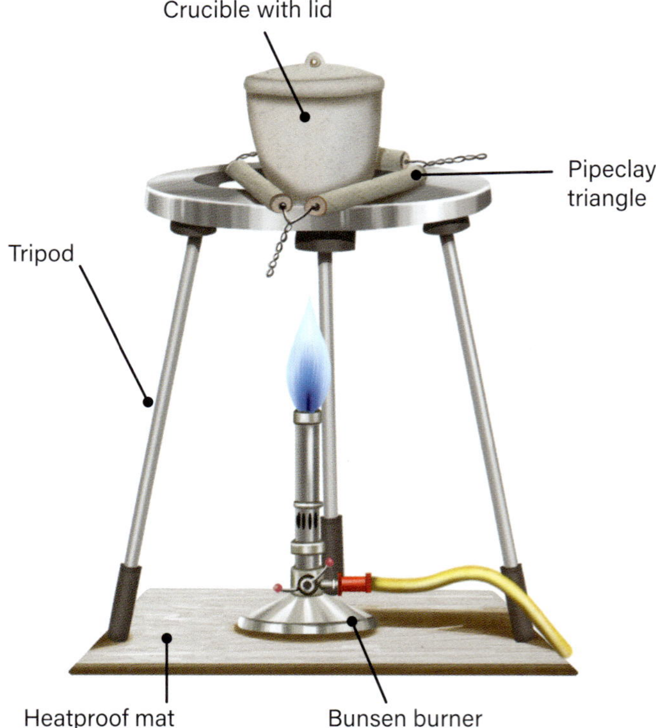

Figure 16.7: For this investigation, set up the materials as shown in this diagram.

Questions

1 How did the appearance of the magnesium differ from that of the magnesium oxide?

2 What was the purpose of regularly lifting the crucible lid?

3 Where did the oxygen in the magnesium oxide come from?

Conclusion

Copy and complete:
'The data shows that ... *(respond to the aim)*.
This makes sense because ...'

AN OPEN FLAME IS A HAZARD. BE CAREFUL. IF YOU BURN YOURSELF, TELL YOUR TEACHER IMMEDIATELY AND PLACE THE BURNT AREA UNDER COLD RUNNING WATER FOR 20 MINUTES.

MAGNESIUM RIBBON IS A HAZARD. IT IS FLAMMABLE AND REACTIVE. BE EXTREMELY CAREFUL. NEVER LOOK DIRECTLY AT A BURNING MAGNESIUM RIBBON.

Investigation 6.9

Modifying metals

Process: Analysing data and information
Focus: Evaluating experimental methods
When conducting investigations, it is important to ensure that the method followed is valid, so that any collected data can be considered of high quality. 'Evaluating' means weighing up the benefits and limitations of the method, before making a judgement on the quality of the data collected.
Hint 1: Is there evidence of precise measurements?
Hint 2: Is there evidence of quantities of materials being controlled?
Hint 3: Is there evidence of repetition?

Aim

To investigate how different heating and cooling techniques alter the properties of metals

Materials

- 4 identical, sharp sewing needles
- wooden peg
- Bunsen burner
- heatproof mat
- matches
- 100 mL beaker, filled with cold water
- pliers

Method

1 Copy the results table into your notebook, adding a title.
2 Stick one of the sewing needles into the wooden peg; the eye of the needle should be in the jaws of the peg. This is needle A.
3 Hold needle A over the blue flame of the Bunsen burner until the whole needle is red hot. Avoid exposing the peg to the flame. Then allow the needle to cool slowly in the air.
4 Mount needle B into the wooden peg and hold it over the blue flame until it is red hot. Remove from the flame and immediately put the needle into a 100 mL beaker filled with cold water.
5 For needle C, repeat the process as for needle B. Then, dry and rewarm needle C over the orange flame of the Bunsen burner. Make sure no part of the needle becomes red hot. Then, allow the needle to cool in the air.
6 Needle D is your control.
7 Use the pliers to bend each needle. Record the properties you observe (for example, hardness and malleability) in the table in your notebook.

Table 16.9: Results

Needle	Treatment	Hardness	Malleability	Additional observations
A				
B				
C				
D				

Questions

1 a Which metal treatment (A, B or C) resulted in the strongest needle?
  b Explain these results.
2 Explain why the original needle is suitable for its intended purpose (sewing).
3 a Without using the internet, make a list of common items made from metal that you think may have been modified by heat treatments.
  b Do an online search to see if your list from Question 3a is accurate.
  c What interesting things did you learn from Question 3b?
4 Evaluate the method followed in this investigation. Use your evaluation to make a judgement about the quality of the data relating to the sewing needles. Suggest one way this method could be improved.

Conclusion

Copy and complete:
'The results show that: *(respond to the aim)*.'

AN OPEN FLAME IS A HAZARD. BE CAREFUL. IF YOU BURN YOURSELF, TELL YOUR TEACHER IMMEDIATELY AND PLACE THE BURNT AREA UNDER COLD RUNNING WATER FOR 20 MINUTES.

Investigation 7.1

Rolling balls

Process: Observing

Focus: Using observations and measurements to answer questions

During first-hand investigations, you can use your senses (sight, hearing, touch, smell) to make observations and to collect qualitative data (written descriptions). You can also use scientific equipment to collect precise quantitative (numerical) data. When answering questions about an investigation, you can draw on both types of data.

Hint: Refer to 'Observing' in the Science how-to section on page 316.

Aim

To investigate the relationship between gravitational potential energy and kinetic energy

Materials

- ramp
- milk carton or small, empty box
- 3 balls or marbles with different masses
- electronic balance
- ruler or tape measure

Method

1 Copy the results table into your notebook, adding a title and as many rows as needed.

2 During the investigation, in your notebook, record the observations you make with your senses.

3 Set up the ramp and the milk carton (see Figure 17.1).

4 Use the electronic balance to measure the mass of each ball. Record this data in the results table.

Write the numerical data you collect in Steps 5–8 in the results table.

5 Roll the first ball down the ramp. Using a ruler, measure the distance the carton moved when the ball hit it.

6 Repeat the test four times with the first ball. Make sure the point of release is the same each time.

7 Use the results of the five tests to calculate the average distance the carton moved when the first ball hit it.

8 Repeat Steps 5–7 for the other two balls.

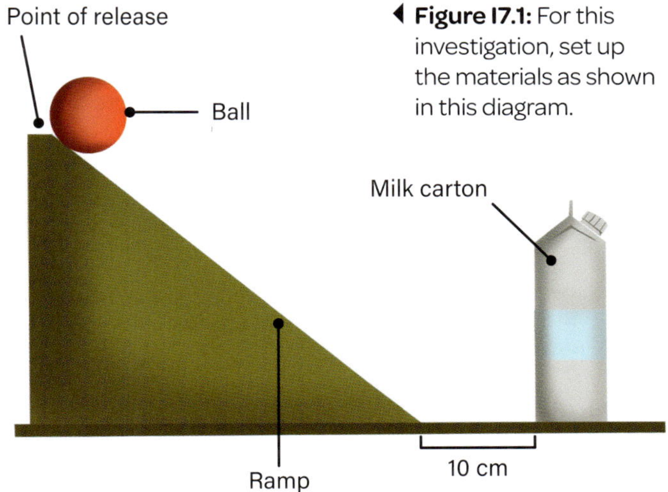

Figure 17.1: For this investigation, set up the materials as shown in this diagram.

Questions

1 Which ball displayed the most kinetic energy at the bottom of the ramp? Suggest a reason for this.

2 Which ball had the most gravitational potential energy at the start? Use the observations you made with your senses to support your response.

3 Compare the numerical data in the results table to the observations you made with your senses. Are there any trends or patterns? Does any part of the data not make sense?

4 Describe how changing the angle of the ramp would affect the gravitational potential energy and kinetic energy.

Conclusion

Copy and complete:

'The results show that: *(respond to the aim)*.'

Table 17.1: Results

Ball mass (g)	Distance (cm) (test 1)	Distance (cm) (test 2)	Distance (cm) (test 3)	Distance (cm) (test 4)	Distance (cm) (test 5)	Average distance (cm)

Investigation 7.2A

Conduction: Heat energy transfer in a solid

Process: Planning investigations

Focus: Constructing investigation plans

As you carry out investigations with a given method, think about how you might write a method for an investigation to address an aim. The first step is to identify a variety of laboratory equipment, what each item is used for, and how to use the equipment. For Investigations 7.2A and 7.2B, consider the required materials. What purpose do they serve in relation to the investigation method? Can you think of other ways to address the investigation aims by using different materials and methods?

Hint: Refer to 'Planning investigations' in the Science how-to section on page 323.

Aim

To investigate how heat energy is conducted in a solid

Materials

- heatproof mat
- tray
- Bunsen burner
- retort stand with bosshead and clamp
- metal bar
- wax beads
- matches
- safety glasses

Method

1 Set up the materials as shown in Figure I7.2A.

2 Wearing safety glasses, light the Bunsen burner and turn it to the blue flame.

3 Observe the metal bar and wax beads, and record your observations in your notebook.

Questions

1 Explain how heat is conducted in a solid, using your observations from this investigation.

2 Plan an investigation to demonstrate that heat energy is transferred through different metals at different rates.

3 Plan an investigation to determine which materials are good at conducting heat energy.

Conclusion

Copy and complete:

'The results show that: *(respond to the aim)*.'

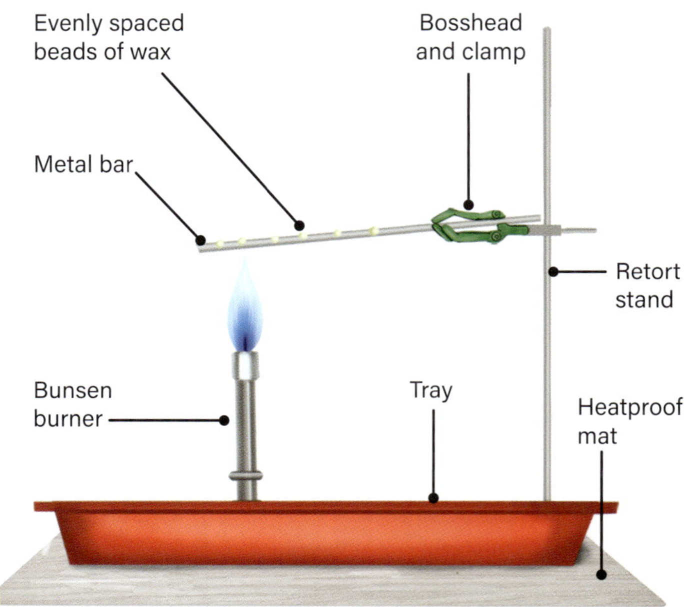

Figure I7.2A: For this investigation, set up the materials as shown in this diagram.

AN OPEN FLAME IS A HAZARD. BE CAREFUL. IF YOU BURN YOURSELF, TELL YOUR TEACHER IMMEDIATELY AND PLACE THE BURNT AREA UNDER COLD RUNNING WATER FOR 20 MINUTES.

Investigation 7.2B

Convection: Heat energy transfer in a liquid

Process: Planning investigations

Focus: Identifying variables

When conducting an investigation, it is important to identify the variables. Being able to identify the independent (changed), dependent (measured) and controlled (same) variables allows you to make sure that the results of an investigation are valid.

Hint: *To help you identify variables, see the Science how-to section on page 321.*

Aim

To investigate how heat energy is convected in a liquid

Materials

- cold water
- 600 mL beaker
- red food colouring
- 50 mL conical flask
- hot tap water (40–50 °C)
- long-handled flask tongs (or crucible tongs with bow)

Method

1. Add cold water to the beaker until it is 3 cm from the top.
2. Place a few drops of red food colouring into the conical flask, then fill it almost to the top with hot water.
3. Using long-handled tongs, carefully lower the small conical flask into the beaker. Take care not to spill the hot water or disturb the flask too much.
4. Record your observations.

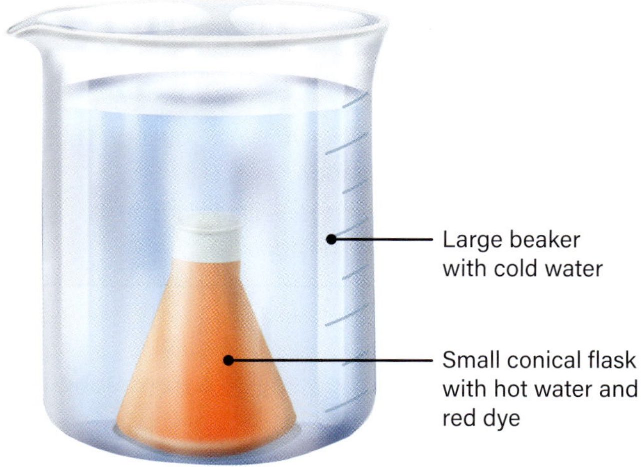

Figure I7.2B: For this investigation, set up the materials as shown in this diagram.

Questions

1. Draw a diagram that shows the movement of the hot water in the cold water.
2. Explain how convection works in a liquid, using your observations from this investigation.
3. Identify the independent, dependent and controlled variables of this investigation.

Conclusion

Copy and complete:
'The results show that: *(respond to the aim).*'

Investigation 7.3  

Energy transformations

Process: Conducting investigations

Focus: Identifying and managing risks

When conducting investigations, it is important that you can identify hazards and implement safe practices to keep you safe. **Answer Question 1 before you start this investigation.**

Hint: *Refer to 'Conducting investigations' in the Science how-to section on page 326.*

Aim

To observe and record how energy is transformed in a variety of familiar systems

Materials

- 250 mL beaker
- sand or gravel (to fill the beaker)
- sparkler
- safety glasses
- matches
- torch
- tennis ball
- elastic band
- peg
- damp rag
- retort stand
- hair dryer

Method

1 Copy the results table into your notebook, adding a title and as many rows as needed. After you complete each of Steps 2–6, fill in the row of the results table for that step.

2 Fill the beaker with sand. Stand a sparkler in the sand. Wearing safety glasses, use a match to light the sparkler.

3 Turn the torch on and off.

4 Roll a tennis ball off the edge of a bench.

5 Hold an elastic band between your finger and thumb. Stretch the elastic band back with your other hand and let it go.

6 Peg a damp rag to a retort stand. Turn on the hair dryer and blow warm air over the rag.

Questions

1 a Work with a partner to identify three hazards in this investigation.

b Suggest one way that the risk of each hazard causing harm could be reduced.

c What other aspects of investigations are important to consider from a safety point of view?

2 What type of potential energy did each system begin with?

3 Explain whether each system still had potential energy once the experiment was complete.

4 Draw an energy flowchart to represent the energy transformations that took place in each system.

Conclusion

Copy and complete:

'The results show that: *(respond to the aim).*'

MATCHES, SPARKLERS AND HAIR DRYERS ARE ALL HAZARDOUS. BE CAREFUL. IF YOU BURN YOURSELF, TELL YOUR TEACHER IMMEDIATELY AND PLACE THE BURNT AREA UNDER COLD RUNNING WATER FOR 20 MINUTES.

Table 17.3: Results

Step	System	Type of energy that initiated the change	Observations (dot points)	Energy transformations observed
2	A sparkler burns			
3	A torch is turned on			
4	A tennis ball falls			
5	An elastic band is stretched and released			
6	A hair dryer blows warm air over a damp rag			

Investigation 7.6

Law of conservation of mass

Process: Conducting investigations

Focus: Using scientific equipment to conduct investigations

To conduct investigations safely and effectively, you must be able to identify the required materials and understand how to correctly use those materials.

Hint: *Refer to 'Conducting investigations' in the Science how-to section on page 326.*

Aim

To observe the mass of reactants before, during and after a chemical reaction

Materials

- 2 × 10 mL measuring cylinders (or plastic pipettes)
- 10 mL of 0.1 mol/L sodium sulfate solution
- 2 funnels
- 100 mL conical flask and stopper
- electronic scales (that weigh masses to two decimal points)
- string
- micro test tube
- 3 mL of 0.1 mol/L barium chloride solution

Method

1 Copy the results table into your notebook, adding a title and as many rows as needed.

2 Working in groups, use a measuring cylinder to measure 10 mL of sodium sulfate solution. Using a funnel, pour the solution into a conical flask. Place the flask on the scales.

3 Tie the string around the micro test tube.

4 Use a different measuring cylinder to measure 3 mL of barium chloride solution. Using a different funnel, pour the solution into the micro test tube.

5 Carefully lower the micro test tube into the conical flask. Make sure the two solutions do not mix.

6 Place the stopper on the flask, using it to pinch the string to hold the test tube upright. Record the mass in the 'before reaction' column of the results table.

7 Tilt the flask just enough to allow the solutions to mix. Observe. Re-weigh and record the mass in the 'during reaction' column of the results table.

8 Tilt the flask again and swirl the solutions together for 10 seconds, or until the reaction is complete. Re-weigh and record the mass in the 'after reaction' column of the results table.

9 Calculate the change in mass and write this in the results table.

10 Share your results with other groups in the class. Add the other groups' results to your results table.

Table I7.6: Results

Class results	Mass of substance before reaction (g)	Mass of substance during reaction (g)	Mass of substance after reaction (g)	Change in mass (g)
Student group 1				
Student group 2				

Questions

1 a Draw a diagram of how to set up the materials for this investigation. Label your diagram.

b Annotate the diagram to include safety precautions and how chemical waste should be managed.

2 The two solutions of sodium sulfate and barium chloride react to form a solid (barium sulfate) and a liquid (sodium chloride).

a Identify the white product of the reaction.

b Write a word equation to represent this reaction.

3 a Describe the purpose of having a stopper on the flask throughout the reaction.

b How might the result have been different without the stopper?

4 What are the benefits of sharing results rather than each group conducting multiple trials?

5 Explain any variations (or lack of variations) in the different groups' results.

6 Explain how it is possible that new substances formed, yet there was no increase in mass.

Conclusion

Copy and complete:

'The results show that: *(respond to the aim).*'

Investigation 7.7A

Modelling respiration and photosynthesis

Process: Planning investigations

Focus: Identifying scientific materials

To plan an investigation, you must understand the function and purpose of a variety of scientific materials. Completing investigations with a given method is a great way to become more familiar with scientific equipment. **Answer Question 1 before you start this investigation.**

Hint 1: Read the method to help you figure out how each item will be used.

Hint 2: Ask a classmate if you are unsure what a piece of equipment is or how to use it.

Hint 3: Refer to 'Planning investigations' in the Science how-to section on page 323.

Aims

- To model cellular respiration using a non-living system
- To observe photosynthesis by a living *Elodea* plant

Materials

- 50 mL water
- 250 mL beaker
- 25–30 drops of bromothymol blue
- test tube rack
- 2 test tubes
- 2 test tube stoppers
- rubber tube to collect gas
- 10 mL measuring cylinder
- marble chips (calcium carbonate)
- 2 mL of 1 mol/L hydrochloric acid
  (or an Alka-Seltzer tablet)
- spatula
- approx. 12 cm *Elodea* plant (or another aquatic plant, such as *Ambulia*)
- light source (the Sun)
- safety glasses
- nitrile gloves

Figure I7.7A1: An *Elodea* plant

Method

1 Copy the results table (on the next page) into your notebook, adding a title and as many rows as needed.

Part A: Modelling cellular respiration by generating carbon dioxide (CO_2)

2 a Pour 50 mL of water into a 250 mL beaker.

b Add 25–30 drops of bromothymol blue to the water.

c In the results table, record your observations of the colour of the solution (see Figure I7.7A3).

3 In the test tube rack, set up one test tube and have a test tube stopper and a delivery tube ready.

4 Place the marble chips into the test tube (about 1 cm depth).

5 a Use the measuring cylinder to add 2 mL of hydrochloric acid to the marble chips and then quickly place the stopper over the test tube.

b Put the end of the tube into the 250 mL beaker to bubble the gas product (carbon dioxide) through the water in the beaker.

6 When the water in the beaker has changed colour, remove the tube. In the results table, record your observations of the colour of the solution (see Figure I7.7A3). Retain the water in the beaker for Part B.

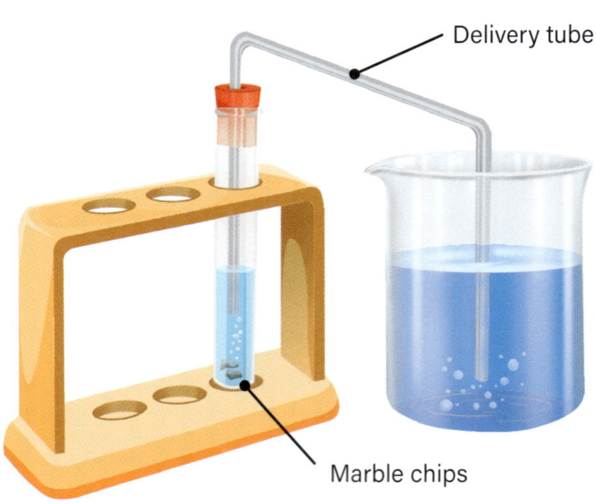

Figure I7.7A2: For Part A of this investigation, set up the materials as shown in this diagram.

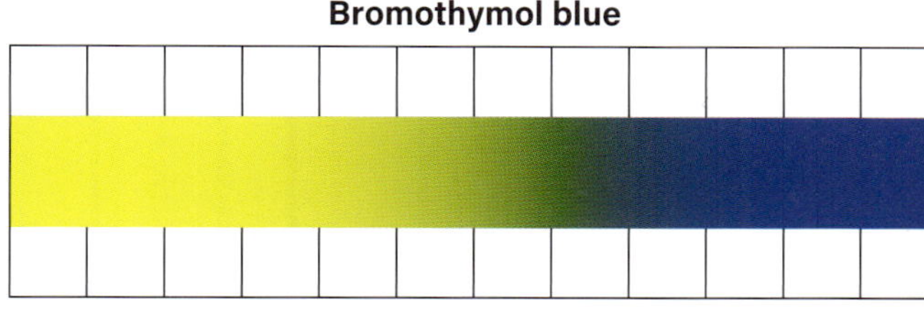

Increasing acidity
Increasing dissolved CO_2 (carbonic acid)

Increasing alkalinity
Decreasing dissolved CO_2 (carbonic acid)

Figure I7.7A3: Indicator colour range for bromothymol blue

Part B: Observing *Elodea* undergo photosynthesis

7 Use the end of a spatula to squash the lower centimetre of the stem of an *Elodea* plant.

8 Place the *Elodea* plant into an empty test tube. Pour the carbonic acid solution from Step 6 into the test tube. Ensure the *Elodea* is submerged in water.

9 Place the test tube rack directly in front of a light source. In the results table, record your observations of any chemical change indicators (see Figure I7.7A3).

10 Place a stopper on the test tube containing the *Elodea* and invert the test tube 5 times to mix the solution.

Table I7.7A: Results

Method step	Observations	Inferences (What do your observations suggest?)
Step 2: After bromothymol blue is added to the water		
Step 5: As the carbon dioxide is bubbled through the water		
Step 6: After the carbon dioxide is bubbled through the water		
Step 9: When the *Elodea* plant first begins to undergo photosynthesis		
Step 11: As the *Elodea* plant undergoes photosynthesis over time (hourly and/or daily)		

11 In the results table, record your observations of how the colour of the solution changes over time (up to several hours or days) (see Figure I7.7A3).

Questions

1 a Identify each item in the materials list by drawing a sketch to represent the item.

b Use dot points to summarise how each item is used in this investigation.

2 Write a word equation for:

a cellular respiration.

b photosynthesis.

3 In your own words, describe the difference between cellular respiration and photosynthesis.

4 What was the purpose of bubbling the gas from the marble chips into the beaker of water?

5 Describe the importance of adding the bromothymol blue to the beaker of water.

6 List any indicators of chemical change that you observed during this investigation.

7 Marble chips do not undergo cellular respiration as they are not living organisms.

a Justify why the reaction between the marble chips and hydrochloric acid was a suitable simulation for this investigation.

b Can you think of another way that carbon dioxide could be introduced into the water?

Conclusion

Copy and complete:

'The results show that: *(respond to the aim).*'

HYDROCHLORIC ACID IS A HAZARD. WEAR SAFETY GLASSES AND GLOVES DURING THIS INVESTIGATION TO PREVENT ACID FROM BURNING YOUR SKIN OR SPLASHING IN YOUR EYES. DISPOSE OF ALL LIQUIDS AS DIRECTED BY YOUR TEACHER.

Investigation 7.7B

Corrosion of iron

Process: Planning investigations

Focus: Identifying and managing risks

All investigations have a degree of risk. **Answer Question 1 before you start this investigation.**

Hint 1: What could happen with sharp equipment, or equipment that might break?

Hint 2: Be specific when describing how to reduce the risk of a hazard causing harm.

Aim

To identify the conditions that cause iron to rust (corrode)

Materials

- 5 test tubes
- test tube rack
- 5 iron nails (non-galvanised or zinc-coated)
- 5 test tube stoppers
- 10 mL vegetable oil
- safety glasses
- 10 mL salt solution
- labelling materials (sticky notes or masking tape and a pen)
- 10 mL distilled water
- 10 mL tap water

Method

1. Place all the test tubes in the test tube rack.
2. Place a nail in each test tube.
3. There are four liquids (vegetable oil, salt solution, distilled water and tap water); place each liquid in a separate test tube to cover each nail. The fifth test tube will not contain any liquid, only a nail and air. Label the test tubes.
4. Place a stopper on each of the test tubes.
5. Over the next week, observe the nails in the test tubes and look for signs of rusting. Record your observations in your notebook.

Table 17.7B: Risk assessment

Hazard	Risk (low, medium, high)	Management strategy

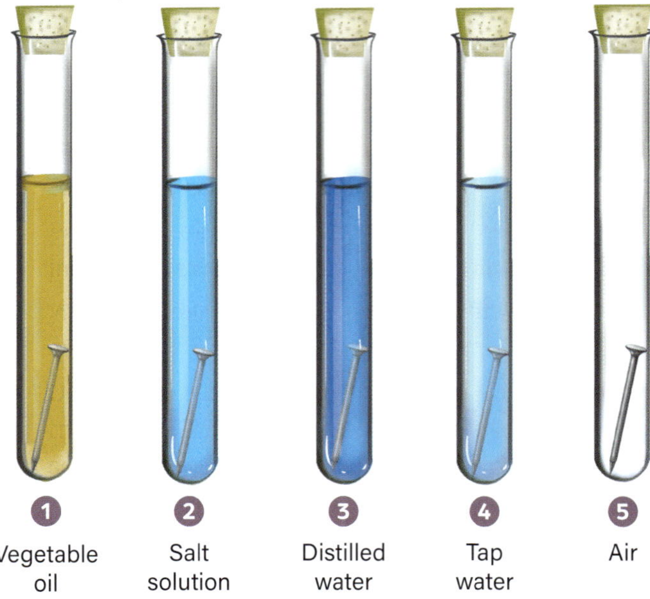

Figure 17.7B: For this investigation, set up the materials as shown in this diagram.

Questions

1. Copy the risk assessment table into your notebook. Fill in the table for this investigation:
   a. Identify two pieces of equipment that could cause an injury. Write these in the hazards column.
   b. Decide whether the chance of each piece of equipment causing an injury is low, medium or high.
   c. Come up with a way to reduce the chances of each piece of equipment causing an injury. This is your management strategy.
2. Rank the test tubes from the one in which the nail shows the most corrosion to the one in which the nail shows the least amount of corrosion.
3. Suggest why some nails showed more signs of corrosion than others.
4. How could you make this investigation more reliable?

Conclusion

Copy and complete:

'The results show that: *(respond to the aim)*.'

CHEMICALS OR SALT SOLUTION MAY GET INTO YOUR EYES. WEAR SAFETY GLASSES THROUGHOUT THE LESSON TO PREVENT THIS.

Investigation 7.7C

Preventing corrosion

Process: Planning investigations

Focus: Identifying investigation variables

In scientific investigations, variables are used to determine the effect of making a change. Consider what is being changed and measured in this investigation, and which factors are remaining the same.

Hint: *To help you identify variables, refer to 'Planning investigations' in the Science how-to section on page 323.*

Aim

To investigate which materials can prevent or slow rust (corrosion)

Materials

- butchers paper
- safety glasses
- nitrile gloves
- lab coats
- 5 iron nails
- vegetable oil
- petroleum jelly
- paint (non-toxic, such as acrylic paint)
- anti-rust spray
- 5 test tubes
- test tube rack
- tap water

Method

1 Line your bench with some butchers paper to protect the surface.

2 Wearing safety glasses, gloves and lab coats, prepare the nails:
   - Cover nail 1 in oil.
   - Cover nail 2 in petroleum jelly.
   - Cover nail 3 in paint.
   - Cover nail 4 in anti-rust spray.
   - Leave nail 5 untreated.

3 Place the 5 test tubes into the test tube rack.

4 Place a nail in each test tube.

5 Fill each test tube with water so the nail is covered.

6 Draw a diagram that shows the nails, covered in water in the test tubes. Label your diagram.

7 Leave the test tubes in the test tube rack for 2 days. In your notebook, record your observations.

Questions

1 What was the purpose of putting the untreated nail in a test tube of water?

2 Identify the:
   a independent variable.
   b dependent variable.
   c three controlled variables.

3 Which treatment provided the best protection from corrosion?

4 What other corrosion-prevention substances could you test?

5 List some other ways of preventing rust.

Conclusion

Copy and complete:
'The results show that: *(respond to the aim)*.'

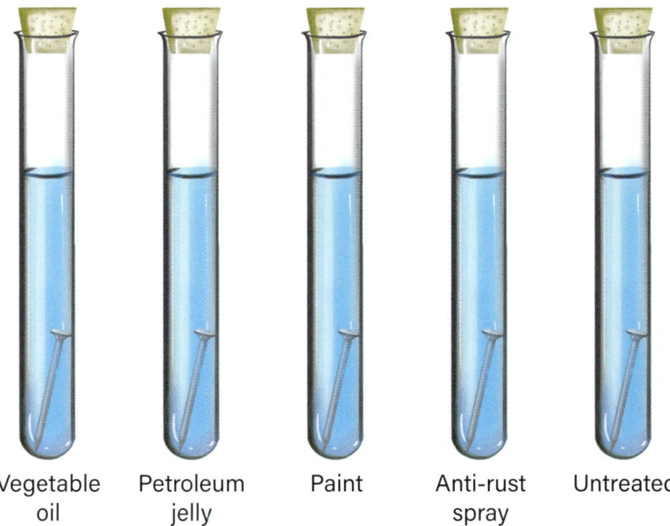

Figure I7.7C: For this investigation, set up the materials as shown in this diagram.

THE PAINT AND ANTI-RUST SPRAY ARE HAZARDOUS. BE CAREFUL. WEAR SAFETY GLASSES, LAB COAT AND GLOVES.

Investigation 7.8

Exothermic and endothermic reactions

Process: Conducting investigations
Focus: Collecting accurate first-hand and second-hand data

When collecting results from an investigation, scientists compare their data to existing data to make sure that it can be considered accurate. Using existing data means that any sources used must be acknowledged and referenced.

Hint 1: To find out what makes data accurate, *refer to the Science how-to section on page 328.*

Hint 2: For help with writing references, *refer to the Science how-to section on page 356.*

Aim

To investigate whether a chemical reaction is exothermic or endothermic

Materials

- 5 test tubes
- test tube rack
- safety glasses
- nitrile gloves
- 10 mL measuring cylinder
- funnel
- thermometer
- glass stirring rod
- spatula
- 15 mL vinegar
- sodium bicarbonate
- 5 mL of 1 mol/L hydrochloric acid
- 5 mL of 1 mol/L sodium hydroxide
- 5 mL of 1 mol/L copper sulfate solution
- 2 × 3 cm pieces of magnesium ribbon
- 5 mL of 1 mol/L sulfuric acid
- steel wool

Method

1. Copy the results table into your notebook, adding a title.
2. Place the 5 test tubes into the test tube rack.
3. Wearing safety glasses and gloves, use the measuring cylinder to measure the correct amount of reactant 1 for Part A (see Table I7.8A). Using a funnel, add reactant 1 to a test tube. Rinse the measuring cylinder and the funnel.
4. Use a thermometer to measure the temperature of reactant 1. Record the temperature in the results table in the 'Initial temperature' column. Rinse the thermometer.
5. Use the measuring cylinder to measure the correct amount of reactant 2 for Part A. Using a funnel, add reactant 2 to the same test tube. Rinse the measuring cylinder and the funnel.
6. a Gently stir the mixture with a stirring rod. Be careful not to break the base of the test tube. Rinse the stirring rod.
   b After a minute, record the temperature of the mixture. Record the temperature in the results table in the 'Final temperature' column.
7. Record your observations in the results table.
8. Repeats Steps 3–7 for Parts B, C, D and E.

Table I7.8A: Reactants

Part	Reactant 1	Reactant 2
A	5 mL vinegar	1 spatula sodium bicarbonate
B	5 mL hydrochloric acid	5 mL 1 mol/L sodium hydroxide
C	5 mL copper sulfate	3 cm piece of magnesium ribbon (cut into small pieces)
D	5 mL sulfuric acid	3 cm piece of magnesium ribbon (cut into small pieces)
E	10 mL vinegar	Piece of steel wool

45 min

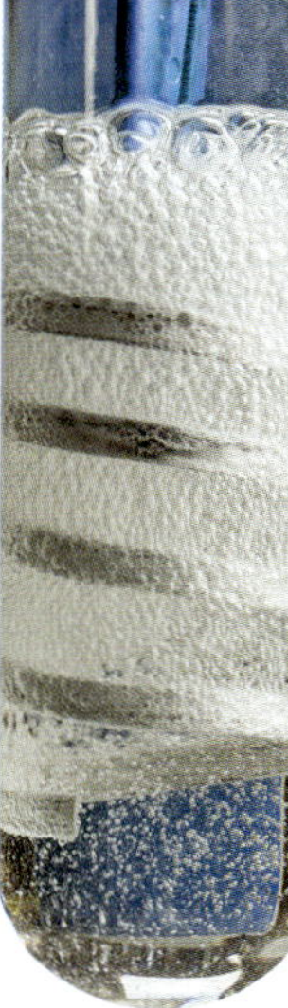

Figure 17.8: In this investigation, you will observe chemical reactions. This is an example of magnesium ribbon reacting to three different acids.

Table 17.8B: Results

Part	Reactants	Initial temp. (°C)	Final temp. (°C)	Observations
A	Vinegar + sodium bicarbonate			
B	Hydrochloric acid + sodium hydroxide			
C	Copper sulfate + magnesium			
D	Sulfuric acid + magnesium			
E	Vinegar + steel wool (iron)			

HYDROCHLORIC AND SULFURIC ACID ARE HAZARDS. BE CAREFUL. WEAR SAFETY GLASSES AND GLOVES WHILE CONDUCTING THIS INVESTIGATION TO PREVENT ACID FROM BURNING YOUR SKIN OR SPLASHING IN YOUR EYES, AND TO PREVENT METAL GETTING IN YOUR EYES.

DISPOSE OF WASTES AS INSTRUCTED BY YOUR TEACHER.

Questions

1. a Read the materials and method carefully.
   b Draw four diagrams: one diagram each for Steps 3–6. Label the diagrams.
   c Annotate the diagrams to include safety precautions and how chemical waste should be managed.
2. a Which reactions were exothermic? Explain your choices.
   b Which reactions were endothermic? Explain your choices.
3. a Find one secondary source that classifies one of the exothermic reactions. How does the source classify the reaction? Does this match your data?
   b Find one secondary source that classifies one of the endothermic reactions. How does the source classify the reaction? Does this match your data?
4. Based on the information you found, is your collected data accurate? Give a reason for your response.
5. Create a reference to acknowledge the use of the sources of information you found.

Conclusion

Copy and complete:
'The results show that: *(respond to the aim)*.'

Investigation 7.9

Modelling Earth's structure

Process: Observing

Focus: Using scientific tools to enhance observations

We can use different scientific equipment to enhance our observations.

Hint 1: When using a measuring cylinder, remember to read from the bottom of the meniscus of the liquid.

Hint 2: Note that $1\ cm^3 = 1\ mL$

Aim

To demonstrate how liquids of different densities interact and to use their properties to model the different layers of Earth

Materials

- electronic balance
- 5 liquids of different densities (for example, water, dishwashing detergent, vegetable oil, honey, light corn syrup) (50 mL of each liquid)
- 3 measuring cylinders (one each of 10, 50 and 100 mL)
- 5 × 100 mL beakers

Method

1 Copy the results table into your notebook, adding a title.

2 Use the 10 mL measuring cylinder to measure 10 mL of each liquid. Pour each liquid into a separate beaker. Rinse the measuring cylinder between each liquid.

3 Use the electronic balance to find the mass of 10 mL of each liquid. Record this information in the results table.

4 Work out the density of each liquid and add this information to the results table.

5 Use your knowledge of density to work out which liquid represents each of Earth's layers. (Hint: The most dense liquid represents Earth's inner core, and the least dense liquid represents the crust.) Check your answer with your teacher. Record this information in the results table.

6 a Use the 50 mL measuring cylinder to measure 19 mL of the liquid that represents Earth's inner core.

b Pour this liquid into the 100 mL measuring cylinder.

c Rinse the 50 mL measuring cylinder.

7 a Measure 35 mL of the liquid that represents the Earth's outer core.

b Carefully pour this liquid into the 100 mL measuring cylinder so that it forms a layer on top of the first liquid. (Hint: Hold the measuring cylinder at an angle.)

c Rinse the measuring cylinder.

8 Measure 42 mL of the liquid that represents Earth's lower mantle. Carefully add this layer to the 100 mL measuring cylinder.

9 Measure 3 mL of the liquid that represents Earth's upper mantle. Carefully add this layer to the 100 mL measuring cylinder.

10 Measure 1 mL of the liquid that represents Earth's crust. Carefully add this layer to the 100 mL measuring cylinder.

Table 17.9: Results

Name of liquid	Mass of 10 mL (g)	Density (g/cm^3)	Earth layer represented	Volume required for scale model (mL)

Questions

1 Draw a diagram of your model of Earth's structure. Label your diagram.

2 Why is it important to know the density of the different substances when making this model?

3 a How could making errors in measuring the density of the liquids affect this investigation?

b What could you do to make these errors less likely to happen?

4 Explain how the model you have created represents Earth's layers and the density of the substances in each layer.

5 With a partner, brainstorm how this investigation would be different if you did not have access to measuring cylinders with graduations marked on them.

Conclusion

Copy and complete:
'The results show that: *(respond to the aim)*.'

Investigation 7.10

Modelling seafloor spreading

Process: Observing

Focus: Making inferences based on your observations

When you write an investigation report, you must include your analysis and explanation of your observations. This is where you explain your results by linking them to what you already know about the science of what you are studying.

Hint: *You can use the following sentence to write about your results: 'My observations show ... and this makes sense because ...'*

Aim

To model the process of seafloor spreading

Materials

- 2 sheets of A4 paper
- ruler
- pencil
- scissors
- sticky tape
- coloured pencils

Method

1 Place a piece of A4 paper in landscape orientation. Use the ruler and pencil to draw a straight line down the centre of the paper. Label the line 'Mid-ocean ridge'.

2 Carefully use the scissors to cut along the line so there is a slit in the paper. Be careful not to cut to the ends of the paper. Leave about 3 cm either side of the slit.

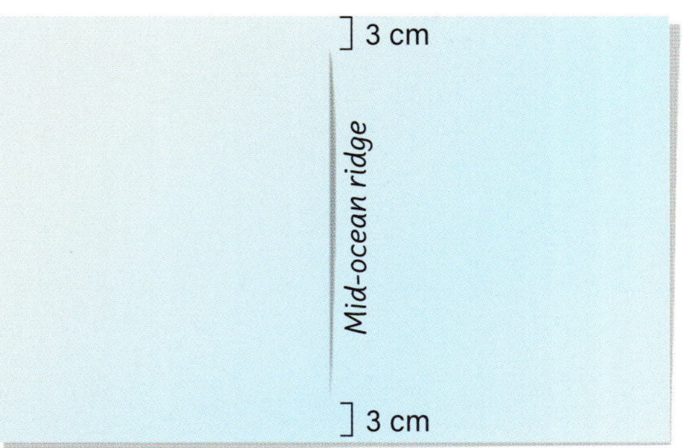

▲ **Figure I7.10A:** Steps 1 and 2

3 Cut the second piece of A4 paper in half lengthwise all the way.

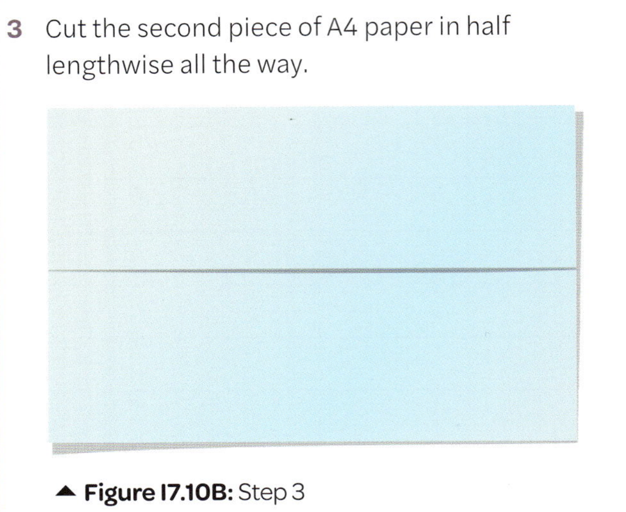

▲ **Figure I7.10B:** Step 3

4 Tape the two pieces of the second piece of paper together along one of the short ends. This represents the oceanic crust being formed at the mid-ocean ridge.

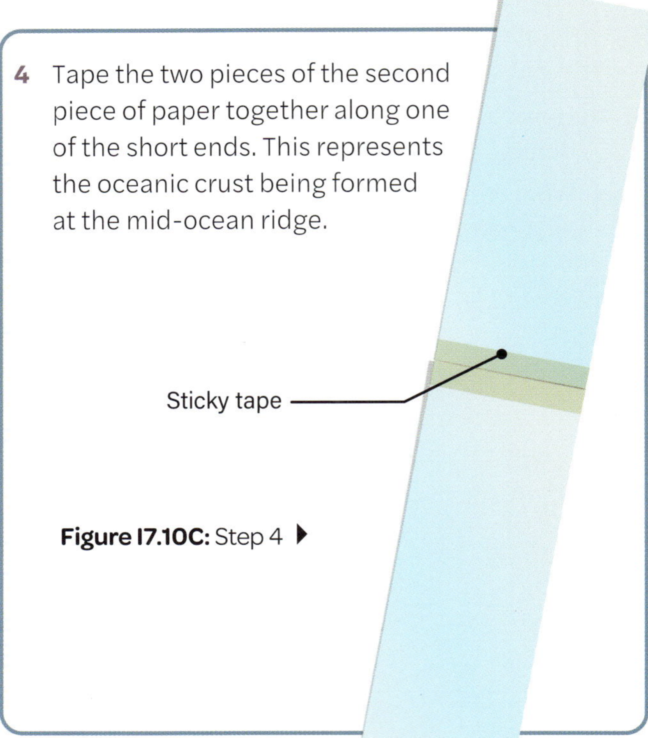

Figure I7.10C: Step 4 ▶

5 Carefully insert the two ends of the second piece of paper into the mid-ocean ridge slit in the first piece of paper (from underneath). Separate the pieces so that one will move towards the left and the other towards the right. Leave about 2 cm exposed.

6 Write 'Continent' on both sides of the exposed paper and draw a line where the paper comes out of the mid-ocean ridge.

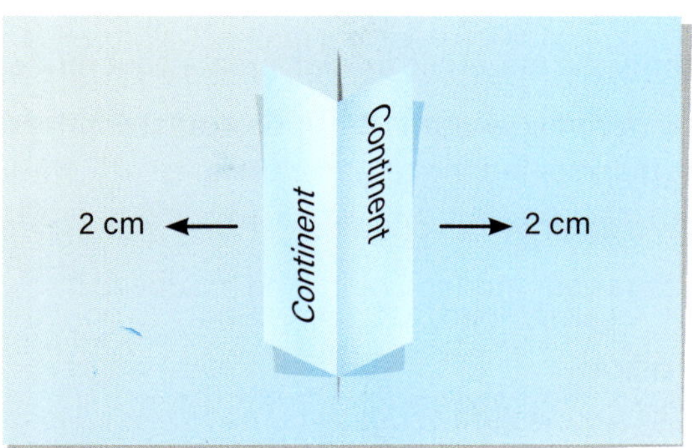

▲ **Figure I7.10D:** Steps 5 and 6

7 Evenly pull the paper outwards 1–3 cm and colour the white paper with the first colour.

8 Repeat Step 7, using different colours until all the paper has been pulled through the mid-ocean ridge.

▲ **Figure I7.10E:** Steps 7 and 8

9 Remove the coloured strip of paper for analysis.

Questions

1 Identify what the different coloured stripes in your model represent.

2 Describe any patterns that you observe in the stripes.

3 Your stripes may not all be the same width. What does this represent?

4 Critique how well this model demonstrates seafloor spreading. How could it be improved?

Conclusion

Copy and complete:
'The results show that: *(respond to the aim).*'

Investigation 7.11A

Modelling slab pull

Process: Conducting investigations

Focus: Implementing safe practices

Different scientific equipment requires different safe-handling procedures. You should always take the time to review the materials you will be using and remind yourself how to use them correctly. **Answer Question 1 before you start this investigation.**

Aim

To create a model that represents the force of slab pull on a tectonic plate

Materials

- variety of materials provided by your teacher

Method

1. Review the information about how the force of slab pull moves a tectonic plate (see Figure I7.11A). You could also use the internet to undertake some further research.
2. Brainstorm how you could use the materials provided to create a model that represents slab pull moving a tectonic plate.
3. Construct your model.
4. Share your model with the class.

Questions

1. a Identify three hazards in this investigation. (Hint: Which materials could cause injuries?)
   b For each hazard you identified in Question 1a, suggest one way that the risk of the hazard causing harm could be reduced. (How could you prevent injuries from occurring?)
2. Identify the materials used to create your model.
3. Describe how your model represents the force of slab pull moving the tectonic plate.
4. Draw a diagram of your model. Label your diagram.
5. Critique your model. Could it be improved?

Conclusion

Copy and complete:

'The results show that: *(respond to the aim)*.'

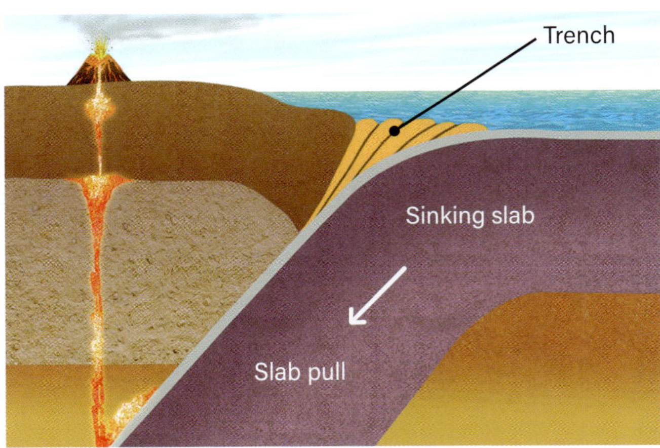

Figure I7.11A: Slab pull happens when the denser lithosphere slowly sinks underneath the less dense asthenosphere. This is like a rock slowly sinking in water.

Investigation 7.11B

Observing convection currents

Process: Conducting investigations

Focus: Using scientific equipment

Scientific equipment is used to collect a range of first-hand data, including qualitative observations from modelling processes.

Hint: *Refer to the Science how-to section on page 339 for more information on the types of data that can be found in an investigation.*

Aim

To investigate the movement of convection currents

Materials

- safety glasses
- nitrile gloves
- Bunsen burner
- heatproof mat
- tripod
- wire gauze
- 500 mL beaker
- water
- glass tube (or a paper straw)
- potassium permanganate crystals
- matches

Method

1 Wearing safety glasses and gloves, set up the Bunsen burner, heatproof mat, tripod and wire gauze (see Figure I7.11B).

2 Fill the beaker with cold water and place it on the wire gauze.

3 Carefully drop 1 or 2 crystals into the glass tube while it is sitting in the beaker, so that they drop and stay on the bottom on one side of the beaker, then carefully remove the tube.

4 Light the Bunsen burner and turn to the orange flame.

5 Observe how the coloured water moves.

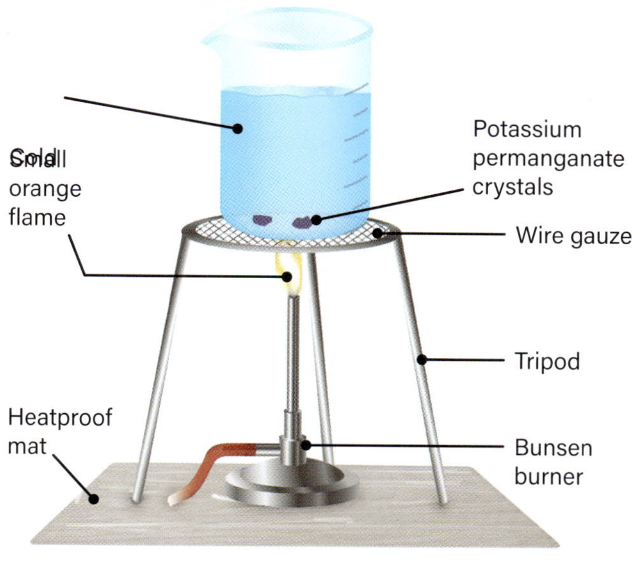

Figure I7.11B: For this investigation, set up the materials as shown in this diagram.

Questions

1 Draw a diagram of the movement of the coloured water. Label your diagram.

2 Describe what happened to cause the pattern of movement. Use the terms ‘heat’ and ‘density’ in your explanation.

3 Relate the movement of the coloured water in the beaker to the movement of rock in the mantle.

4 Explain how collecting qualitative first-hand data can assist our understanding of scientific phenomena.

Conclusion

Copy and complete:

‘The results show that: *(respond to the aim).*’

AN OPEN FLAME IS A HAZARD. BE CAREFUL. IF YOU BURN YOURSELF, TELL YOUR TEACHER IMMEDIATELY AND PLACE THE BURNT AREA UNDER COLD RUNNING WATER FOR 20 MINUTES. WEAR SAFETY GLASSES AND GLOVES.

DISPOSE OF WASTES AS INSTRUCTED BY YOUR TEACHER.

Investigation 7.12

20 min

Slinky waves

Process: Planning investigations

Focus: Constructing scientific aims

To make sure models are appropriate, an aim should be constructed so the purpose of the model is identified. This allows scientists to effectively communicate which concept a model is demonstrating.

Hint 1: What does a scientific aim start with?

Hint 2: What variables are needed to construct a scientific aim?

Aim

To use slinkies to model different seismic waves

Materials

- video camera (optional)
- 2 slinkies
- 5 cm piece of string
- paperclip

Method

1 Copy the results table into your notebook, adding a title.

2 Set up the video camera (if you have one) to record the movement of the slinky.

3 One person holds one end of a slinky on the ground. A second person holds the other end of the slinky in the air above the first person.

4 Tie the string in a bow around one of the coils near the middle of one of the slinkies.

5 a The first person pulls some coils directly down and releases them.

b Observe how the string moves. Record your observations in the results table.

6 a When the slinky is still again, the first person moves the bottom of the slinky from side to side.

b Observe how the string moves. Record your observations in the results table.

7 Use a paperclip to create a hook to connect the end of one slinky to the middle of the other slinky. People hold each of the three ends. Hold one slinky vertically and hold the other slinky horizontally.

8 Tie the string in a bow to the second slinky that is being held horizontally.

9 a The person holding the bottom of the vertical slinky pulls some coils down and to the side before releasing them.

b Observe how the string moves. Record your observations in the results table.

Table I7.12: Results

Movement	Observations (description of how the string on the slinky moves)
Compression	
Side to side	
Both	

Questions

1 Identify what you were modelling at the following steps:

a Step 5 b Step 6 c Step 9

2 For this investigation, identify the:

a independent variable.

b dependent variable.

3 Use your answers to Question 2 to write an aim for this investigation.

4 Consider how you might improve this investigation to more effectively model seismic waves. Rewrite the method to include your improvements.

Conclusion

Copy and complete:

'The results show that: *(respond to the aim).*'

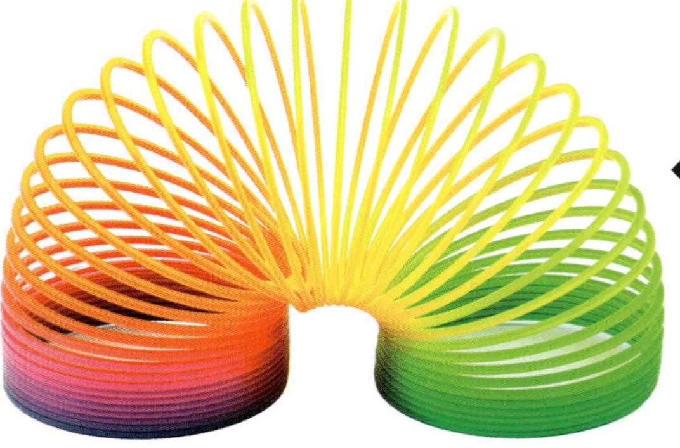

Figure I7.12: In this investigation, you will use two slinkies to model seismic waves.

Investigation 7.13A

Viscosity of lava

Process: Conducting investigations

Focus: Collecting accurate data

To make sure the data collected from first-hand investigations is accurate, we need to make sure we are using the best methods to take a measurement. This could mean choosing a piece of equipment with smaller graduations, or positioning ourselves in a way that allows us to take a more accurate reading.

Hint: *What equipment will enable you to record the most precise measurements?*

Aim

To investigate how the viscosity (thickness) of a substance affects how well it flows over a distance

Materials

- 2 spatulas
- whipped cream
- tray
- thickened cream
- 2 stopwatches
- ruler

Method

1 Copy the results table into your notebook, adding a title.

2 Measure the length of the tray.

3 a Use a spatula to place one dollop (about the size of a 20-cent coin) of whipped cream at the end of a tray.

  b Use a different spatula to place one dollop (about the size of a 20-cent coin) of thickened cream at the same end of the tray as the whipped cream.

4 Start both stopwatches and immediately lift the end of the tray with the cream on it so the tray is at a 45-degree angle.

5 Time how long it takes for each dollop of cream to reach the opposite end of the tray. Record this information in the results table.

6 Clean and dry the tray and repeat Steps 2–4 two more times.

7 Calculate the average time each type of cream took to travel the length of the tray. Record this information in the results table.

8 Calculate the flow rate of each type of cream. To do this, divide the distance travelled by the average time. Record this information in the results table.

Questions

1 a Which cream had the fastest flow rate?

  b Which cream had the slowest flow rate?

2 Describe the relationship between flow rate and the viscosity of the cream.

3 Explain how viscosity affected the time it took for the creams to travel down to the end of the tray.

4 a Which type of cream has a similar viscosity to the lava from a hotspot volcano?

  b Which type of cream has a similar viscosity to the lava from a strato volcano?

5 When taking and recording measurements during this investigation, how did you make sure they were accurate?

Conclusion

Copy and complete:
'The results show that: *(respond to the aim).*'

NEVER EAT ANYTHING IN A SCIENCE LABORATORY.

Table I7.13A: Results

	Time taken to travel the length of the tray (seconds)				Distance (cm)	Flow rate
	Trial 1	Trial 2	Trial 3	Average		
Whipped cream						
Thickened cream						

Investigation 7.13B

Wax volcano

Process: Conducting investigations
Focus: Implementing safe practices
Safety is critical in science because all investigations carry some degree of risk. Things can go wrong if you do not use scientific equipment safely. **Answer Question 1 before you start this investigation.**

Aim

To model a volcanic eruption using wax

Materials

- safety glasses
- coloured candle wax
- 500 mL beaker
- washed sand
- water
- hotplate
- video camera (if available)

Method

1. Wearing safety glasses, put the candle wax in the beaker. Heat it on the hotplate until it melts and forms a 1 cm layer in the bottom of the beaker. Allow the wax to set.
2. Cover the wax with a 1–2 cm layer of sand.
3. Carefully fill the beaker with water.
4. Place the beaker on the hot plate and turn on the heat.

HOT WAX CAN BURN. BE CAREFUL. IF YOU BURN YOURSELF, TELL YOUR TEACHER IMMEDIATELY AND PLACE THE BURNT AREA UNDER COLD RUNNING WATER FOR 20 MINUTES. WEAR SAFETY GLASSES.

5. Observe from a safe distance. Record your observations by writing notes in your notebook or by recording a short video.

Questions

1. Copy the risk assessment table into your notebook. Write your answers to the following questions in the table.
   a. **Hazards**: Identify three hazards in this investigation. (Hint: Which materials could cause injuries?)
   b. **Risks**: Decide whether you think there is a low, medium or high chance that each hazard will harm the person conducting this investigation.
   c. **Management strategies**: For each hazard you identified in Question 1a, suggest one way that the risk of the hazard causing harm could be reduced. (How could you prevent injuries from occurring?)
2. Identify the structures of Earth that are represented by the different layers in the volcanic model you have made with:
   a. wax. b. sand. c. water.
3. Draw a diagram of what your volcano model looks like after you have completed Step 3. Label your diagram.
4. Review the video of the eruption. Determine an effective way to present the result of the investigation shown in the video.
5. Critique this model. How does it reflect real volcanic eruptions? How is it different?

Conclusion

Copy and complete:
'The results show that: *(respond to the aim)*.'

Table I7.13B: Risk assessment

Hazard	Risk (low, medium or high)	Management strategy

Investigation 7.15A

Observing minerals

Process: Observing

Focus: Using senses to make observations

The observations we make using our senses during an investigation can help us to answer questions.

Hint: Identify the types of observations you will need to make before you start this investigation.

Aim

To investigate the properties of some common minerals

Materials

- selection of mineral samples (mineral kit)
- magnifying lens
- streak plate (or an unglazed white ceramic tile)
- nail (or a fingernail, copper coin, glass file)

Method

1 Copy the results table into your notebook, adding a title and as many rows as needed.

2 Complete Steps 3–6 for each mineral sample. Write your observations in the results table. Use the magnifying lens to examine the samples closely.

3 **Lustre**: Determine if each mineral has a metallic or a non-metallic lustre.

4 **Colour**: Describe the colour of each mineral.

5 **Streak**: Scratch each mineral along the streak plate. Describe the colour of the powder that is left on the plate. Note: If a mineral has a measurement of more than 6.5, it will not make a streak. In this case, write 'n/a' ('not applicable') in the results table.

6 **Hardness**: Refer to Table 7.2 on page 259 and use the nail and the streak plate to determine the approximate hardness of each mineral.

Questions

1 a Which mineral is the softest?

b Which material is the hardest?

2 Which property was the easiest to test? Why?

3 Explain why we observe both the colour and streak of minerals.

4 Is there one property that you could use on its own to tell the difference between the mineral samples? Explain your response.

Conclusion

Copy and complete:

'The results show that: *(respond to the aim).*'

Table I7.15A: Results

Mineral name	Mineral property			
	Lustre	Colour	Streak	Hardness

Investigation 7.15B

Extracting copper

Process: Observing

Focus: Making inferences based on observations

An inference is a conclusion based on evidence and reasoning. You need to use both your scientific knowledge and evidence to make an inference.

Hint: *What do the things you see tell you about what is happening in the investigation?*

Aim

To investigate how chemical reactions are used to extract copper from a copper ore

Materials

- safety glasses
- nitrile gloves
- 100 mL beaker
- glass stirring rod
- 5 g malachite or copper sulfate crystals
- 50 mL distilled water
- zinc metal strip
- camera (optional)

Method

1. Wearing safety glasses and gloves, in a beaker, use the stirring rod to carefully dissolve 5 g of malachite or copper sulfate crystals in 50 mL of distilled water.
2. Add the zinc metal strip to the beaker. Make sure the solution covers the zinc strip.
3. Write the heading 'Start of the investigation' in your notebook. Record your observations, including the colour of the solution and the colour of the zinc strip.
4. Draw or take a photo of the materials at this stage of the investigation.
5. Leave the solution and the zinc strip in the beaker overnight.
6. Write the heading 'End of the investigation' in your notebook. Record your observations, including the colour of the solution and the colour of the zinc strip.
7. Draw or take a photo of the materials at this stage of the investigation.

Questions

1. Describe the change to the colour of the solution in the beaker.
2. Describe the change to the colour of the zinc strip.
3. a. Which chemical reaction occurred in this investigation? (Note: You may need to conduct some research to find out this information.)
   b. Write a word equation for the chemical reaction that occurred in this investigation.
4. Where were the copper atoms located:
   a. at the start of the investigation?
   b. at the end of the investigation?
5. What does this investigation demonstrate about how chemical reactions can be used to extract minerals from ores?
6. How can you tell that not all the copper had been extracted from the solution?

Conclusion

Copy and complete:
'The results show that: *(respond to the aim)*.'

COPPER SULFATE IS HAZARDOUS. BE CAREFUL. WEAR SAFETY GLASSES AND GLOVES.

Figure I7.15B: Copper sulfate crystals

Investigation 7.16

Modelling the rock cycle

Process: Conducting investigations

Focus: Implementing safe practices when using scientific equipment

Safety is critical in science because all investigations carry some degree of risk. Things can go wrong if you do not use scientific equipment safely. **Answer Question 1 before you start this investigation.**

Aim

To model the process of the rock cycle using crayons

Materials

- camera
- safety glasses
- butter knife (or plastic knife)
- 2–4 wax crayons of different colours (or plasticine or lollies)
- mortar and pestle
- cutting mat
- rolling pin
- crucible
- Bunsen burner
- heatproof mat
- tripod
- pipeclay triangle
- matches
- brass (crucible) tongs

Method

1. Use the camera to take photographs of each step of this process.
2. Wearing safety glasses, use the knife to cut the crayons into very small pieces. Keep the colours separate.
3. Layer the different colours of crayon in the mortar.
4. Use the pestle to gently push down on the crayon layers so they stick together.
5. Carefully take the crayon layers out of the mortar and place them on the cutting mat.
6. Use the rolling pin to further press down and combine the crayon layers.
7. Use the knife to carefully cut out a section of the crayon layers. Place this section into the crucible.
8. Set up the Bunsen burner, heatproof mat, tripod and pipeclay triangle.

**AN OPEN FLAME AND HOT WAX ARE HAZARDS.
BE EXTREMELY CAREFUL. IF YOU BURN YOURSELF, TELL YOUR TEACHER IMMEDIATELY AND PLACE THE BURNT AREA UNDER COLD RUNNING WATER FOR 20 MINUTES.
WEAR SAFETY GLASSES.**

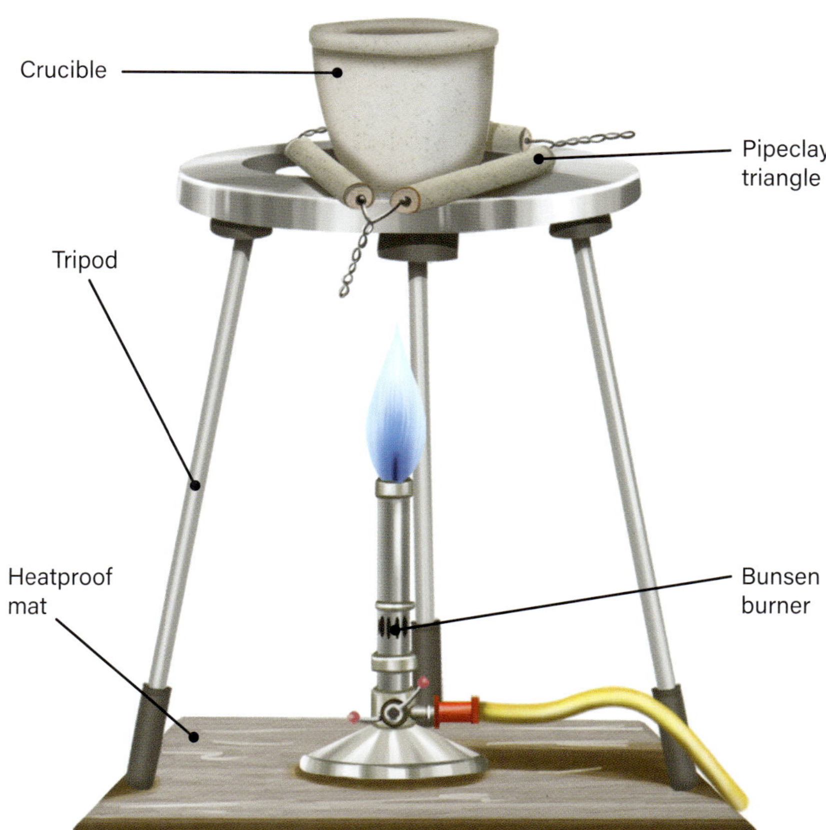

▲ **Figure I7.16:** For this investigation, set up the Bunsen burner, heatproof mat, tripod, pipeclay triangle and crucible as shown in this diagram.

continues ▶

9 Carefully place the crucible without its lid on the pipeclay triangle.

10 Light the Bunsen burner and turn it to a blue flame.

11 Observe how the wax changes. Once all the wax has melted, turn off the Bunsen burner.

12 Remove the crucible with the tongs and allow the melted wax to cool.

Questions

1 a Work with a partner to identify three hazards in this investigation. (Hint: What equipment could cause injuries?)

b For each hazard you identified in Question 1a, suggest one way that the risk of the hazard causing harm could be reduced. (How could you prevent injuries from occurring?)

2 Identify the rock cycle processes that are modelled by Steps 2, 3, 4, 6, 11 and 12.

3 Using the processes outlined in Figure 7.57 (page 261), along with the photos you took, create your own rock cycle diagram.

4 Evaluate the method of this investigation for its effectiveness in modelling the rock cycle. How could you improve the method?

Conclusion

Copy and complete:
'The results show that: *(respond to the aim).*'

Investigation 7.17A

Cooling rate and crystal size

Process: Conducting investigations

Focus: Using scientific equipment to conduct investigations

When conducting an investigation, it is important to use scientific equipment carefully and appropriately. Take time to carefully read the materials list and use each item as described in the method.

Hint: Read and complete the instructions one step at a time to make sure you understand how to use each piece of equipment properly.

Aim

To investigate how cooling rate affects crystal size

Materials

- safety glasses
- nitrile gloves
- 3 test tubes
- 3 × 250 mL beakers
- ice
- cotton wool
- Bunsen burner
- heatproof mat
- tripod
- gauze mat
- 40 mL saturated copper sulfate (or alum) solution
- 100 mL beaker
- matches
- thermometer
- 10 g pentahydrate
- glass stirring rod
- beaker tongs
- magnifying lens

Method

1 Wearing safety glasses and gloves, place a test tube into each of the three 250 mL beakers.

   a Beaker 1: Surround the test tube with ice up to the top of the beaker.

   b Beaker 2: Surround the test tube with cotton wool up to the top of the beaker.

   c Beaker 3: Leave the test tube surrounded by air.

2 Set up the Bunsen burner, heatproof mat, tripod and gauze mat.

3 Add 40 mL of saturated copper sulfate solution to the 100 mL beaker.

4 Light the Bunsen burner and turn to a blue flame.

5 Place the beaker of copper sulfate solution onto the Bunsen burner. Heat the solution to about 40 °C. (Use a thermometer to measure the temperature of the solution.) Turn off the Bunsen burner.

6 Add 10 g of pentahydrate to the copper sulfate solution, stirring with the stirring rod until the pentahydrate dissolves. Keep adding the pentahydrate until it no longer dissolves.

7 Using the beaker tongs, carefully remove the beaker of the solution from the Bunsen burner. Pour one-third of the solution into each test tube.

8 Plug the top of each test tube with some cotton wool. Leave the test tubes to cool overnight.

9 Use a magnifying lens to observe the size of the crystals in each test tube. Write your observations of the crystals in your notebook.

Questions

1 Explain why it was important to plug all three test tubes with cotton wool.

2 Identify the speed of cooling (for example, slow or fast) for the crystals in each beaker.

3 a Describe the relationship that you observed between the rate of cooling and the size of the crystals formed.

   b Did you expect to see this relationship? Why or why not?

4 Which test tube represented the formation of an intrusive igneous rock? Justify your choice.

5 Which test tube represented the formation of an extrusive igneous rock? Justify your choice.

Conclusion

Copy and complete:
'The results show that: *(respond to the aim)*.'

AN OPEN FLAME IS A HAZARD. BE CAREFUL. IF YOU BURN YOURSELF, TELL YOUR TEACHER IMMEDIATELY AND PLACE THE BURNT AREA UNDER COLD RUNNING WATER FOR 20 MINUTES. WEAR SAFETY GLASSES AND GLOVES.

Investigation 7.17B

Observing igneous rocks

Process: Conducting investigations

Focus: Collecting accurate data

When conducting a first-hand investigation, it is important that the data you collect is accurate. One way to do this is to use scientific equipment such as magnifying lenses and measuring devices.

Hint: *Bring your eyes down to the level of the measuring device to avoid parallax errors.*

Aim

To observe the key characteristics of some igneous rocks

Materials

- selection of igneous rocks (for example, basalt, granite, pumice, rhyolite, gabbro, obsidian)
- magnifying lens
- ruler

Method

1 Copy the results table into your notebook, adding a title and as many rows as needed.

2 Use the magnifying lens to carefully observe each igneous rock sample.

3 Fill in the results table. Use the ruler to make sure the information you record in the table is accurate.

Questions

1 a Identify the extrusive igneous rocks.

b What feature enabled you to identify them?

2 a Identify the intrusive igneous rocks.

b What feature enabled you to identify them?

3 You are walking through the bush and find a rock outcrop. What observations do you need to make to confirm that it is an igneous rock?

Conclusion

Copy and complete:

'The results show that: *(respond to the aim).*'

Table 17.17B: Results

Rock name	Approximate percentage of light and dark minerals	Crystal size (mm)	Other observations	Rate of cooling

Investigation 7.18

Modelling contact metamorphism

Process: Planning investigations

Focus: Constructing scientific methods

When conducting an investigation, it is important to make sure that you have an appropriate method to follow. Is the method for this investigation as good as it could be?

Hint: *What are the features of a good method?*

Aim

To investigate the process of contact metamorphism

Materials

- safety glasses
- vinyl gloves
- raw egg white
- 100 mL Petri dish
- 100 mL beaker
- tap water
- spatula
- salt
- Bunsen burner
- heatproof mat
- tripod
- gauze mat
- matches
- beaker tongs
- camera (optional)

Method

1 Wearing safety glasses and vinyl gloves, place the egg white in the Petri dish, making sure there is enough to cover the base of the dish.

2 Half fill the beaker with water. Add a spatula of salt to the water.

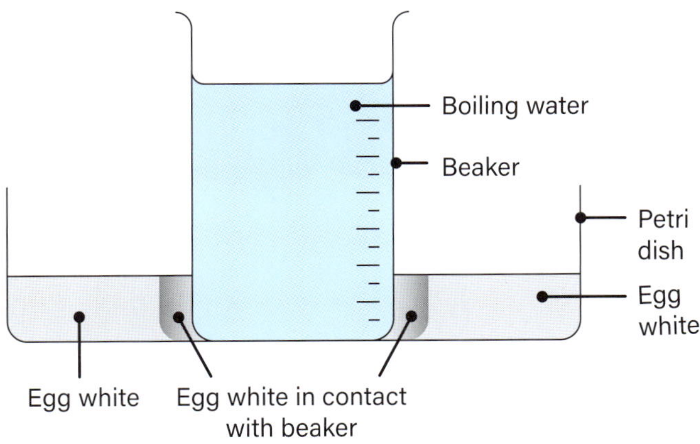

▲ **Figure 17.18:** For Step 6 of this investigation, set up the materials as shown in this diagram.

3 Set up the Bunsen burner, heatproof mat, tripod and gauze mat.

4 Light the Bunsen burner and set it to the blue flame.

5 Place the beaker containing the salty water on the gauze mat and bring the water to the boil.

6 When the water is boiling, use beaker tongs to carefully remove the beaker from the gauze mat and place it in the centre of the Petri dish.

7 Observe what happens over 10 minutes. Record your observations in your notebook. You may like to take photos to support your observations.

Questions

1 This investigation modelled contact metamorphism.

  a Explain what the egg white represented.

  b Explain what the beaker of hot water represented.

2 a What would happen if you used water that was a different temperature?

  b Would the water be a controlled, dependent or independent variable?

3 Explain how the change to the egg white is similar to the change to rocks that have undergone contact metamorphism.

4 Brainstorm some improvements to this investigation. Rewrite the method to include these improvements.

Conclusion

Copy and complete:

'The results show that: *(respond to the aim).*'

BOILING WATER, STEAM AND AN OPEN FLAME ARE HAZARDS. BE CAREFUL. IF YOU BURN YOURSELF, TELL YOUR TEACHER IMMEDIATELY AND PLACE THE BURNT AREA UNDER COLD RUNNING WATER FOR 20 MINUTES.

IF YOU ARE ALLERGIC TO EGG WHITE, TELL YOUR TEACHER BEFORE STARTING THIS INVESTIGATION.

NEVER EAT ANYTHING IN A SCIENCE LABORATORY. WEAR SAFETY GLASSES AND GLOVES.

Investigation 7.19A

30 min

Modelling the formation of sandstone

Process: Observing

Focus: Recording observations accurately

When conducting investigations, it is important to record your observations and to do so accurately. When deciding how to do this, it is a good idea to read the method and think about what observations could be made.

Aim

To investigate, using a model, the formation of sandstone

Materials

- sand: separate samples of different colours
- 0.5 cup dry plaster of Paris
- 2 clear disposable cups
- spatula (or spoon)
- water
- magnifying lens

Method

1. Mix the first sample of sand with the dry plaster of Paris.
2. Place a layer of sand of one colour into a cup.
3. a. Add a little water to the sand so it is moist but not saturated.
   b. Use a spatula to combine the sand and water.
4. Press the second cup into the first cup to compress the layer of sand.
5. Repeat Steps 2–4 for sands of different colours.
6. Allow your 'sandstone' to dry overnight before removing it from the cup.

Questions

1. Describe the look and feel of your 'sandstone'.
2. Use a magnifying lens to observe your 'sandstone'. Draw a diagram of your rock, showing the different sediments.
3. Create a table or a Venn diagram to compare your 'sandstone' to a real sample of sandstone.
   a. What are the similarities?
   b. What are the differences?
4. Identify the steps in the method that represent deposition, compaction and cementation.
5. What does this investigation demonstrate about the formation of sandstone?

Conclusion

Copy and complete:
'The results show that: *(respond to the aim)*.'

Figure I7.19A: Sandstone ▶

Investigation 7.19B

Observing sedimentary rocks

Process: Observing

Focus: Using senses to make observations

When conducting first-hand investigations, scientists can make observations using any of the five senses: sight, touch, smell, sound and taste.

Aim

To observe samples of sedimentary rocks

Materials

- selection of sedimentary rocks (for example, sandstone, conglomerate, mudstone, shale, limestone)
- magnifying lens
- ruler

Method

1 Copy the results table into your notebook, adding a title and as many rows as needed.

2 Use a magnifying lens to carefully observe each rock. Record your observations of each rock in the 'Observations' column of the results table.

3 Use a ruler to measure the size of the sediments in each rock. Record these measurements in the results table.

4 Does each rock have obvious layers? Record this information in the results table.

Questions

1 a With a partner, brainstorm different types of observations that could be made during this investigation.

b Suggest one way that each type of observation could be recorded.

2 If one of the rock samples did not have obvious layers, what observation did you make that told you the sample is a sedimentary rock?

3 a Which rock contained the largest sediments?

b Which rock contained the smallest sediments?

4 Large sediments are deposited in high-energy environments and small sediments are deposited in low-energy environments. Identify the rock that was deposited in the:

a highest energy environment.

b lowest energy environment.

5 You are walking through the bush and find a rock outcrop. What observations do you need to make to confirm that it is a sedimentary rock?

Conclusion

Copy and complete:

'The results show that: *(respond to the aim)*.'

Table 17.19B: Results

Rock name	Observations (e.g. colour, presence of fossils)	Size of sediments (mm)	Obvious layers present? (Yes or no)

Investigation 7.20

Making 'fossils'

Process: Planning investigations

Focus: Identifying variables

Being able to identify the independent (changed), dependent (measured) and controlled (same) variables of an investigation is an important scientific skill.

Aim

To model the formation of a fossil

Materials

- 0.5 cup plaster of Paris
- plastic cup
- water
- popstick
- plasticine
- plastic container (for example, a take-away food container)
- shell (or other item)
- camera (optional)

Method

1. Place the plaster of Paris into a plastic cup.
2. Carefully add water to the plaster of Paris. Stir with a popstick until the mixture is the same consistency as toothpaste.
3. Press the plasticine into the base of a plastic container and smooth the top.
4. Press the shell into the plasticine to make an imprint, then remove it.
5. Carefully add a 1–2 cm layer of plaster of Paris on top of the plasticine. Tap the container to remove any air bubbles.
6. Leave the plaster to dry overnight.
7. Carefully separate the plaster from the plasticine.
8. Record your observations in your notebook. You may like to photograph the plaster and the plasticine to support your observations.

Questions

1. Identify materials that you could substitute for the ones listed for this investigation.
2. Imagine you want to compare alternatives to plasticine in the creation of a fossil. Identify the:
   a. independent variable.
   b. dependent variable.
   c. controlled variable.
3. Write an aim for the investigation in Question 2.
4. Use the terms *deposition*, *compaction* and *cementation* to compare the method you followed in this investigation to the process of fossilisation.

Conclusion

Copy and complete:

'The results show that: *(respond to the aim)*.'

Figure 17.20: Clam fossils in limestone

Investigation 8.1

Pin drop

Process: Analysing data and information

Focus: Identifying trends in data

After data has been gathered in an investigation, it needs to be reviewed to see if there are any relationships or trends in the information. Trends and relationships help us to explain what data is telling us and to make conclusions about the results of an investigation.

Hint: Is there a link between the number of times a pin is dropped and the way the pin lands?

Aim

To determine the likelihood of a dropped pin landing point side up

Materials

- 20 drawing pins
- small plastic container

Method

1. Copy the results table into your notebook, adding a title.
2. Place the drawing pins into the plastic container. Hold the container over a bench.
3. Gently toss the container so the pins move up into the air and land on the bench top.
4. In the results table, record the number of pins that landed point side up.
5. Repeat the test four times.

Table 18.1: Results

Test number	Number of pins that landed point side up
1	
2	
3	
4	
5	

Questions

1. What type of data does this investigation collect?
2. How was the data sourced in this investigation?
3. In which test did the most pins land point side up?
4. a. Describe any trends in the data about how many pins landed point side up.
   b. Why do you think this was the case?
5. a. Compare your data to that of the students sitting on either side of you.
   b. Is their data similar or different?
   c. What are the reasons for the differences?
6. Estimate the likelihood of a pin landing point side up when it is dropped, based on the data you have collected.
7. How might the data from this investigation be applied in a different context?

Conclusion

Copy and complete:

'The results show that: *(respond to the aim).*'

Figure 18.1: In this investigation, you will estimate the likelihood of a drawing pin landing point side up when it is dropped.

Investigation 8.2

Have you been influenced?

Process: Problem-solving

Focus: Solving scientific problems using cause-and-effect relationships

There is a lot of good information on the internet. However, there is also a lot of misinformation, including false claims made by advertisers.

Hint: Refer to the Science how-to section on page 354 for tips on how to find high-quality information.

Aim

To use secondary sources to determine the accuracy of a claim

Materials

- device with internet access

Background information

Some influencers promote anti-wrinkle creams and claim that when applied, the extremely small molecules allow these creams to easily penetrate the subcutaneous layer of the skin. They claim that you will notice a reduction of wrinkles after just one application. Influencers are usually paid to promote products. However, does scientific evidence support the claims they make?

Method

1 Search the internet to find an example of a specific anti-wrinkle cream that is being promoted by an influencer.

2 Record information about the cream and the influencer in your notebook. Include what the influencer claims the product does.

3 a Find and read reviews of the product.

b Find and read scientific articles that relate to the product.

4 Decide if the product is likely to work or not, based on the information you have read.

5 Compare your research to that of other students in your class. Are their findings similar or different?

Questions

1 Summarise three of the reviews you read about the product and its effectiveness.

2 Consider the information in the scientific articles you found. Does the scientific evidence support or not support the likelihood of this product working?

3 Compare the product claims with the scientific evidence. Based on this, would you expect this product to work? Give reasons to support your decision.

4 Write a review that you could post online relating to this product. Remember to consider how to safely protect your digital footprint when writing and posting a review online. (Note: Do not actually post the review; just create it.)

Conclusion

Copy and complete:

'The results show that: *(respond to the aim)*.'

Figure I8.2: Search engines like Google Scholar can help you to find high-quality information.

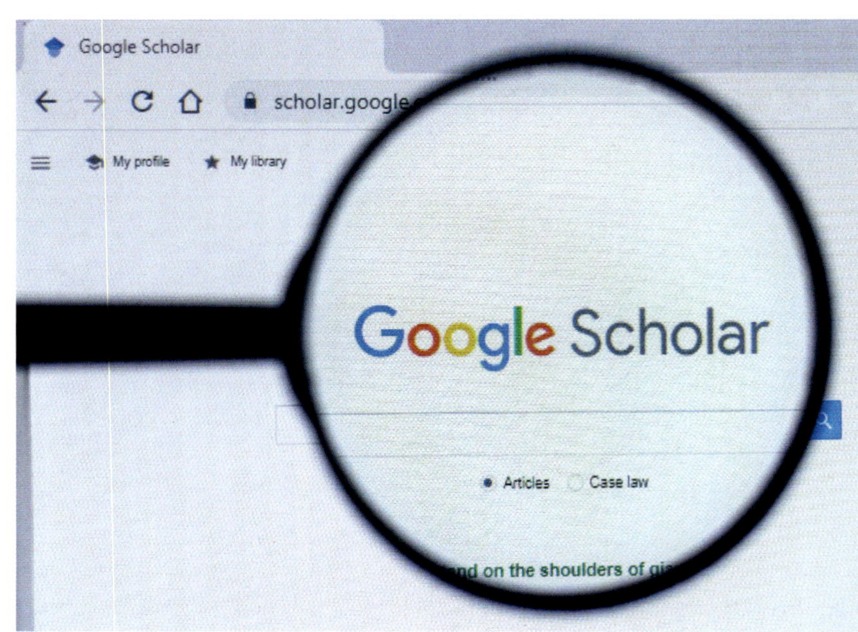

Investigation 8.3

30 min

Can you believe your eyes?

Process: Problem-solving

Focus: Suggesting solutions to scientific problems

Data collected from investigations can be used to make inferences about larger populations. Could you use your data to do this?

Hint: *Comparing your data to other investigations can help you decide if your investigation effectively addresses the problem of identifying the percentage of people in a population that have each eye colour.*

Figure I8.3: Human eyes come in many different colours.

Aim

To determine the number of individuals who have each eye colour in a school

Materials

- class members
- magnifying lens

Method

1. Copy the results table into your notebook, adding a title.
2. Use the magnifying lens to inspect the eye colour of each of your classmates.
3. Record the eye colour of each of your classmates as a tally mark in the results table.
4. a Add the tally marks for each eye colour.
   b Calculate the percentage of classmates with each eye colour.
5. Use the percentages to calculate how many students in your school should have each different eye colour.

Questions

1. a Which eye colour is the most common among your classmates?
   b Which eye colour is the least common among your classmates?
2. Describe how you calculated the percentage of classmates with each eye colour.
3. Construct a pie chart showing the percentages of students in your class that have each eye colour. Use page 371 of the Science how-to section to help you.
4. Explain how you used the data in your results table to calculate the number of students in your school with each eye colour.
5. Worldwide, eye percentages are as follows: 5% amber, 9% blue, 75% brown, 2% green, 3% grey, 5% hazel and 1% other.
   a Do these percentages match your data?
   b Provide a reason for any differences between your data and the worldwide percentages.
   c Apply the worldwide percentages to all the students at your school. Re-calculate the number of students in your school with each eye colour.

Conclusion

Copy and complete:

'The results show that: *(respond to the aim).*'

Table I8.3: Results

Eye colour	Tally	Total	Percentage of classmates
Amber			
Blue			
Brown			
Green			
Grey			
Hazel			
Other			

Investigation 8.4

Modelling bridges

Process: Problem-solving

Focus: Using criteria to solve problems

In investigations, the aim is often to find a solution to a problem. Sometimes, data makes the criteria for a problem. How can you use your understanding of bridges to solve the problem of a bridge holding the most mass?

Hint: *What shapes do you normally see used in construction to help stabilise bridges?*

Aim

To construct a model of a bridge

Materials

- wooden blocks
- 50 popsticks
- roll of sticky tape
- 25 g masses

Method

1 Copy the results table into your notebook, adding a title.

2 Set up the wooden blocks so there is a 50 cm distance between them.

3 Use the popsticks and the sticky tape to construct a model of a bridge that spans a 50 cm distance. Think about the shapes you are using to build your bridge.

4 a Suspend your bridge across the 50 cm space you created using wooden blocks.

b Add the 25 g masses to the bridge, one at a time, until the bridge can no longer support the mass. Record your findings in the results table.

Table 18.4: Results

Mass (g)	Mass supported or not supported?
0	
25	
50	
75	
100	
125	
150	

Questions

1 a What overall shape did you use for your bridge?

b Why did you use this shape?

2 a What was the maximum mass your bridge was able to support without breaking or falling?

b How did this compare to the other bridges constructed by your classmates?

3 What is the link between the shape of a bridge and its ability to hold mass?

4 How did you construct your bridge to solve the problem of crossing an open space, with the criterion of holding a set mass?

Conclusion

Copy and complete:

'The results show that: *(respond to the aim)*.'

Figure 18.4: This is one of the many different bridge shapes.

Investigation 8.5

Trends in the weather

Process: Analysing data and information

Focus: Describing trends from data and information

Being able to see and interpret patterns in data is an important scientific skill. Computer models use these patterns to make predictions for the future.

Hint 1: Is there a general increase or decrease in your data?

Hint 2: Refer to the Science how-to section on page 334.

Aim

To analyse the weather patterns that are used in computer models

Materials

- device with internet access
- spreadsheet

Method

1. Copy the results table into your notebook, adding a title and as many rows as needed.
2. Find online weather data for New South Wales from last year. (This information is available on the website of the Australian Bureau of Meteorology.)
3. Identify the average maximum temperature for each month of the previous year for Sydney. Record this information in the results table.
4. Identify the total rainfall for each month of the previous year for Sydney. Record this information in the results table.
5. a Construct a second results table.
   b Find the New South Wales weather data for 2016.
   c Identify the average maximum temperature for each month of 2016 for Sydney. Record this information in the second results table.
   d Identify the total rainfall for each month of 2016 for Sydney. Record this information in the second results table.
6. a Construct a third results table.
   b Find the New South Wales weather data for 1916.
   c Identify the average maximum temperature for each month of 1916 for Sydney. Record this information in the third results table.
   d Identify the total rainfall for each month of 1916 for Sydney. Record this information in the third results table.

Table I8.5: Results

Month	Average maximum temperature (°C)	Total monthly rainfall (mm)
January		
February		
March		
April		

Questions

1. a Were there any differences between the maximum temperature data for last year and 2016?
   b Were there any differences between the monthly rainfall data for last year and 2016?
   c Suggest a reason for the similarities or differences in the data.
2. Based on the data you have collected, suggest what a computer model might predict about future temperature and rainfall patterns for Sydney in:
   a 10 years' time.
   b 100 years' time.
3. Justify your reasoning for your responses to Questions 2a and 2b.

Conclusion

Copy and complete:

'The results show that: *(respond to the aim)*.'

Investigation 8.6

30 min

Modelling the big bang

Process: Problem-solving

Focus: Identifying scientific problems

Scientists often use models to explain and provide supporting evidence for scientific theories. However, models can have limitations. What are some problems with the model in this investigation? How could these problems be fixed?

Aim

To construct a model that demonstrates how the universe is expanding

Materials

- balloon
- 2 marker pens (blue and red)
- fabric tape measure
- balloon pump

Method

1 Copy the results table into your notebook, adding a title.

2 Use the blue pen to draw 3 stars on the uninflated balloon. Label one star 'Star 1'.

3 Use the red pen to draw 3 spirals (representing galaxies) on the uninflated balloon. Label one galaxy 'Galaxy 1'.

4 a Use a tape measure to measure the widest point of each of the 3 stars and the widest point of each of the 3 galaxies.

  b Calculate the average size of the stars and the average size of the galaxies. Record this information in the results table.

5 a Use a tape measure to measure the distance between Star 1 and the nearest object (a star or galaxy).

  b Use a tape measure to measure the distance between Galaxy 1 and the nearest object (a star or galaxy).

  c Record these distances in the results table.

6 Use the balloon pump to add 3 pumps of air into the balloon so it is partially inflated.

7 Repeat Steps 4 and 5.

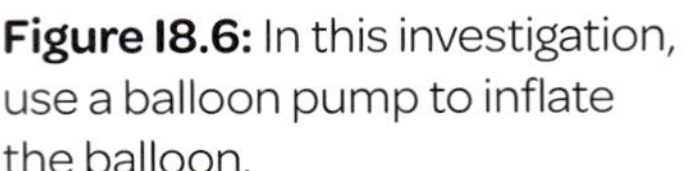

▲ **Figure 18.6:** In this investigation, use a balloon pump to inflate the balloon.

8 Use the balloon pump to fully inflate the balloon.

9 Repeat Steps 4 and 5.

Table 18.6: Results

Uninflated balloon		
Object	**Average size (cm)**	**Distance from nearest object (cm)**
Star		
Galaxy		
Partially inflated balloon		
Object	**Average size (cm)**	**Distance from nearest object (cm)**
Star		
Galaxy		
Fully inflated balloon		
Object	**Average size (cm)**	**Distance from nearest object (cm)**
Star		
Galaxy		

Questions

1 What happened to the sizes of the stars and galaxies as the balloon inflated?

2 What happened to the distances between the stars and galaxies as the balloon inflated?

3 How does this model represent what scientists believe is happening to our universe?

4 What is one problem with this model?

5 Explain a way you could create a different model that better represents how the universe is expanding.

Conclusion

Copy and complete:
'The results show that: *(respond to the aim).*'

Investigation 8.7A

30 min

Reaching great heights

Process: Problem-solving

Focus: Providing solutions to problems

There can be many problems that need to be overcome during a scientific investigation. Sometimes these problems involve physical limitations.

Hint 1: A piece of paper is 0.1 mm thick. Each time it is folded, the thickness doubles.

Hint 2: How could you figure out how many folds are needed when you can no longer physically fold the paper?

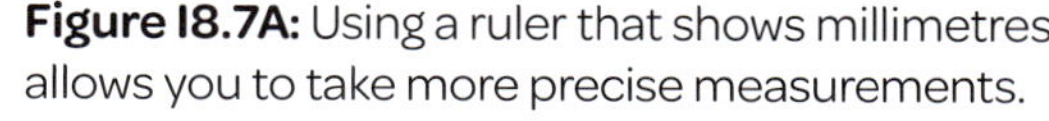

Figure 18.7A: Using a ruler that shows millimetres allows you to take more precise measurements.

Aim

To determine how many folds in a piece of paper you would need to cover the distance between Earth and the Moon

Materials

- A4-sized piece of paper
- ruler showing millimetres

Method

1 Copy the results table into your notebook, adding a title and enough rows to record the thickness of 10 folds.

2 Use the ruler to measure the thickness of the A4-sized piece of paper.

3 Fold the paper in half. Measure the total thickness of the paper. Record this measurement in the results table.

4 Repeat Step 3 until you can no longer fold the paper in half.

Table 18.7A: Results

Number of folds	Paper thickness (mm)
0	0.1
1	

Questions

1 How many times could you fold the paper in half before it was no longer possible?

2 How could you calculate the thickness of the paper, even when you could no longer fold the paper in half?

3 If you could fold the paper in half as many times as you liked, how many folds would it take to make it as tall as Centrepoint Tower in Sydney (which is 309 m or 309 000 mm tall)?

4 How many folds would it take to reach the Moon from Earth (384 400 km)? Remember: Convert the distance from Earth to the Moon to millimetres. (See the Science how-to section on page 360.)

Conclusion

Copy and complete:

'The results show that: *(respond to the aim).*'

Investigation 8.7B

Height and arm span

Process: Problem-solving

Focus: Explaining scientific problems using cause-and-effect relationships

Many investigations examine the relationship between two specific factors. In this investigation, you will collect data to see if there is a relationship between a person's height and their arm span.

Aim

To determine the relationship between a person's height and their arm span

Materials

- 5-metre tape measure

Method

1 Copy the results table into your notebook, adding a title and as many rows as needed.

2 Use the tape measure to work out the height of each person in your class. Record this information in the results table.

3 Use the tape measure to work out the arm span of each person in your class. Record this information in the results table.

4 Use the height and arm span of each person to calculate their height-to-arm span ratio. Record this information in the results table. (See 'Understanding ratios' in the Science how-to section on page 361.)

Table I8.7B: Results

Individual height (cm)	Individual arm span (cm)	Individual height-to-arm span ratio

Questions

1 What were the shortest and tallest heights you recorded in the results table?

2 What was the range of height data in your class?

3 What was the mean arm span length in your class?

4 Explain the relationship between height and arm span, using data from the results table to support your explanation.

5 Convert the data in the results table into a graph to clearly show any relationships in the data.

6 How could you change the results table to show multiple individuals of the same height and their different arm spans in one row?

7 Does recording the data in a different way make relationships clearer? Explain your answer.

Conclusion

Copy and complete:

'The results show that: *(respond to the aim)*.'

Figure I8.7B: The height-to-arm span ratio of swimmers is often different from other people's.

Investigation 8.8

30 min

Analysing rainfall

Process: Analysing data and information
Focus: Explaining relationships between datasets and information

Scientists analyse the data they collect to find and assess any trends or patterns. These trends can help to explain the relationship between variables.

Figure 18.8: Climate weather stations like this record the amount of rainfall in different areas.

Aim

To analyse the link between the number of rainy days and the amount of rainfall

Materials

- device with internet access
- weather website (for example, the Australian Bureau of Meteorology website)

Method

1. Copy the results table into your notebook, adding a title.
2. On the internet, find the following information for your area for last year:
   a. monthly rainfall (mm).
   b. number of rainy days per month.
3. Record the data for each month in the results table.
4. Calculate the mean, mode, median and range for both the amount of rain and the number of rainy days. Record this information in the results table.

Questions

1. What type of data is the amount of rainfall?
2. Construct a column graph showing the total rainfall in millimetres for each month of last year in your local area.
3. Construct a column graph showing the number of rainy days for each month of last year in your local area.
4. Do the graphs show any links between the data? Explain any relationships you see.
5. Do these correlations equate to causation? Give a reason for your answer.

Conclusion

Copy and complete:
'The results show that: *(respond to the aim)*.'

Table 18.8: Results

Month	Total rainfall (mm)	Number of rainy days
January		
February		
March		
April		
May		
June		
July		
August		
September		
October		
November		
December		
Mean		
Mode		
Median		
Range		

Investigation 8.9

Modelling the yowie

Process: Problem-solving
Focus: Explaining scientific phenomena
Scientists use their observations and the data they collect to help them explain the world around them.

Aim

To use data to construct a model of a yowie (a large, mythical creature said to live in the Australian wilderness)

Materials

- 1-metre ruler
- 3-metre tape measure
- A3-sized piece of paper
- coloured pencils

Method

1 Copy the results table into your notebook, adding a title.
2 Break up into groups of 3 people.
3 Use the ruler to measure the length of the left foot of each person in your group. Record this information in the results table.
4 Use the tape measure to measure the height of each person in your group. Record this information in the results table.
5 a Calculate the average foot length of the members of your group. Record this information in the results table.
b Calculate the average height of the members of your group. Record this information in the results table.
6 Calculate the average height-to-foot length ratio by dividing the average height by the average foot length.

Questions

1 State the height-to-foot length ratio for your group.

Observational data: The yowie

There has been a reported sighting of a yowie in the Blue Mountains. Based on observations collected by different hikers, it has been determined that the yowie is covered in dark brown fur and has two sharp top fangs and long, gorilla-like limbs. Hikers have also found a left footprint that is 48 cm long.

2 Use the height-to-foot length ratio for your group to calculate the approximate height of the yowie.
3 Construct a labelled, two-dimensional model of the yowie. Base your model on the observational data and the data you have collected in this investigation.
4 Suggest one way the model could be improved. Give a reason for your suggestion.

Conclusion

Copy and complete:
'The results show that: *(respond to the aim)*.'

Table 18.9: Results

Feature	Measurement (cm)			
	Person 1	Person 2	Person 3	Average
Left foot length				
Height				

Figure 18.9: The Yowie is reported to roam Australia's densely forested wilderness, from southern Queensland to southeastern Victoria. The Blue Mountains have been the site of numerous Yowie reported sightings.

Glossary

A

abiotic: non-living

acceleration: the rate at which speed changes over time, or the rate at which you fall due to gravity

accuracy: how close a measured value is to the true, exact value; how closely a recorded value matches the expected outcome of an investigation

adaptation: a feature that an organism is born with that helps it to survive in its environment

aerobic respiration: the process of turning glucose into energy where the cells take in oxygen and release carbon dioxide

algorithm: a rule or set of rules that specifies how to solve a problem, often used in mathematics or information technology

alloy: a metal mixed with other metals or non-metals

alpha particle: a positively charged particle that is four times larger than a proton

anaerobic respiration: how living things produce energy without oxygen

Ancestors: the people we are descended from

angiosperm: a plant that produces seeds in fruit or flowers

aorta: the main artery that leads from the heart to the rest of the body

aqueous: containing water

artery: a type of blood vessel that carries blood away from the heart

asthenosphere: the portion of Earth's mantle underneath the lithosphere that can flow

atmosphere: the layer of gas that surrounds Earth

atmospheric nitrogen: nitrogen in the atmosphere

atom: a particle that makes up all matter; made up of protons, neutrons and electrons

atomic mass: the average mass of the atoms in a sample of an element

atomic number: the number of protons in an atom

atria: filling chambers of the heart

attractive force: a force that pulls objects towards each other

axis: a real or imaginary line through the centre of an object

axle: the pin, post or rod that a wheel can spin around to transfer force

B

big bang: when the universe began as an incredibly hot, dense point roughly 13.8 billion years ago

big data: large, diverse sets of information that grow at increasing rates

bile: a salty solution stored in the gallbladder, which aids digestion in the upper small intestine

biodiversity: the variety of organisms in an ecosystem

biotic: living

blood vessel: a tube such as a vein or artery that carries blood in the body

body system: two or more organs that are connected and working together

branch: a type of science, such as biology or chemistry

brittle: not able to be bent; will break or shatter if stressed

bronze: an alloy made of copper and tin

brumation: the state of dormancy (inactivity) reptiles undergo in winter (similar to hibernation)

bryophyte: a plant that does not contain vascular tissue

Bunsen burner: a piece of scientific equipment that produces a single open gas flame

buoyancy: the upward force that opposes the downward force of gravity on an object in a liquid or gas

C

capillary: the smallest type of blood vessels

carbon dioxide: a colourless and odourless gas containing carbon and oxygen that is produced during respiration

cardiovascular: related to the heart (cardio) and blood vessels (vascular)

cast: an object created when sediment or minerals fill a mould

cathode ray: a stream of electrons observed in a high-vacuum tube

causation: a change in one factor being the reason for a change in another factor

cell: the smallest functional unit of an organism

cell membrane: a thin layer around a cell that controls which substances enter and leave the cell

cell wall: a stiff layer around a plant cell that supports it

cellular respiration: the process of turning glucose into energy for cells to use

census: a complete count of everything

centrifugation: a technique for separating the components of a mixture on the basis of their different densities

chemical bond: a force that holds atoms together

chemical change: a change in properties with a new substance formed

chemical digestion: breaking down food in a chemical reaction that forms new molecules

chemical formula: an expression of the elements that make up a chemical compound, usually presented as a ratio using chemical symbols and numbers; for example, H_2O

chemical property: a property of a substance that is observed during a chemical reaction

chemical reaction: a process in which one or more substances are changed to form new substances

chemical sedimentary rock: sedimentary rock formed from layers of mineral crystals that have crystallised from water

chemical symbol: a symbol of one or two letters used to represent an element

chlorophyll: the green pigment in plant cells; this pigment absorbs sunlight and enables photosynthesis

chloroplast: a small organelle in a plant cell that makes food for the plant

chyme: a soupy mixture of partially digested food in the stomach and small intestine

citation: a way of giving credit to a source, usually in the same text where the information appears

classification: the process of sorting things into groups or classes

classification key: a tool to help classify an organism by identifying important characteristics such as its features and behaviours

clastic sedimentary rock: sedimentary rock made of sediments cemented together

closed system: a system from which energy can flow in and out but matter cannot

collaboration: working cooperatively with other people to complete a task or solve a problem

colloid: a mixture made of tiny insoluble particles

combustion: the reaction of a substance with oxygen to give off heat

community: a naturally occurring group of animals, plants and other organisms

compound: a substance containing atoms of two or more elements; the atoms are chemically bonded together and are present in the compound in a fixed ratio

compress: squash something into a smaller shape

computer model: a model generated using computer software to show a scientific concept

concave: hollowed or rounded inwards

concentrated: when a solution has a large amount of solute in a certain volume

concentration: the amount of a dissolved solute in a defined amount of solvent

condensation: the cooling of a gas to become a liquid

condense: change state from gas to liquid

condenser: a glass tube cooled by water that cools a gas to become a liquid

conduction: the transfer of energy through a substance

conductor: a substance that allows the transfer of energy

conglomerate rock: sedimentary rock made of large, rounded pebbles and fragments cemented together

conservation of energy: the scientific law that states that energy cannot be created or destroyed

conservation of mass: the scientific law that states that mass cannot be created or destroyed

consumer: an organism that gains energy by consuming other living organisms

contact metamorphism: the process of change that happens to rock over small areas, often near volcanoes

continental drift: the theory that the continents have moved over time

continuous data: data that can take any value along a continuum or between fixed values

controlled variable: a thing that needs to stay the same during an investigation

convection: the transfer of energy by movement of a liquid or gas

convergent boundary: where two tectonic plates are moving towards each other and colliding

conversion factor: a number used to change one unit of measurement to another

coolamon: a shallow dish traditionally used by First Nations Peoples to carry water and food and for winnowing seeds

core: Earth's central layer, made up of a liquid outer core and a solid inner core

correlation: a relationship between two factors or variables that shows them changing together, in either the same or opposite way

corrosion: the reaction of a metal with oxygen

Country: a specific area of a First Nations People that includes physical, linguistic and spiritual features

crude oil: oil that has not been separated into usable petroleum products

crust: Earth's thin outer layer, made up of continental crust and oceanic crust

crystal: a solid substance made up of very ordered microscopic parts

crystallisation: the separation of a solution by evaporating the solvent, leaving behind solute crystals

culture: a shared system of customs, habits, beliefs/spirituality, social organisation and ways of life that characterise different groups and communities

cytoplasm: a jelly-like fluid inside a cell

D

dark energy: a hypothetical form of energy that expands the fabric of the universe

dark matter: particles that do not absorb, reflect or emit light, so they cannot be detected by observing electromagnetic radiation

data: (1) facts and information collected for reference or analysis; (2) a collection of information gathered through observations, measurements, study or analysis

data point (datum): a single identified element in a dataset

dataset: a collection of data, often from numerous trials related to a single factor

decant: to carefully pour off the liquid from a mixture, leaving the sediment behind

decomposer: an organism that breaks down and recycles decaying matter

deforestation: the removal of trees to make land available for other uses

density: how heavy something is for its size; mass divided by volume

density formula: $\rho = \frac{m}{V}$ (unit: g/cm^3)

dependent variable: the thing that is measured in a first-hand investigation

deposition: (1) changing state from gas to solid; (2) a process in which sediment is left in a new place

diaphragm: the band of muscle under the lungs that enables the physical action of inhalation and exhalation

dichotomous: divided into two parts

differentiate: to change to have a particular function

diffuse: (1) when light becomes spread out as it passes through the atmosphere; (2) to move from an area of high concentration to an area of low concentration

digestion: the physical and chemical processes that break down food in the body

digital footprint: information about a particular person that exists on the internet as a result of their online activity

dilute: when a solution has a small amount of solute in a certain volume

direct force: a force that acts on an object in contact with another object; also known as contact force

discrete data: data that can only take certain values; fixed values; for example, a quantity count

disinfection: a method of destroying bacteria, often using special light or chlorine

dissolve: particles separating and spreading out, so that they seem to disappear, when added to another substance

distillation: the separation of liquids with different boiling points by heating

divergent boundary: where two tectonic plates are moving away from one another

DNA: the genetic information inside a body's cells

Dreaming stories: important stories for First Nations Peoples; these stories often contain important knowledge of the world

ductile: can be drawn out into a wire

E

ebb: the phase of the tide when the water flows towards the ocean

eclipse: the blocking of the Sun's light from Earth

ecosystem: a community of living and non-living things interacting with one another and with their environment

effort: the force applied to a simple machine to move another object

Elder: a First Nations person who is a recognised custodian of knowledge, who has permission to pass on that knowledge, and who has specific responsibilities to young people in a community

electric field: an area around a charged particle in which it exerts a force on other charged particles

electrically neutral: having an equal number of protons (positively charged) and electrons (negatively charged)

electromagnet: a device that uses electricity to make a magnetic field

electron: a subatomic particle that orbits the nucleus of an atom; it is negatively charged

electron configuration: the number of electrons in each shell of an atom, starting with the smallest; for example, 2, 8, 8, 2

electron shell: the space around an atom's nucleus, in which electrons circulate

electrostatic charge: the electric charge on the surface of an object

electrostatic force: an indirect force between any objects with an electric charge

electrostatic shock: the rapid movement of charged particles from one object to another

element: a pure substance made of one type of atom

endangered: at risk of becoming extinct

endemic: only found in a certain location or community

endothermic: a reaction that absorbs energy in the form of heat

energy: the ability to do work

enzyme: a substance that enables or speeds up a chemical reaction

epicentre: the point on Earth's surface directly above the focus of an earthquake

epidermis: the outer layer of cells

equation: a mathematical statement that shows that two things are equal; for example, $2x + 6 = 14$ is an equation that needs to be solved so that $2x + 6$ does actually equal 14

equinox: the two times of the year when night and day are about the same length

erosion: a process in which sediments are moved from one place to another

ethical: in science, minimising harm to those involved, and ensuring investigations are conducted honestly and data is collected and recorded accurately

evaluate: judge value based on scientific evidence

evaporate: change state from liquid to gas

evaporation: the separation of a solid from a solution by heating to remove the liquid

evidence: facts and observations that can be used to support or oppose a theory

excretion: the elimination of cellular waste from the body through urine

exothermic: a reaction that releases energy in the form of heat

experiment: a repeatable investigation carried out under controlled conditions to test a hypothesis

experimental data: information collected by conducting an investigation and recording the results

exponent: the superscript value to the right of a number that says how many times to use the number in a multiplication; for example, when we write 10^3, '3' is the exponent; it means we need to multiply 10 by itself 3 times

extinct: having no living members of the species; died out

extrapolate: to estimate or guess what could happen by using information that is already known

extrusive igneous rock: igneous rock formed at Earth's surface

F

fair test: a test where all variables are kept the same, except for the independent variable and the dependent variable

fault: a break or an area of breaks between two blocks of rock

fauna: animals

field: an area of space where objects are affected by an indirect force

field line: a line used to show the direction of a force within a field

filtrate: the liquid that passes through a filter on filtration

filtration: the process of separating a mixture of solid particles from a liquid by using a filter

first-hand investigation: an investigation where the data is collected by the investigator

flora: plants and microorganisms

flow: the phase of the tide when the water flows towards the land

focus: the origin of an earthquake

food chain: a path of energy through an ecosystem

food web: a system of interlocking food chains

force: a push or pull on an object when it interacts with another object

force multiplier: a machine that amplifies the effect of an effort applied in a task

formulate: to invent something, usually in scientific or mathematical terms

fossil: preserved remains or traces of once-living things

fossil record: the history of life on Earth as documented by fossils

fossilisation: the process of the formation of a fossil

free-body diagram: a simplified diagram of the forces acting on an object

freeze: change state from liquid to solid

friction: a force between two surfaces that are sliding, or trying to slide, across each other

fulcrum: the point on which a lever turns when moving an object

G

galaxy: a large system of stars

gas exchange: the exchange of oxygen and carbon dioxide between an organism and the environment

geocentric model: a model of the solar system with Earth at the centre

geological history: how Earth has changed over time

germination: the process in which previously dormant seeds put out shoots for new growth

glomerulus: a bundle of capillaries; the filtering unit within each nephron

glucose: a type of sugar that is the energy source for cells

gravitational field: a region in which an object with mass can experience a gravitational force

gravitational force: an indirect force that attracts physical objects with mass towards each other

gravity: the force of attraction between two objects

grinding: the action of rubbing objects together to break up materials into smaller particles

group: a column in the periodic table

gymnosperm: a plant that produces seeds in cones

H

habitat: the place where an animal or a plant naturally lives

hand picking: a simple method of separating a mixture of objects

hazard: something that can harm living things, objects or the environment

heat: the flow of thermal energy between objects

heating flame: the blue (very hot) flame of a Bunsen burner (approx. 1500 °C); used for heating substances

heliocentric model: a model of the solar system with the Sun at the centre

heterogeneous: a mixture with an uneven (non-uniform) composition

hiccups: when the diaphragm contracts involuntarily and repeatedly to produce a vocal 'hic' sound

homogeneous: a mixture with an even (uniform) composition

hormone: a chemical substance produced by the body that controls the activity of certain cells or organs

hotspot (shield) volcano: a volcano formed by magma upwelling underneath a tectonic plate

hypothesis: a suggested explanation or prediction of a scientific problem that can be tested with an investigation

I

igneous rock: rock that has solidified from lava or magma

immiscible: cannot be mixed

inclined plane: a tilted flat surface; also called a ramp

independent variable: the thing that is deliberately changed in a first-hand investigation

indirect force: a force that acts on an object without the need for physical contact; also known as non-contact force

inference: a conclusion that is based on evidence and observations

insight: applying inferences to real-world scenarios

insoluble: does not dissolve

insulator: a substance that resists the transfer of energy

intensity: how much heat is released from a fire front

interact: act upon or have an effect on

interdisciplinary: combining more than one branch of knowledge

intraplate earthquake: an earthquake that takes place in the middle of tectonic plates

intrusive igneous rock: igneous rock formed under Earth's surface

invertebrate: an organism without a backbone or spinal cord

isolated system: a system from which neither energy nor matter can flow in or out

J, K

kinetic energy: the energy of movement

knapping: the action of applying force to a specific type of rock to create a sharp shard

L

lattice: a three-dimensional shape made of a repeating pattern of atoms

lava: molten (melted) rock above Earth's surface

law of superposition: older rocks are found beneath younger rocks in a sequence

leach: to remove soluble substances by the action of water passing through the material

lever: a bar acted upon at different points by two forces

leverage: the action of a lever

like poles: magnetic poles that are the same (both north or both south); they repel each other

lithosphere: Earth's rigid outer zone (crust and most of the upper mantle), made up of tectonic plates

living thing: an organism that displays specific characteristics of being alive

load: an object that is moved or lifted by a simple machine

load force: the force from the object that is being moved

M

magma: molten (melted) rock under Earth's surface

magnetic force: an indirect force that affects any object made of certain metals

magnetic pole: one of the two ends of a magnet

magnitude: the size of a force, an object or an amount of energy

malleable: able to be bent and shaped

mantle: Earth's middle layer, made up of an upper mantle and a lower mantle

mass: the amount of matter that a physical body contains

mathematical formula: a rule or principle that helps you find the answer to a question or understand the relationship between variables

matter: any substance that has mass and takes up space

mean: a measure of centre (an average) calculated by adding all the numbers together and dividing by how many numbers there are

mechanical digestion: physically breaking down food into smaller pieces

median: the middle number in a set of numbers when they are arranged in order

melt: change state from solid to liquid

meniscus: the curve seen at the top of a liquid in its container

metalloid: an element with properties of both metals and non-metals

metamorphic rock: rock formed from another rock that has been changed by heat or pressure, or both

metamorphism: the process of change that happens to a rock because of heat or pressure or both

mid-ocean ridge: a long chain of mountains under the ocean formed by plate tectonics

mineral: a naturally occurring inorganic (non-living) substance

mitochondria: the organelles where respiration happens

mixture: two or more substances combined that can be physically separated

mode: the number that appears most frequently in a set of numbers

model: a physical or mathematical representation of a scientific idea or concept

molecule: a unit made up of two or more atoms chemically bonded to each other

motion: the change in position of an object over time

mould: a hollow impression formed by an imprint of an organism, or when the original bone or shell has dissolved

multicellular: consisting of many specialised cells that form tissues and organs, which together make one functional organism

mutualism: a relationship between two organisms in which both organisms benefit

N

naturalist: someone who studies nature and its history

nebula: an enormous cloud of dust and gas in space

negative correlation: as one variable increases, the other decreases

nephron: a microscopic filtration structure in the kidneys that removes waste and excess water

net force: the resultant force when all forces acting on an object are added together and cancelled out

neutron: a subatomic particle located in the nucleus of an atom; it is neutrally charged

Newer Volcanics Province: an area covering roughly 15 000 square kilometres in south-east Australia; the most recent active volcanic area in Australia

newton: the unit for measuring force, SI unit N; 1 kg on Earth is equivalent to 10 N

nocturnal: active during the night

non-biodegradable: does not break down in the environment

non-scientific approach: an approach relying on tradition, intuition and personal belief to arrive at a conclusion

normal force: the force that acts from a surface pushing against an object

nucleus: (1) the centre of an atom, which contains protons and neutrons; (2) the control centre of a cell; contains DNA

O

observation: noticing something you can see, touch, smell, hear or taste, and know to be true

oil slick: a thin layer of oil on the surface of water

open system: a system that allows energy and matter to transfer in and out

orbit: the path of an object around a star or planet; for example, Earth around the Sun, or the Moon around Earth

organ: a structure made up of two or more tissues that has a specific function

organelle: a structure within a cell

organic sedimentary rock: sedimentary rock formed from the remains of plants or animals

organism: an individual animal, plant or other living thing

P

paid promotion: when someone is paid to say they like a product (even if they do not)

palaeontologist: a scientist who studies fossils

paper chromatography: a technique that separates the components of a mixture based on their solubilities

parallax error: the apparent shift in something's position when it is viewed from different angles

parasite: an organism that lives in or on another organism, causing it harm

particle: a very small amount of matter

particle theory: a model used to describe the properties of solids, liquids and gases

pattern (data): when data repeats in a predictable way

pendulum: a mass that is attached to a fixed point by a string and can move backwards and forwards freely

penumbra: the outer part of the Moon's shadow on Earth

period: a row in the periodic table

periodic table: a table of all known elements and their chemical symbols

peristalsis: the involuntary muscle action that pushes food through the digestive tract

personal data: potentially sensitive, private information connected to a particular individual

pesticide: a chemical used to kill insects or other organisms that feed on plants

phenomenon (plural **phenomena**)**:** an observable fact or event

phloem: a tubular structure that transports food around a plant

photosynthesis: the chemical reaction, powered by sunlight, in plants that converts carbon dioxide and water into sugar and oxygen

physical change: a change in appearance with no new substance formed

physical property: a characteristic or attribute of a substance that can be observed and measured, such as colour, texture, melting and boiling points, density and hardness

pistil: the female reproductive part of a flower (the stigma, style and ovary)

plagiarise: to copy someone else's work and present it as your own

plausible: could be reasonably accepted based on available evidence

polarity: the direction of a force, such as a magnetic force

pollen: the fine powder produced by the male part of a flower; contains male sex cells

pollination: the movement of pollen from the male part of a flower to the female part (from the anther to the stigma)

pollinator: an organism that transfers pollen from the male part of a plant to the female part

positive correlation: as one variable increases, so does the other

potential energy: stored energy; has the ability to cause movement

power: the end product obtained by multiplying a quantity by itself one or more times; for example, 2 to the *power* of 3 is 2 × 2 × 2, which is 8, so the *power* of 2^3 is 8

precision: how close measured values are to each other within a dataset

predator: an organism that kills and feeds on prey

prey: an organism that is killed by a predator

primary data: data that you have collected from your own investigations

producer: an organism that makes its own food using energy from the Sun

product: a substance formed in a chemical reaction

property: a characteristic of a substance that affects what it looks like and how it behaves

proton: a subatomic particle located in the nucleus of an atom; it is positively charged

pseudoscience: beliefs mistakenly regarded as being based on scientific method

pulley: a wheel with a grooved rim for carrying a cable

Q

qualitative data: data with qualities or characteristics that can be observed and described

quantitative data: information based on numbers that can be counted, measured or represented by numbers

R

radial field: a field in which the field lines radiate from a centre

radiation: the transfer of energy that does not require contact with matter

range: in a set of numbers, a measure of spread between the highest number and the lowest number

ratio: a way of comparing like quantities without units

reactant: a substance that undergoes a chemical change

reactive: undergoes chemical reactions easily

regional metamorphism: the process of change that happens to rock over large areas

relationship: a link between two factors

relative age: the approximate age of a rock determined by comparing it to another rock

reliability: the extent to which the results or measurements can be reproduced when the research is repeated

reliable: provides consistent results when repeated

renal: an adjective that means 'related to the kidney'

repulsive force: a force that pushes objects away from each other

residue: the solid that does not pass through a filter on filtration

respiration: a chemical reaction that converts glucose to energy

reversible: can be changed back to its previous state

revolve: move in a circular path around another object

rift valley: a valley formed when a continent was pulled apart

risk: the chance that a hazard will cause harm

rotate: spin on an axis

rotation: turning around as if on an axis

Rube Goldberg machine: a machine or contraption that uses a chain reaction to carry out a simple task

S

safety flame: the orange (cooler) flame of a Bunsen burner (approx. 300 °C); used between heating substances

satellite: an object that orbits a planet; Earth has a natural satellite called the Moon and also has many purpose-built satellites (referred to as artificial satellites)

saturated: a solution that cannot dissolve any more solute

saturation point: the point at which a solution has dissolved the maximum amount of solute

scale factor: the ratio between corresponding measurements of an object and a copy of that object

scientific approach: the process of using testing and experiments to objectively establish facts

scientific law: an explanation of a natural phenomenon that usually uses a mathematical equation

scientific notation: a way to write very large or very small numbers in a simple form

scientific theory: an explanation of a natural phenomenon that is supported by evidence and the results of repeated tests

secondary data: data that has been collected by someone else

second-hand investigation: an investigation that uses data previously collected in a separate investigation

sediment: (1) the solid that settles to the bottom of a liquid in a mixture; (2) small particles of rock, such as clay, sand and pebbles

sedimentary rock: rock formed by sediments that have been pressed together

seismic wave: a wave of energy that passes through Earth's layers that is caused by an earthquake

seismometer: a scientific instrument that detects seismic waves

sequence: the order of something

sewage: semi-liquid human waste

sewerage: a system of pipes that carry sewage

sieving: using a sieve to separate larger particles from smaller particles

simulations: experimental results generated by a computer model

sky Country: in First Nations knowledge, the aspects of Country that are not linked to the land; all the observable celestial phenomena and the spirits and lore linked to these events

social media: the means of interacting and sharing information with other people via virtual networks

solidify: become a solid

solstice: the two times of the year when night and day are the most different in length

solubility: the amount of solute that will dissolve in a solvent

soluble: able to dissolve

solute: a substance that is dissolved by a solvent

solution: a mixture made up of a solvent and a solute

solvent: a substance that a solute dissolves in

songlines or **song series:** complex ways that First Nations Peoples record important information relating to many different topics; a body of songs sung sequentially and which are intended to convey knowledge

species: a single, specific type of living thing

spectral line: a dark or bright line in an otherwise uniform spectrum

speed multiplier: a machine that increases the speed of a task

spirits: the supernatural beings that many First Nations Peoples believe exist; can be associated with particular places, objects and rituals; are often connected with Dreaming stories

spore: a tiny part of some plants that is used to reproduce

stamen: the male reproductive part of a flower (the anther and filament)

statistics: the analysis of large collections of data to make inferences

steel: an alloy made of carbon and iron

stimuli: changes in the environment that cause responses in living things

stomata: pores on the surface of a leaf; the site of gas exchange in plants

strato volcano: a volcano formed at a subduction zone

structure: the construction and arrangement of tissues, parts or organs

subatomic particles: particles that make up atoms: protons, neutrons and electrons

subduction: when one tectonic plate moves underneath another

sublimation: changing state from solid to gas

subscript: a letter or number written slightly below and to one side of another; for example, '2' in H_2O

substance: matter that has a fixed chemical make-up

supersaturated: when a solution contains more than the maximum amount of solute it can normally dissolve

superscript: a letter or number written slightly above and to one side of another; for example, '2' in 8^2

surface tension: where the molecules at the surface of a liquid are more attracted to each other than to the air above the liquid

suspension: a mixture made of large insoluble particles

system: a set of simple things that work together to perform a function

T

taxonomy: the science of classifying organisms

tectonic plate: a section of Earth's lithosphere

telescope: a device that uses mirrors or lenses to focus on distant objects

tension: a pulling force exerted by each end of an object

threatened: likely to become endangered in the near future

tide: the rise and fall of the waters in the ocean, caused by forces from the Moon and the Sun

tissue: a group of cells with a similar structure and function

toxin: a poisonous substance produced by a living thing

tracheophyte: a plant that contains vascular tissue

transdisciplinary: collaboratively working across multiple branches to learn or solve problems

transfer: move from one place or object to another

transform: change from one type to another

transform boundary: where two tectonic plates slide past each other

trend (data): when data moves in a general direction, usually up or down

trophic level: the position of an organism in a food chain

tubule: a small tube made of epithelial cells

U

umbra: the inner part of the Moon's shadow on Earth

understorey: ground-level vegetation

unicellular: consisting of one cell that carries out all the functions itself

universal solvent: a solvent that dissolves most substances

unlike poles: magnetic poles that are different (south–north or north–south); they attract each other

unsubstantiated: not supported with any evidence

urbanisation: the creation of urban areas such as cities

urea: a product of the chemical breakdown of food, specifically proteins

V

valence electron: an electron in the valence shell

valence shell: the electron shell furthest away from the nucleus of an atom; occupied by electrons

valid: when an investigation meets its intended purpose

validity: whether an investigation measured what it intended to measure

variable: a factor in an investigation or a model that can be changed, measured or controlled

vascular tissue: tissue that transports fluid and nutrients throughout a plant

vein: a type of blood vessel that returns blood to the heart

vena cava (plural **venae cavae**)**:** the main vein that leads to the heart from the rest of the body

ventricles: pumping chambers of the heart

vertebrate: an organism with a backbone or spinal cord

villi: the tiny finger-like projections that line the walls of the small intestine

volcano: a point in Earth's crust where lava erupts

volume: the amount of space an object takes up

W

waning: when the Moon looks as though it is getting smaller

waxing: when the Moon looks as though it is getting bigger

weathering: a process in which rocks are worn down into smaller particles

weight: the force of a gravitational field on the mass of a body

winnowing: separating a mixture of seeds and husks by blowing air through the mixture

word equation: a representation of a chemical change in words

work: when a force applied to an object causes the object to move

X

xylem: a tubular structure that transports water and nutrients from the roots to various parts of a plant

Y

yandying: separating less dense particles from denser ones

Z

zone of tolerance: the range of an abiotic factor that an organism can survive in

Index

A

D

E

F

G

H

I

K

L

M

T

U

Acknowledgements

The authors and publisher are grateful to the following for permission to reproduce material:

Photographs

aAP/Dave Wethey, **278-79**; **Aboriginal Artists Agency** for the artwork Possum Nungurrayi Anmatyerre people, Northern Territory born Mt Allen, Northern Territory 1967, *Seven Sisters, Milky Way Dreaming* 2002, Melbourne, synthetic polymer paint on canvas, 115.0 x 204.0 cm, Santos Fund for Aboriginal Art 2002, Art Gallery of South Australia, Adelaide, © Gabriella Possum Nungurrayi /Aboriginal Artists Agency, 20022P8, **38**; **Adobe Stock**/169169, **95**, **108** (bottom), /3dsculptor, **242** (bottom), /7activestudio, **152,** /9dreamstudio, **219** (bottom), /Aastels, **118** (top), /alchena, **84,** /Alecander Raths, **312,** /Alexandr Yurtchenko, **60,** /Ali, **197** (top right), /Alizada Studios, **268** (left), /alphaspirit, **496,** /anankkml, **36** (left), **43** (bottom left), **256** (top), /Andrea Danti, **12,** /angkhan, **345,** /Ayudh Karanwal, **499**, /beau, **151,** /blueringmedia, **168** (bottom), **187** (bottom right), /brgfx, **134,** /Chadchai, **316,** /chaiviewfinder, **304-5**, **308,** /concept w, **208-209**, **221** (centre right), /Craig, **97** (goanna), /Dabarti, **234** left, /Daria Nipot, **263,** /David Bokuchava, **284,** /DenPhoto, **282** (bottom), /desdemona72, **287** (bottom), /designer_things, **353** (top), /designua, **105** (left), /designua, **166,** /dinozzaver, **107** (top), /eBsimplicity, **199** (silver), /Elena, **403,** /Elisa, **200** (top), /Emil Lime, **19** (left), /Eric Isselée, **94** (bottom), **109** (bottom left), **80**, /Eskymaks, 194 (bottom), /exopixel, **251,** /Ekaterina Kriminskaya, **260** (sedimentary), /fotomatrix, **489,** /freshidea, **159,** /gguy, **280,** /ghoststone, **51** (top), /Gred Epperson, **196** (top), /Greg Meland, **319,** /Henri Koskinen, **265** (bottom), /HollyHarry, **340,** /Ian Scott, **97** (stingray), /IB Photography, **355,** /ijp2726, **199** (diamond), /Imagevixen, **260-61,** /Imogen, **97** (kingfisher), /Inna, **156** (bottom), /ISENGARD, **89** (centipede), /James Steidl, **27** (top), /janmiko, **121,** /Jens Metschurat, **174** (top), /Jo Sak Roc, **216**, **221** (bottom left), /Jose Luis Stephens, **410,** /Juraiwan, **199** (gold), /Justlight, **311**, /KOTO, **15**, **42** (bottom left), /ktdesign, **285** (right), /lauritta, **37** (bottom), /Lena Balk, **199** (aluminium), /Liaurinko, 365, /Lightfield Studios, **195** (bottom), /lorenza62, **168** (top), /Lubos Chlubny, **199** (ruby), /luchschenF, **111**, **480**, /Macronatura.es, **89** (spider), /magann, **8** (top), /matimix, **iii** (bottom), **48** (top), **76** (top left), /matis75, **150,** /Mauricio G, **61**, /mdurson, **44**, /Meng_Dakara, **443**, /Microgen, **21,** /Milan, **200** (centre), **220** (top left), /mnauliady, **333** (top left), /Monkey Business, **285** (left), /Narsil, **199** (sulphur), /nataliazakharova, **129** (top), /NatureStock, **89** (millipede), /Naypong Studio, **197** (top left), /New Africa, **112,** /nobeastsofierce, **289** (bottom), /Nurachmadi, **333** (bottom right), /Oleh, **161**, **186** (centre right), /Only Fabrizio, **260** (igneous), /pat_hastings, **119** (top), /peter, **198** (bottom), **220** (bottom), /Phillip Minnis, **270-71** (bottom), /phototrip.cz, **iv**, **97** (white cockatoo), **106**, **343**, /photoworld, **196** (bottom), /PixieME, **302** (left), /pxl.store, **142**, **147** (centre), /rh2010, **282** (top), /Roger, **174** (centre), /RomanVX, **269** (bottom), /ronstik, **215** (centre), /ryanking999, **105** (right), /SAMYA, **360,** /sarahdoow, **224,** /schankz, **88** (lizard), /Seventyfour, **295** (bottom), /siimsepp, **486,** /Skfoto, **226,** /Sonja, **219** (top), /steheap, **16** (right), /StockPhotoPro, **19** (right), /stockshoppe, **154**, **158** (right), **186** (bottom & centre left), /studioDG, **199** (copper), /Syda Productions, **53** (bottom), /TADDEUS, **195** (top), /TOimages, **293** (bottom), /Tomasz Zajda, **298-99,** /tonaquatic, **100** (left), /trekandphoto, **199** (brass), /TSpider, **176** (bottom), /Uros Petrovic, **258** (top), /Uwe, **317,** /ValentinaKru, **26** (left), /Valery Voennyy, **260** (metamorphic), /VectorMine, **363,** /verdateo, **199** (iron), /ViniSouza128, **199** (quartz), /Visualmind, **293** (top), /vladim_ka, **425,** /wektorygrafika, **117** (bottom), **146** (top), /Winston Link, **379** (left), /Wollwerth Imagery, **143,** /xamtiw, **405,** /YB, **124** (left), /Yeti Studio, **295** (top), /zefein, **404,** /Zorro Stock images, **34; Airview Online**, **257**; **Alamy Stock Photo**/Abe Smith Photography, **491,** /Alex Mustard, **247,** /Andrew Unangst, **238** (bottom left), /Bill Bachman, **iii** (top), **41**, **110** (right), **141** (top), **147** (top right), /Björn Wylezich, **258** (bottom), /blickwinkel, **406,** /Botany Vision, **296,** /BrazilPhotos, **69** (bottom), /Budimier Jevtic, **267,** /Carol Dembinsky / Dembinsky Photo Associates, **13** (right), **42** (top right), /Christina Gandolfo, **88** (cat), /David Arky, **289** (top), /David J. Green, **422,** /David R. Frazier Photolibrary, Inc, **116** (left), /Dorling Kindersley ltd, **339**, **324** (top), **417**, /EarlyBird, **181,** /Editorial Image, LLC, **74** (top), **492**, /Eye Ubiquitous, **497,** /Gerault Grefory/hemis.fr, **119** (bottom), /GoodIdeas, **320,** /GRANGER, **273,** /Grant Heilman Photography, **10** (right), /Greg Balfour Evans, **28** (top), /Hans Blossey, **180** (top), /Iain Sarjeant, **194** (top), /Image Source Limited, **326** (left), /Imagebroker, **87,** /ImageBROKER.com GmbH & Co. KG, **253** (bottom), /imageBROKER/David Talukdar, **303**, /Jack Tuszynski, **225,** /Jackek Laskowski, **264,** /Javier Jaime, **11,** /Jean-Paul Ferrero/AUSCAPE, **184** (left), 189 (bottom), /Jeff J Daly, **117** (top), /Jim West, **9**, **42** (top left), **45**, /John Eveson, **102,** /Julee Ashmeed, **274** (bottom), /KamilSD, **193,** /krystynaSzulecka, **141** (bottom), /Lee Wonyeop, **347,** /Louise George, **83** (bottom), /Mark Garlick/Science Photo Library, **63**, **292**, /Martin Harvey, **145,** /Martin Shields, **464** (left), /Maurice Savage, **338** (bottom), /Mediadrumimages/Don Marx, **173** (top), /NASA, **346** (top), /NASA / Dembinsky Photo Associates, **13** (left), **64** (right), /NG Images, **245,** /NOAA, **144-45,** /Oksana Adigamova, **92** (bottom), /Outback Australia, **96,** /Pacific Imagica, **148** (right), /Paul Ridsdale, **68** (bottom), **77** (top right), /Penny Tweedie, **4-5,** /Peter Casolino, **75,** /Peter Lenk, **148** (left), /Phil Degginger, **204** (top), /Photography By Marco, **39** (bottom), /Picturebank, **58** (top), /Pulsar Imagens,

185, /Rolf Hickler, **190-91,** /Romans Klevcovs, **299** (top), /Ron Giling, **128** (left), /Ruslan Nesterenko, **346,** /Sara Sadler, **409,** /Scenics & Science, **156** (top), /Science History Images, **205,** **229**, /sciencephotos, **66** (right), **213** (bottom), **318**, /Scott Camazine, **288,** /Simon Anders, **94** (top), /Simon Belcher, **18** (left), **42** (bottom right), /Sipa USA, **490,** /Trevor Clifford Photography/SPL, **211** (top), /Universal Images Group North America LLC / DeAgostini, **24,** /UrbanZone, **20**, **324** (bottom), /Valery Voennyy, **262** (left), /View Stock, **78,** /Westend61 GmbH, **16** (left), /Yuri Arcurs, **351,** /Zoonar GmbH, **8** (bottom), **494**; **Copyright Agency** for 'Baidam Tithuyil' by Brian Robinson, © Brian Robinson/Copyright Agency, 2024, **6**; **CSIRO**/Andrea Wild (CC BY 3.0), **183**; **Fairfax Photos**/ Wolter Peeters, **73** (bottom right), /Andrew Taylor, **256** (bottom); **Getty Images**/Alfred Pasieka/Science Photo Library, **192,** /Anadolu, **291** (bottom), /Andrew Merry, **17** (left), /antonioiacobelli, **113,** /Auscape, **182** (bottom), **189** (top), /Bettmann, **40,** /Brownie Harris, **290,** /coberschneider, **42,** /DEA/A.DAGLI ORTI, **218** (left & right), **221** (bottom right), /Grant Faint, **7,** /Jeffrey Coolidge, **272** (right), /Jens Schlueter, **57** (top), /Mark Garlick/Science Photo Library, **49**, **76** (bottom), /Matt Anderson Photography, **31**, **43** (top right), /Michael Uthe / 500px, **3,** /NASA, **10**, **42** (top centre), /Philipp Boettcher/500px, **81,** /shells1, **36** (right), **43** (bottom right), /Star Tribune, **272** (left), /Tex Image/Science Photo Library, **83** (top), /Winfried Wisniewski, **172,** /Yulia Reznikov, **236** (bottom), **275** (centre left); **iStockphoto**/AGorohov , **99** (left), , /ai_yoshi, **242** (top), **275** (bottom left), /airdone, **228,** /Aleksandr Grechanyuk, **495,** /aluxum, **214** (plant), /Andreas Häuslbetz, **89** (beetles), /Andrew Peacock, **114** (top), /Antagain, **93** (top), /Antic Zlatko, **414,** /Artem Stephanov, **199** (graphite), /atakan, **202** (top), /Ayakochun, **379** (right), /BanksPhotos, **130,** /Betka82, **222** (left), /bitontawan, **332** (bottom), /c12, **250** (bottom), /Capstoc, **97** (red-winged parrot), /chameleonseye, **72**, **77** (bottom right), /Chansom Pantip, **240** (right), /Children of Dune, **376** (warning pictograms), /Chris Ryan, **51** (bottom), /chrisdorney, **476,** /ChrisGorgio, **62,** /Christopher Freeman, **85** (cat), /ClaraNila, **93** (bottom), /claylib, **92** (top), /coddy, **199** (titanium), /colematt, **98**, **109** (bottom right), /Colematt & Nandalal Sarkar, **164**, **187** (top right), /CraigRJD, **97** (black cockatoo), /CribbVisuals, **252** (left), /Davizro, **358,** /defun, **89** (dragonfly), /Dmitr1ch, **350** (background), /Dragon Claws, **46,** /elenabs, **333** (bottom left), /Emilija Randjelovic, **353** (bottom), **204** (bottom), /fcafotodigital, **115** (top), /Fotocam, **199** (steel), /fotoVoyager, **212** (top), /GlobalP, **vii**, **85** (lion), **82**, /Good Life Studio, **71** (top), /Hipanolistic, **375,** /illustratrice, **27** (bottom), /imv, **91,** /industryview, **126-27,** /jacus, **v**, **240** (left), /Jazziell, **39** (top), /jessicahyde, **97** (stringybark), /jessicaphoto, **26** (bottom), /Jose Gonzalez Buenaposada, **322,** /Jurkos, **182** (top), /kali9, **327,** /kazakovmaksim, **169** (leaf), /Ken Griffiths, **97** (cycad palm), /Kirillm, **259** (bottom), /komta, **59** (bottom, /Lamaip, **306,** /Liana Manukyan, **350,** /Liliboas, **28** (bottom), /lukpedclub, **214** (watering can), /manpuku7, **35,** /Maor Winetrob, **100** (right), **109** (top right), /Mari_C, **333** (top right), /mashabuba, **89** (crab), /mediaphotos, **325,** /megaflopp, **445,** /Meindert van der Haven, **442,** /metamorworks, **281,** /microgen, **392,** /MJ_Prototype, **236** (top), **275** (bottom right), /Mlenny, **174** (bottom), /Moof, **126** (left), /Nina Borisova, **330,** /NNehring, **176** (top), /Outback to Coast, **169** (top), /Paolo Cipriani, **180** (bottom), /pavlinec, **309,** /PeopleImages, **22,** /PeterHermesFurian, **99** (right), **215** (right), /PeterJamesSampson, **90,** /Pgiam, **2,** /pidjoe, **287** (top), **326** (right), /RainvonBrandis, **175,** /Rawpixel.com, **313,** /rediguana_ nz, **251** (bottom), /ricardoreitmeyer, **321,** /scisettialflo, **384,** /SebastianGauert, **59** (top), **394**, /Sergly Zamureenko, **120** (bottom), **146** (bottom left), /sharrocks, **334** (bottom), /spukkato, **138-39**, **147** (bottom), /suphakit73, **368,** /supsktiypumpy, **334** (top), /t_kimura, **133** (top), /Tetiana Lazunova, **85** (top right), /themacx, **114** (bottom), /TomekD76, **237,** /ttsz, **25** (top), **116** (right), **342**, /VectorMine, **153**, **167**, **186** (top), **187** (bottom left), /VeselovaElena, **155** (top), /volschenkh, **14** (bottom left), /VvoeVale, **253** (top), /WebSubstance, **361,** /Wirestock, **173** (bottom), /Wittayayut, **54** (left), /Wojciech Kozielczyk, **439; National Library of Australia** for Aboriginal Woman Winnowing Grass Seed, Arnhem Land, Northern Territory, 1948 [Picture]/Harrison Howell Walker, **140**; **NOAA**/ GFDL, 291 (top); Queensland Museum/ Peter Waddington, **269** (top); **Science Photo Library**, **14** (top right & top left), **115** (bottom), **217**, **434**, /Alexandre Dotta/Science Source, **132-33**, /Andrew Lambert Photography, **197** (bottom), **323**, /B.G. Thomson, **110** (left), /CNRI, **223** (right), /DK IMAGES, **411**, **488**, /Dr Jeremy Burgess, **169** (somata), /Geoff Tompkinson, **135** (bottom), /GIPHOTOSTOCK, **25** (bottom), /Gunilla Elam, **101**, /Joe Arem, **259** (top), /John Sanford, **32**, /Martyn F. Chillmaid, **17** (right), **157** (top), **158** (top), **159** (top), **210**, **211** (bottom), **376** (bottom right), **397**, **469**, /Maurizio De Angelis, **120** (top, /Nancy Kedersha / UCLA, **104**, /Petr Jan Juracka, **206** (top), /Sheila Terry, **132** (top), /Tex Image, **14** (right); **Simon Slattery**/Vallery Imagery, **cover**; **Turtle Rock Scientific**/Science Source, **452**.

OTHER

Aipki-em permut bepiti-e (The Twinkling Stars) by George Passi reproduced with the permission of Doug Passi, **35**; CRAAP, Meriam Library, California State University, Chico, https://library.csuchico.edu/sites/default/files/craap-test.pdf, CC BY 4.0 https://creativecommons.org/licenses/by/4.0/, **354**; Ray Norris for 'Dreaming story', **33**.

Science 7-10 Syllabus and Glossary © NSW Education Standards Authority for and on behalf of the Crown in right of the State of New South Wales, 2023.